D0926414

Directional Statistics

WILEY SERIES IN PROBABILITY AND STATISTICS

Established by WALTER A. SHEWHART and SAMUEL S. WILKS

A complete list of the titles in this series appears at the end of this volume

Directional Statistics

Kanti V. Mardia
University of Leeds, UK

Peter E. Jupp
University of St Andrews, UK

JOHN WILEY & SONS, LTD
Chichester · New York · Weinheim · Brisbane · Singapore · Toronto

First published under the title *Statistics of Directional Data* by
Academic Press Inc. (London) Ltd.
Baffins Lane, Chichester,
West Sussex, PO19 1UD England

National 01243 779777
International (+44) 1243 779777

e-mail (for orders and customer service enquiries): cs-books@wiley.co.uk

Visit our Home Page on http://www.wiley.co.uk or http://www.wiley.com

Other Wiley Editorial Offices

John Wiley & Sons, Inc., 605 Third Avenue,
New York, NY 10158-0012, USA

Wiley-VCH Verlag GmbH, Pappelallee 3,
D-69469 Weinheim, Germany

Jacaranda Wiley Ltd, 33 Park Road, Milton,
Queensland 4064, Australia

John Wiley & Sons (Asia) Pte Ltd, 2 Clementi Loop #02-01,
Jin Xing Distripark, Singapore 129809

John Wiley & Sons (Canada) Ltd, 22 Worcester Road,
Rexdale, Ontario, M9W 1L1, Canada

Library of Congress Cataloging-in-Publication Data
Mardia, K.V. (Kanti V.)
Directional Statistics / K.V. Mardia and P.E. Jupp
p. cm. — (Wiley series in probability and statistics)
Includes bibliographical references and index.
ISBN 0-471-95333-4 (cased : alk. paper)
1. Mathematical statistics. 2. Distribution (Probability theory)
3. Sampling (Statistics) I. Jupp, P. E. II. Title
III. Series.
QA276, J864 1999 99-33679
519.5—dc21 CIP

British Library Cataloguing in Publication Data

A catalogue record for this book is available from the British Library

ISBN 0 471 95333 4

Produced from files supplied by the authors and processed by Alden Bookset
Printed and bound in Great Britain by Antony Rowe Ltd, Chippenham
This book is printed on acid-free paper responsibly manufactured from sustainable forestry in which at least two trees are planted for each one used for paper production.

TO

PAVAN AND STEPHANIE

Ah, but my Computations, People say,
Have squared the Year to human compass, eh?
If so, by striking from the Calendar
Unborn To-morrow and dead Yesterday.

Edward FitzGerald, *Rubáiyát of Omar Khayyám*, 2nd edition, Quatrain 49

The theory of errors was developed by Gauss primarily in relation to the needs of astronomers and surveyors, making rather accurate angular measurements. Because of this accuracy it was appropriate to develop the theory in relation to an infinite linear continuum, or, as multivariate errors came into view, to a Euclidean space of the required dimensionality. The actual topological framework of such measurements, the surface of a sphere, is ignored in the theory as developed, with a certain gain in simplicity.

It is, therefore, of some little mathematical interest to consider how the theory would have had to be developed if the observations under discussion had in fact involved errors so large that the actual topology had had to be taken into account. The question is not, however, entirely academic, for there are in nature vectors with such large natural dispersions.

R. A. Fisher

Contents

Preface

The aim of this book

Directional statistics is concerned mainly with observations which are unit vectors (possibly with sign unknown) in the plane or in three-dimensional space. Thus the sample space is typically a circle or a sphere, so that standard methods for analysing univariate or multivariate measurement data cannot be used. Special directional methods are required which take into account the structure of these sample spaces. The aim of this book is to give a comprehensive, systematic and unified treatment of the theory and methodology of directional statistics. We give the theory underlying each technique, and illustrate applications by working through real-life examples. Three basic approaches (the embedding, wrapping and intrinsic approaches) to directional statistics are highlighted throughout. Our original intention was to produce a minor revision of *Statistics of Directional Data* (Mardia, 1972a), which has been out of print. However, it soon became clear that the considerable advances in the subject in the last quarter century (some of which were summarised in our review paper, Jupp & Mardia, 1989) meant that drastic revision and major extension were required. This book is the result.

Relationship with other books

The relationship between this book and Mardia (1972a) is firstly that the material of the latter has been updated to reflect the advances in the area since 1972, and secondly that we have added the following material:

(i) spheres of arbitrary dimension are considered (Chapter 9);
(ii) a full chapter on correlation and regression analysis is given (Chapter 11);
(iii) modern topics such as robust techniques, bootstrap methods and density estimation have been added (Chapter 12);
(iv) Stiefel manifolds, Grassmann manifolds and other more general manifolds are considered (Chapter 13);
(v) shape analysis is described in detail (Chapter 14) and its relationship to directional statistics is discussed.

Excellent accounts of some of the developments in directional statistics in the last 25 years are given in the following important books, none of which was intended to be comprehensive, in contrast to this book. Watson's (1983a) *Statistics on Spheres* is concerned mainly with the theory of inference for distributions on spheres of arbitrary dimension. *Statistical Analysis of Circular Data* by Fisher (1993) and *Statistical Analysis of Spherical Data* by Fisher, Lewis & Embleton (1987) concentrate on modern methods of data analysis on the unit circle and on the unit sphere in three-dimensional space, respectively. A wide variety of biological applications can be found in Batschelet's (1981) *Circular Statistics in Biology.* Theory and applications of circular and spherical statistics are considered also in Chapters 9 and 10 of *Spatial Data Analysis by Example* by Upton & Fingleton (1989).

Survey of contents

The book has three parts. The first part (Chapters 1–8) is concerned with statistics on the circle. Following a general discussion of circular data in Chapter 1, various summary statistics are introduced in Chapter 2. Chapter 3 presents the basic concepts and models for circular data. Chapter 4 gives the fundamental theorems and distribution theory for the uniform and von Mises distributions. Chapter 5 treats point estimation, mainly for the von Mises and wrapped Cauchy distributions. The uniform distribution plays a central role in circular statistics, and so Chapter 6 is devoted to tests of uniformity and to related tests of goodness-of-fit. Chapter 7 gives a detailed account of tests on von Mises distributions. Non-parametric methods for circular data are considered in Chapter 8.

The second part (Chapters 9–12) considers statistics on spheres of arbitrary dimension. The basic models and distribution theory for distributions on spheres are presented in Chapter 9. A detailed account of inference on the main distributions on spheres is given in Chapter 10. Correlation, regression and time series are considered in Chapter 11. Chapter 12 describes some modern methodology, in particular robust techniques, bootstrap methods, Bayesian methods, density estimation, and curve fitting and smoothing.

The third part (Chapters 13–14) considers extensions to statistics on more general sample spaces. Chapter 13 treats extensions to general manifolds, in particular to rotation groups, Stiefel manifolds and Grassmann manifolds. Chapter 14 treats the latest advances in the fast-growing field of shape analysis from the perspective of directional statistics. In particular, statistics on complex projective spaces is considered.

Since Bessel functions and Kummer functions play a major role in directional statistics, some relevant formulae have been placed in Appendix 1. Tables for use in the analysis of circular and spherical data are given in Appendices 2 and 3, respectively.

We have not given any general historical account of directional statistics, since the history of the subject has been covered well by Fisher, Lewis

& Embleton (1987) and Fisher (1993). Although the package *Oriana for Windows* (Kovach Computing Services) performs some analyses of circular data, as far as we are aware there are not yet any commercial computer packages for handling general directional data. However, there is a forthcoming package *DDSTAP – A Statistical Package for the Analysis of Directional Data* by Professor Ashis SenGupta, Indian Statistical Institute, Calcutta.

Acknowledgements
We are grateful to many people for their permission to reproduce figures and tables.

Toby Lewis, who inspired *Statistics of Directional Data*, has also been a great source of inspiration for this book. Thanks are due also to Ian Dryden, John Kent, Paul McDonnell, Kevin de Souza and Alistair Walder for their comments.

Special thanks are due to Catherine Thomson, Valerie Sturrock and Christine Rutherford for preparing large sections of the text. Rachel Fewster and David Kemp kindly provided valuable solutions to our more technical problems with LaTeX. We thank Helen Ramsey, Sharon Clutton and Juliet Booker of Wiley for their patience, encouragement and assistance. It seems inevitable that, in spite of our best efforts, various errors and obscurities will remain. We shall be most grateful to readers who draw our attention to these, or suggest any other improvements.

Kanti Mardia
Peter Jupp
March 1999

1

Circular Data

1.1 INTRODUCTION

Circular data arise in various ways. The two main ways correspond to the two principal circular measuring instruments, the *compass* and the *clock*. Typical observations measured by the compass include wind directions and directions of migrating birds. Data of a similar type arise from measurements by spirit level or protractor. Typical observations measured by the clock include the arrival times (on a 24-hour clock) of patients at a casualty unit in a hospital. Data of a similar type arise as times of year (or times of month) of appropriate events.

A circular observation can be regarded as a point on a circle of unit radius, or a unit vector (i.e. a *direction*) in the plane. Once an initial direction and an orientation of the circle have been chosen, each circular observation can be specified by the angle from the initial direction to the point on the circle corresponding to the observation.

Circular data are usually measured in degrees. However, it is sometimes useful to measure in radians. Recall that angular measurements are converted from degrees to radians by multiplying by $\pi/180$.

Closely related to circular data are *axial* data, i.e. observations of *axes*. They are usually given as observations on the circle for which each direction is considered as equivalent to the opposite direction, so that the angles θ and $\theta + 180°$ are equivalent. The standard way of handling axial data is to convert them to circular data by 'doubling the angles', i.e. transforming θ to 2θ and so removing the ambiguity in direction.

1.2 DIAGRAMMATICAL REPRESENTATION

1.2.1 Ungrouped Data

The simplest representation of circular data is a *circular raw data plot*, in which each observation is plotted as a point on the unit circle. Figure 1.1 illustrates this method for the following example.

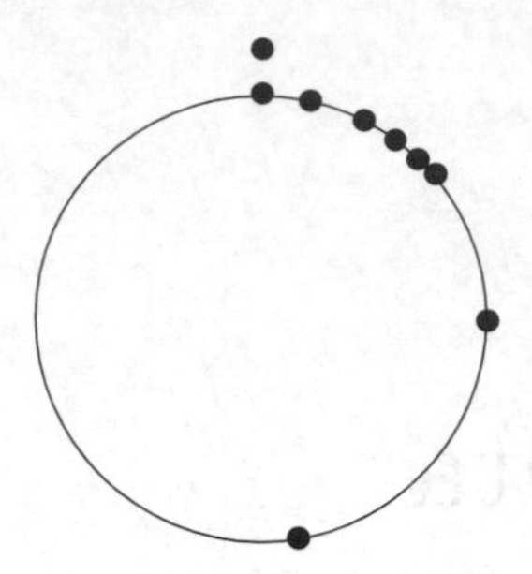

Figure 1.1 Circular plot of the roulette data of Example 1.1.

Example 1.1
A roulette wheel was spun and the positions at which it stopped were measured. The stopping positions in 9 trials were 43°, 45°, 52°, 61°, 75°, 88°, 88°, 279°, 357°. The circular raw data plot in Fig. 1.1 suggests that there is a preferred direction.

1.2.2 Grouped Data

Circular Histograms

Grouped circular data can be represented by circular histograms, which are analogous to histograms on the real line. Each bar in a circular histogram is centred at the midpoint of the corresponding group of angles, and the *area* of the bar is proportional to the frequency in that group. As with histograms on the line, the visual impression given by a circular histogram may be sensitive to the grouping used.

Table 1.1 shows the frequencies of the vanishing angles of 714 non-migratory British mallards. The birds were displaced from Slimbridge, Gloucestershire, to various sites between 30 km and 250 km away. When the birds were released from a site, they flew off, vanishing from sight in a direction with bearing known as the *vanishing angle*. A circular histogram of this data set is given in Fig. 1.2. Note that, because these vanishing angles have been measured clockwise from north, the circular histogram in Fig. 1.2 follows this convention, whereas the usual convention for circular histograms measures angles anticlockwise and takes the x-axis as the zero direction.

Linear Histograms

Because statisticians have acquired expertise in interpreting histograms on the line, it can be useful to transform a circular histogram into a linear histogram. This is done by cutting the circular histogram at a suitably chosen point on the circle and then 'unrolling' the circular histogram to a linear histogram on an interval of width 360°.

Table 1.1 Vanishing angles of 714 British mallards (adapted from Matthews, 1961, reproduced by permission of British Ornithologists' Union). Directions are measured clockwise from north.

Direction (in degrees)	Number of birds
0–19	40
20–39	22
40–59	20
60–79	9
80–99	6
100–119	3
120–139	3
140–159	1
160–179	6
180–199	3
200–219	11
220–239	22
240–259	24
260–279	58
280–299	136
300–319	138
320–339	143
340–359	69
Total	714

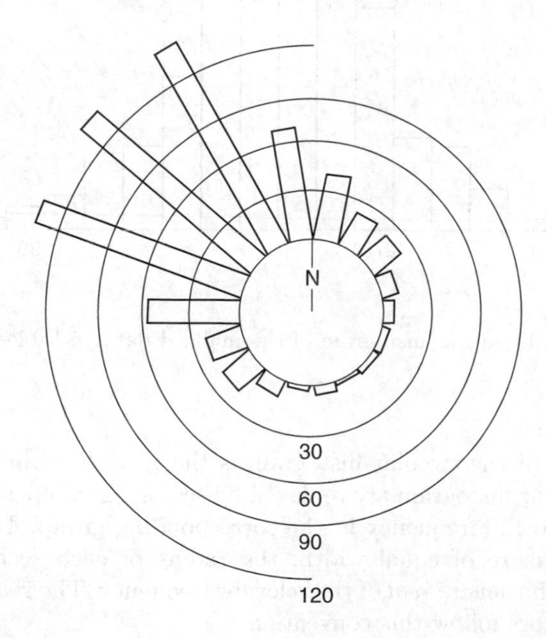

Figure 1.2 Circular histogram of the mallard data of Table 1.1.

The visual impression given by a linear histogram can be sensitive to the point at which the circle is cut. If the data have a single mode (preferred direction) then it is wise to use a cut almost opposite this mode. Then the centre of the linear histogram will be near the mode. A cut near the mode would give the misleading impression that the data are bimodal. For data sets which do not have a single pronounced mode, it is useful to modify the circular histogram by repeating a complete cycle of the data, to give a linear histogram on an interval of width 720°.

Figure 1.3 shows a linear histogram of the data in Table 1.1.

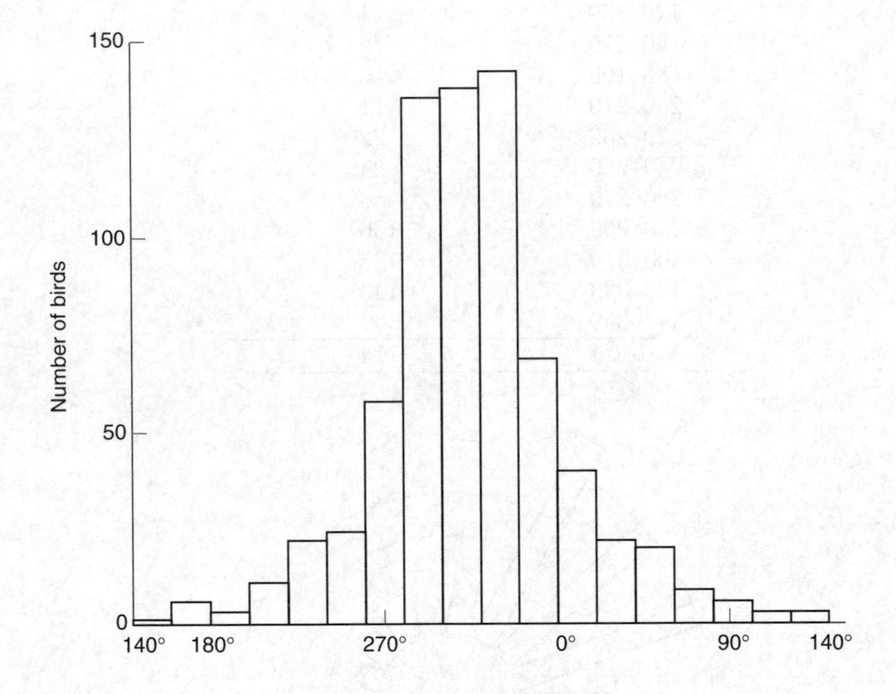

Figure 1.3 Linear histogram of the mallard data of Table 1.1.

Rose Diagrams

A useful variant of the circular histogram is the *rose diagram*, in which the bars of the circular histogram are replaced by sectors. The *area* of each sector is proportional to the frequency in the corresponding group. To achieve this when the groups are of equal width, the radius of each sector should be proportional to the *square root* of the relevant frequency. The reader is warned that not all authors follow this convention.

Figure 1.4 shows a rose diagram of the data in Table 1.1.

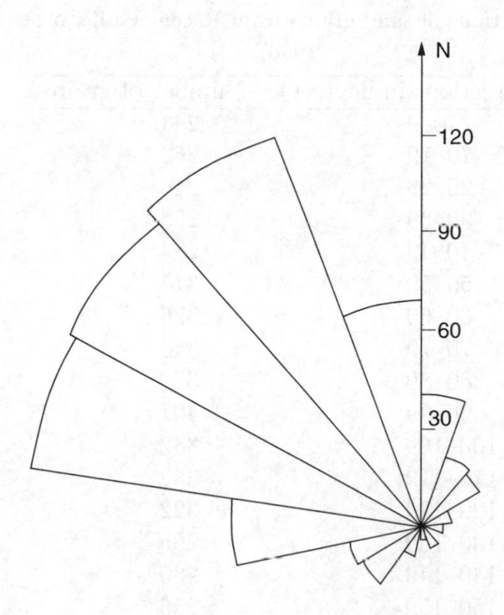

Figure 1.4 Rose diagram of the mallard data of Table 1.1.

1.2.3 Axial Data

Axial data can be represented diagrammatically by first transforming them to circular data by 'doubling the angles' and then constructing circular histograms or rose diagrams of the circular data.

Table 1.2 shows the orientations of the shortest projection elongations of sand grains from Recent Gulf Coast beach. Note that these observations have been recorded as angles in the range $(0°, 180°)$. Since elongations are axes, the data are axial.

1.3 FORMS OF FREQUENCY DISTRIBUTIONS

The general forms of the circular distributions appearing in practice can be identified roughly from the linear histograms, just as in the case of linear data.

1.3.1 Unimodal Distributions

The data given in Tables 1.1 and 1.2 are unimodal. The frequencies tail off steadily, giving rise to a minimum in the corresponding circular histogram. This point of minimum is called the *antimode* (and the point of maximum is called the *mode*). The data in Tables 1.1 and 1.2 are fairly symmetrical. The distribution of the directions of birds in Table 1.1 has a high peak and the frequencies tail off steadily to zero. The birds have a preferred direction

Table 1.2 Orientations of sand grains from Recent Gulf Coast beach (Curray, 1956)

Direction (in degrees)	Number of grains
0–9	244
10–19	262
20–29	246
30–39	290
40–49	284
50–59	314
60–69	326
70–79	340
80–89	371
90–99	401
100–109	382
110–119	332
120–129	322
130–139	295
140–149	230
150–159	256
160–169	263
170–179	281
Total	5439

between 280° and 340°, i.e. there is a tendency for the birds to select a north-westerly course.

Not all data sets on the circle are symmetrical. Table 1.3 gives an example of an asymmetrical distribution which is 'negatively skewed'. The data give azimuths (angles measured clockwise from north) of cross-beds in the upper Kamthi river, India.

Table 1.4 shows the month of onset of lymphatic leukaemia in the UK during 1946–1960. The distribution is unimodal.

1.3.2 Multimodal Distributions

In Section 1.3.1 we have discussed unimodal distributions. Multimodal distributions also occur. An example of these is given in Table 1.5, which shows the directions in which 76 female turtles moved after laying their eggs on a beach.

The circular raw data plot of this data set given in Fig. 1.5 shows that the distribution is bimodal and the two modes are roughly 180° apart. The dominant mode is in the interval 60°–90° and the subsidiary mode is in the interval 240°–270°. The data indicate that the turtles have a preferred direction (towards the sea) but a substantial minority seem to prefer the opposite direction.

Table 1.3 Azimuths of cross-beds in the upper Kamthi river (Sengupta & Rao, 1966, reproduced by permission of the Indian Statistical Institute)

Azimuth (in degrees)	Frequency
0–19	75
20–39	75
40–59	15
60–79	25
80–99	7
100–119	3
120–139	3
140–159	0
160–179	0
180–199	0
200–219	21
220–239	8
240–259	24
260–279	16
280–299	36
300–319	75
320–339	90
340–359	107
Total	580

1.4 FURTHER EXAMPLES OF DIRECTIONAL DATA

Some examples giving rise to circular data have been described in Sections 1.2–1.3. We now give further examples involving circular or spherical data. Although these examples come from widely differing scientific disciplines, they all involve directional data and give rise to similar statistical questions.

1.4.1 Earth Sciences

Since the surface of the earth is approximately a sphere, spherical data arise readily in the earth sciences. For example, an important feature of an earthquake is its epicentre (the point on the earth's surface vertically above the origin of the earthquake). Other spherical data in which the observations are points on the earth's surface occur in the estimation of relative rotations of tectonic plates (Chang, 1993).

Spherical observations on a smaller scale include directions of remnant magnetism in rocks. These are used to infer directions of palaeomagnetic fields and hence to infer the wander path of the north magnetic pole. The study of palaeomagnetism was a major stimulus in the development the analysis of spherical data.

Directional data arise in investigations of various geological processes, since these involve transporting matter from one place to another. In particular, orientations of cross-bedding structures and of the long axes of particles in undeformed sediments are used in inference about the directions of palaeocurrents. See Curray (1956) and Pincus (1953, p. 584). Tables 1.2 and 1.3 give examples from this area.

A wide-ranging account of directional statistics in the earth sciences was given by Watson (1970), who gave an excellent description of geological terms and concepts.

Further examples of spherical data arising in the earth sciences are used in Chapters 9 and 10.

Table 1.4 Month of onset of cases of lymphatic leukaemia in the UK, 1946–1960 (Lee, 1963, reproduced by permission of the BMJ)

Month	Number of cases
January	40
Febuary	34
March	30
April	44
May	39
June	58
July	51
August	55
September	36
October	48
November	33
December	38
Total	506

1.4.2 Meteorology

Wind directions provide a natural source of circular data. A distribution of wind directions may arise either as a marginal distribution of the wind speed and direction, or as a conditional distribution for a given speed. Other circular data arising in meteorology include the times of day at which thunderstorms occur and the times of year at which heavy rain occurs.

Table 1.5 Orientations of 76 turtles after laying eggs (Gould's data cited by Stephens, 1969e)

Direction (in degrees) clockwise from north									
8	9	13	13	14	18	22	27	30	34
38	38	40	44	45	47	48	48	48	48
50	53	56	57	58	58	61	63	64	64
64	65	65	68	70	73	78	78	78	83
83	88	88	88	90	92	92	93	95	96
98	100	103	106	113	118	138	153	153	155
204	215	223	226	237	238	243	244	250	251
257	268	285	319	343	350				

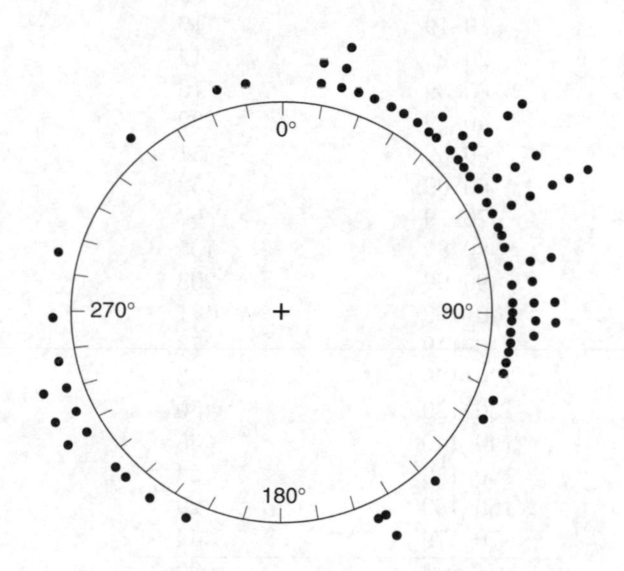

Figure 1.5 Circular plot of the turtle data of Table 1.5.

1.4.3 Biology

Studies of animal navigation lead to circular data. Typical questions of interest are (i) whether the directions of the animals are uniformly distributed on the circle and (ii) whether the animals tend to head in a specified direction. The answer to the latter question is useful in attempts to determine whether or not the animals use navigational clues such as the direction of the sun or the earth's magnetic field.

Discussions of various investigations in the field of bird navigation are given in Schmidt-Koenig (1965) and Batschelet (1981).

The mallard data in Table 1.1 and the turtle data in Table 1.5 provide examples from the area of animal movement. Further examples will be given in Chapters 6, 7 and 8.

Table 1.6 Angles between the swimming directions of *Daphnia* and the plane of polarisation of light (data of Waterman & Jander, cited in Waterman, 1963, reproduced by permission of Springer-Verlag)

Direction (in degrees)	Frequency
0–9	65
10–19	17
20–29	12
30–39	16
40–49	22
50–59	51
60–69	58
70–79	67
80–89	105
90–99	208
100–109	81
110–119	73
120–129	43
130–139	50
140–149	35
150–159	24
160–169	29
170–179	44
Total	1000

An example of axial data in animal navigation is given in Table 1.6, which consists of angles between the swimming directions of *Daphnia* and the plane of polarisation of the surrounding light. The distribution is bimodal and the modes are roughly 90° apart — as happens for many sets of bimodal axial data.

Circular data arise also in the study of circadian rhythms, typically as times of day at which peak activity occurs.

1.4.4 Physics

A set of circular data which led to the introduction of one of the basic distributions in circular statistics consists of the fractional parts of atomic weights. Before the discovery of isotopes, it was hypothesised that measured atomic weights are integers subject to error. Von Mises (1918) proposed testing this by testing whether or not the corresponding distribution on the circle has a mode at 0°. (See Example 6.4.)

Axial data occur as directions of optical axes of crystals.

The distribution of the resultant of a random sample of unit vectors arises in representations of sound waves (Rayleigh, 1919), interference among oscillations with random phases (Beckmann, 1959) and in molecular links (Rayleigh, 1919; Kuhn & Grün, 1942).

The distribution on the celestial sphere of sources of high-energy cosmic rays can be investigated using distributions on rotating spherical caps (Mardia & Edwards, 1982).

1.4.5 Psychology

Directional data appear in the perception of direction under various conditions, such as simulated zero gravity (Ross *et al.*, 1969). Circular data occur also in studies of the mental maps which people use to represent their surroundings (Gordon, Jupp & Byrne, 1989).

1.4.6 Image Analysis

Circular data occur in machine vision, as transformed versions of cross-ratios of sets of four collinear points (Mardia, Goodall & Walder, 1996). Axial data occur in the orientation of textures (Blake & Marinos, 1990). In particular, they occur as planar orientation fields representing the orientation of ridges on fingerprints (Mardia *et al.*, 1997).

1.4.7 Medicine

The incidence of onsets of a particular disease (or of deaths due to the disease) at various times of year provides circular data. The data in Table 1.4 are such an example.

Spherical data occur in vector cardiology. In vector cardiograms, information about the electrical activity in a heart during a heartbeat is described in terms of a near-planar orbit in three-dimensional space. See Downs & Liebman (1969) and Gould (1969).

1.4.8 Astronomy

Since information on distance is often not readily available for astronomical objects, many astronomical observations are of points on the celestial sphere, and so provide spherical data.

It is an interesting historical point that Gauss developed the theory of errors primarily for the analysis of astronomical measurements. Since the measurements involved were concentrated in a small region of the celestial sphere, it was reasonable to approximate the sphere locally by a tangent plane. This led to the development of the theory of statistics on Euclidean spaces, rather than on the sphere.

Several hypotheses have been considered about the distribution on the celestial sphere of various astronomical objects. For example, Pólya (1919) enquired whether or not the stars are distributed uniformly over the celestial sphere, and uniformity of visual binary stars has been considered by Jupp (1995).

Orbits of planets (with known sense of rotation) can be regarded as points on the sphere (see Example 10.5). Bernoulli (1735) enquired whether or not the close coincidence of the orbital planes of the six planets then known could have arisen by chance. His hypothesis can be interpreted as the statement that the corresponding points come from the uniform distribution on the sphere. Similar questions have been asked about the orbits of comets and asteroids.

1.5 Wrapping and Projecting

Looking back at the examples in Sections 1.2–1.4, we see that they fall into two groups. Data in the first group are obtained by taking data on the line and *wrapping* the line round the circle, e.g. by reducing times to times of day. Examples of this are given in Example 1.1 and Table 1.4. Such data are obtained typically from measurements with a clock. Data in the second group are obtained by taking data in the plane and *projecting* the plane radially onto the unit circle. Examples of this are given in Tables 1.1–1.3 and 1.5, and by wind directions. Such data are obtained typically from measurements with a compass.

2

Summary Statistics

2.1 INTRODUCTION

After data have been plotted (as described in Section 1.2), it is useful to summarise them by appropriate descriptive statistics. At first sight, it is tempting to cut the circle at a suitable point and to use conventional summary statistics on the resulting observations on the line. The drawback of this approach is that the resulting summary statistics depend strongly on the point at which the circle is cut. To see this, consider a sample of size 2 on the circle consisting of the angles 1° and 359°. Cutting the circle at 0° would give the sample mean as 180° and the sample standard deviation (using the 'biased' version of the sample variance with divisor n) as 179°, whereas cutting the circle at $180^\circ = (-180^\circ)$ would give the sample mean as 0° and the sample standard deviation as 1°.

It turns out that the appropriate way of constructing summary statistics for circular data is to regard points on the circle as unit vectors in the plane and then to take polar coordinates of the sample mean of these vectors. Measures of location and of dispersion obtained in this way are considered in Sections 2.2 and 2.3, respectively. The idea of 'multiplying angles by p' (for $p = 1, 2, \ldots$) leads to the more general trigonometric moments, which are introduced in Section 2.4.

2.1.1 *Preliminaries and Notation*

Directions in the plane can be regarded as unit vectors $\mathbf{x}$, or equivalently as points on the unit circle (i.e. the circle of unit radius centred at the origin). There are two other useful ways of regarding such directions – as angles and as unit complex numbers. Choose an initial direction and an orientation for the unit circle. (This is equivalent to choosing an orthogonal coordinate system on the plane.) Then each point $\mathbf{x}$ on the circle can be represented by an angle θ or equivalently by a unit complex number z. These are related to $\mathbf{x}$ by

$$\mathbf{x} = (\cos\theta, \sin\theta)^T \quad \text{and} \quad z = e^{i\theta} = \cos\theta + i\sin\theta$$

(see Fig. 2.1).

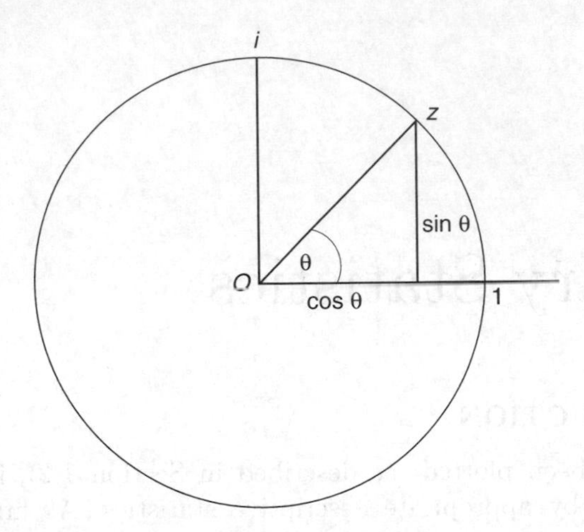

Figure 2.1 Representation of the direction **x** by the angle θ and by the complex number $z = \cos\theta + i\sin\theta$.

From now on, all angles will be measured in radians, unless stated otherwise. Note that the angles θ and $\theta + 2\pi$ (radians) give the same point on the circle. Thus all arithmetic that we shall do on the circle will be modulo 2π, so that we shall write without further comment e.g. '$\theta = \theta_1 + \theta_2$' instead of '$\theta = \theta_1 + \theta_2 \bmod 2\pi$'.

The representation of directions by angles can be regarded as an *intrinsic approach* to directional statistics, since the directions are considered as points on the circle itself, whereas the representation by unit complex numbers can be regarded as an *embedding approach*, since the directions are considered as special points in the plane. In the embedding approach, distributions on the unit circle will be considered as singular distributions on the plane which have their mass concentrated on the unit circle.

Note that the representations of directions by angles and by unit complex numbers depend on the choice of initial direction and orientation. In most contexts there are no preferred choices and so it is essential that statisticians who have made different choices (but observed the same data) make the same inferences. This requirement of coordinate-independent inference (which goes almost unnoticed in inference about ordinary real-valued random variables) dominates the theory and practice of inference in directional statistics.

2.2 MEASURES OF LOCATION

2.2.1 The Mean Direction

Suppose that we are given unit vectors $\mathbf{x}_1, \ldots, \mathbf{x}_n$ with corresponding angles θ_i, $i = 1, \ldots, n$. The *mean direction* $\bar{\theta}$ of $\theta_1, \ldots, \theta_n$ is the direction of the resultant $\mathbf{x}_1 + \ldots + \mathbf{x}_n$ of $\mathbf{x}_1, \ldots, \mathbf{x}_n$. It is also the direction of the centre of mass $\bar{\mathbf{x}}$ of $\mathbf{x}_1, \ldots, \mathbf{x}_n$. Since the Cartesian coordinates of $\mathbf{x}_j$ are $(\cos\theta_j, \sin\theta_j)$ for $j = 1, \ldots, n$, the Cartesian coordinates of the centre of mass are $(\bar{C}, \bar{S})$, where

$$\bar{C} = \frac{1}{n}\sum_{j=1}^{n} \cos\theta_j, \qquad \bar{S} = \frac{1}{n}\sum_{j=1}^{n} \sin\theta_j. \tag{2.2.1}$$

Therefore $\bar{\theta}$ is the solution of the equations

$$\bar{C} = \bar{R}\cos\bar{\theta}, \qquad \bar{S} = \bar{R}\sin\bar{\theta} \tag{2.2.2}$$

(provided that $\bar{R} > 0$), where the *mean resultant length* $\bar{R}$ is given by

$$\bar{R} = (\bar{C}^2 + \bar{S}^2)^{1/2}. \tag{2.2.3}$$

Note that $\bar{\theta}$ is not defined when $\bar{R} = 0$. When $\bar{R} > 0$, $\bar{\theta}$ is given explicitly by

$$\bar{\theta} = \begin{cases} \tan^{-1}(\bar{S}/\bar{C}) & \text{if } \bar{C} \geq 0, \\ \tan^{-1}(\bar{S}/\bar{C}) + \pi & \text{if } \bar{C} < 0, \end{cases} \tag{2.2.4}$$

where the inverse tangent function '$\tan^{-1}$' (or 'arctan') takes values in $[-\pi/2, \pi/2]$. Note that in the context of circular statistics $\bar{\theta}$ does *not* mean $(\theta_1 + \ldots + \theta_n)/n$ (which is not well defined, as it depends on where the circle is cut).

Example 2.1

For the roulette data in Example 1.1, $\bar{C} = 0.447$ and $\bar{S} = 0.553$, so the mean direction $\bar{\theta}$ and mean resultant length $\bar{R}$ are

$$\bar{\theta} = 51^\circ \quad \text{and} \quad \bar{R} = 0.711.$$

Figure 2.2 shows $\bar{\theta}$ and $\bar{R}$ for this data set and indicates the preferred direction of 51°.

It follows from (2.2.1) and (2.2.2) that

$$\frac{1}{n}\sum_{j=1}^{n} \cos(\theta_j - \bar{\theta}) = \bar{R} \tag{2.2.5}$$

and (for $\bar{R} > 0$)

$$\sum_{j=1}^{n} \sin(\theta_j - \bar{\theta}) = 0. \tag{2.2.6}$$

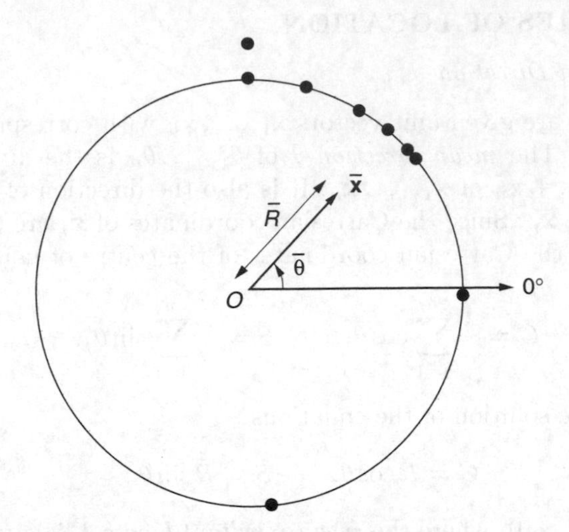

Figure 2.2 The mean direction $\bar{\theta}$ and the mean resultant length $\bar{R}$ for the roulette data in Example 1.1.

Equation (2.2.6) is analogous to

$$\sum_{j=1}^{n}(x_j - \bar{x}) = 0, \tag{2.2.7}$$

for observations $x_1, \ldots, x_n$ on the line with sample mean $\bar{x}$. Equations (2.2.6) and (2.2.7) state that the sums of deviations about the mean are zero. We shall show in Section 2.3.2 that the mean direction minimises a suitable measure of dispersion.

We now consider the effect of rotations on the sample mean direction. Suppose that a new initial direction is chosen, making angle α with the original initial direction. Then the data points correspond to angles

$$\theta'_j = \theta_j - \alpha, \qquad j = 1, \ldots, n, \tag{2.2.8}$$

in this new coordinate system. Put

$$\bar{C}' = \frac{1}{n}\sum_{j=1}^{n}\cos\theta'_j, \qquad \bar{S}' = \frac{1}{n}\sum_{j=1}^{n}\sin\theta'_j.$$

Then

$$\bar{C}' = \bar{R}\cos(\bar{\theta} - \alpha), \qquad \bar{S}' = \bar{R}\sin(\bar{\theta} - \alpha). \tag{2.2.9}$$

If the polar coordinates of $(\bar{C}', \bar{S}')$ are $(\bar{R}', \bar{\theta}')$ then

$$\bar{C}' = \bar{R}'\cos\bar{\theta}', \qquad \bar{S}' = \bar{R}'\sin\bar{\theta}'. \tag{2.2.10}$$

Comparison of (2.2.9) and (2.2.10) gives

$$\bar{\theta}' = \bar{\theta} - \alpha, \qquad \bar{R}' = \bar{R}. \tag{2.2.11}$$

Thus the mean direction of $\theta_1 - \alpha, \ldots, \theta_1 - \alpha$ is $\bar{\theta} - \alpha$, i.e. the sample mean direction is *equivariant* under rotation. This equivariance is analogous to the equivariance under translation of the sample mean of observations on the line, i.e. if $x_1, \ldots, x_n$ are data points on the line with sample mean $\bar{x}$ then the sample mean of $x_1 - a, \ldots, x_n - a$ is $\bar{x} - a$. The importance of such equivariance on the circle (or the line) is that statisticians using different coordinate systems will agree on *where* the sample mean is, even though they may use different numbers to describe its position.

For axial data, the mean axis can be defined as follows. Let $\theta_1, \ldots, \theta_n$ be angles representing axes, and let $\bar{x}$ be the mean direction of the *doubled* angles $2\theta_1, \ldots, 2\theta_n$. Then the *mean axis* is the axis given by $\bar{x}/2$ and $\bar{x}/2 + 180°$.

2.2.2 The Median Direction

For the purpose of robust estimation (see Section 12.4) it is useful to have a version for circular data of the sample median. A *sample median* direction $\tilde{\theta}$ of angles $\theta_1, \ldots, \theta_n$ is any angle ϕ such that (i) half of the data points lie in the arc $[\phi, \phi + \pi)$, and (ii) the majority of the data points are nearer to ϕ than to $\phi + \pi$. When the sample size n is odd, the sample median is one of the data points. When n is even, it is convenient to take the sample median as the midpoint of two appropriate adjacent data points. Population medians are considered in Section 3.4.2.

Example 2.2

For the roulette data in Example 1.1, the observations (in degrees) are 43, 45, 52, 61, 75, 88, 88, 279, 357. From Fig. 1.1, we see that the median direction is 52°. This is close to the mean direction of 51° given in Example 2.1.

2.3 MEASURES OF CONCENTRATION AND DISPERSION

2.3.1 The Mean Resultant Length and the Circular Variance

The mean resultant length $\bar{R}$ was introduced in (2.2.3) as the length of the centre of mass vector $\bar{\mathbf{x}}$, and is given by

$$\bar{R} = (\bar{C}^2 + \bar{S}^2)^{1/2}.$$

Since $\mathbf{x}_1, \ldots, \mathbf{x}_n$ are *unit* vectors,

$$0 \leq \bar{R} \leq 1. \tag{2.3.1}$$

If the directions $\theta_1, \ldots, \theta_n$ are tightly clustered then $\bar{R}$ will be almost 1. On the other hand, if $\theta_1, \ldots, \theta_n$ are widely dispersed then $\bar{R}$ will be almost

0. Thus $\bar{R}$ is a measure of *concentration* of a data set. Note that *any* data set of the form $\theta_1, \ldots, \theta_n, \theta_1 + \pi, \ldots, \theta_n + \pi$ has $\bar{R} = 0$. It follows that $\bar{R} \simeq 0$ does *not* imply that the directions are spread almost evenly round the circle. Equation (2.2.11) shows that $\bar{R}$ is invariant under rotation.

The *resultant length* R is the length of the vector resultant $\mathbf{x}_1 + \ldots + \mathbf{x}_n$. Thus

$$R = n\bar{R}. \tag{2.3.2}$$

For most descriptive and inferential purposes, the mean resultant length $\bar{R}$ is more important than any measure of dispersion. However, for purposes of comparison with data on the line, it is sometimes useful to consider measures of dispersion of circular data. The simplest of these is the sample *circular variance* defined as

$$V = 1 - \bar{R}. \tag{2.3.3}$$

Note that some authors, e.g. Batschelet (1981) use 'circular variance' to refer to $2(1 - \bar{R})$ (compare (2.4.9) with $p = 1$). It follows from (2.3.1) that

$$0 \leq V \leq 1. \tag{2.3.4}$$

2.3.2 Decomposition of Dispersion

A useful measure of the distance between two angles θ and ξ is

$$1 - \cos(\theta - \xi). \tag{2.3.5}$$

Thus one way of measuring the dispersion of angles $\theta_1, \ldots, \theta_n$ about a given angle α is by

$$D(\alpha) = \frac{1}{n} \sum_{i=1}^{n} \{1 - \cos(\theta_i - \alpha)\}. \tag{2.3.6}$$

It follows from (2.2.5) that

$$D(\bar{\theta}) = V. \tag{2.3.7}$$

From the decomposition

$$n - \sum_{i=1}^{n} \cos(\theta_i - \alpha) = (n - n\bar{R}) + \left(n\bar{R} - \sum_{i=1}^{n} \cos(\theta_i - \alpha) \right), \tag{2.3.8}$$

together with (2.3.7) and (2.2.5), we have

$$D(\alpha) = V + 2\bar{R} \left\{ \sin\left(\frac{\bar{\theta} - \alpha}{2} \right) \right\}^2. \tag{2.3.9}$$

This is analogous to the identity

$$\frac{1}{n} \sum_{i=1}^{n} (x_i - u)^2 = \frac{1}{n} \sum_{i=1}^{n} (x_i - \bar{x})^2 + (\bar{x} - u)^2, \tag{2.3.10}$$

for observations $x_1, \ldots, x_n$ on the line with sample mean $\bar{x}$. Equations (2.3.9) and (2.3.10) provide decompositions of the total variation about α or u into the sum of (i) the deviation of the sample about the sample mean (direction) and (ii) the deviation of the sample mean (direction) from α or u. We shall see in Sections 4.8.2 and 7.4.1 that the decomposition (2.3.8) plays an important role in the analysis of variance of circular data.

It follows from (2.3.9) that the dispersion D has a minimum of V at $\bar{\theta}$. This is analogous to the familiar result that for observations $x_1, \ldots, x_n$ on the line,

$$\frac{1}{n}\sum_{i=1}^{n}(x_i - u)^2$$

is minimised when $u = \bar{x}$ and that the minimal value is the (biased version of the) sample variance (with divisor n).

2.3.3 The Circular Standard Deviation

It is sometimes useful to have an analogue for circular data of the standard deviation of data on the line. One way of obtaining such a summary statistic is by transformation of the sample variance V. We shall show later (in Section 3.4.2) that an appropriate transformed statistic is the sample *circular standard deviation* given by

$$v = \{-2\log(1 - V)\}^{1/2} = \{-2\log \bar{R}\}^{1/2}. \tag{2.3.11}$$

Note that v takes values in $[0, \infty]$, whereas V takes values in $[0, 1]$.

For small V, (2.3.11) reduces to

$$v \simeq (2V)^{1/2} = \{2(1 - \bar{R})\}^{1/2}. \tag{2.3.12}$$

2.3.4 Other Measures of Dispersion

The *sample circular dispersion* $\hat{\delta}$ is

$$\hat{\delta} = \frac{1 - \bar{R}_2}{2\bar{R}^2},$$

where $\bar{R}_2$ denotes the mean resultant length of the doubled angles $2\theta_1, \ldots, 2\theta_n$ (see (2.4.4)).

An alternative to (2.3.5) for measuring the distance between two angles θ and ξ is

$$\min(\theta - \xi, 2\pi - (\theta - \xi)) = \pi - |\pi - |\theta - \xi||. \tag{2.3.13}$$

The corresponding measure of the dispersion of angles $\theta_1, \ldots, \theta_n$ about a given angle α is

$$d_0(\alpha) = \frac{1}{n}\sum_{i=1}^{n}\{\pi - |\pi - |\theta - \alpha||\}. \tag{2.3.14}$$

The function d_0 has a minimum at the sample median $\tilde{\theta}$. The *circular mean deviation* is $d_0(\tilde{\theta})$.

The *circular mean difference* is

$$\bar{D}_0 = \frac{1}{n^2}\sum_{i=1}^{n}\sum_{j=1}^{n}\{\pi - |(\pi - |\theta_i - \theta_j|)|\}, \tag{2.3.15}$$

i.e. the mean distance between pairs of data points. It follows from (6.3.46) and (6.3.48) that $\bar{D}_0 \leq \pi/2$. The analogue of (2.3.15) which uses (2.3.5) instead of (2.3.13) to measure distance is

$$\frac{1}{n^2}\sum_{i=1}^{n}\sum_{j=1}^{n}\{1 - \cos(\theta_i - \theta_j)\},$$

which reduces to $1 - \bar{R}^2$.

The *circular range* is the length of the smallest arc which contains all the observations. One way of calculating the circular range is to cut the circle at the initial direction and consider $\theta_1, \ldots, \theta_n$ in the range $0 \leq \theta_i \leq 2\pi$. Let $\theta_{(1)} \leq \ldots \leq \theta_{(n)}$ be the linear order statistics of $\theta_1, \ldots, \theta_n$. The arc lengths between adjacent observations are

$$T_i = \theta_{(i+1)} - \theta_{(i)}, \quad i = 1, \ldots, n-1; \qquad T_n = 2\pi - \theta_{(n)} + \theta_{(1)}.$$

The circular range w is

$$w = 2\pi - \max(T_1, \ldots, T_n). \tag{2.3.16}$$

The circular range is used in a non-parametric test presented in Section 6.3.4.

Example 2.3
For the roulette data in Example 1.1,

$$\begin{array}{rl} \theta_{(i)}: & 43, 45, 52, 61, 75, 88, 88, 279, 357 \\ T_i: & 2, 7, 9, 14, 13, 13, 191, 78, 46 \end{array}$$

so $w = 360° - 191° = 169°$. This is clear from Fig. 1.1, since the longest arc which does not contain any observation is $(88°, 279°)$.

2.4 TRIGONOMETRIC MOMENTS

2.4.1 Definitions

We have seen in Sections 2.2–2.3 that the moments

$$\bar{C} = \frac{1}{n}\sum_{i=1}^{n}\cos\theta_i, \qquad \bar{S} = \frac{1}{n}\sum_{i=1}^{n}\sin\theta_i$$

play key roles in defining the sample mean direction and the sample circular variance. It is useful to combine them into the *first trigonometric moment about the zero direction*

$$m_1' = \bar{C} + i\bar{S}.$$

Then

$$m_1' = \bar{R}e^{i\bar{\theta}}. \tag{2.4.1}$$

Extending this notion, we define the *pth trigonometric moment about the zero direction* for $p = 1, 2, \ldots$ as

$$m_p' = a_p + ib_p, \tag{2.4.2}$$

where

$$a_p = \frac{1}{n}\sum_{j=1}^{n} \cos p\theta_j, \qquad b_p = \frac{1}{n}\sum_{j=1}^{n} \sin p\theta_j. \tag{2.4.3}$$

Then

$$m_p' = \bar{R}_p e^{i\bar{\theta}_p}, \tag{2.4.4}$$

where $\bar{\theta}_p$ and $\bar{R}_p$ denote the sample mean direction and sample mean resultant length of $p\theta_1, \ldots, p\theta_n$.

The *pth trigonometric moment about the mean direction* is

$$m_p = \bar{a}_p + i\bar{b}_p, \tag{2.4.5}$$

where

$$\bar{a}_p = \frac{1}{n}\sum_{i=1}^{n} \cos p(\theta_i - \bar{\theta}), \qquad \bar{b}_p = \frac{1}{n}\sum_{i=1}^{n} \sin p(\theta_i - \bar{\theta}). \tag{2.4.6}$$

In particular, it follows from (2.2.5) and (2.2.6) that

$$m_1 = \bar{R}. \tag{2.4.7}$$

The population versions of the trigonometric moments (introduced in Section 3.4.1) play an important role in the theory of distributions on the circle.

In order to relate the pth trigonometric moments for concentrated data on the circle to moments on the line, consider points $x_1, \ldots, x_n$ on the line with sample mean $\bar{x}$. The line can be wrapped onto the circle to give angles

$$\theta_i = \mu + cx_i \bmod 2\pi, \quad i = 1, \ldots, n.$$

It follows from the expansions

$$\sin\theta = \theta - \frac{1}{3!}\theta^3 + O(\theta^5), \qquad \cos\theta = 1 - \frac{1}{2!}\theta^2 + \frac{1}{4!}\theta^4 + O(\theta^6)$$

that, for c small,

$$\bar{\theta}_p = p\mu + cp\bar{x} + O(c^3) \tag{2.4.8}$$

$$\bar{R}_p = 1 - \frac{c^2p^2}{2}\frac{1}{n}\sum_{i=1}^{n}(x_i - \bar{x})^2 + O(c^4) \tag{2.4.9}$$

$$\bar{b}_2 = -\frac{4c^3}{3}\frac{1}{n}\sum_{i=1}^{n}(x_i - \bar{x})^3 + O(c^5) \tag{2.4.10}$$

$$\bar{a}_2 = 1 - 4(1 - \bar{R}) + \frac{2c^4}{3}\frac{1}{n}\sum_{i=1}(x_i - \bar{x})^4 + O(c^6). \tag{2.4.11}$$

2.4.2 Measures of Skewness and Kurtosis

The interpretations (2.4.8)–(2.4.10) of trigonometric moments of concentrated distributions on the circle motivate the use of

$$\hat{s} = \frac{\bar{R}_2 \sin(\bar{\theta}_2 - 2\bar{\theta})}{(1 - \bar{R})^{3/2}} \tag{2.4.12}$$

as a measure of skewness of circular data. Similarly, (2.4.11) and the fact that $\bar{a}_2 \simeq \bar{R}^4$ for samples from wrapped normal distributions (see the sentence after (3.5.65)) motivate the use of

$$\hat{k} = \frac{\bar{R}_2 \cos(\bar{\theta}_2 - 2\bar{\theta}) - \bar{R}^4}{(1 - \bar{R})^2} \tag{2.4.13}$$

as a measure of kurtosis of circular data. For symmetric unimodal data sets, $\hat{s}$ is nearly zero. For unimodal data sets with a peak which is fitted well by a wrapped normal distribution, $\hat{k}$ is nearly zero.

Example 2.4

For the mallard data in Table 1.1, the first and second trigonometric moments are

$$m_1' = 0.500 - 0.513i \qquad m_2' = -0.040 - 0.382i.$$

Thus

$$\bar{\theta} = 314°, \qquad \bar{R} = 0.716,$$

and

$$m_2 = 0.383 + 0.491i, \qquad \bar{R}_2 = 0.384,$$

and so the measures of skewness and kurtosis are

$$\hat{s} = 0.322 \quad \text{and} \quad \hat{k} = 1.488,$$

indicating some skewness and moderate kurtosis (compared with a wrapped normal distribution). The value of $\bar{R}$ indicates moderate concentration about the mean direction. Such concentration about 314° is in agreement with the histogram in Fig. 1.2. The median direction and the mean deviation of this data set are 314.3° and 48.3°, respectively.

2.5 CORRECTIONS FOR GROUPING

When grouped data are summarised, the calculations proceed as if all data points in an interval were at the midpoint of that interval. This can affect the values of some statistics. If all the class intervals have the same length, then it follows from calculations similar to those in Stuart & Ord (1987, pp. 93–94) that $\bar{\theta}$ and m_p do not require correction for grouping, whereas $\bar{R}_p$ does need such a correction. The corrected value of R_p is

$$\bar{R}_p^* = a(ph)\bar{R}_p, \tag{2.5.1}$$

where

$$a(h) = \frac{h/2}{\sin(h/2)}$$

and h is the length in radians of each class interval. In particular,

$$\bar{R}^* = a(h)\bar{R}. \tag{2.5.2}$$

Since $a(h) \leq 1.03$ for $h \leq 45°$, the corrections (2.5.1) and (2.5.2) are important only when the grouping is coarse.

3

Basic Concepts and Models

3.1 INTRODUCTION

This chapter gives some basic theoretical concepts and various models for circular data. Distribution functions and characteristic functions of distributions on the circle are introduced in Sections 3.2–3.3. Trigonometric moments and various measures of location and dispersion are considered in Section 3.4. The key model for circular data consists of the von Mises distributions, which are described fully in Section 3.5.4. The von Mises distributions can be obtained by conditioning bivariate normal distributions. Various other distributions on the circle arise by radial projection of distributions in the plane, or by wrapping distributions from the line to the circle. Some of these distributions are considered in Sections 3.5.6–3.5.7. Some models on the torus and on the cylinder are given in Section 3.7.

We shall use the following notation:

$\simeq$ means 'is approximately equal to';

$\sim$ usually means 'is distributed as' (sometimes it indicates an asymptotic expansion but the usage should be clear from the context);

$\overset{.}{\sim}$ means 'is approximately distributed as'.

3.2 THE DISTRIBUTION FUNCTION

One way of specifying a distribution on the unit circle is by means of its distribution function. Suppose that an initial direction and an orientation of the unit circle have been chosen. Then the distribution can be regarded as that of a random angle θ, and its *distribution function* F is defined as the function on the whole real line given by

$$F(x) = \Pr(0 < \theta \leq x), \quad 0 \leq x \leq 2\pi,$$

and

$$F(x + 2\pi) - F(x) = 1, \quad -\infty < x < \infty. \tag{3.2.1}$$

Equation (3.2.1) just states that any arc of length 2π on the unit circle has probability 1 (since such an arc is the whole of the circumference of the circle).

For $\alpha \leq \beta \leq \alpha + 2\pi$,

$$\Pr(\alpha < \theta \leq \beta) = F(\beta) - F(\alpha) = \int_{\alpha}^{\beta} dF(x), \tag{3.2.2}$$

where the integral is a Lebesgue-Stieltjes integral. The distribution function F is a right-continuous function. In contrast to distribution functions on the real line,

$$\lim_{x \to \infty} F(x) = \infty, \quad \lim_{x \to -\infty} F(x) = -\infty.$$

By definition,

$$F(0) = 0, \quad F(2\pi) = 1.$$

Note that, although the function F depends on the choice of the zero direction, (3.2.2) shows that $F(\beta) - F(\alpha)$ is independent of this choice. Thus changing the zero direction simply adds a constant to F.

If the distribution function F is absolutely continuous then it has a probability density function f such that

$$\int_{\alpha}^{\beta} f(\theta) d\theta = F(\beta) - F(\alpha), \quad -\infty < \alpha \leq \beta < \infty.$$

A function f is the probability density function of an absolutely continuous distribution if and only if

(i) $f(\theta) \geq 0$ almost everywhere on $(-\infty, \infty)$,
(ii) $f(\theta + 2\pi) = f(\theta)$ almost everywhere on $(-\infty, \infty)$,
(iii) $\int_0^{2\pi} f(\theta) d\theta = 1$.

3.3 THE CHARACTERISTIC FUNCTION

3.3.1 *Definition*

Analogy with distributions on the line suggests that a useful tool for handling the distribution of a random angle θ would be the function $t \mapsto \mathrm{E}[e^{it\theta}]$. Since θ and $\theta + 2\pi$ represent the same direction, it is necessary to restrict t to integer values. The *characteristic function* of a random angle θ is the doubly-infinite sequence of complex numbers $\{\phi_p : p = 0, \pm 1, \ldots\}$ given by

$$\phi_p = \mathrm{E}[e^{ip\theta}] = \int_0^{2\pi} e^{ip\theta} dF(\theta), \quad p = 0, \pm 1, \pm 2, \ldots. \tag{3.3.1}$$

Then

$$\phi_0 = 1, \quad \bar{\phi}_p = \phi_{-p}, \quad |\phi_p| \leq 1, \tag{3.3.2}$$

where $\bar{\phi}_p$ denotes the complex conjugate of ϕ_p. We shall write

$$\phi_p = \alpha_p + i\beta_p, \tag{3.3.3}$$

where

$$\alpha_p = \mathrm{E}[\cos p\theta] = \int_0^{2\pi} \cos p\theta \, dF(\theta) \tag{3.3.4}$$

and

$$\beta_p = \mathrm{E}[\sin p\theta] = \int_0^{2\pi} \sin p\theta \, dF(\theta). \tag{3.3.5}$$

Then

$$\alpha_{-p} = \alpha_p, \quad \beta_{-p} = -\beta_p, \quad |\alpha_p| \leq 1, \quad |\beta_p| \leq 1. \tag{3.3.6}$$

Note that ϕ_p, α_p and β_p are population versions of the pth sample trigonometric moments m'_p, a_p and b_p defined in (2.4.2)–(2.4.3).

3.3.2 Fourier Series

The complex numbers $\{\phi_p : p = 0, \pm 1, \ldots\}$ are the Fourier coefficients of F (see Feller, 1966, p. 595; or Zygmund, 1959, p. 11). When the ϕ_p are related to F by formula (3.3.1), it is usual to write

$$dF(\theta) \sim \frac{1}{2\pi} \sum_{p=-\infty}^{\infty} \phi_p e^{-ip\theta}. \tag{3.3.7}$$

The relationship (3.3.7) does not carry any implication that the series is convergent, still less that it converges to F. However, as we shall see in Section 4.2, if $\sum_{p=1}^{\infty}(\alpha_p^2 + \beta_p^2)$ is convergent then the random variable θ has a density f which is defined almost everywhere by

$$f(\theta) = \frac{1}{2\pi} \sum_{p=-\infty}^{\infty} \phi_p e^{-ip\theta} \tag{3.3.8}$$

(where convergence of the sum is in the L^2 sense). This result is an analogue on the unit circle of the inversion theorem for continuous random variables on the real line. Equation (3.3.8) can be written as

$$f(\theta) = \frac{1}{2\pi} \left\{ 1 + 2 \sum_{p=1}^{\infty} (\alpha_p \cos p\theta + \beta_p \sin p\theta) \right\}. \tag{3.3.9}$$

3.3.3 Independence and Convolution

Let θ_1 and θ_2 be two angular random variables (so that the variable (θ_1, θ_2) takes values on the unit torus). The characteristic function of (θ_1, θ_2) is defined as the double sequence $\{\phi_{p,q} : p, q = 0, \pm 1, \pm 2, \ldots\}$ given by

$$\phi_{p,q} = \mathrm{E}[e^{ip\theta_1 + iq\theta_2}].$$

The variables θ_1 and θ_2 are independent if and only if

$$\phi_{p,q} = \phi_p \phi'_q, \tag{3.3.10}$$

where ϕ and ϕ' denote the marginal characteristic functions of θ_1 and θ_2.

Let $\theta_1, \ldots, \theta_n$ be random variables on the circle. The sum

$$S_n = \theta_1 + \ldots + \theta_n$$

is the analogue of the sum of random variables on the line. If $\theta_1, \ldots, \theta_n$ are distributed independently then the characteristic function of S_n is the product of the characteristic functions of $\theta_1, \ldots, \theta_n$. Further, if $\theta_1, \ldots, \theta_n$ are identically distributed with the common characteristic function $\{\phi_p\}_{0,\pm1,\pm2,\ldots}$ then the characteristic function of S_n is $p \mapsto \phi_p^n$. If the series $\sum_{p=-\infty}^{\infty} |\phi_p|^2$ is convergent then S_n has probability density function

$$\frac{1}{2\pi} \sum_{p=-\infty}^{\infty} \phi_p^n e^{-ip\theta}.$$

Let θ and ξ be two independent random variables with corresponding distribution functions F_1 and F_2. A calculation similar to that used for convolution of random variables on the line shows that the distribution function F of $\theta + \xi$ is given by

$$dF(\zeta) = \int_0^{2\pi} dF_2(\zeta - \theta) dF_1(\theta), \tag{3.3.11}$$

where $\zeta = \theta + \xi$. If one of the random variables has a density then so does the convolution. If θ and ξ have densities f_1 and f_2 then (3.3.11) shows that the density f of $\theta + \xi$ is

$$f(\zeta) = \int_0^{2\pi} f_1(\theta) f_2(\zeta - \theta) d\theta. \tag{3.3.12}$$

It is shown in Section 3.5.3 that if one of the two random variables is uniformly distributed then so is their convolution.

Various other properties of the characteristic function are considered in Sections 4.2.1 and 4.2.3.

3.4 MOMENTS AND MEASURES OF LOCATION AND DISPERSION

In this section we consider population versions of various sample quantities given in Chapter 2.

3.4.1 *Trigonometric Moments*

The *trigonometric moments*

$$\alpha_p = \mathrm{E}[\cos p\theta], \quad \beta_p = \mathrm{E}[\sin p\theta] \tag{3.4.1}$$

have already been defined in (3.3.4)–(3.3.5). Note that the sequence $\{(\alpha_p, \beta_p) : p = 0, \pm 1, \ldots\}$ of trigonometric moments of a random angle θ is equivalent to the characteristic function of θ. It follows from the uniqueness property (key property (i) of Section 4.2.1) that, in contrast to distributions on the line, any distribution on the circle is determined by its moments.

For $p \geq 0$, we write

$$\phi_p = \rho_p e^{i\mu_p}, \quad \rho_p \geq 0 \tag{3.4.2}$$

as the population version of (2.4.4). As the case $p = 1$ is used frequently, we shall write

$$\alpha_1 = \alpha, \quad \beta_1 = \beta, \quad \rho_1 = \rho, \quad \mu_1 = \mu. \tag{3.4.3}$$

The *pth trigonometric moment about the mean direction* is defined by analogy with (2.4.5) as

$$\bar{\alpha}_p + i\bar{\beta}_p, \tag{3.4.4}$$

where

$$\bar{\alpha}_p = \mathrm{E}[\cos p(\theta - \mu)], \qquad \bar{\beta}_p = \mathrm{E}[\sin p(\theta - \mu)]. \tag{3.4.5}$$

3.4.2 *Measures of Location and Dispersion*

Recall from (3.4.3) that

$$\phi_1 = \rho e^{i\mu}. \tag{3.4.6}$$

The direction μ is called the *mean direction* and ρ is called the *mean resultant length.* Note that μ and ρ are the population versions of $\bar{\theta}$ and $\bar{R}$, respectively. The effect of a rotation by $-\psi$ is to map θ to θ^*, where

$$\theta^* = \theta - \psi. \tag{3.4.7}$$

The mean direction of θ^* is

$$\mu^* = \mu - \psi, \tag{3.4.8}$$

i.e. mean direction is equivariant under rotation. Further,

$$\phi_p^* = e^{-ip\psi}\phi_p. \tag{3.4.9}$$

The mean resultant length ρ is invariant under rotation (and reflection). A straightforward calculation shows that

$$\mathrm{E}[\sin(\theta - \mu)] = 0, \tag{3.4.10}$$

which is the population analogue of (2.2.6).

There are various useful measures of dispersion of a distribution on the circle, analogous to the sample measures of dispersion considered in Sections 2.3.2–2.3.4. The *circular variance* ν of a random angle θ is defined as

$$\nu = 1 - \rho = 1 - \mathrm{E}[\cos(\theta - \mu)] \tag{3.4.11}$$

and is the population analogue of V as defined in (2.3.3). Then

$$0 \leq \nu \leq 1.$$

Using $1 - \cos(\theta - \xi)$ to measure the distance between angles θ and ξ as in (2.3.5) leads to the use of

$$V(\alpha) = 1 - \mathrm{E}[\cos(\theta - \alpha)] \tag{3.4.12}$$

as a measure of the variability of θ about any direction α. Then calculations similar to those in Section 2.3.2 give

$$V(\alpha) = \nu + 2\rho \left\{ \sin\left(\frac{\mu - \alpha}{2} \right) \right\}^2 . \tag{3.4.13}$$

Thus V has a minimum of ν at μ.

Since

$$\nu = \int_0^{2\pi} \{1 - \cos(\theta - \mu)\} dF(\theta)$$

and the integrand is continuous and non-negative, $\nu = 0$ if and only if the distribution is concentrated at the point μ. In this respect ν is analogous to the variance of a random variable on the line. If $\nu = 1$ then the distribution can be regarded as so scattered that there is no concentration around any particular direction.

The *circular standard deviation* σ is defined as

$$\sigma = \{-2 \log(1 - \nu)\}^{1/2} = \{-2 \log \rho\}^{1/2} \tag{3.4.14}$$

and is the population analogue of the sample circular standard deviation defined in (2.3.11).

The motivation for this definition is that wrapping the normal distribution $N(\mu, \sigma^2)$ round the circle gives the wrapped normal distribution $WN(\mu, \rho)$ with $\rho = \exp\{-\sigma^2/2\}$ as in (3.5.63), so that

$$1 - \nu = e^{-\sigma^2/2}. \tag{3.4.15}$$

For small ν, (3.4.14) reduces to

$$\sigma \simeq (2\nu)^{1/2}. \tag{3.4.16}$$

The population version of the sample circular dispersion $\hat{\delta}$ is the *circular dispersion* δ defined by

$$\delta = \frac{1 - \rho_2}{2\rho^2}. \tag{3.4.17}$$

The population version of the sample median direction $\tilde{\theta}$ is the *median direction* $\tilde{\mu}$, defined as a direction ϕ which minimises

$$\mathrm{E}[\pi - |\pi - |\theta - \phi||]. \tag{3.4.18}$$

Note that (3.4.18) is the population analogue of (2.3.14). A median direction satisfies

$$\Pr(\theta \in [\tilde{\mu}, \tilde{\mu}+\pi)) \geq \tfrac{1}{2}, \qquad \Pr(\theta \in (\tilde{\mu}-\pi, \tilde{\mu}]) \geq \tfrac{1}{2} \tag{3.4.19}$$

A unimodal distribution has a unique median direction.

Asymmetry of a circular distribution can be measured by the *skewness*

$$s = \frac{\bar{\beta}_2}{(1-\rho)^{3/2}}, \tag{3.4.20}$$

which is the population version of the sample skewness $\hat{s}$. Peakedness can be measured by the *kurtosis*

$$k = \frac{\bar{\alpha}_2 - \rho^4}{(1-\rho)^2}, \tag{3.4.21}$$

which is the population version of the sample kurtosis $\hat{k}$.

3.4.3 A Chebyshev Inequality

Applying Chebyshev's inequality to the random variable $\sin([\theta-\mu]/2)$ gives

$$\Pr\left(\left|\sin\frac{1}{2}(\theta-\mu)\right| \geq \varepsilon\right) \leq \frac{\nu}{2\varepsilon^2}, \quad 0 < \varepsilon \leq 1. \tag{3.4.22}$$

The inequality (3.4.22) cannot be sharpened, as is shown by the symmetric discrete distribution with

$$\Pr(\theta = 0) = 1 - \frac{\nu}{2\varepsilon^2}, \quad \Pr(\theta = \pm 2\sin^{-1}\varepsilon) = \frac{\nu}{2\varepsilon^2}.$$

Various other inequalities for distributions on the circle are given by Marshall & Olkin (1961).

3.4.4 Symmetrical Distributions

The distribution of θ is symmetrical about μ if the distribution is invariant under the transformation

$$\theta \mapsto \mu - \theta, \tag{3.4.23}$$

i.e. under reflection in μ. If θ is symmetrical about μ and has density f, then

$$f(\theta - \mu) = f(\mu - \theta). \tag{3.4.24}$$

If a distribution is symmetrical about $\theta = \mu$ then it is also symmetrical about $\theta = \mu + \pi$. Further, if the distribution is unimodal then the mean direction, the median direction and the mode are all equal. Then the sine moments of $\theta - \mu$ are zero, so that for $\mu = 0$ the Fourier expansion (3.3.9) simplifies to

$$f(\theta) = \frac{1}{2\pi}\left\{1 + 2\sum_{p=1}^{\infty} \alpha_p \cos p\theta\right\}. \tag{3.4.25}$$

3.5 CIRCULAR MODELS

3.5.1 Introduction

In this section we describe some of the more important families of distributions on the circle. The most basic distribution on the circle is the uniform distribution. The von Mises distributions studied in Section 3.5.4 play a key role in statistical inference on the circle, analogous to that of the normal distributions on the line. Other useful distributions are the cardioid, wrapped normal, wrapped Cauchy, and projected normal distributions, considered in Sections 3.5.5–3.5.7. Some discrete models are considered in Sections 3.5.2 and 3.5.7.

Most models for directional data belong to the two main classes of parametric models: exponential models and transformation models. Because the major inferential properties of the models for directional data come from their structure as exponential models or transformation models, it seems worth summarising here the main properties of these two classes of models.

An *exponential model* has probability density functions of the form

$$f(x;\omega) = b(x)\exp\{\boldsymbol{\phi}(\boldsymbol{\omega})^T\mathbf{t}(x) - \psi(\boldsymbol{\omega})\} \tag{3.5.1}$$

with respect to some dominating measure λ, where $\mathbf{t}$ and $\boldsymbol{\phi}$ are $\mathbb{R}^m$-valued functions on the sample space $\mathcal{X}$ and the parameter space Ω, respectively. The function $\mathbf{t}$ is called the *canonical statistic* and $\boldsymbol{\phi}(\boldsymbol{\omega})$ is called the *canonical parameter*. If d is the dimension of $\phi(\Omega)$ and the representation (3.5.1) is minimal then the model is called an (m,d) *exponential model.* If $d<m$ then the model is called a *curved exponential model.* If $d=m$ then (3.5.1) can be written as

$$f(x;\boldsymbol{\theta}) = b(x)\exp\{\boldsymbol{\theta}^T\mathbf{t}(x) - \psi(\boldsymbol{\theta})\}, \tag{3.5.2}$$

where $\boldsymbol{\theta}$ runs through a subset Θ of $\mathbb{R}^m$. If Θ is open in $\mathbb{R}^m$ and

$$\Theta = \left\{\boldsymbol{\theta} : \int b(x)\exp\{\boldsymbol{\theta}^T\mathbf{t}(x)\}d\lambda(x) < \infty\right\}$$

then the model (3.5.2) is called a *regular exponential model.* The key properties of regular exponential models are as follows:

(i) the first two moments of the canonical statistic are given by

$$\mathrm{E}_{\boldsymbol{\theta}}[\mathbf{t}] = \frac{\partial\psi}{\partial\boldsymbol{\theta}}, \tag{3.5.3}$$

$$\mathrm{var}_{\boldsymbol{\theta}}(\mathbf{t}) = \frac{\partial^2\psi}{\partial\boldsymbol{\theta}\partial\boldsymbol{\theta}^T}; \tag{3.5.4}$$

(ii) the Fisher information matrix is $\mathrm{var}_{\boldsymbol{\theta}}(\mathbf{t})$;

(iii) if $x_1, \ldots, x_n$ are independently distributed with probability density function (3.5.2) then the sample mean $\bar{\mathbf{t}}$ has probability density function proportional to

$$\exp\left\{n\left(\boldsymbol{\theta}^T\bar{\mathbf{t}} - \psi(\boldsymbol{\theta})\right)\right\}, \tag{3.5.5}$$

and

$$\bar{\mathbf{t}} \text{ is sufficient for } \boldsymbol{\theta}; \tag{3.5.6}$$

(iv) the maximum likelihood estimator $\hat{\boldsymbol{\theta}}$ of $\boldsymbol{\theta}$ is unique and is given by

$$\mathrm{E}_{\hat{\boldsymbol{\theta}}}[\mathbf{t}] = \bar{\mathbf{t}}, \tag{3.5.7}$$

where $\bar{\mathbf{t}}$ denotes the sample mean of $\mathbf{t}$.

Further details about exponential models can be found in Chapter 2 of Barndorff-Nielsen (1978a).

A *(composite) transformation model* is a model in which a group G acts on both the sample space $\mathcal{X}$ and the parameter space Ω, such that if the random variable x has probability density function $f(x;\omega)$ then the random variable gx has probability density function $f(gx;g\omega)$. Then

$$f(gx;g\omega) = f(x;\omega)\chi(g,x), \tag{3.5.8}$$

for some function χ on $G \times \mathcal{X}$. A *transformation model* is a composite transformation model in which the group G acts transitively on the parameter space, i.e., for any ω and ω' in Ω, there is a g in G such that $g\omega = \omega'$. The standard examples of transformation models are location models on the line. Here the group is the additive group $\mathbb{R}$ and the underlying measure is Lebesgue measure, so that (3.5.8) becomes

$$f(x+g;\omega+g) = f(x;\omega).$$

For our purposes, the key property of composite transformation models is that the maximum likelihood estimator $\hat{\omega}$ of ω is equivariant, i.e.

$$\hat{\omega}(gx_1, \ldots, gx_n) = g\hat{\omega}(x_1, \ldots, x_n). \tag{3.5.9}$$

Further details about composite transformation models can be found in Chapter 2 of Barndorff-Nielsen (1988) and Section 8 of Barndorff-Nielsen, Blæsild & Eriksen (1989). An *exponential transformation model* is an exponential model which is also a composite transformation model. Exponential transformation models have a very rich structure. See Barndorff-Nielsen *et al.* (1982) or Section 2.5 of Barndorff-Nielsen (1988).

Many of the tests used in directional statistics are either likelihood ratio tests or score tests. Since score tests are not as well known as likelihood ratio tests, we recall their construction and main property. For a parametric statistical model with parameter $\omega = (\psi^T, \chi^T)^T$, denote by $l(\omega; x_1, \ldots, x_n)$ the log-likelihood based on a random sample $x_1, \ldots, x_n$, and let

$$\mathbf{I} = \begin{pmatrix} \mathbf{I}_{\psi\psi} & \mathbf{I}_{\psi\chi} \\ \mathbf{I}_{\chi\psi} & \mathbf{I}_{\chi\chi} \end{pmatrix}$$

be the corresponding Fisher information matrix. If ψ_0 is a given value of ψ then the score test of the null hypothesis $H_0 : \psi = \psi_0$ rejects H_0 for large values of

$$S = \mathbf{U}^T \mathbf{I}_{\psi\psi.\chi}^{-1} \mathbf{U},$$

where

$$\begin{aligned} \mathbf{U} &= \frac{\partial l}{\partial \psi^T}, \\ \mathbf{I}_{\psi\psi.\chi} &= \mathbf{I}_{\psi\psi} - \mathbf{I}_{\psi\chi} \mathbf{I}_{\chi\chi}^{-1} \mathbf{I}_{\chi\psi} \end{aligned}$$

and both $\mathbf{U}$ and $\mathbf{I}_{\psi\psi.\chi}$ are evaluated at the maximum likelihood estimate of ω under H_0. Under suitable regularity conditions, the large-sample asymptotic null distribution of S is

$$S \mathrel{\dot{\sim}} \chi^2_\nu,$$

where ν is the dimension of the interest parameter ψ (see Cox & Hinkley, 1974, Section 9.3.)

3.5.2 Lattice Distributions

Consider a discrete distribution with

$$\Pr\left(\theta = \nu + \frac{2\pi r}{m}\right) = p_r, \quad r = 0, 1, \ldots, m-1, \tag{3.5.10}$$

and

$$p_r \geq 0, \quad \sum_{r=0}^{m-1} p_r = 1.$$

The points $\nu + 2\pi r/m$ are the vertices of an m-sided regular polygon inscribed in the unit circle. If all the weights are equal then

$$p_r = \frac{1}{m}. \tag{3.5.11}$$

This distribution is called a *discrete uniform distribution* on m points. The case $m = 37$ gives the distribution of the stopping position of the ball on an unbiased roulette wheel.

If $\nu = 0$ then the characteristic function of (3.5.10) is given by

$$\phi_p = \sum_{r=0}^{m-1} p_r e^{2\pi r i p/m} \tag{3.5.12}$$

and so

$$\phi_p = 1 \quad \text{for } p = 0 \pmod m. \tag{3.5.13}$$

For the discrete uniform distribution, (3.5.12) reduces to

$$\phi_p = \begin{cases} 1, & p = 0 \pmod m, \\ 0, & \text{otherwise.} \end{cases} \tag{3.5.14}$$

A Poisson distribution on the circle will be presented in Section 3.5.7. A model for the distribution of the first significant digits is given in Example 4.1.

3.5.3 Uniform Distribution

The most basic distribution on the circle is the uniform distribution; this is often used as the null model. It is the unique distribution on the circle which is invariant under rotation and reflection. It has probability density function

$$f(\theta) = \frac{1}{2\pi}. \tag{3.5.15}$$

Thus for $\alpha \le \beta \le \alpha + 2\pi$

$$\Pr(\alpha < \theta \le \beta) = \frac{\beta - \alpha}{2\pi},$$

i.e. probability is proportional to arc length. Integration of $\exp(ip\theta)$ shows that

$$\phi_p = \begin{cases} 1, & p = 0, \\ 0, & p \ne 0. \end{cases} \tag{3.5.16}$$

Thus $\rho = 0$, so $\nu = 1$ and there is no concentration about any particular direction.

Let $\theta_1, \ldots, \theta_n$ be n independent uniform random variables with common characteristic function $\{\phi_p\}_{0,\pm1,\pm2,\ldots}$. The characteristic function of the sum $S_n = \theta_1 + \ldots + \theta_n$ is $\{\phi_p^n\}_{0,\pm1,\pm2,\ldots}$. From (3.5.16),

$$\phi_p^n = \begin{cases} 1, & p = 0, \\ 0, & p \ne 0, \end{cases}$$

which is the characteristic function of the uniform distribution. It follows from the uniqueness property (key property (i) of Section 4.2.1) that S_n is uniformly distributed on the circle. We shall see in Section 4.3.1 that, under a mild condition, for any independent and identically distributed random variables $\theta_1, \ldots, \theta_n$, S_n tends to the uniform distribution as $n \to \infty$.

Furthermore, let θ_1 be distributed uniformly and let θ_2 have any distribution whatsoever. If θ_1 and θ_2 are independently distributed then the characteristic function of $\theta_1 + \theta_2$ is given by (3.5.16). Hence, by the uniqueness property (key property (i) of Section 4.2.1), $\theta_1 + \theta_2$ is distributed uniformly.

3.5.4 Von Mises Distributions

From the point of view of statistical inference, perhaps the most useful distributions on the circle are the von Mises distributions.

Definition

The *von Mises distribution* $M(\mu, \kappa)$ has probability density function

$$g(\theta; \mu, \kappa) = \frac{1}{2\pi I_0(\kappa)} e^{\kappa \cos(\theta - \mu)}, \tag{3.5.17}$$

where I_0 denotes the modified Bessel function of the first kind and order 0, which can be defined by

$$I_0(\kappa) = \frac{1}{2\pi} \int_0^{2\pi} e^{\kappa \cos\theta} \, d\theta \tag{3.5.18}$$

(take $p = 0$ in (A.1) of Appendix 1). The function I_0 has power series expansion

$$I_0(\kappa) = \sum_{r=0}^{\infty} \frac{1}{(r!)^2} \left(\frac{\kappa}{2}\right)^{2r} \tag{3.5.19}$$

(take $p = 0$ in (A.2) of Appendix 1). The parameter μ is the mean direction and the parameter κ is known as the *concentration parameter*. The mean resultant length ρ is $A(\kappa)$, where A is the function defined in (3.5.31) below.

Note that $M(\mu + \pi, \kappa)$ and $M(\mu, -\kappa)$ are the same distribution. To eliminate this indeterminancy of the parameters μ, κ, it is usual to take $\kappa \geq 0$.

This distribution was introduced by von Mises (1918) in order to study the deviations of measured atomic weights from integral values. See Example 6.4.

The Shape of the Distribution

The distribution is unimodal and is symmetrical about $\theta = \mu$. The mode is at $\theta = \mu$ and the antimode is at $\theta = \mu + \pi$. The ratio of the density at the mode to the density at the antimode is given by $e^{2\kappa}$, so that the larger the value of κ, the greater is the clustering around the mode.

Figure 3.1 shows the density for $\mu = 0$ and $\kappa = 0.5, 1, 2, 4$. For $\kappa = 4$, over 99% of the probability lies in the arc $(-90°, 90°)$. Figure 3.2 gives a polar representation of the density for $\mu = 0$ and $\kappa = 1.6$.

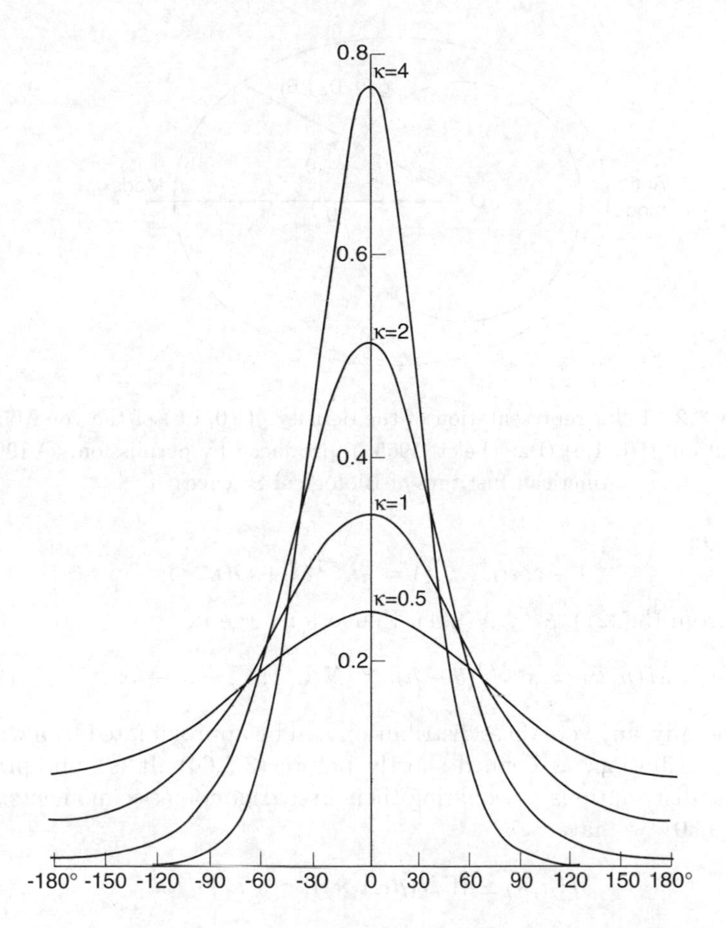

Figure 3.1 Density of the von Mises distribution $M(0, \kappa)$ for $\kappa = 0.5, 1, 2, 4$.

Relationship with other distributions

When $\kappa = 0$, $M(\mu, \kappa)$ is the uniform distribution. The approximation $\exp(x) \simeq 1 + x$ shows that for small κ,

$$M(\mu, \kappa) \simeq C(\mu, \kappa/2), \tag{3.5.20}$$

where $C(\mu, \kappa/2)$ denotes a cardioid distribution, as defined in Section 3.5.5. Thus a von Mises distribution with small concentration parameter can be approximated by the cardioid distribution with the same mean direction and mean resultant length. As $\kappa \to \infty$, the $M(\mu, \kappa)$ distribution becomes concentrated at the point $\theta = \mu$. If κ is large, put $\xi = \kappa^{1/2}(\theta - \mu)$. Then from (3.5.17) the probability density function of ξ is proportional to

$$\exp\{-\kappa[1 - \cos(\kappa^{-1/2}\xi)]\}. \tag{3.5.21}$$

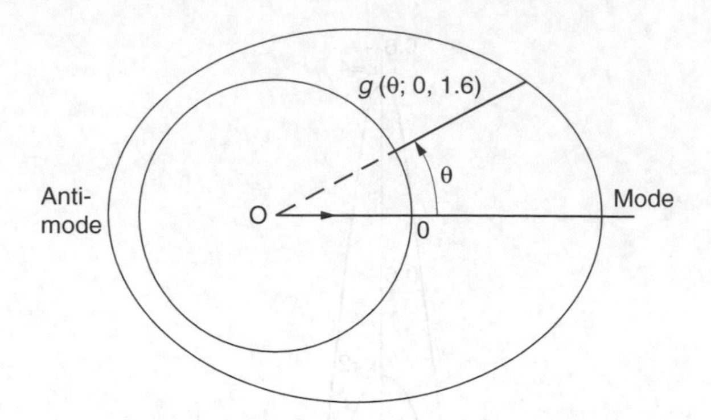

Figure 3.2 Polar representation of the density $g(\cdot; 0, 1.6)$ of the von Mises distribution $M(0, 1.6)$ (Batschelet, 1965, reproduced by permission, © 1999 American Institute of Biological Sciences).

For large κ,

$$1 - \cos(\kappa^{-1/2}\xi) = \tfrac{1}{2}\kappa^{-1}\xi^2 + O(\kappa^{-2}),$$

so that, from (3.5.21), $\xi \mathrel{\dot{\sim}} N(0, 1)$. Hence for large κ,

$$\theta \mathrel{\dot{\sim}} M(\mu, \kappa) \Rightarrow \kappa^{-1/2}(\theta - \mu) \mathrel{\dot{\sim}} N(0, 1), \qquad \kappa \to \infty. \tag{3.5.22}$$

More generally, any von Mises distribution can be approximated by a wrapped normal distribution, as defined shortly before (3.5.64). It is appropriate to match the distributions by equating their first trigonometric moments. Thus, using (3.5.30), we have

$$M(\mu, \kappa) \simeq WN(\mu, A(\kappa)), \qquad \kappa \to \infty. \tag{3.5.23}$$

Although the approximation (3.5.23) was derived here as a first-order approximation for large κ, Kent (1978) has shown that the approximation holds to a higher order in κ. More precisely,

$$f_{VM}(\theta; \mu, \kappa) - f_{WN}(\theta; \mu, A(\kappa)) = O(\kappa^{-1/2}) \qquad \kappa \to \infty, \tag{3.5.24}$$

where $f_{VM}(\cdot; \mu, \kappa)$ and $f_{WN}(\cdot; \mu, A(\kappa))$ denote the densities of the von Mises distribution $M(\mu, \kappa)$ and the approximating wrapped normal distribution $WN(\mu, A(\kappa))$, respectively. Stephens (1963) has verified numerically that the approximation (3.5.23) is satisfactory for intermediate values of κ. The worst match (in terms of maximum absolute difference of probability density functions) between a von Mises distribution and the approximating wrapped normal distribution occurs for $\kappa \simeq 1.4$. Even in this case the two densities are very close. The von Mises distribution $M(\mu, \kappa)$ is also close to the wrapped Cauchy distribution $WC(\mu, A(\kappa))$, as defined in Section 3.5.7, with the same mean direction and mean resultant length. This means that the statistician

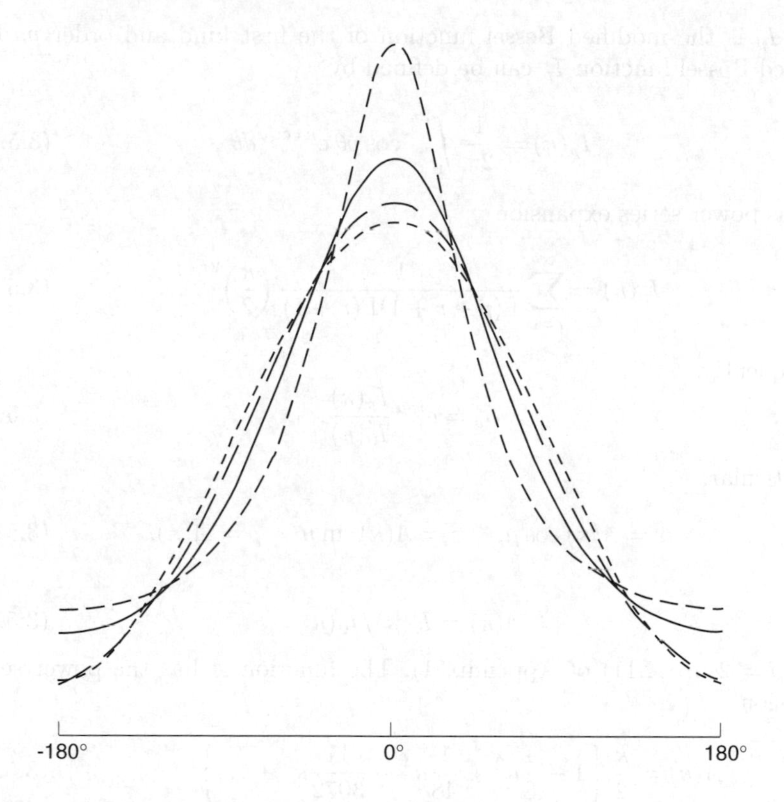

Figure 3.3 Comparison of the probability density functions of the von Mises (————), cardioid (- - - - - - -), wrapped normal (— — — —) and wrapped Cauchy (– – – – –) distributions with $\mu = 0$ and $\rho = 0.45$.

can work with whichever of these distributions is most convenient for his purpose. For inferential purposes, the von Mises distributions are most useful, because of their structure as exponential transformation models.

Figure 3.3 gives a comparison of the cardioid, wrapped normal, wrapped Cauchy and von Mises distributions with same values of μ and ρ.

Characteristic function and moments

Since the distribution is symmetrical about μ,

$$\bar{\beta}_p = \mathrm{E}[\sin p(\theta - \mu)] = 0. \tag{3.5.25}$$

Also,

$$\begin{aligned}\bar{\alpha}_p &= \frac{1}{2\pi I_0(\kappa)} \int_0^{2\pi} \cos p(\theta - \mu)\, e^{\kappa \cos(\theta - \mu)}\, d\theta \\ &= \frac{I_p(\kappa)}{I_0(\kappa)},\end{aligned} \tag{3.5.26}$$

where I_p is the modified Bessel function of the first kind and order p. The modified Bessel function I_p can be defined by

$$I_p(\kappa) = \frac{1}{2\pi}\int_0^{2\pi} \cos p\theta\, e^{\kappa\cos\theta}\, d\theta \tag{3.5.27}$$

and has power series expansion

$$I_p(\kappa) = \sum_{r=0}^{\infty} \frac{1}{\Gamma(p+r+1)\Gamma(r+1)}\left(\frac{\kappa}{2}\right)^{2r+p}. \tag{3.5.28}$$

Consequently,

$$\phi_p = e^{ip\mu}\frac{I_p(\kappa)}{I_0(\kappa)}. \tag{3.5.29}$$

In particular,

$$\alpha = A(\kappa)\cos\mu, \quad \beta = A(\kappa)\sin\mu, \quad \rho = A(\kappa), \tag{3.5.30}$$

where

$$A(\kappa) = I_1(\kappa)/I_0(\kappa). \tag{3.5.31}$$

(take $p = 2$ in (A.11) of Appendix 1). The function A has the power series expansion

$$A(\kappa) = \frac{\kappa}{2}\left\{1 - \frac{1}{8}\kappa^2 + \frac{1}{48}\kappa^4 - \frac{11}{3072}\kappa^6 + \dots\right\}, \tag{3.5.32}$$

which is useful for small κ (take $p = 2$ in (A.12) of Appendix 1). This relation follows upon using the series definitions of $I_0(\kappa)$ and $I_1(\kappa)$ given by (3.5.28). For large κ, I_p has the asymptotic expansion (see (A.4) of Appendix 1 with $p = 2$)

$$\begin{aligned} I_p(\kappa) \quad &\sim \frac{e^\kappa}{\sqrt{2\pi\kappa}}\left\{1 - \frac{m-1}{8\kappa} + \frac{(m-1)(m-9)}{2!(8\kappa)^2}\right. \\ &\left. - \frac{(m-1)(m-9)(m-25)}{3!(8\kappa)^3} + \dots\right\}, \end{aligned} \tag{3.5.33}$$

where $m = 4p^2$. Using this expansion in (3.5.31) gives

$$A(\kappa) \sim 1 - \frac{1}{2\kappa} - \frac{1}{8\kappa^2} - \frac{1}{8\kappa^3} + o(\kappa^{-3}), \tag{3.5.34}$$

as in (A.13) of Appendix 1. Differentiation of the series for $I_1(\kappa)$ given in (3.5.28) gives

$$\frac{d}{d\kappa}\{\kappa I_1(\kappa)\} = \kappa I_0(\kappa), \quad \frac{d}{d\kappa}\left\{\frac{I_1(\kappa)}{\kappa}\right\} = \frac{I_2(\kappa)}{\kappa}. \tag{3.5.35}$$

Elimination of the derivative of $I_1(\kappa)$ yields

$$I_2(\kappa) = I_0(\kappa) - \frac{2I_1(\kappa)}{\kappa}, \tag{3.5.36}$$

and so

$$\alpha_2 = \frac{I_2(\kappa)}{I_0(\kappa)} = 1 - \frac{2A(\kappa)}{\kappa}. \tag{3.5.37}$$

The distribution function

The distribution function of the von Mises distribution $M(0, \kappa)$ is

$$F(\theta; 0, \kappa) = \frac{1}{I_0(\kappa)} \int_0^\theta e^{\kappa \cos u} du \tag{3.5.38}$$

and is not particularly easy to evaluate numerically. A table of $F(\theta; \pi, \kappa)$ for $0 < \kappa \leq 10$ is given in Appendix 2.1. The result (3.5.23) means that the distribution function of $N(\mu, \kappa^{-1})$ provides an approximation to the distribution function of $M(\mu, \kappa)$. For $\kappa > 10$, this approximation is very accurate.

Improved approximations to the distribution functions of von Mises distributions are given by approximate normalising transformations which refine the transformation $\theta \mapsto \kappa^{1/2}(\theta - \mu)$ used in (3.5.21). Upton (1974) gave the approximation

$$\kappa^{\frac{1}{2}} \left\{ \left(1 - \frac{1}{8\kappa}\right) \theta - \frac{1}{24} \left(1 + \frac{1}{4\kappa}\right) \theta^3 \right\} \;\dot{\sim}\; N(0,1),$$

where $\theta \sim M(0, \kappa)$, and showed that this yields the cumulative distribution function of $M(0, \kappa)$ correct to 2 decimal places for $\kappa \geq 6$. A more refined approximation by Hill (1976) is that if $\theta \sim M(0, \kappa)$ then $\chi \;\dot{\sim}\; N(0, 1)$, where

$$\chi = y - \frac{y}{8\kappa} - \frac{2y^3 + 7y}{128\kappa^2} - \frac{8y^5 + 46y^3 + 177y}{3072\kappa^3} - \ldots$$

and

$$y = 2\sqrt{\kappa} \sin\left(\frac{\theta}{2}\right).$$

Computer algorithms for evaluating the von Mises distribution function have been given by Mardia & Zemroch (1975b) and Hill (1977).

Selected quantiles of $M(0, \kappa)$ are given in Appendix 2.2.

Genesis

We now give five ways in which von Mises distributions can arise. The first four of these are analogous to ways in which normal distributions arise on the line.

Conditioning Normal Distributions

Let $\mathbf{x}$ have a bivariate normal distribution with mean $\boldsymbol{\mu} = (\cos\mu, \sin\mu)^T$ and variance matrix $\kappa^{-1}\mathbf{I}_2$. Put $\mathbf{x} = r(\cos\theta, \sin\theta)^T$. Then the probability density function of (r, θ) is proportional to

$$r \exp\left\{-\frac{\kappa}{2}\left[r^2 - 2r\cos(\theta - \mu)\right]\right\}, \tag{3.5.39}$$

and so the conditional distribution of θ given $r = 1$ is $M(\mu, \kappa)$.

Maximum Likelihood Characterisation

Consider a continuous location model on the circle, i.e. a set of probability density functions of the form

$$f(\theta; \mu) = g(\theta - \mu),$$

where g is some given positive function on the circle. Suppose that g has continuous second derivative and that (for $n = 2, 3$) the maximum likelihood estimate of μ based on observations $(\theta_1, \ldots, \theta_n)$ is the sample mean direction $\bar{\theta}$. Then

$$\sum_{i=1}^{n} \frac{g'(\theta_i - \bar{\theta})}{g(\theta_i - \bar{\theta})} = 0 \tag{3.5.40}$$

and

$$\sum_{i=1}^{n} \sin(\theta_i - \bar{\theta}) = 0, \tag{3.5.41}$$

by (2.2.6). Taking $n = 2$ and firstly $\theta_2 = -\theta_1$ and then $\theta_2 = \theta_1 + \pi$ shows that $g'(\theta)/g(\theta) = h(\sin\theta)$ for some continuously differentiable function h. Taking $n = 3$ shows that h satisfies $h(x + y) = h(x) + h(y)$. It follows that $h(x) = \kappa x$ for some κ, and so g is the probability density function of a von Mises distribution. This characterisation is due to von Mises (1918) and is analogous to the characterisation by Gauss of normal distributions as location models on the line for which the sample mean $\bar{x}$ is the maximum likelihood estimate of the population mean.

Maximum Entropy Approach

Another way in which the von Mises distributions (3.5.17) arise is as maximum entropy distributions. The *entropy* of a distribution on the circle with probability density function f is defined as

$$H(f) = -\int_0^{2\pi} f(\theta) \log f(\theta) d\theta \tag{3.5.42}$$

and is one way of measuring the closeness of a distribution to the uniform distribution. Let $\mathbf{t} = (t_1, \ldots, t_d)^T$ be a d-dimensional function on the circle and let $\mathbf{c}$ be a given vector in $\mathbb{R}^d$. Consider the problem of finding the distribution on the circle for which the entropy is maximal subject to $\mathrm{E}[\mathbf{t}(x)] = \mathbf{c}$. It follows from Jensen's inequality and the convexity of the function $h(x) = x \log x$ that this maximum entropy distribution is a member of the (d, d) exponential model with canonical statistic $\mathbf{t}$. (This result holds for any sample space. See Kagan, Linnik & Rao, 1973, p. 409. The choice of this maximum entropy distribution as a distribution when $\mathrm{E}[\mathbf{t}(x)]$ is specified is sometimes called 'the maximum entropy principle'. See Jaynes, 1957; 1963.) In particular, the maximum entropy distribution on the circle with given mean direction μ and given mean resultant length ρ is the von Mises distribution $M(\mu, \kappa)$ with $A(\kappa) = \rho$. This characterisation is due to Mardia (1972a, pp. 65–66).

Connection with a Diffusion Process on the Circle

Consider a diffusion on the circle with infinitesimal variance σ^2 and drift $-\lambda \sin(\theta - \mu)$. Such a diffusion is called a *von Mises process* (Kent, 1975) and is analogous to the Ornstein–Uhlenbeck process on the line. The equilibrium distribution of the von Mises process is the von Mises distribution $M(\mu, 2\lambda\sigma^{-2})$.

Connection with a Diffusion Process in the Plane

Consider particles which move under a Brownian motion in the plane with infinitesimal variance $\mathbf{I}_2$ and drift $\kappa(\cos\mu, \sin\mu)^T$. Gordon & Hudson (1977) showed that if the particles start at the origin then the distribution of the points at which they first hit the unit circle is the von Mises distribution $M(\mu, \kappa)$.

Simulation

The von Mises distribution $M(\mu, \kappa)$ can be simulated efficiently using the following algorithm of Best & Fisher (1979), which is available in the IMSL (1991) library. Put

$$a = 1 + (1 + 4\kappa^2)^{\frac{1}{2}}, \qquad b = \frac{a - (2a)^{\frac{1}{2}}}{2\kappa}, \qquad r = \frac{1 + b^2}{2b}.$$

Let U_1, U_2, U_3 be pseudo-random numbers in $[0, 1]$ (chosen afresh each time step (i), (ii) or (iv) is executed).

(i) Put $z = \cos(\pi U_1), f = (1 + rz)/(r + z), c = \kappa(r - f)$.
(ii) If $c(2 - c) - U_2 > 0$ then go to step (iv).
(iii) If $\log(c/U_2) + 1 - c < 0$ then return to step (i).
(iv) $\theta = \mu + \mathrm{sign}(U_3 - 0.5) \cos^{-1}(f)$.

The algorithm implements an acceptance–rejection method with envelope proportional to the wrapped Cauchy distribution $WC(\mu, b)$. A different algorithm, which may be faster than the above if κ changes from call to call, was given by Dagpunar (1990).

Convolutions of von Mises distributions

Let θ_1 and θ_2 be independently distributed as $M(\mu_1, \kappa_1)$ and $M(\mu_2, \kappa_2)$, respectively. It follows from the convolution formula (3.3.12) that the probability density function of $\theta = \theta_1 + \theta_2$ is given by

$$h(\theta) = \frac{1}{4\pi^2 I_0(\kappa_1) I_0(\kappa_2)} \int_0^{2\pi} \exp\{\kappa_1 \cos(\xi - \mu_1) + \kappa_2 \cos(\theta - \xi - \mu_2)\}\, d\xi.$$

Since

$$\begin{aligned}&\kappa_1 \cos(\xi - \mu_1) + \kappa_2 \cos(\theta - \xi - \mu_2)\\ &= [\kappa_1 \cos\mu_1 + \kappa_2 \cos(\theta - \mu_2)]\cos\xi + [\kappa_1 \sin\mu_1 + \kappa_2 \sin(\theta - \mu_2)]\sin\xi,\end{aligned}$$

an application of (3.5.18) shows that (Mardia, 1972a, p. 67)

$$h(\theta) = \frac{1}{2\pi I_0(\kappa_1) I_0(\kappa_2)} I_0\left(\{\kappa_1^2 + \kappa_2^2 + 2\kappa_1\kappa_2 \cos(\theta - [\mu_1 + \mu_2])\}^{1/2}\right). \tag{3.5.43}$$

Thus, the convolution of two von Mises distributions is not a von Mises distribution. However, (3.5.43) can be approximated by a von Mises distribution, as follows. From (3.5.23), the distributions $M(\mu_1, \kappa_1)$ and $M(\mu_2, \kappa_2)$ can be approximated by the wrapped normal distributions $WN(\mu_1, A(\kappa_1))$ and $WN(\mu_2, A(\kappa_2))$. It follows from (3.5.67) that the convolution of these two distributions is the wrapped normal distribution $WN(\mu_1 + \mu_2, A(\kappa_1)A(\kappa_2))$, which in turn can be approximated by $M(\mu_1 + \mu_2, A^{-1}(A(\kappa_1)A(\kappa_2)))$. Thus

$$\theta_1 + \theta_2 \dot{\sim} M(\mu_1 + \mu_2, A^{-1}(A(\kappa_1)A(\kappa_2))). \tag{3.5.44}$$

Numerical studies (Stephens, 1963) have shown that this approximation is quite satisfactory.

The probability density function of an n-fold convolution of von Mises distributions can be obtained by substituting the characteristic function of the sum into the Fourier series (3.3.9). In the case $n = 2$, equating this expression to (3.5.43) gives

$$I_0(\{\kappa_1^2 + \kappa_2^2 + 2\kappa_1\kappa_2 \cos\theta\}^{1/2}) = I_0(\kappa_1)I_0(\kappa_2) + 2\sum_{p=1}^{\infty} I_p(\kappa_1)I_p(\kappa_2)\cos p\theta. \tag{3.5.45}$$

This is the Neumann addition formula, which is valid for any complex numbers κ_1 and κ_2 (see (A.5) of Appendix 1). For $n > 2$, the Fourier series for the probability density function of the convolution of n von Mises distributions does not have a simplification analogous to that in the case $n = 2$. However, the probability density function can be approximated by a von Mises distribution.

Generalisations of von Mises Distributions

Useful generalisations of von Mises distributions can be obtained by extending the canonical statistic of the exponential model from $(\cos\theta, \sin\theta)$ to $(\cos\theta, \sin\theta, \cos 2\theta, \sin 2\theta, \ldots, \cos q\theta, \sin q\theta)$. Taking $q = 2$ gives the $(4, 4)$ exponential model with probability density functions

$$\frac{1}{2\pi} \times \frac{\exp\{\kappa_1 \cos(\theta - \alpha) + \kappa_2 \cos 2(\theta - \beta)\}}{I_0(\kappa_1)I_0(\kappa_2) + \sum_{p=1}^{\infty} I_{2p}(\kappa_1)I_p(\kappa_2)\cos 2p(\alpha - \beta)} \tag{3.5.46}$$

proposed by Cox (1975) and investigated by Yfantis & Borgman (1982). The distributions can be unimodal or bimodal, and they are not necessarily symmetrical.

3.5.5 Cardioid Distributions

Perturbation of the uniform density by a cosine function produces the *cardioid distribution* $C(\mu, \rho)$, which has probability density function

$$f(\theta) = \frac{1}{2\pi}\{1 + 2\rho\cos(\theta - \mu)\}, \quad |\rho| < \frac{1}{2}. \tag{3.5.47}$$

The name comes from the fact that the curve given in polar coordinates by $r = f(\theta)$ with f as in (3.5.47) is a cardioid curve. The distribution was introduced by Jeffreys (1948, p. 302).

The mean resultant length of $C(\mu, \rho)$ is ρ and (if $\rho > 0$) the mean direction is μ. The distribution is symmetrical and unimodal with mode at μ (if $\rho > 0$). For $\rho = 0$, the cardioid distribution reduces to the uniform distribution. For small ρ, $C(\mu, \rho)$ represents a slight departure from uniformity. The main use of the cardioid distributions is as small-concentration approximations (3.5.20) to the von Mises distributions.

If $\theta \sim C(\mu, \rho)$ and ψ is any given angle then $\theta - \psi \sim C(\mu - \psi, \rho)$, so that the cardioid distributions form a composite transformation model under the group $SO(2)$ of rotations of the circle.

A simple calculation with characteristic functions shows that if θ_1 and θ_2 are independent then

$$\theta_i \sim C(\mu_i, \rho_i) \quad (i = 1, 2) \Rightarrow \theta_1 + \theta_2 \sim C(\mu_1 + \mu_2, \rho_1\rho_2).$$

Thus the set (3.5.47) of cardioid distributions is closed under convolution.

3.5.6 Projected Normal Distributions

Distributions on the circle can be obtained by radial projection of distributions on the plane. Let $\mathbf{x}$ be a random two-dimensional vector such that $\Pr(\mathbf{x} = \mathbf{0}) = 0$. Then $\|\mathbf{x}\|^{-1}\mathbf{x}$ is a random point on the unit circle. (For more on this construction, see the discussion of projected distributions in Section 9.3.3.) An important instance is that in which $\mathbf{x}$ has a bivariate normal distribution $N_2(\boldsymbol{\mu}, \boldsymbol{\Sigma})$, in which case $\|\mathbf{x}\|^{-1}\mathbf{x}$ is said to have a *projected normal* (or *angular Gaussian* or *offset normal*) distribution $PN_2(\boldsymbol{\mu}, \boldsymbol{\Sigma})$. A typical application occurs in meteorology: when wind velocity is modelled by a bivariate normal distribution, the resulting marginal distribution for wind direction is a projected normal distribution.

A tedious calculation shows that the probability density function of the projected normal distribution $PN_2(\boldsymbol{\mu}, \boldsymbol{\Sigma})$ is (Mardia, 1972a, p. 52)

$$p(\theta; \boldsymbol{\mu}, \boldsymbol{\Sigma}) = \frac{\phi(\theta; \mathbf{0}, \boldsymbol{\Sigma}) + |\boldsymbol{\Sigma}|^{-1/2} D(\theta) \Phi(D(\theta)) \phi(|\boldsymbol{\Sigma}|^{-1/2} (\mathbf{x}^T \boldsymbol{\Sigma}^{-1} \mathbf{x})^{-1/2} \boldsymbol{\mu} \wedge \mathbf{x})}{\mathbf{x}^T \boldsymbol{\Sigma}^{-1} \mathbf{x}}, \tag{3.5.48}$$

where $\phi(\cdot; \mathbf{0}, \boldsymbol{\Sigma})$ denotes the probability density function of $N_2(\mathbf{0}, \boldsymbol{\Sigma})$, ϕ and Φ denote the probability density function and cumulative density function of $N(0, 1)$, $\mathbf{x} = (\cos\theta, \sin\theta)^T$,

$$D(\theta) = \frac{\boldsymbol{\mu}^T \boldsymbol{\Sigma}^{-1} \mathbf{x}}{(\mathbf{x}^T \boldsymbol{\Sigma}^{-1} \mathbf{x})^{1/2}}$$

and $\boldsymbol{\mu} \wedge \mathbf{x} = \mu_1 \sin\theta - \mu_2 \cos\theta$ with $\boldsymbol{\mu} = (\mu_1, \mu_2)^T$. In particular,

$$p(\theta; (\mu, 0), \mathbf{I}_2) = \frac{1}{\sqrt{2\pi}} \phi(\mu) + \mu \cos\theta \, \phi(\mu \sin\theta) \Phi(\mu \cos\theta), \tag{3.5.49}$$

where $\mathbf{I}_2$ denotes the 2×2 identity matrix. The distribution $PN_2(\boldsymbol{\mu}, \boldsymbol{\Sigma})$ reduces to the uniform distribution if and only if $\boldsymbol{\mu} = \mathbf{0}$ and $\boldsymbol{\Sigma} = \sigma^2 \mathbf{I}_2$. Projected normal distributions can be bimodal and/or asymmetrical.

Projected normal distributions with $\boldsymbol{\mu} = \mathbf{0}$ are called *angular central Gaussian distributions.* The probability density function of the angular central Gaussian distribution $PN_2(\mathbf{0}, \boldsymbol{\Sigma})$ is

$$p(\theta; \boldsymbol{\Sigma}) = \frac{(1 - b^2)^{1/2}}{2\pi(1 - b\cos 2(\theta - \mu))}, \tag{3.5.50}$$

where

$$b = \frac{2(\mathrm{tr}(\boldsymbol{\Sigma}) - 2|\boldsymbol{\Sigma}|^{1/2})^{1/2} (\mathrm{tr}(\boldsymbol{\Sigma}) + 2|\boldsymbol{\Sigma}|^{1/2})^{3/2}}{(2\mathrm{tr}\boldsymbol{\Sigma})^2 + 4|\boldsymbol{\Sigma}|}, \qquad \tan\mu = \frac{2\sigma_{12}}{\sigma_{11} - \sigma_{22}}$$

with

$$\boldsymbol{\Sigma} = \begin{pmatrix} \sigma_1^2 & \sigma_{12} \\ \sigma_{12} & \sigma_2^2 \end{pmatrix}.$$

Note that $p(\theta; c\mathbf{\Sigma}) = p(\theta; \mathbf{\Sigma})$ for any non-zero c, so we may assume without loss of generality that $|\mathbf{\Sigma}| = 1$. A connection between angular central Gaussian distributions and wrapped Cauchy distributions is given in (3.5.72). Because $p(\theta; \mathbf{\Sigma}) = p(\theta+\pi; \mathbf{\Sigma})$, the angular central Gaussian distributions provide useful models for axial data.

Each invertible linear transformation $\mathbf{A}$ of the plane gives rise to an invertible transformation $\varphi_{\mathbf{A}}$ of the unit circle by

$$\varphi_{\mathbf{A}}(\mathbf{x}) = \frac{1}{||\mathbf{A}\mathbf{x}||}\mathbf{A}\mathbf{x}. \qquad (3.5.51)$$

Then

$$\theta \sim PN_2(\boldsymbol{\mu}, \mathbf{\Sigma}) \quad \Rightarrow \quad \mathbf{A}\mathbf{x} \sim PN_2(\mathbf{A}\boldsymbol{\mu}, \mathbf{A}\mathbf{\Sigma}\mathbf{A}^T). \qquad (3.5.52)$$

Thus the set of projected normal distributions is closed under these transformations. Since $\varphi_{\mathbf{A}} = \varphi_{|\mathbf{A}|^{-1}\mathbf{A}}$, we may assume without loss of generality that $|\mathbf{A}| = 1$, i.e. that $\mathbf{A}$ is a unimodular matrix. Thus the projected normal distributions form a composite transformation model with group $SL_2(\mathbb{R})$, the set of 2×2 unimodular matrices. This was exploited by Cairns (1975) and Fraser (1979, pp. 219–231). By (3.5.72), maximum likelihood estimation of the parameters in angular central Gaussian distributions (with $|\mathbf{\Sigma}| = 1$) is equivalent to maximum likelihood estimation in wrapped Cauchy distributions. This is considered in Section 5.4.

It follows from (3.5.52) that

$$\theta \sim PN_2(\mathbf{0}, \mathbf{\Sigma}) \quad \Rightarrow \quad \mathbf{A}\mathbf{x} \sim PN_2(\mathbf{0}, \mathbf{A}\mathbf{\Sigma}\mathbf{A}^T), \qquad (3.5.53)$$

so that the angular central Gaussian distributions form a transformation model with group $SL_2(\mathbb{R})$. In particular, each angular central Gaussian distribution can be obtained from the uniform distribution by a suitable transformation $\varphi_{\mathbf{A}}$. This is why the angular central Gaussian distributions occur in image analysis, when 'textures' (fields of axes) are projected from one plane to another (see Blake & Marinos, 1990).

3.5.7 Wrapped Distributions

Definition

Given a distribution on the line, we can wrap it around the circumference of the circle of unit radius. That is, if x is a random variable on the line, the corresponding random variable x_w of the wrapped distribution is given by

$$x_w = x \,(\mathrm{mod}\, 2\pi). \qquad (3.5.54)$$

If the circle is identified with the set of complex numbers with unit modulus then the wrapping map $x \mapsto x_w$ can be written as

$$x \mapsto e^{2\pi i x}. \qquad (3.5.55)$$

If x has distribution function F then the distribution function F_w of x_w is given by

$$F_w(\theta) = \sum_{k=-\infty}^{\infty} \{F(\theta + 2\pi k) - F(2\pi k)\}, \quad 0 \leq \theta \leq 2\pi. \tag{3.5.56}$$

In particular, if x has a probability density function f then the corresponding probability density function f_w of x_w is

$$f_w(\theta) = \sum_{k=-\infty}^{\infty} f(\theta + 2\pi k). \tag{3.5.57}$$

Properties

(a) Perhaps the most important property of wrapping is that

$$(x + y)_w = x_w + y_w. \tag{3.5.58}$$

(In algebraic language, wrapping is a homomorphism from $\mathbb{R}$ to the circle group S^1.)

(b) If the characteristic function of x is ϕ then the characteristic function $\{\phi_p : p = 0, \pm 1, \ldots\}$ of x_w is given by

$$\phi_p = \phi(p). \tag{3.5.59}$$

To see this, note that

$$\phi_p = \int_0^{2\pi} e^{ip\theta} dF_w(\theta) = \sum_{k=-\infty}^{\infty} \int_{2\pi k}^{2\pi(k+1)} e^{ip\theta} dF(\theta) = \int_{-\infty}^{\infty} e^{ipx} dF(x) = \phi(p).$$

(c) If ϕ is integrable then x has a density and

$$f_w(\theta) = \sum_{k=-\infty}^{\infty} f(\theta + 2\pi k) = \frac{1}{2\pi}\left[1 + 2\sum_{p=1}^{\infty}(\alpha_p \cos p\theta + \beta_p \sin p\theta)\right], \tag{3.5.60}$$

where $\phi(p) = \alpha_p + i\beta_p$. To see this, note that since ϕ is integrable, we have

$$\sum_p^{\infty} |\phi_p|^2 \leq \sum_p^{\infty} |\phi_p| \leq \int_{-\infty}^{\infty} |\phi(t)| dt.$$

Therefore the series $\sum_{p=1}^{\infty}(\alpha_p^2 + \beta_p^2)$ is convergent and the result follows from (3.5.57) and (3.5.59).

(d) If x is infinitely divisible then x_w is infinitely divisible. (This follows from the homomorphism property (3.5.58).)

(e) There are (infinitely) many distributions on the line which can be wrapped onto any given distribution on the circle. To see this, let g be the probability density function of a distribution on the circle and define a probability density function on the line by

$$f(x) = p_r g(x), \quad 2\pi r < x \leq 2\pi(r+1), \quad r = 0, \pm 1, \pm 2, \ldots,$$

where p_r are any non-negative numbers such that $\sum_{r=-\infty}^{\infty} p_r = 1$. Then $f_w = g$.

We now consider some important particular cases.

Wrapped Poisson Distribution

Just as reduction modulo 2π wraps the line onto the circle, so (if m is a positive integer) reduction modulo $2\pi m$ wraps the integers onto the group of mth roots of 1, regarded as a subgroup of the circle. More precisely, if x is a random variable on the integers, then x_w, defined by

$$x_w = 2\pi x \pmod{2\pi m} \tag{3.5.61}$$

is a random variable on the lattice $\{2\pi r/m : r = 0, 1, \ldots, m-1\}$ on the circle. The probability function of x_w is given by

$$\Pr\left(x_w = \frac{2\pi r}{m}\right) = \sum_{k=-\infty}^{\infty} p(r + km), \quad r = 0, 1, \ldots, m-1, \tag{3.5.62}$$

where p is the probability function of x. In particular, if x has the Poisson distribution with mean λ then, from (3.5.62), x_w has the *wrapped Poisson distribution* with probability function

$$\Pr\left(\theta = \frac{2\pi r}{m}\right) = e^{-\lambda} \sum_{k=0}^{\infty} \frac{\lambda^{r+km}}{(r+km)!}, \quad r = 0, 1, \ldots, m-1.$$

Ball & Blackwell (1992) showed that these probabilities can be written in the finite form

$$\Pr\left(\theta = \frac{2\pi r}{m}\right) = \frac{1}{m} \sum_{j=0}^{m-1} \exp\{\omega^j \lambda\} \omega^{-rj},$$

where ω is a complex mth root of 1.

From (3.5.59) the characteristic function of θ is

$$\phi_p = \exp\{\lambda(1 - e^{2\pi i p/m})\}.$$

It follows that the convolution of wrapped Poisson distributions with parameters λ_1 and λ_2 is wrapped Poisson with parameter $\lambda_1 + \lambda_2$. (This is also a consequence of the homomorphism property (3.5.58).) This distribution appears in the study of distributions of triangular arrays on the circle (Lévy, 1939).

Wrapped Normal Distribution

The *wrapped normal distribution* $WN(\mu, \rho)$ is obtained by wrapping the $N(\mu, \sigma^2)$ distribution onto the circle, where

$$\sigma^2 = -2 \log \rho,$$

i.e.

$$\rho = e^{-\sigma^2/2}. \tag{3.5.63}$$

From (3.5.57), the probability density function of $WN(\mu, \rho)$ is

$$\phi_w(\theta; \mu, \rho) = \frac{1}{\sigma\sqrt{2\pi}} \sum_{k=-\infty}^{\infty} \exp\left\{\frac{-(\theta - \mu + 2\pi k)^2}{2\sigma^2}\right\}. \tag{3.5.64}$$

Since the characteristic function of $N(\mu, \sigma^2)$ is given by $\phi(t) = \exp(i\mu t - t^2\sigma^2/2)$, (3.5.59) gives

$$\phi_p = e^{i\mu p - p^2\sigma^2/2}, \quad \alpha_p = e^{-p^2\sigma^2/2} \cos p\mu, \quad \beta_p = e^{-p^2\sigma^2/2} \sin p\mu. \tag{3.5.65}$$

In particular, the mean direction is $\mu \pmod{2\pi}$, the mean resultant length is ρ, and $\bar{\alpha}_2 = \rho^4$. Using (3.5.65) in (3.5.60) gives a useful representation of the density (3.5.64) as

$$\phi_w(\theta; \mu, \rho) = \frac{1}{2\pi}\left\{1 + 2\sum_{p=1}^{\infty} \rho^{p^2} \cos p(\theta - \mu)\right\}, \quad 0 \leq \rho \leq 1. \tag{3.5.66}$$

For practical purposes, the density ϕ_w can be approximated adequately by the first three terms of (3.5.66) when $\sigma^2 \geq 2\pi$, while for $\sigma^2 \leq 2\pi$ the term with $k = 0$ of (3.5.64) gives a reasonable approximation. The density ϕ_w can be expressed in terms of the theta function ϑ_3 as

$$\phi_w(\theta; \mu, \rho) = \frac{1}{2\pi}\vartheta_3(\theta - \mu, \rho)$$

(see Abramowitz & Stegun, 1965, p. 576, 16.27.3.)

The distribution $WN(\mu, \rho)$ is unimodal and symmetric about its mode μ. As $\rho \to 0$, $WN(\mu, \rho)$ tends to the uniform distribution, while as $\rho \to 1$ it tends to a point distribution at μ. The distribution function can be obtained on integrating the series (3.5.66) term by term.

If $\theta \sim WN(\mu, \rho)$ then $\theta - \psi \sim WN(\mu - \psi, \rho)$, so that the wrapped normal distributions form a composite transformation model under the group $SO(2)$ of rotations of the circle.

It follows from the homomorphism property (3.5.58) and from (3.5.63) (or from consideration of the characteristic function) that if θ_1 and θ_2 are independent then

$$\theta_i \sim WN(\mu_i, \rho_i) \quad (i = 1, 2) \Rightarrow \theta_1 + \theta_2 \sim WN(\mu_1 + \mu_2, \rho_1\rho_2). \tag{3.5.67}$$

The wrapped normal distribution appears in Brownian motion on the circle. More precisely, consider a continuous-time Markov process on the circle, regarded as a random walk in which (i) a particle starts at $\theta = \mu$ at time $t = 0$, (ii) the particle moves infinitesimal distances in infinitesimal time periods, and (iii) at time t the infinitesimal displacement has mean zero and variance $\sigma^2 = ct$. Then the distribution of the position of the particle at time t is $WN(\mu, \exp(-ct/2))$ (see de Hass-Lorentz, 1913, pp. 24–25; Stephens, 1963; M. S. Bingham, 1971, Theorem 6.2). The entire process is obtained by wrapping Brownian motion (on the line) onto the circle.

Wrapped Cauchy Distribution

Consider the Cauchy distribution on the real line with density

$$f(x;\mu,a) = \frac{1}{\pi}\frac{a}{a^2+(x-\mu)^2}, \qquad -\infty < \mu < \infty, \quad a > 0.$$

Its characteristic function is $e^{-a|t|-it\mu}$. Consequently, from (3.5.60), we find that the corresponding wrapped distribution is the *wrapped Cauchy distribution* with density

$$c(\theta;\mu,\rho) = \sum_{k=-\infty}^{\infty} f(\theta+2\pi k;\mu,a) = \frac{1}{2\pi}\left\{1+2\sum_{p=1}^{\infty}\rho^p \cos p(\theta-\mu)\right\}, \tag{3.5.68}$$

where $\rho = e^{-a}$. We shall denote this distribution by $WC(\mu,\rho)$. It follows from considering the real part of the geometric series

$$\sum_{p=1}^{\infty}\rho^p e^{-ip(\theta-\mu)}$$

that (3.5.68) reduces to

$$c(\theta;\mu,\rho) = \frac{1}{2\pi}\frac{1-\rho^2}{1+\rho^2-2\rho\cos(\theta-\mu)}. \tag{3.5.69}$$

Further,

$$\phi_p = \rho^{|p|}, \quad \alpha_p = \rho^{|p|}\cos\mu, \quad \beta_p = \rho^{|p|}\sin\mu. \tag{3.5.70}$$

In particular, the mean direction is μ (mod 2π) and the mean resultant length is ρ. The $WC(\mu,\rho)$ distribution is unimodal and symmetric about μ. As $\rho \to 0$, it tends to the uniform distribution and as $\rho \to 1$ it becomes concentrated at the point μ. Its distribution function is given by

$$F(\theta) - F(\mu) = \frac{1}{2\pi}\cos^{-1}\left\{\frac{(1+\rho^2)\cos(\theta-\mu)-2\rho}{1+\rho^2-2\rho\cos(\theta-\mu)}\right\} \qquad \mu \le \theta \le \mu+\pi. \tag{3.5.71}$$

It follows from the homomorphism property (3.5.58) that the convolution of the wrapped Cauchy distributions $WC(\mu_1, \rho_1)$ and $WC(\mu_2, \rho_2)$ is the wrapped Cauchy distribution $WC(\mu_1 + \mu_2, \rho_1\rho_2)$.

There is a close connection between the wrapped Cauchy and the projected normal distributions. Kent & Tyler (1988) and Mardia (1972a, p. 52) showed that

$$\theta \sim PN_2(\mathbf{0}, \mathbf{\Sigma}) \Rightarrow 2\theta \sim WC(\mu, \rho), \tag{3.5.72}$$

where

$$\rho^2 = \frac{\operatorname{tr}(\mathbf{\Sigma}) - 2|\mathbf{\Sigma}|^{1/2}}{\operatorname{tr}(\mathbf{\Sigma}) + 2|\mathbf{\Sigma}|^{1/2}}, \qquad \tan\mu = \frac{2\sigma_{12}}{\sigma_{11} - \sigma_{22}}, \tag{3.5.73}$$

with

$$\mathbf{\Sigma} = \begin{pmatrix} \sigma_1^2 & \sigma_{12} \\ \sigma_{12} & \sigma_2^2 \end{pmatrix}.$$

The wrapped Cauchy distribution was introduced by Lévy (1939) and has been studied by Wintner (1947). McCullagh (1996) showed that wrapped Cauchy distributions can be obtained by mapping Cauchy distributions onto the circle by the transformation $x \mapsto 2\tan^{-1} x$, and that they form a transformation model under the corresponding action of the Möbius group $SL_2(\mathbb{R})$ on the circle. He also observed that the wrapped Cauchy distributions have the following *harmonic property*: if θ has distribution given by (3.5.69) then

$$\mathrm{E}[g(e^{i\theta})] = g(\rho e^{i\mu}),$$

for any complex function g which is analytic on the open unit disc and continuous on the closed disc.

The wrapped Cauchy distributions generalise to wrapped stable distributions. The characteristic function of a general stable distribution on the line has the form

$$\phi(t) = \exp\left\{i\mu t - \lambda|t|^a \exp\left\{-i\,\mathrm{sgn}(t)\frac{\pi\gamma}{2}\right\}\right\}.$$

(see Lukacs, 1970, p. 136). It follows from (3.5.59) and (3.5.60) that the probability density function of the corresponding wrapped distribution is

$$\frac{1}{2\pi}\left\{1 + 2\Sigma\rho^{p^a}\cos(p(\theta - \mu) + bp^a)\right\}, \quad 0 < a \leq 2.$$

In the case $b = 0$, Wintner (1947) proved that the distribution has a unique mode at μ. The wrapped stable distributions with $b = 0$ include the wrapped normal ($a = 2$) and the wrapped Cauchy ($a = 1$) distributions.

3.6 MULTIPLY-WRAPPED DISTRIBUTIONS

3.6.1 Wrapping the Circle onto Itself

In various contexts (some of which were considered in Chapter 1) it is appropriate to consider distributions with k-fold rotational symmetry, i.e.

distributions which are invariant under

$$\theta \mapsto \theta + \frac{2\pi}{k}.$$

In particular, if $k = 2$ these are the distributions with antipodal symmetry. Antipodally symmetric distributions on the circle are appropriate for modelling axial data, i.e. data where each observation is a unit vector with unknown sign. (In terms of unit complex numbers, axial data are unit complex numbers z such that we cannot distinguish between z and $-z$.)

For any positive integer k, any distribution on the circle gives rise to a correponding distribution with k-fold rotational symmetry by 'putting k copies of the original distribution end-to-end'. More precisely, if the original distribution has probability density function f then the new distribution has probability density function f^* given by

$$f^*(\theta) = f(k\theta).$$

All distributions with k-fold rotational symmetry can be obtained by this construction. The reverse construction consists of *k-fold wrapping of the circle onto itself*, i.e. the transformation $\theta \mapsto k\theta$. Then f can be recovered from f^* by this transformation. More precisely,

$$f(\theta) = \sum_{r=0}^{k-1} f^* \left(\theta + \frac{2r\pi}{k} \right).$$

3.6.2 Mixtures

One context in which wrapping of the circle onto itself is useful is that of distributions with probability density functions of the form

$$g(\theta) = \lambda f(\theta) + (1 - \lambda) f(\theta + \pi), \tag{3.6.1}$$

for some probability density function f and some λ with $0 \leq \lambda \leq 1$. If f is unimodal then g is bimodal with (in general) modes π radians apart. By double wrapping of the circle onto itself, i.e. by the transformation $\theta \mapsto \phi = 2\theta$, we obtain a distribution with probability density function

$$h(\phi) = f\left(\frac{\phi}{2}\right) + f\left(\frac{\phi}{2} + \pi\right), \tag{3.6.2}$$

which does not depend on the nuisance parameter λ.

If f is the probability density function of $M(\mu, \kappa)$ and $\lambda = 1/2$ then the probability density function given by (3.6.1) reduces to

$$g(\theta) = \frac{1}{2\pi I_0(\kappa)} \cosh(\kappa \cos(\theta - \mu)). \tag{3.6.3}$$

In order to make inferences about κ and μ, it is worth approximating this distribution by a doubly-wrapped von Mises distribution with probability density function

$$\frac{1}{2\pi I_0(\kappa_1)} e^{\kappa_1 \cos 2(\theta-\mu)}, \tag{3.6.4}$$

where $\kappa_1$is defined by matching mean resultant lengths, i.e. by

$$A(\kappa) = I_2(\kappa_1)/I_0(\kappa_1).$$

To see that this is a good approximation for large κ, note that the characteristic functions ϕ_p and ϕ'_p of (3.6.3) and (3.6.4) are given by

$$\phi_p = \begin{cases} 0, & p \text{ odd}, \\ I_p(\kappa)/I_0(\kappa), & p \text{ even}, \end{cases}$$

$$\phi'_p = \begin{cases} 0, & p \text{ odd}, \\ I_{p/2}(\kappa_1)/I_0(\kappa_1), & p \text{ even}. \end{cases}$$

Define σ and σ_1 by

$$A(\kappa) = e^{-\sigma^2/2}, \quad I_2(\kappa_1)/I_0(\kappa_1) = e^{-\sigma_1^2}. \tag{3.6.5}$$

Then using (3.5.23) to approximate von Mises distributions by wrapped normal distributions yields

$$I_p(\kappa)/I_0(\kappa) \simeq e^{-p\sigma^2/2} \simeq e^{-p\sigma_1^2} \simeq I_{2p}(\kappa_1)/I_0(\kappa_1), \tag{3.6.6}$$

so that the distributions (3.6.3) and (3.6.4) are close.

3.7 DISTRIBUTIONS ON THE TORUS AND THE CYLINDER

3.7.1 Distributions on the Torus

Sometimes it is necessary to consider the joint distribution of two circular random variables θ_1 and θ_2. Then (θ_1, θ_2) takes values on the unit torus. In the *uniform distribution* on the torus, θ_1 and θ_2 are independent and uniformly distributed. One application of the uniform distribution on the torus occurs in Buffon's needle problem (see Feller, 1966, p. 61). In this problem a needle of unit length is thrown onto a plane partitioned into parallel strips of unit width and the quantity of interest is the probability that the needle does not lie entirely within a strip. Let θ_1 and θ_2 denote respectively the direction of the needle and 2π times the fractional part of the position of the centre of the needle from the edge of any specified strip. A suitable model takes (θ_1, θ_2) to be uniformly distributed on the torus.

One useful set of distributions on the torus is the bivariate von Mises model (Mardia, 1975a; 1975c) with probability density functions proportional to

$$\exp\{\kappa_1 \cos(\theta_1 - \mu_1) + \kappa_2 \cos(\theta_2 - \mu_2) + (\cos\theta_1, \sin\theta_1)^T \mathbf{A}(\cos\theta_2, \sin\theta_2)\}, \tag{3.7.1}$$

where $\mathbf{A}$ is a 2×2 matrix. The marginal distributions of θ_1 and θ_2 are von Mises if and only if either $\mathbf{A} = \mathbf{0}$ (so that θ_1 and θ_2 are independent) or $\kappa_1 = \kappa_2 = 0$ and $\mathbf{A}$ is a multiple of an orthogonal matrix (so that θ_1 and θ_2 are uniformly distributed). An important submodel of (3.7.1) is obtained by imposing the constraint that $\mathbf{A}$ is a multiple of a rotation matrix. A series expansion for the normalising constant in this case was given by Jupp & Mardia (1980).

3.7.2 Distributions on the Cylinder

There are various practical situations which involve both a linear random variable x and a circular random variable θ. Then (x, θ) takes values in the cylinder $\mathbb{R} \times S^1$. Examples in rhythmometry, medicine and demography can be found in Batschelet *et al.* (1973) and Batschelet (1981). A suitable model for most of these situations is that proposed by Mardia & Sutton (1978) in which

$$\begin{aligned} \theta &\sim M(\mu_0, \kappa), \\ x|\theta &\sim N(\mu(\theta), \sigma^2(1-\rho^2)), \end{aligned}$$

where

$$\mu(\theta) = \mu + \sigma\sqrt{\kappa}\rho\cos(\theta - \nu).$$

Here μ_0 and ν are angles, μ is a real number, $\kappa \geq 0$, and $0 \leq \rho < 1$. The joint density of (x, θ) is

$$f(x, \theta) = \frac{1}{(2\pi)^{\frac{3}{2}} I_0(\kappa)\sigma^2(1-\rho^2)} \exp\left\{\kappa\cos(\theta - \mu) - \frac{(x - \mu(\theta))^2}{2\sigma^2(1-\rho^2)}\right\}. \tag{3.7.2}$$

Maximum likelihood estimates of the parameters and a practical example are given in Mardia & Sutton (1978). A related regression model is considered in Section 11.3.1.

4

Fundamental Theorems and Distribution Theory

4.1 INTRODUCTION

Characteristic functions of distributions on the circle were introduced in Section 3.3. In Section 4.2 we present key properties of these characteristic functions. We also show how the usual characteristic function of a random vector in the plane can be used to obtain the distributions of the polar coordinates of this vector. This provides a method of calculating the distributions of the sample mean direction and resultant length of a sample from a distribution on the circle. Some limit theorems on the circle are considered in Section 4.3. In the subsequent sections, we use results from Section 4.2 to obtain the distributions of the sample resultant length and related statistics for samples from the uniform distribution and von Mises distributions.

4.2 PROPERTIES OF CHARACTERISTIC FUNCTIONS

4.2.1 Key Properties

The key properties of characteristic functions of distributions on the circle are as follows:

(i) a probability distribution on the circle is determined by its characteristic function;
(ii) weak convergence of distributions is equivalent to pointwise convergence of characteristic functions, i.e. a sequence $F_1, F_2, \ldots$ of distribution functions converges weakly to F if and only if $\phi_p^{(n)} \to \phi_p$ for $p = 0, \pm1, \ldots$, where $\phi^{(n)}$ and ϕ denote the characteristic functions of F_n and F.

A proof of (i) can be found in Feller (1966, pp. 591–592). The continuity property (ii) can be proved using Helly's selection theorem as in the proof for random variables on the line given in Lukacs (1970, pp. 49–50). Note that property (i) is in marked contrast to the behaviour of characteristic functions on the line. One intuitive explanation for this is that on the circle it is not possible for the probability mass to 'escape to infinity'.

The neatest connection between probability distributions on the circle and their characteristic functions occurs for the distributions with square-summable density functions. Consider a probability distribution on the circle with density function f and with characteristic function $\{\phi_p : p = 0, \pm 1, \ldots\}$. Then

$$\int_0^{2\pi} f(\theta)^2 d\theta < \infty \Leftrightarrow \sum_{p=-\infty}^{\infty} |\phi_p|^2 < \infty. \tag{4.2.1}$$

If (4.2.1) holds then

$$f(\theta) = \frac{1}{2\pi} \sum_{p=-\infty}^{\infty} e^{-ip\theta} \phi_p = \frac{1}{2\pi} \left\{ 1 + 2 \sum_{p=1}^{\infty} (\alpha_p \cos p\theta + \beta_p \sin p\theta) \right\} \tag{4.2.2}$$

almost everywhere, where $\phi_p = \alpha_p + i\beta_p$.

The connection (4.2.2) between square-summable density functions and characteristic functions has the pleasant property that it 'preserves length', in the following sense. Let f and g be two square-summable density functions with characteristic functions ϕ_p and ϕ'_p, respectively. Then Parseval's formula (Titchmarsh, 1958, p. 425) states that

$$2\pi \int_0^{2\pi} f(\theta) g(\theta) d\theta = \sum_{p=-\infty}^{\infty} \phi_p \bar{\phi'}_p. \tag{4.2.3}$$

In particular,

$$2\pi \int_0^{2\pi} f(\theta)^2 d\theta = 1 + 2 \sum_{p=1}^{\infty} (\alpha_p^2 + \beta_p^2).$$

4.2.2 Polar Distributions and Characteristic Functions

In various contexts it is necessary to obtain the distribution of the polar coordinates (r, θ) of a continuous two-dimensional random variable (x, y) from the characteristic function of (x, y). The polar coordinates (r, θ) are defined by

$$x = r\cos\theta, \quad y = r\sin\theta. \tag{4.2.4}$$

If the characteristic function ψ of (x, y) is integrable, it follows from the inversion theorem that the probability density function of (x, y) is given by

$$f(x, y) = \frac{1}{4\pi^2} \int_{-\infty}^{\infty} \int_{-\infty}^{\infty} e^{-it_1 x - it_2 y} \psi(t_1, t_2) dt_1 dt_2. \tag{4.2.5}$$

Using (4.2.4) and defining ρ, Φ and $\tilde{\psi}$ by

$$t_1 = \rho\cos\Phi, \quad t_2 = \rho\sin\Phi,$$

$$\tilde{\psi}(\rho, \Phi) = \mathrm{E}[\exp\{i\rho r\cos(\theta - \Phi)\}] = \psi(t_1, t_2), \tag{4.2.6}$$

we find that the joint density of r and θ is given by

$$p(r, \theta) = \frac{1}{4\pi^2} r \int_{\rho=0}^{\infty} \int_{\Phi=0}^{2\pi} e^{-i\rho r\cos(\theta-\Phi)} \rho\tilde{\psi}(\rho, \Phi) d\rho d\Phi. \tag{4.2.7}$$

On integrating over θ and interchanging of the order of integration, it follows that the density of r is

$$p_1(r) = \frac{1}{2\pi} r \int_{\rho=0}^{\infty} \int_{\Phi=0}^{2\pi} \left\{ \frac{1}{2\pi} \int_0^{2\pi} e^{-i\rho r\cos(\theta-\Phi)} d\theta \right\} \rho\tilde{\psi}(\rho, \Phi) d\rho d\Phi.$$

The inner integral is

$$I_0(i\rho r) = J_0(\rho r), \tag{4.2.8}$$

where J_0 is the Bessel function of the first kind and order zero, which can be defined by

$$J_0(x) = \sum_{k=0}^{\infty} \frac{(-1)^k}{(k!)^2} \left(\frac{x}{2}\right)^{2k} \tag{4.2.9}$$

(take $p = 0$ in (A.15) of Appendix 1). Hence

$$p_1(r) = \frac{1}{2\pi} r \int_{\rho=0}^{\infty} \int_{\Phi=0}^{2\pi} J_0(\rho r)\tilde{\psi}(\rho, \theta)\rho d\rho d\theta. \tag{4.2.10}$$

Formula (4.2.10) may be described as an inversion formula for the distribution of r. We can rewrite it as

$$p_1(r) = r \int_0^{\infty} J_0(\rho r)\tilde{\psi}_1(\rho)\rho d\rho, \tag{4.2.11}$$

where

$$\tilde{\psi}_1(\rho) = \frac{1}{2\pi} \int_0^{2\pi} \psi(\rho, \Phi) d\Phi. \tag{4.2.12}$$

We now invert (4.2.11) to express $\tilde{\psi}_1(\rho)$ in terms of $p_1(r)$. From (4.2.6), we have

$$\tilde{\psi}(\rho, \Phi) = \int_{r=0}^{\infty} \int_{\theta=0}^{2\pi} e^{i\rho r\cos(\theta-\Phi)} p(r, \theta) dr d\theta.$$

On substituting for $\tilde{\psi}$ in (4.2.12) and integrating over Φ (in the same way that we integrated over θ in (4.2.7)), we obtain

$$\tilde{\psi}_1(\rho) = \int_0^{\infty} J_0(r\rho) p_1(r) dr, \tag{4.2.13}$$

i.e. $\tilde{\psi}_1$ is the Hankel transform of p_1. Applications of the Hankel transform to statistics are given in Lord (1954).

On using the expansion (4.2.9) for J_0 in (4.2.13), we obtain

$$\tilde{\psi}_1(\rho) = \sum_{k=0}^{\infty}(-1)^k \frac{\mu'_{2k}\rho^{2k}}{2^{2k}k!^2},$$

where μ'_k denotes the kth moment of r. Thus the moments of r^2 can be derived from ψ_1. In view of this property, $\tilde{\psi}_1$ is described as the *polar moment generating function.*

The marginal density of θ can be obtained from (4.2.7) by integration over r. The joint characteristic function of r and θ can be expressed as

$$\mathrm{E}[e^{itr+ip\theta}] = \int_{r=0}^{\infty}\int_{\theta=0}^{2\pi} e^{itr+ip\theta}\psi_2(r,\theta)dr d\theta,$$

where

$$\psi_2(r,\theta) = r\int_0^{\infty} I_p(i\rho r)\tilde{\psi}(\rho,\theta)\rho d\rho,$$

and the Bessel function I_p is defined by (3.5.27). It is interesting to note that the moment generating function of $\cos\theta$ is simply

$$\phi(t) = \frac{1}{2\pi}\left\{I_0(t) + 2\sum_{p=1}^{\infty}\alpha_p I_p(t)\right\}, \tag{4.2.14}$$

where α_p is the real part of the component ϕ_p of the characteristic function of θ. This result follows on using the Fourier expansion for the probability density function of θ together with the characteristic function of the von Mises distribution. It can be used to obtain the moments of $\sum_{i=1}^{n}\cos\theta_i$, where $\theta_1,\ldots,\theta_n$ are independent observations on a von Mises distribution.

The Distribution of R

Suppose that $\theta_1,\ldots,\theta_n$ are distributed independently on the circle and that θ_j has probability density function f_j for $j = 1,\ldots,n$. We give a general method of obtaining the distribution of R, where

$$R^2 = C^2 + S^2,$$

with

$$C = \sum_{j=1}^{\infty}\cos\theta_j, \quad S = \sum_{j=1}^{\infty}\sin\theta_j. \tag{4.2.15}$$

The joint characteristic function of (C, S) is

$$\prod_{j=1}^{n}\tilde{\psi}_j(\rho,\Phi), \tag{4.2.16}$$

where $\tilde{\psi}_j(\rho, \Phi)$ is the joint characteristic function of $(\cos\theta_j, \sin\theta_j)$, i.e.

$$\tilde{\psi}_j(\rho, \Phi) = \mathrm{E}[\exp\{i\rho\cos(\theta_j - \Phi)\}]. \tag{4.2.17}$$

Hence, from the inversion formula (4.2.10) the probability density function of R is

$$p(R) = \frac{1}{2\pi} R \int_{\rho=0}^{\infty} \int_{\Phi=0}^{2\pi} J_0(\rho R) \left\{ \prod_{j=1}^{n} \tilde{\psi}_j(\rho, \Phi) \right\} \rho d\rho d\Phi. \tag{4.2.18}$$

This method will be used in Sections 4.4–4.6 to obtain the distribution of R in various particular cases.

4.2.3 Further Properties of the Characteristic Function

(a) A distribution on the circle with distribution function F is called *stable* if for all angles c_1 and c_2 there is a c such that the convolution $F_{c_1} * F_{c_2}$ of F_{c_1} and F_{c_2} satisfies

$$F_{c_1} * F_{c_2} = F_c,$$

where

$$F_c(x) = F(x - c),$$

etc. Let $\{\phi_p\}_{0,\pm1,\pm2,\ldots}$ be the characteristic function of F and put $\nu = c - c_1 - c_2$. Then

$$\phi_p^2 = \phi_p e^{ip\nu},$$

and so

$$\phi_p = 0 \quad \text{or} \quad \phi_p = e^{ip\nu}.$$

Hence we deduce from Sections 3.5.2 and 3.5.3 that the stable circular distributions are precisely (i) the uniform distribution on the circle, and (ii) the discrete uniform distributions concentrated on $\{\nu + 2\pi r/m : r = 0, 1, \ldots, m-1\}$ for some ν and m.

(b) It follows from the Riemann–Lebesgue theorem (Titchmarsh, 1958, p. 403) that for absolutely continuous distributions,

$$\lim_{p\to\infty} \phi(p) = 0.$$

From Lukacs (1970, pp. 17–18), we deduce that $\{\phi_p\}_{0,\pm1,\pm2,\ldots}$ is the characteristic function of a lattice distribution if and only if

$$|\phi_p| = 1 \qquad \text{for some } p \neq 0. \tag{4.2.19}$$

4.3 LIMIT THEOREMS

4.3.1 Central Limit Theorems

The main distributions on the circle which arise from convolutions of independent and identically distributed random angles are the uniform distribution and the discrete uniform distributions.

Let $\theta_1, \ldots, \theta_n$ be independent and identically distributed random variables on the circle. We show that the distribution of the sum

$$S_n = \theta_1 + \ldots + \theta_n \tag{4.3.1}$$

converges to the uniform distribution, provided that the parent distribution is not a lattice distribution.

Let ϕ_p be the characteristic function of the distribution of $\theta_1, \ldots, \theta_n$. Since this is not a lattice distribution, we have from (4.2.19)

$$|\phi_p| < 1 \qquad \text{for all } p \neq 0.$$

Hence the characteristic function ϕ_p^n of S_n tends to zero for all $p \neq 0$. From (3.5.16), this limiting characteristic function is the characteristic function of the uniform distribution. Therefore the result follows on using the continuity property (property (ii) of Section 4.2.1).

We now show that if θ has a lattice distribution on the circle with zero as a lattice point then the distribution of the sum S_n converges to a discrete uniform distribution.

Consider a lattice distribution assigning probability p_r to the point $2\pi r/m$ for $r = 0, 1, \ldots, m-1$, where $0 < p_0 < 1$. We may take m to be the smallest possible positive integer such that the support of the distribution is contained in these points. Then there is an s which is coprime to m and with $p_s > 0$. From (3.5.12), the characteristic function of the distribution is

$$\phi_p = \sum_{r=0}^{m-1} p_r \omega^{rp}, \tag{4.3.2}$$

where $\omega = \exp\{2\pi i/m\}$. We have

$$\phi_p = 1 \quad \text{for } p = 0 \ (\text{mod } m).$$

If $p \neq 0$ (mod m) then $\omega^{sp} \neq 1$. Since the coefficients p_0 and p_s of ω^0 and ω^{sp} in (4.3.2) are positive, it follows from convexity of the unit disc that $|\phi_p| < 1$. Then

$$\lim_{n\to\infty} \phi_p^n = \begin{cases} 0 & \text{if } p \neq 0 \bmod m, \\ 1 & \text{if } p = 0 \bmod m. \end{cases}$$

From (3.5.14), this is the characteristic function of the discrete uniform distribution on m points. The result now follows from property (ii) of Section 4.2.1.

To see that convergence need not occur if the point $\theta = 0$ is not a lattice point, consider a distribution with

$$\Pr(\theta = c\pi) = q, \quad \Pr(\theta = c\pi + \pi) = 1 - q,$$

where c is irrational and $q > 0$. Then $\phi_{2p} = e^{2pci}$, so that $\lim_{n\to\infty} \phi_{2p}^n$ does not exist.

4.3.2 Poincaré's Theorem

Intuition suggests that if a continuous distribution is spread out over a large region of the line then the corresponding wrapped distribution will be almost uniform on the circle. Poincaré (1912) formalised this intuitive idea and gave the following example to illustrate it. Consider a needle which is free to rotate about the centre of a unit disc. After the needle is given an initial push, it spins and eventually stops. Let x denote the total distance covered by a given end of the needle before it stops. The stopping position x_w of the needle is

$$x_w = x \,(\mathrm{mod}\, 2\pi).$$

Thus x_w is obtained by wrapping x round the circle. Poincaré's result states that the variable x_w is distributed nearly uniformly if the spread of the variable x is large.

For a given number c, define

$$x' = cx. \tag{4.3.3}$$

We show that the distribution of the wrapped variable

$$x'_w = x' \,(\mathrm{mod}\, 2\pi) \tag{4.3.4}$$

tends to the uniform distribution as $c \to \infty$. Let $\phi(t)$ be the characteristic function of x. Then the characteristic function of x_w is given by $\phi_p = \phi(cp)$. Since x is a continuous random variable, the Riemann–Lebesgue theorem (Titchmarsh, 1958, p. 403) gives

$$\lim_{|t|\to\infty} \phi(t) = 0.$$

This implies that the characteristic function of x'_w satisfies

$$\lim_{c\to\infty} \phi_p = 0, \qquad \text{for all } p \neq 0.$$

Consequently, the distribution of x'_w tends to the uniform distribution.

This result has already been demonstrated in Section 3.5.7 for the special cases of the wrapped normal distribution $WN(\mu, \exp(-\sigma^2/2))$ as $\sigma \to \infty$ and the wrapped Cauchy distribution $WN(\mu, e^{-a})$ as $a \to \infty$. Another version of

Poincaré's result, in which it is assumed that the maximum of the density of x tends to zero, is given by Feller (1966, pp. 62–63).

Example 4.1: The Distribution of First Significant Digits
It has been observed that for a wide variety of large data sets, the distribution of the first significant digits is not uniform on $1, \ldots, 9$ (as one might naively suppose) but is a good fit to the model

$$\Pr\{\text{first significant digit of } x = i\} = \log_{10}(i+1) - \log_{10} i \qquad (4.3.5)$$

for $i = 1, 2, \ldots, 9$, which is often known as the 'significant digit law', or 'Benford's law'. This model was proposed by Newcomb (1881). Benford (1938) arrived at (4.3.5) independently through empirical studies of many data sets. Table 4.1 reproduces one of Benford's examples, which gives a complete count (except for dates and page numbers) of first significant digits from large numbers in an issue of the *Reader's Digest.* The expected frequencies are obtained from (4.3.5). The observed value of χ^2 is 3.27 and $\Pr(\chi^2_8 > 3.27) = 0.92$, so that the fit is satisfactory.

Table 4.1 The frequencies of first significant digits from large numbers in an issue of the *Reader's Digest* (Benford, 1938, reproduced by permission of American Philosophical Society)

First digit	Observed frequency	Expected frequency
1	103	92.7
2	57	54.2
3	38	38.5
4	23	29.8
5	22	24.4
6	20	20.6
7	17	17.9
8	15	15.8
9	13	14.1
Total	308	308

One heuristic argument for (4.3.5) is the following. Let x be the random variable which generates the data. Then it follows from Poincaré's theorem that, if the distribution of $\log x$ is very spread out along the line, the distribution of the wrapped random variable

$$x_w = \log_{10} x \,(\text{mod}\, 1)$$

is almost uniform on the circle of unit circumference. The first significant digit of x is i, $i = 1, 2, \ldots, 9$, if and only if

$$i \times 10^r \leq x < (i+1) \times 10^{r+1},$$

for some integer r. Consequently,

$$\begin{aligned}\Pr\{\text{first significant digit of } x = i\} &= \Pr\{\log_{10} i \leq x_w < \log_{10}(i+1)\} \\ &\simeq \log_{10}(i+1) - \log_{10} i,\end{aligned}$$

for $i = 1, 2, \ldots, 9$.

Careful derivations of (4.3.5), from both (i) an argument using scale invariance and (ii) a scheme of random sampling from randomly chosen distributions, were given by Hill (1995).

4.4 THE DISTRIBUTION OF $\bar{\theta}$ AND $\bar{R}$ FROM THE UNIFORM DISTRIBUTION

The distribution of the mean direction $\bar{\theta}$ and that of the resultant length R for random samples from the uniform distribution are important in inference on the circle. They also arise naturally in random walks in the plane which start from the origin and have steps of unit length in directions which are distributed independently and uniformly on the circle. Such random walks occur in the studies of random migration (Pearson, 1905) and the superposition of random vibrations (Rayleigh, 1880; 1905). It is straightforward to generalise the results given in this section to the case where the lengths of successive steps of the random walk are not necessarily equal.

4.4.1 The Distribution of $\bar{\theta}$ and $\bar{R}$

If $\theta_1, \ldots, \theta_n$ are independently and uniformly distributed then, for any constant angle c, $(\theta_1 + c, \ldots, \theta_n + c)$ has the same joint distribution as $(\theta_1, \ldots, \theta_n)$, i.e.

$$(\theta_1 + c, \ldots, \theta_n + c) \sim (\theta_1, \ldots, \theta_n).$$

By (2.2.11), the sample mean direction $\bar{\theta}$ is equivariant under rotation (i.e. the mean direction of $\theta_1 + c, \ldots, \theta_n + c$ is $\bar{\theta} + c$), while the sample mean resultant length $\bar{R}$ is invariant (i.e. the mean resultant length of $\theta_1 + c, \ldots, \theta_n + c$ is $\bar{R}$), and so

$$(\bar{\theta} + c, \bar{R}) \sim (\bar{\theta}, \bar{R}),$$

and

$$\bar{\theta} + c | \bar{R} \sim \bar{\theta} | \bar{R},$$

i.e. the conditional distribution of $\bar{\theta}$ given $\bar{R}$ is invariant under rotation. Since the uniform distribution is the unique distribution which is invariant under rotation, it follows that

$$\bar{\theta} \text{ is uniformly distributed,} \tag{4.4.1}$$

$$\bar{\theta} \text{ and } \bar{R} \text{ are independent.} \tag{4.4.2}$$

Result (4.4.2) has an interesting converse. Kent, Mardia & Rao (1979) showed that independence of $\bar{\theta}$ and $\bar{R}$ (for some $n \geq 2$) characterises the uniform distribution (among distributions with density continuous almost everywhere). This characterisation is in contrast to Geary's (1936) characterisation of normal distributions by independence of the sample mean $\bar{x}$ and the sample variance s^2.

Since $\bar{R} = R/n$, the distribution of $\bar{R}$ can be obtained readily from that of R. To obtain the distribution of R, we follow the characteristic function method of Section 4.2.2. Using

$$f(\theta) = \frac{1}{2\pi}$$

in (4.2.17) shows that the characteristic function of $(\cos\theta_j, \sin\theta_j)$ is

$$\tilde{\psi}_j(\rho, \Phi) = \frac{1}{2\pi}\int_0^{2\pi} e^{i\rho\cos(\theta-\Phi)}d\theta, \tag{4.4.3}$$

where we have used the notation of (4.2.6). Since θ has a distribution on the circle, the integral in (4.4.3) does not depend on Φ. On substituting its value from (4.2.8), we have

$$\tilde{\psi}_j(\rho, \Phi) = J_0(\rho).$$

Hence the characteristic function of (C, S) is given by

$$J_0(\rho)^n, \tag{4.4.4}$$

which does not depend on Φ. On substituting (4.4.4) into the inversion formula (4.2.10) for R, we find that the probability density function h_n of R is given by

$$h_n(R) = R\int_0^\infty uJ_0(Ru)J_0(u)^n du, \tag{4.4.5}$$

$$= R\phi_n(R^2), \tag{4.4.6}$$

where

$$\phi_n(R^2) = \frac{h_n(R)}{R}. \tag{4.4.7}$$

Since $p(R) = 0$ for $R > n$, the integral on the right of (4.4.5) vanishes for $R > n$. A few particular cases of the probability density function h_n are discussed below.

We now obtain the distribution function of R. Let J_1 denote the Bessel function of the first kind and order 1, defined by (A.15) of Appendix 1 as

$$J_1(x) = \sum_{k=0}^{\infty}\frac{(-1)^k}{k!(k+1)!}\left(\frac{1}{2x}\right)^{2k+1}.$$

We have

$$\frac{d}{dx}\{xJ_1(x)\} = xJ_0(x) \tag{4.4.8}$$

(cf. (3.5.35)), so that the distribution function corresponding to h_n can be written as

$$\begin{aligned} H_n(R) &= \int_0^\infty \left[\int_0^R \frac{d}{d(u\rho)}\{J_1(u\rho)u\rho\}d\rho\right] J_0(u)^n du \\ &= R\int_0^\infty J_1(uR)J_0(u)^n du. \end{aligned} \tag{4.4.9}$$

For $n = 2, 3$, it is possible to get comparatively simple expressions for h_n. In the case $n = 2$, we can exploit the fact that $R^2 = 2(1+\cos\theta)$, where the angle θ between the two steps is uniformly distributed, to obtain

$$h_2(R) = \frac{2}{\pi(4-R^2)^{1/2}}, \quad 0 < R < 2. \tag{4.4.10}$$

For $n = 3$, the probability density function can be expressed in terms of elliptic functions as

$$h_3(R) = \begin{cases} \pi^{-2}mR^{1/2}K(m) & \text{if } R < 1, \\ \pi^{-2}R^{1/2}K(1/m) & \text{if } R > 1, \end{cases} \tag{4.4.11}$$

where

$$m = 4\left\{\frac{R}{(3-R)(1+R)^3}\right\}^{1/2}, \quad K(m) = \int_0^1 \frac{1}{\{(1-t^2)(1-m^2t^2)\}^{1/2}}dt.$$

As $R \to 1, h_3(R) \to \infty$. For details of the derivation, see, e.g. Stephens (1962a). For $n = 4, \ldots, 7$, the probability density function has been calculated by Pearson (1906), who also gave a series expansion which was used later by Greenwood & Durand (1955) to tabulate the distribution for $n = 6, \ldots, 24$. Durand & Greenwood (1957) have examined the adequacy of some approximations by truncating Pearson's series expansion and compared various methods of approximation.

A useful approximation for large n is

$$2n\bar{R}^2 \doteq \chi_2^2,$$

which is derived below just before (4.8.14).

A more refined approximation, suitable even for small values of n, is the saddlepoint approximation

$$\Pr(\bar{R} > z) \simeq e^{-nt^2/2}\left\{\frac{z}{\hat{\kappa}}\left(1 - z^2 - \frac{z}{\hat{\kappa}}\right)\right\}^{-1/2} + O\left(\frac{1}{n}e^{-nt^2/2}\right), \tag{4.4.12}$$

where

$$t^2 - 2\{\hat{\kappa}z - \log I_0(\hat{\kappa})\}$$

and

$$\hat{\kappa} = A^{-1}(z)$$

(see Jensen (1995, p. 165)). The approximation (4.4.12) holds uniformly in κ.

History

The above problem of the random walk on the circle was proposed by K. Pearson (1905) in a letter to *Nature*. Its asymptotic solution had already been obtained by Rayleigh (1880) and was reported in response to Pearson's letter in Rayleigh (1905). The exact solution was obtained by Kluyver (1906) and Pearson (1906) gave another proof of Kluyver's result. Later Markov (1912) and Rayleigh (1919) studied this problem. The solution given here (see also Mardia, 1972a, Section 4.4.2) seems to require a minimum of Bessel function theory.

4.4.2 The Distribution of C and S

We now obtain the joint probability density function of

$$C = \sum_{i=1}^{n} \cos\theta_i \quad \text{and} \quad S = \sum_{i=1}^{n} \sin\theta_i.$$

From Section 4.4.1, the joint probability density function of $(\bar{\theta}, R)$, where $C = R\cos\bar{\theta}$ and $S = R\sin\bar{\theta}$, is

$$\frac{1}{2\pi} R\phi_n(R^2), \quad R > 0, \tag{4.4.13}$$

where ϕ_n is defined by (4.4.7). Consequently, the joint probability density function of (C, S) is

$$g_0(C, S) = \frac{1}{2\pi}\phi_n(C^2 + S^2). \tag{4.4.14}$$

We shall obtain the marginal distributions of C and S in Section 4.5.3.

4.5 DISTRIBUTION OF C, S AND R FOR A VON MISES POPULATION

We now assume that $\theta_1, \ldots, \theta_n$ is a random sample from the von Mises distribution $M(\mu, \kappa)$.

4.5.1 The Joint Distribution of C and S

The joint probability density function $g(\cdot, \cdot; \mu, \kappa)$ of (C, S) can be obtained by integrating the density function of $\theta_1, \ldots, \theta_n$ keeping C and S fixed. Thus

$$g(C, S; \mu, \kappa) = \frac{1}{I_0(\kappa)^n} e^{\kappa(C \cos \mu + S \sin \mu)} \int \cdots \int \left(\frac{1}{2\pi}\right)^n d\theta_1 \ldots d\theta_n, \quad (4.5.1)$$

where the integral is taken over all values of $\theta_1, \ldots, \theta_n$ satisfying

$$\sum_{i=1}^{n} \cos \theta_i = C, \quad \sum_{i=1}^{n} \sin \theta_i = S.$$

Taking $\kappa = 0$ gives

$$g(C, S; \mu, 0) = \int \cdots \int \left(\frac{1}{2\pi}\right)^n d\theta_1 \ldots d\theta_n,$$

and so

$$\begin{aligned} g(C, S; \mu, \kappa) &= \frac{1}{I_0(\kappa)^n} e^{\kappa(C \cos \mu + S \sin \mu)} g(C, S; \mu, 0) \\ &= \frac{1}{\{2\pi I_0(\kappa)\}^n} e^{\kappa(C \cos \mu + S \sin \mu)} \phi_n(C^2 + S^2), \quad (4.5.2) \end{aligned}$$

using (4.4.6). The marginal distributions of C and S cannot be obtained in a useful form from (4.5.2). We shall obtain them in Section 4.5.3 by a different approach.

4.5.2 Distributions of $\bar{\theta}$ and $\bar{R}$

On transforming (C, S) to $(\bar{\theta}, R)$ by $C = R \cos \bar{\theta}$ and $S = R \sin \bar{\theta}$ in (4.5.2), the joint probability density function of $\bar{\theta}$ and R is seen to be

$$g(\bar{\theta}, R; \mu, \theta) = \frac{1}{2\pi I_0(\kappa)^n} e^{\kappa R \cos(\bar{\theta} - \mu)} h_n(R), \quad 0 < R < n, \quad (4.5.3)$$

where h_n is the probability density function of R for the uniform case and is given by (4.4.5). Integration with respect to $\bar{\theta}$ shows that the probability density function of R is given by

$$p(R) = \frac{1}{I_0(\kappa)^n} I_0(\kappa R) h_n(R), \quad 0 < R < n, \quad (4.5.4)$$

a result due to Greenwood & Durand (1955). The large-sample limiting distribution of $(\bar{\theta}, \bar{R})$ is considered in Section 4.8.1.

The marginal probability density function of $\bar{\theta}$ does not have a particularly simple form. However, using (4.5.3) and (4.5.4), we obtain the important result (Mardia, 1972a, p. 98) that the conditional distribution of $\bar{\theta}$ given R is $M(\mu, \kappa R)$, i.e.

$$f(\bar{\theta}|R; \mu, \theta) = \frac{1}{2\pi I_0(\kappa R)} \exp\{\kappa R \cos(\bar{\theta} - \mu)\}. \tag{4.5.5}$$

4.5.3 Marginal Distributions of C and S

The characteristic function of $\cos\theta$ is given by

$$\phi(t) = \frac{1}{2\pi I_0(\kappa)} \int_0^{2\pi} \exp\{it\cos\theta + \kappa\cos(\theta - \mu)\}d\theta. \tag{4.5.6}$$

To evaluate this integral, we note that for complex a and b

$$\int_0^{2\pi} \exp(a\cos\theta + ib\sin\theta)d\theta = 2\pi I_0(\{a^2 + b^2\}^{1/2}), \tag{4.5.7}$$

which can be verified by expanding the exponential term and using

$$\int_0^{2\pi} (a\cos\theta + ib\sin\theta)^r d\theta = \begin{cases} 2B(\frac{1}{2}, m + \frac{1}{2})(a^2 - b^2)^m & \text{if } r = 2m, \\ 0 & \text{if } r = 2m+1. \end{cases}$$

Thus (4.5.6) reduces to

$$\phi(t) = \frac{J_0(\{(t - i\kappa\cos\mu)^2 - (\kappa\sin\mu)^2\}^{1/2})}{I_0(\kappa)}. \tag{4.5.8}$$

The characteristic function of C is $\phi(t)^n$, so that by the inversion theorem the probability density function of C is

$$g(C; \mu, \kappa) = \frac{1}{2\pi(I_0(\kappa))^n} \int_{-\infty}^{\infty} e^{-iCt} J_0(\{(t - i\kappa\cos\mu)^2 - (\kappa\sin\mu)^2\}^{1/2})^n dt.$$

By contour integration of the integrand around the rectangle with vertices $(\pm c, 0), (\pm c, \kappa\mu)$ with $c \to \infty$, we find that

$$g(C; \mu, \kappa) = \frac{1}{2\pi I_0(\kappa)^n} e^{\kappa C\cos\mu} \int_{-\infty}^{\infty} e^{-iCt} J_0(\{t^2 - (\kappa\sin\mu)^2\}^{1/2})^n dt. \tag{4.5.9}$$

Since the second part of the integrand is an even function of t, we have

$$g(C; \mu, \kappa) = \frac{1}{\pi I_0(\kappa)^n} e^{\kappa C\cos\mu} \int_0^{\infty} \cos(Ct) J_0(\{t^2 - (\kappa\sin\mu)^2\}^{1/2})^n dt. \tag{4.5.10}$$

For $\mu = 0$, (4.5.10) reduces to

$$g(C; 0, \kappa) = \frac{1}{\pi I_0(\kappa)^n} e^{\kappa C} \int_0^{\infty} \cos(Ct)\, J_0(t)^n dt, \tag{4.5.11}$$

as given by Greenwood & Durand (1955). For the isotropic case, we take $\kappa = 0$ to get

$$g(C; \mu, 0) = \frac{1}{\pi}\int_0^\infty \cos(Ct) J_0(t)^n dt, \tag{4.5.12}$$

as obtained by Lord (1948). Large-sample and high-concentration approximations to (4.5.11)–(4.5.12) are discussed in Sections 4.8.1 and 4.8.2, respectively.

Replacing μ by $\pi/2 - \mu$ in (4.5.10) gives the probability density function of S as (Mardia, 1972a, Section 4.5.4)

$$g(S; \mu, \kappa) = \frac{1}{\pi I_0(\kappa)^n} e^{\kappa S \sin\mu} \int_0^\infty \cos(St)\, J_0(\{t^2 - (\kappa\cos\mu)^2\}^{1/2})^n dt. \tag{4.5.13}$$

Putting $\kappa = 0$ gives the probability density function of S in the uniform case as

$$g(S; \mu, \kappa) = \pi^{-1}\int_0^\infty \cos(St) J_0(t)^n dt. \tag{4.5.14}$$

4.6 DISTRIBUTIONS RELATED TO THE MULTI-SAMPLE PROBLEM FOR VON MISES POPULATIONS

Let $(\theta_{11}, \ldots, \theta_{1n_1}), \ldots, (\theta_{q1}, \ldots, \theta_{qn_q})$ be q independent random samples of sizes $n_1, \ldots, n_q$ from $M(\mu_j, \kappa_j)$, for $j = 1, \ldots, q$. Let $\bar{\theta}_j$ and R_j denote the mean direction and resultant of the jth sample. We shall write

$$\boldsymbol{\mu} = (\mu_1, \ldots, \mu_q), \qquad \boldsymbol{\kappa} = (\kappa_1, \ldots, \kappa_q),$$

$$C_j = \sum_{k=1}^{n_j} \cos\theta_{jk}, \qquad S_j = \sum_{k=1}^{n_j} \sin\theta_{jk},$$

$$C = \sum_{j=1}^{q} C_j, \qquad S = \sum_{j=1}^{q} S_j,$$

$$\mathbf{R} = (R_1, \ldots, R_q), \qquad n = \sum_{j=1}^{q} n_j. \tag{4.6.1}$$

Let $\bar{\theta}$ and R be the mean direction and resultant of the combined sample. Then

$$R^2 = C^2 + S^2, \quad C = \sum_{j=1}^{q} R_j \cos\bar{\theta}_j, \quad S = \sum_{j=1}^{q} R_j \sin\bar{\theta}_j. \tag{4.6.2}$$

4.6.1 *The Distribution of R*

Again we utilise the characteristic function method used in Section 4.4 for the distribution of R in the uniform case. In the notation of (4.2.6), the characteristic function of $(\cos\theta, \sin\theta)$ is defined by

$$\tilde{\psi}(\rho, \Phi) = \mathrm{E}[\exp\{i\rho\cos(\theta - \Phi)\}].$$

After using the transformation $\theta' = \theta - \Phi$, this reduces to

$$\tilde{\psi}(\rho, \Phi) = \frac{1}{2\pi I_0(\kappa)} \int_0^{2\pi} \exp\{i\rho\cos\theta + \kappa\cos(\theta - \mu + \Phi)\}d\theta.$$

On substituting the value of this integral from (4.5.7), we obtain

$$\tilde{\psi}(\rho, \Phi) = \frac{J_0(\{\rho^2 - \kappa^2 - 2i\rho\kappa\cos(\Phi - \mu)\}^{1/2})}{I_0(\kappa)}. \tag{4.6.3}$$

Hence the joint characteristic function of C and S is given by

$$\prod_{j=1}^{q} \left\{\frac{J_0(w_j)}{I_0(\kappa_j)}\right\}^{n_j}, \tag{4.6.4}$$

where

$$w_j = \left\{\rho^2 - \kappa_j^2 - 2i\rho\kappa_j\cos(\Phi - \mu_j)\right\}^{1/2}. \tag{4.6.5}$$

On using (4.6.4) in the inversion formula (4.2.10), we obtain the probability density function of R as

$$f(R) = 2\pi \left\{\prod_{j=1}^{q} I_0(\kappa_j)^{n_j}\right\}^{-1} R\Psi_1(R; \mu, \kappa), \tag{4.6.6}$$

where

$$\Psi_1(R; \mu, \kappa) = \int_{\rho=0}^{\infty} \int_{\Phi=0} 2\pi J_0(R\rho) \left\{\prod_{j=1}^{q} J_0(w_j)^{n_j}\right\} \rho d\rho d\Phi. \tag{4.6.7}$$

By using the method of Section 4.6.3 it can be shown that, for $q = 1$, (4.6.6) reduces to the probability density function of R for the von Mises distribution given by (4.5.4).

4.6.2 The Joint Distribution of $(R, \mathbf{R})$

First we obtain the conditional distribution of R given $\mathbf{R}$. Since, by (4.5.5), the conditional distribution of $\bar{\theta}_j$ given R_j is $M(\mu_j, \kappa_j R_j)$, the characteristic function of $(R_j\cos\bar{\theta}_j, R_j\sin\bar{\theta}_j)$ can be obtained from (4.6.3) on replacing μ, κ and ρ by $\mu_j, \kappa_j R_j$ and ρR_j, respectively. Consequently, using the representation (4.6.2) of C and S, it follows that the conditional characteristic function of (C, S) given $\mathbf{R}$ is

$$\prod_{j=1}^{q} \frac{J_0(w_j R_j)}{I_0(\kappa_j R_j)}. \tag{4.6.8}$$

Substituting (4.6.8) into the inversion formula (4.2.18) shows that the conditional distribution of R given $\mathbf{R}$ is

$$f(R|\mathbf{R}) = \frac{1}{2\pi \prod_{j=1}^{q} I_0(\kappa_j R_j)} R\Psi_2(R, \mathbf{R}; \mu, \kappa), \tag{4.6.9}$$

where

$$\Psi_2(R, \mathbf{R}; \mu, \kappa) = \int_{\rho=0}^{\infty} \int_{\Phi=0}^{2\pi} J_0(R\rho) \left\{ \prod_{j=1}^{q} J_0(w_j R_j) \right\} \rho d\rho d\Phi, \tag{4.6.10}$$

with $w_1, \ldots, w_q$ defined by (4.6.5).

From (4.5.4) and the independence of $R_1, \ldots, R_q$,, the probability density function of $\mathbf{R}$ is

$$f(\mathbf{R}; \mu, \kappa) = \prod_{j=1}^{q} \frac{I_0(\kappa_j R_j) h_{n_j}(R_j)}{I_0(\kappa_j)^{n_j}}. \tag{4.6.11}$$

Multiplying this by (4.6.9) gives the joint probability density function of R and $\mathbf{R}$ as

$$f(R, \mathbf{R}; \mu, \kappa) = \frac{1}{2\pi} \left\{ \prod_{j=1}^{q} \frac{h_{n_j}(R_j)}{I_0(\kappa_j R_j)^{n_j}} \right\} R\Psi_2(R, \mathbf{R}; \mu, \kappa). \tag{4.6.12}$$

On dividing (4.6.12) by (4.6.6), it follows that the conditional probability density function of $\mathbf{R}$ given R is

$$f(\mathbf{R}|R, \mu, \kappa) = \left\{ \prod_{j=1}^{q} h_{n_j}(R_j) \right\} \frac{\Psi_2(R, \mathbf{R}; \mu, \kappa)}{\Psi_1(R; \mu, \kappa)}, \tag{4.6.13}$$

where Ψ_1 and Ψ_2 are given by (4.6.7) and (4.6.10), respectively.

4.6.3 Distributions for the Homogeneous Case

We now consider the homogeneous case, in which $\mu_1 = \ldots = \mu_q = \mu$ and $\kappa_1 = \ldots = \kappa_q = \kappa$. In this case, the distribution of R is given by (4.5.4). Comparing this with (4.6.6) shows that

$$\Psi_1(R; \mu\mathbf{1}, \kappa\mathbf{1}) = \frac{2\pi I_0(\kappa R) h_n(R)}{R}, \tag{4.6.14}$$

where $\mathbf{1} = (1, \ldots, 1)$ denotes a q-vector of 1s. Next we consider Ψ_2 defined by (4.6.10). Without loss of generality, we may rotate coordinates so that $\mu = 0$. Further, on transforming (ρ, Φ) to Cartesian coordinates (x, y) and performing the same contour integration with respect to x as carried out for (4.5.9), we obtain

$$\Psi_2(R, \mathbf{R}; \mu\mathbf{1}, \kappa\mathbf{1})$$
$$= \int_{-\infty}^{\infty}\int_{-\infty}^{\infty} J_0(\{(x+i\kappa)^2 + y^2\}^{1/2}) \left\{\prod_{j=1}^{q} J_0(R_j\{x^2+y^2\}^{1/2})\right\} dxdy.$$

On transforming (x, y) to polar coordinates and using the result

$$\int_0^{2\pi} J_0(R\{\rho^2 - \kappa^2 + 2i\kappa\rho\cos\Phi\}^{1/2})d\Phi = 2\pi J_0(R\rho)I_0(\kappa R), \tag{4.6.15}$$

obtained from the Neumann addition formula (3.5.45), we find that

$$\Psi_2(R, \mathbf{R}; \mu\mathbf{1}, \kappa\mathbf{1}) = 2\pi I_0(\kappa R)\int_0^{\infty} J_0(R\rho)\left\{\prod_{j=1}^{q} J_0(R_j\rho)\right\}\rho d\rho. \tag{4.6.16}$$

With the help of the simplified versions of Ψ_1 and Ψ_2 given by (4.6.14) and (4.6.16), various distributions related to $(R, \mathbf{R})$ can be obtained. In particular, from (4.6.13), the conditional probability density function of $\mathbf{R}$ given R is

$$f(\mathbf{R}|R) = \frac{\prod_{j=1}^{q} h_{n_j}(R_j)}{h_n(R)}\int_0^{\infty} RJ_0(R\rho)\left\{\prod_{j=1}^{q} J_0(R_j\rho)\right\}\rho d\rho. \tag{4.6.17}$$

Note that the conditional probability density function (4.6.17) does not involve the parameters μ and κ. This is of considerable practical interest, and it forms the basis of most of the two-sample and multi-sample tests presented in Sections 7.3 and 7.4.

The case $q = 2$ is particularly important in practice. In this case (4.6.17) simplifies nicely. Simple trigonometry gives

$$R^2 = R_1^2 + R_2^2 + 2R_1R_2\cos\lambda, \tag{4.6.18}$$

where λ (with $0 \leq \lambda \leq \pi$) is the angle between $\bar{\theta}_1$ and $\bar{\theta}_2$. If $\kappa = 0$ then $\bar{\theta}_1$ and $\bar{\theta}_2$ are uniformly distributed, and so λ is distributed uniformly on $[0, \pi]$. Hence, on using the transformation (4.6.18) from λ to R, we find that the conditional probability density function of R given R_1 and R_2 is

$$f(R|R_1, R_2) = \frac{R}{\pi R_1 R_2 \sin\lambda}, \tag{4.6.19}$$

on $|R_1 - R_2| \leq R \leq R_1 + R_2$, $0 \leq R_1 \leq n_1$, $0 \leq R_2 \leq n_2$, where, from (4.6.18),

$$\sin\lambda = \frac{\{[(R_1+R_2)^2 - R^2][R^2 - (R_1-R_2)^2]\}^{1/2}}{2R_1R_2}. \tag{4.6.20}$$

The fact that the probability density function of R given by (4.6.19) is identical to (4.6.9) with $q = 2$ and $\kappa = 0$ implies that

$$\int_0^\infty J_0(R\rho)J_0(R_1\rho)J_0(R_2\rho)\rho d\rho = \frac{1}{\pi R_1 R_2 \sin\lambda}. \tag{4.6.21}$$

Substituting (4.6.21) into (4.6.17) with $q = 2$ shows that the conditional probability density function of R_1 and R_2 given R reduces to

$$f(R_1, R_2|R) = \frac{R}{h_n(R)}\frac{h_{n_1}(R_1)}{R_1}\frac{h_{n_2}(R_2)}{R_2}\frac{1}{\pi\sin\lambda}, \tag{4.6.22}$$

where $0 < |R_1 - R_2| < R, 0 < R_1 + R_2 < n$ and $\sin\lambda$ is given by (4.6.20).

Result (4.6.22) was obtained by Watson & Williams (1956). The generalisation of this result to (4.6.17) is due to Rao (1969).

4.7 MOMENTS OF $\bar{R}$

Let θ be a random variable on the circle. Put $\mathbf{x} = (\cos\theta, \sin\theta)^T$ and denote the variance matrix of $\mathbf{x}$ by $\boldsymbol{\Sigma}$. Then

$$\operatorname{tr}\boldsymbol{\Sigma} = 1 - \rho^2. \tag{4.7.1}$$

Similarly, if $\mathbf{S}$ denotes the variance matrix of a random sample from this distribution, we have

$$\operatorname{tr}\mathbf{S} = \frac{n}{n-1}(1 - \bar{R}^2). \tag{4.7.2}$$

Since $\mathbf{S}$ is an unbiased estimator of $\boldsymbol{\Sigma}$, taking the expectation of (4.7.2) yields

$$1 - \rho^2 = \frac{n}{n-1}(1 - \mathrm{E}[\bar{R}^2])$$

and so

$$\mathrm{E}[\bar{R}^2] = \rho^2 + \frac{1}{n}(1 - \rho^2). \tag{4.7.3}$$

For the uniform distribution $\rho = 0$, and so (4.7.3) gives

$$\mathrm{E}[\bar{R}^2] = \frac{1}{n}. \tag{4.7.4}$$

A calculation based on (9.6.2) shows that

$$n^2\operatorname{var}(\bar{R}^2) = 1 - \frac{1}{n}. \tag{4.7.5}$$

4.8 LIMITING DISTRIBUTIONS OF CIRCULAR STATISTICS

4.8.1 Large-Sample Approximations

The Joint Distribution of $\bar{C}$ and $\bar{S}$

By the central limit theorem, the joint distribution of $\bar{C} = C/n$ and $\bar{S} = S/n$ is asymptotically normal. Their means, variances and covariance are given by

$$\mathrm{E}[(\bar{C},\bar{S})] = (\alpha, \beta) \tag{4.8.1}$$

$$n\begin{pmatrix} \mathrm{var}(\bar{C}) & \mathrm{cov}(\bar{C},\bar{S}) \\ \mathrm{cov}(\bar{S},\bar{C}) & \mathrm{var}(\bar{S}) \end{pmatrix} = \frac{1}{2}\begin{pmatrix} 1+\alpha_2-2\alpha^2 & \beta_2-2\alpha\beta \\ \beta_2-2\alpha\beta & 1-\alpha_2-2\beta^2 \end{pmatrix}. \tag{4.8.2}$$

In particular, the asymptotic marginal distributions of $\bar{C}$ and $\bar{S}$ are normal. For the uniform case $\alpha = \beta = \alpha_2 = \beta_2 = 0$, and so for n large, $\sqrt{2n}(\bar{C},\bar{S})$ is distributed asymptotically as $N(0, \mathbf{I}_2)$. However, the distribution of $(\bar{\theta}, \bar{R})$ is not so simple and depends on whether ρ is zero or not.

The Distributions of $\bar{\theta}$ and $\bar{R}$

Since

$$\bar{R} = (\bar{S}^2 + \bar{C}^2)^{1/2}$$

and

$$\bar{\theta} = \begin{cases} \tan^{-1}(\bar{S}/\bar{C}) & \text{if } \bar{C} \geq 0, \\ \pi + \tan^{-1}(\bar{S}/\bar{C}) & \text{if } \bar{C} < 0 \end{cases}$$

(where $\tan^{-1}(\theta)$ is measured in the interval $[-\pi/2, \pi/2]$), the asymptotic distributions of $\bar{\theta}$ and $\bar{R}$ can be obtained from those of $\bar{C}$ and $\bar{S}$.

Case I: $\rho > 0$. If $\rho > 0$ then the transformation from $(\bar{C},\bar{S})$ to $(\bar{\theta},\bar{R})$ is invertible at (α,β) and it follows from the asymptotic normality of $(\bar{C},\bar{S})$ together with Taylor expansion (the 'δ method': see Rao, 1973, p. 388), that the joint distribution of $\bar{\theta}$ and $\bar{R}$ is asymptotically bivariate normal with

$$\mathrm{E}[\bar{\theta}] \simeq \mu, \qquad \mathrm{E}[\bar{R}] \simeq \rho, \tag{4.8.3}$$

$$n\mathrm{var}(\bar{\theta}) \simeq \frac{\rho^2 + \alpha_2(\beta^2 - \alpha^2) - 2\alpha\beta\beta_2}{2\rho^4}, \tag{4.8.4}$$

$$n\mathrm{var}(\bar{R}) \simeq \frac{\rho^2(1-2\rho^2) + \alpha_2(\alpha^2 - \beta^2) + 2\alpha\beta\beta_2}{2\rho^2}, \tag{4.8.5}$$

$$n\mathrm{cov}(\bar{\theta},\bar{R}) \simeq \frac{\beta_2(\alpha^2 - \beta^2) - 2\alpha\beta\alpha_2}{2\rho^3}. \tag{4.8.6}$$

For distributions which are symmetrical about zero,

$$\mathrm{E}[\bar{\theta}] = 0 \tag{4.8.7}$$

and

$$n\text{cov}(\bar{\theta}, \bar{R}) = 0. \tag{4.8.8}$$

Also, $\beta = \beta_2 = 0$, so (4.8.3)–(4.8.5) simplify to

$$\text{E}[\bar{R}] \simeq \alpha, \tag{4.8.9}$$

$$n\text{var}(\bar{\theta}) \simeq \frac{1-\alpha_2}{2\alpha^2}, \tag{4.8.10}$$

$$n\text{var}(\bar{R}) \simeq \frac{1-2\alpha^2+\alpha_2}{2}. \tag{4.8.11}$$

A more detailed calculation gives

$$\text{E}[\bar{R}] = \alpha + \frac{1-\alpha_2}{4\alpha n} + O(n^{-3/2}). \tag{4.8.12}$$

If the parent distribution is $M(0, \kappa)$ then it follows from the general result (9.6.8) below due to Hendriks, Landsman & Ruymgaart (1996) that

$$2n\kappa A(\kappa)\left(1-\cos\bar{\theta}\right) \stackrel{\cdot}{\sim} \chi_1^2, \qquad n \to \infty. \tag{4.8.13}$$

Case II: $\rho = 0$. When $\rho = 0$, the transformation from $(\bar{C}, \bar{S})$ to $(\bar{\theta}, \bar{R})$ is not invertible at (α, β), so the above approach is not applicable. For simplicity, we consider only samples from the uniform distribution. In this case, (4.4.1) states that the distribution of $\bar{\theta}$ is uniform. Since $\sqrt{2n}(\bar{C}, \bar{S})$ is distributed asymptotically as $N(0, \mathbf{I}_2)$,

$$2n\bar{R}^2 \stackrel{\cdot}{\sim} \chi_2^2, \qquad n \to \infty. \tag{4.8.14}$$

The von Mises Distribution

We now assume that the parent distribution is $M(0, \kappa)$ with $\kappa > 0$.

Since $\rho > 0$, the joint distribution of $\bar{\theta}$ and R can be obtained from case I above. The asymptotic means and variances are given by (4.8.7)–(4.8.12). From (3.5.30) and (3.5.37),

$$\alpha = A(\kappa), \qquad \alpha_2 = 1 - \frac{2A(\kappa)}{\kappa}, \tag{4.8.15}$$

where

$$\alpha = A(\kappa) = \frac{I_1(\kappa)}{I_0(\kappa)}.$$

Consequently,

$$n\text{var}(\bar{\theta}) \simeq \frac{1}{\kappa A(\kappa)}. \tag{4.8.16}$$

Further, from (4.8.12),

$$\mathrm{E}[\bar{R}] \simeq A(\kappa) + \frac{1}{2n\kappa}. \tag{4.8.17}$$

From (4.7.3) and (4.8.17), we have

$$n\mathrm{var}(\bar{R}) = \left(1 - A(\kappa)^2 - \frac{A(\kappa)}{\kappa}\right) - \frac{1}{4n\kappa^2} + O(n^{-2}). \tag{4.8.18}$$

By more detailed calculations, we can refine (4.8.16) to

$$n\mathrm{var}(\bar{\theta}) = \frac{1}{\kappa A(\kappa)} + \frac{3\kappa(1 - A(\kappa)^2) - 5A(\kappa)}{n\kappa^2 A(\kappa)^3} + O(n^{-2}). \tag{4.8.19}$$

If κ is large, (4.8.17)–(4.8.19) can be simplified by using the expansion for A given by (3.5.34), i.e.

$$A(\kappa) = 1 - \frac{1}{2\kappa} - \frac{1}{8\kappa^2} + \cdots \tag{4.8.20}$$

For example,

$$n\mathrm{var}(\bar{\theta}) \simeq \frac{1}{\kappa}\left\{1 + \frac{1}{2\kappa}\right\}. \tag{4.8.21}$$

4.8.2 High-Concentration Approximations

Distribution of $1 - \bar{C}$

Recall from (3.5.22) that if $\theta \sim M(0,\kappa)$ and κ is large then $\kappa^{1/2}\theta \stackrel{\cdot}{\sim} N(0,1)$. Since $2\kappa(1 - \cos\theta) \simeq \kappa\theta^2$, we have

$$2\kappa(1 - \cos\theta) \stackrel{\cdot}{\sim} \chi_1^2, \qquad \kappa \to \infty. \tag{4.8.22}$$

It follows from (4.8.22) and the additive property of χ^2 distributions that if $\theta_1, \ldots \theta_n$ are independent and distributed as $M(0,\kappa)$ then

$$2n\kappa(1 - \bar{C}) \stackrel{\cdot}{\sim} \chi_n^2. \tag{4.8.23}$$

In practice, the asymptotic result (4.8.23) is not adequate for moderately large values of κ. One way of improving the approximation (4.8.23) is to multiply $2n\kappa(1 - \bar{C})$ by a suitable constant so that that its mean is almost exactly the limiting value n, i.e. to replace κ by γ such that

$$\mathrm{E}[2n\gamma(1 - \bar{C})] = n,$$

either exactly or with error of order $O(\kappa^{-2})$. Such a multiplicative correction is a form of Bartlett correction for highly concentrated distributions. Then

$$2n\gamma(1 - \bar{C}) \stackrel{\cdot}{\sim} \chi_n^2. \tag{4.8.24}$$

Using $\mathrm{E}[\bar{C}] = A(\kappa)$ and the expansion (4.8.20) for large values of κ gives

$$\gamma^{-1} = \kappa^{-1} + \tfrac{1}{4}\kappa^{-2}. \tag{4.8.25}$$

An alternative expression for γ can be obtained by identifying the expression for $n\mathrm{var}(\bar{\theta})$ given by (4.8.21) with $1/\gamma$, giving

$$\gamma^{-1} = \kappa^{-1} + \tfrac{1}{2}\kappa^{-2}. \tag{4.8.26}$$

Stephens (1969a) found that taking

$$\gamma^{-1} = \kappa^{-1} + \tfrac{3}{8}\kappa^{-2} \tag{4.8.27}$$

(the average value of $1/\gamma$ from (4.8.25) and (4.8.26)) ensures that approximation (4.8.24) is reasonable for $\kappa \geq 2$.

Distribution of $\bar{R} - \bar{C}$

Recall from (4.5.5) that $\bar{\theta}|\bar{R} \sim M(\mu, \kappa R)$. Since

$$\bar{R} - \bar{C} = \bar{R}(1 - \cos\bar{\theta}),$$

it follows that

$$2n\kappa(\bar{R} - \bar{C})|\bar{R} \mathrel{\dot{\sim}} \chi_1^2,$$

for large κ. Since this approximation to the conditional distribution does not depend on $\bar{R}$, it follows that

$$2n\kappa(\bar{R} - \bar{C}) \mathrel{\dot{\sim}} \chi_1^2, \qquad \kappa \to \infty. \tag{4.8.28}$$

A more refined approximation is

$$2n\gamma(\bar{R} - \bar{C}) \mathrel{\dot{\sim}} \chi_1^2, \qquad \kappa \to \infty, \tag{4.8.29}$$

where γ is defined in (4.8.27).

Distribution of $1 - \bar{R}$

In the identity

$$2n\kappa(1 - \bar{C}) = 2n\kappa(1 - \bar{R}) + 2n\kappa(\bar{R} - \bar{C}), \tag{4.8.30}$$

each term is approximately quadratic in $\theta_1, \ldots, \theta_n$. It follows from Cochran's theorem (see, e.g. Stuart & Ord, 1991, p. 1490) that

$$2n\kappa(1 - \bar{R}) \mathrel{\dot{\sim}} \chi_{n-1}^2, \qquad \kappa \to \infty, \tag{4.8.31}$$

and the random variables $2n\kappa(1 - \bar{R})$ and $2n\kappa(\bar{R} - \bar{C})$ are approximately independent. This important result will be extended to higher dimensions in

Section 9.6.3. It was obtained by a different method by Watson & Williams (1956). The approximation (4.8.31) can be refined to

$$2n\gamma(1-\bar{R}) \ \dot{\sim}\ \chi_1^2, \qquad \kappa \to \infty, \tag{4.8.32}$$

where γ is defined in (4.8.27). Replacing the decomposition (4.8.30) by

$$2n\gamma(1-\bar{C}) = 2n\gamma(1-\bar{R}) + 2n\gamma(\bar{R}-\bar{C}),$$

where γ is given by (4.8.27), and using the refinement (4.8.24) of (4.8.23) yields

$$2n\gamma(1-\bar{C}) \ \dot{\sim}\ \chi_n^2, \quad 2n\gamma(1-\bar{R}) \ \dot{\sim}\ \chi_{n-1}^2, \quad 2n\gamma(\bar{R}-\bar{C}) \ \dot{\sim}\ \chi_1^2, \quad \kappa \to \infty. \tag{4.8.33}$$

Note the analogy between the decomposition (4.8.30) and the decomposition

$$\frac{1}{\sigma^2}\sum_{i=1}^{n}(x_i-\mu)^2 = \frac{1}{\sigma^2}\sum_{i=1}^{n}(x_i-\bar{x})^2 + n\frac{(\bar{x}-\mu)^2}{\sigma^2} \tag{4.8.34}$$

on the line (see (2.3.10)). If $x_1, \ldots, x_n$ are distributed independently as $N(\mu, \sigma^2)$ and κ is large then corresponding terms in (4.8.30) and (4.8.34) have approximately the same distributions, which are those in the decomposition

$$\chi_n^2 = \chi_{n-1}^2 + \chi_1^2.$$

Approximations in the Multi-sample Case

Suppose that we have q independent random samples of sizes $n_1, \ldots, n_q$ from $M(\mu, \kappa)$, $j = 1, \ldots, q$, summarised as in (4.6.1). From the above discussion, it follows that for large κ

$$2\kappa(n-R) \ \dot{\sim}\ \chi_{n-1}^2, \quad 2\kappa(n_j-R_j) \ \dot{\sim}\ \chi_{n_j-1}^2, \quad j = 1, \ldots, q. \tag{4.8.35}$$

Therefore

$$2\kappa\left(n - \sum_{j=1}^{q} R_j\right) \ \dot{\sim}\ \chi_{n-q}^2, \quad 2\kappa\left(\sum_{j=1}^{q} R_j - R\right) \ \dot{\sim}\ \chi_{q-1}^2, \tag{4.8.36}$$

and these two statistics are approximately independent. Further approximations will be discussed in Section 7.4.1.

4.8.3 Further Approximations to the Distribution of $\bar{R}$

We now obtain approximations to the distribution of $\bar{R}$ when (i) $0 < \kappa < 1$ and (ii) $1 \leq \kappa \leq 2$. We know from Section 4.8.1 that $\bar{R}$ is asymptotically normal with

$$\mathrm{E}[\bar{R}] = \rho, \quad n\mathrm{var}\bar{R} = \left\{1 - \rho^2 - \frac{\rho}{\kappa}\right\}, \tag{4.8.37}$$

where $\rho = A(\kappa)$. By using variance-stabilising transformations (see Rao, 1973, p. 426), we obtain, for each range of κ, a function of $\bar{R}$ with asymptotic variance independent of ρ, and so of κ. After suitable modifications of the asymptotic variances, these transformations provide adequate normal approximations to the distribution of R.

Case I. $0 < \kappa < 1$. On using the approximation

$$A(\kappa) = \frac{\kappa}{2}\left(1 - \frac{\kappa^2}{8}\right)$$

from (3.5.32) in (4.8.37) we find that for small κ (Mardia, 1972a, (6.3.27))

$$2n(\bar{R} - \kappa) \; \dot{\sim} \; N(0, 2(1 - a^2\kappa^2)), \tag{4.8.38}$$

where

$$a = \sqrt{\frac{3}{8}}. \tag{4.8.39}$$

Note that the same constant a appears in (4.8.38) and Stephens's approximation (4.8.27). The variance-stabilising transformation g_1 is (Mardia, 1972a, (6.3.29))

$$g_1(x) = \sin^{-1}(ax). \tag{4.8.40}$$

Therefore the statistic

$$g_1(2\bar{R}) = \sin^{-1}(2a\bar{R}) \tag{4.8.41}$$

has mean approximately equal to $g_1(\kappa)$ and

$$n \text{var} g_1(2\bar{R}) \simeq 3/4.$$

On comparing the exact tail area of $\bar{R}$ with various approximations obtained on modifying the variance of $g_1(2\bar{R})$, it is found that the approximation with

$$n \text{var} g_1(2\bar{R}) \simeq \frac{3}{4(1 - 4/n)} \tag{4.8.42}$$

is quite satisfactory for $n \geq 8$. For κ very near 1, the approximation given below in case II is recommended. For $n > 40$, the normal approximation with mean and variance given by (4.8.37) is adequate.

The asymptotic mean given above can be improved. By expanding g_1 about κ and using the asymptotic variance of $\bar{R}$ given by (4.8.38), we find that

$$\text{E}[g_1(\bar{R})] = g_1(\kappa) + \frac{1}{n}\frac{a^3\kappa}{(1 - a^2\kappa^2)^{1/2}}.$$

For small κ, this reduces to

$$\text{E}[g_1(2\bar{R})] \simeq g_1(\kappa) + \frac{1}{n}a^3\kappa. \tag{4.8.43}$$

Case II. $1 \leq \kappa \leq 2$. Using (4.8.37), the variance-stabilising transformation g_2 is seen to be (Mardia, 1972a, (6.3.32))

$$g_2(x) = cn^{1/2} \int_0^x \left\{ 1 - r^2 - \frac{r}{A^{-1}(r)} \right\}^{-1/2} dr, \tag{4.8.44}$$

where c is given below. By expanding $r/A^{-1}(r)$ about $A^{-1}(1.5)$, we find after some simplification that

$$g_2(\kappa) = \sinh^{-1} \frac{\rho - c_1}{c_2} = \log \left\{ \frac{\rho - c_1}{c_2} + \left(1 + \frac{\rho - c_1}{c_2}^2 \right)^{1/2} \right\}, \tag{4.8.45}$$

where

$$c = c_3/n^{1/2}, \qquad c_3 = 0.893 \qquad c_1 = 1.089, \qquad c_2 = 0.258.$$

Hence the variance-stabilising transformation is

$$g_2(\bar{R}) = \sinh^{-1} \frac{\bar{R} - c_1}{c_2} \tag{4.8.46}$$

with

$$n \text{var} g_2(\bar{R}) \simeq c_3^2.$$

Again, on comparing the exact tail area of $\bar{R}$ with the approximation obtained after modifying the variance to

$$n \text{var} g_2(\bar{R}) = \frac{c_3^2}{1 - 3/n}, \tag{4.8.47}$$

it is found that the modified approximation is quite adequate for $n \geq 8$. Expanding g_2 about κ and using the asymptotic variance of $\bar{R}$ given by (4.8.38) shows that

$$\text{E}[g_2(\bar{R})] = g_2(\kappa) - \frac{1}{n} \frac{c_3^2(\rho - c_1)}{2\{c_2^2 + (\rho - c_1)^2\}^{1/2}}. \tag{4.8.48}$$

The variance-stabilising transformations (4.8.40) and (4.8.46) will be used in Sections 7.4.2 and 7.4.4 to construct tests of equal concentration of von Mises distributions.

5

Point Estimation

5.1 INTRODUCTION

Because of the special nature of the circle, the concept of unbiased estimation of circular parameters requires careful definition. Section 5.2 provides such a definition and presents an appropriate analogue of the Cramér-Rao bound. Point estimation for von Mises and wrapped Cauchy distributions is discussed in Sections 5.3 and 5.4, respectively. Interval estimation will be considered in Section 7.2. Robust estimation on the circle will be considered in Section 12.4 in the more general context of robust estimation on spheres. Estimation in mixtures of von Mises distributions is considered in Section 5.5.

5.2 UNBIASED ESTIMATORS AND A CRAMÉR–RAO BOUND

Circular distributions often involve *circular parameters*, i.e. parameters taking values on the unit circle. For example, the population mean direction μ is a circular parameter of von Mises distributions. Because we cannot directly take expectations of circular variables, it is not immediately obvious how to define unbiasedness of estimators of circular parameters. The embedding approach of regarding each point θ on the circle as the unit vector $\mathbf{x} = (\cos\theta, \sin\theta)^T$ in the plane enables us to take expectations and so to define unbiasedness.

Let ω be a circular parameter of some family of circular distributions and let t be a statistic taking values in the unit circle. At first sight, it is tempting to call t an unbiased estimator of ω if

$$\mathrm{E}[(\cos t, \sin t)] = (\cos\omega, \sin\omega). \tag{5.2.1}$$

However, convexity of the unit disc means that if t satisfies (5.2.1) then the distribution of t is concentrated at ω. Thus it is appropriate to use the weaker definition that a statistic t taking values on the unit circle is an *unbiased* estimator of ω if the mean direction of t is ω, i.e.

$$||\mathrm{E}[(\cos t, \sin t)]||^{-1}\mathrm{E}[(\cos t, \sin t)] = (\cos\omega, \sin\omega). \tag{5.2.2}$$

In this case, it follows from (3.4.10) that

$$\mathrm{E}[\sin(t-\omega)] = 0. \tag{5.2.3}$$

There is a lower bound on the variability of unbiased estimators of circular parameters, analogous to the usual Cramér–Rao bound for real-valued unbiased estimators. It can be obtained by the following variant of the usual derivation. Let t be an unbiased estimator of ω based on random samples $\theta_1, \ldots, \theta_n$ from a circular distribution with probability density function $f(\cdot;\omega)$ which is positive everywhere. Differentiation of (5.2.3) with respect to ω gives

$$\mathrm{E}[\cos(t-\omega)] = \mathrm{cov}\left(\sin(t-\omega), \frac{\partial l}{\partial \omega}\right), \tag{5.2.4}$$

where

$$l = l(\omega; \theta_1, \ldots, \theta_n) = \sum_{i=1}^{n} \log f(\theta_i; \omega)$$

is the log-likelihood. Using the Cauchy–Schwarz inequality

$$\mathrm{cov}(X,Y)^2 \leq \mathrm{var}(X)\mathrm{var}(Y)$$

in (5.2.4) gives

$$\mathrm{var}(\sin(t-\omega)) \geq \frac{\rho_\omega(t)^2}{I_\omega}, \tag{5.2.5}$$

where $\rho_\omega(t)$ is the mean resultant length of t and I_ω denotes the Fisher information

$$I_\omega = \mathrm{E}\left[\left(\frac{\partial l}{\partial \omega}\right)^2\right] = \mathrm{E}\left[-\frac{\partial^2 l}{\partial \omega^2}\right],$$

which is assumed to be positive. Inequality (5.2.5) is due to Mardia (1972a, Section 5.1). A generalisation by Hendriks (1991) to the context of manifolds is outlined in Section 13.4.2. Equality holds in (5.2.5) if and only if

$$\frac{\partial l}{\partial \omega} = c \sin(t-\omega), \tag{5.2.6}$$

for some constant c.

As an example, consider estimation of μ by a single observation θ from the von Mises distribution $M(\mu, \kappa)$, where $\kappa > 0$. Take $t = \theta$ as the estimator of μ. Then (5.2.6) holds with $c = \kappa$, so equality holds in (5.2.5). To verify this, note that it follows from (3.5.30) and (3.5.37) that

$$\begin{aligned}
\mathrm{E}[\cos(\theta-\mu)] &= A(\kappa), \\
\mathrm{var}(\sin(\theta-\mu)) &= \frac{A(\kappa)}{\kappa}, \\
I_\omega &= \kappa A(\kappa).
\end{aligned}$$

5.3 VON MISES DISTRIBUTIONS

5.3.1 Maximum Likelihood Estimation

Let $\theta_1, \ldots, \theta_n$ be a random sample from $M(\mu, \kappa)$. The log-likelihood is

$$\begin{aligned} l(\mu, \kappa; \theta_1, \ldots, \theta_n) &= n \log 2\pi + \kappa \sum_{i=1}^{n} \cos(\theta_i - \mu) - n \log I_0(\kappa) \\ &= n \left\{ \log 2\pi + \kappa \bar{R} \cos(\bar{\theta} - \mu) - \log I_0(\kappa) \right\}. \quad (5.3.1) \end{aligned}$$

Since $\cos x$ has its maximum at $x = 0$, the maximum likelihood estimate $\hat{\mu}$ of μ is

$$\hat{\mu} = \bar{\theta}. \quad (5.3.2)$$

Differentiating (5.3.1) with respect to κ and using

$$I_0'(\kappa) = I_1(\kappa) \quad (5.3.3)$$

gives

$$\frac{\partial l}{\partial \kappa} = n \left\{ \bar{R} \cos(\bar{\theta} - \mu) - A(\kappa) \right\}, \quad (5.3.4)$$

where $A(\kappa) = I_1(\kappa)/I_0(\kappa)$, as in (3.5.31). Then the maximum likelihood estimate $\hat{\kappa}$ of κ is the solution of

$$A(\hat{\kappa}) = \bar{R}, \quad (5.3.5)$$

i.e.

$$\hat{\kappa} = A^{-1}(\bar{R}). \quad (5.3.6)$$

Expressions (5.3.2) and (5.3.6) for the maximum likelihood estimates can be obtained alternatively from general theory, using the fact that the von Mises distributions form a regular exponential model with natural parameter $\kappa(\cos \mu, \sin \mu)^T$. By (3.5.7), the maximum likelihood estimate of $\kappa(\cos \mu, \sin \mu)^T$ is obtained by equating the sample mean of the canonical statistic $(\cos \theta, \sin \theta)$ to its population mean, so that

$$\bar{R}(\cos \bar{\theta}, \sin \bar{\theta}) = A(\hat{\kappa})(\cos \hat{\mu}, \sin \hat{\mu}),$$

giving (5.3.2) and (5.3.6).

Some selected values of the functions A and A^{-1} are given in Appendix 2.3 and Appendix 2.4, respectively. Approximate solutions of (5.3.6) can be obtained by inverting the expansions (3.5.32) and (3.5.34) to obtain

$$\hat{\kappa} \simeq 2\bar{R} + \bar{R}^3 + \tfrac{5}{6}\bar{R}^5 \quad (5.3.7)$$

for small $\bar{R}$ and

$$\begin{aligned} \hat{\kappa} &\simeq 1/\{2(1 - \bar{R}) - (1 - \bar{R})^2 - (1 - \bar{R})^3\} \quad (5.3.8) \\ &\simeq \frac{1}{2(1 - \bar{R})} \quad (5.3.9) \end{aligned}$$

for large $\bar{R}$. The approximations (5.3.7) and (5.3.8) are reasonable for $\bar{R} < 0.53$ and $\bar{R} \geq 0.85$, respectively. For $0.53 \leq \bar{R} < 0.85$, the approximation

$$\hat{\kappa} \simeq -0.4 + 1.39\bar{R} + 0.43/(1 - \bar{R}) \tag{5.3.10}$$

is adequate (see Fisher, 1993, p. 88). The approximation

$$\hat{\kappa} \simeq (1.28 - 0.53\bar{R}^2) \tan\left(\frac{\pi\bar{R}}{2}\right) \tag{5.3.11}$$

has maximum relative error of 0.032 (Dobson, 1978). A routine for numerical calculation of $\hat{\mu}$ and $\hat{\kappa}$ was given by Mardia & Zemroch (1975a).

Example 5.1

For the roulette data in Example 1.1, the mean direction and mean resultant length were calculated in Example 2.1 as $\bar{\theta} = 51°$ and $\bar{R} = 0.711$. Then $\hat{\mu} = 51°$ and $\hat{\kappa} = 2.08$.

The distributions of $\hat{\mu}$ and $\hat{\kappa}$ are complicated. However, it follows from (4.5.5) and (5.3.2) that the conditional distribution of $\hat{\mu}$ given $\hat{\kappa}$ is

$$\hat{\mu}|\hat{\kappa} \sim M(\mu, n\kappa\bar{R}) \tag{5.3.12}$$

as noted first by Mardia (1972a).

Asymptotic Properties

Large-Sample Asymptotics

Standard theory of maximum likelihood estimators (Cox & Hinkley, 1974, pp. 294–296) shows that the large-sample asymptotic distribution of $(\hat{\mu}, \hat{\kappa})$ is

$$\sqrt{n}(\hat{\mu} - \mu, \hat{\kappa} - \kappa) \overset{\cdot}{\sim} N(0, \mathbf{I}^{-1}), \tag{5.3.13}$$

where $\mathbf{I}$ denotes the Fisher information matrix

$$\mathbf{I} = \mathrm{E}\left[-\begin{pmatrix} \dfrac{\partial^2 l}{\partial \mu^2} & \dfrac{\partial^2 l}{\partial \mu \partial \kappa} \\ \\ \dfrac{\partial^2 l}{\partial \kappa \partial \mu} & \dfrac{\partial^2 l}{\partial \kappa^2} \end{pmatrix} \right] \tag{5.3.14}$$

based on a single observation. (In (5.3.13) $\hat{\mu}$ is regarded as unwrapped onto the line.) Repeated differentiation of (5.3.1) with respect to μ and κ, together with (3.4.10) and (A.14) of Appendix 1 (with $p = 2$) shows that

$$\mathbf{I} = \begin{pmatrix} \kappa A(\kappa) & 0 \\ 0 & 1 - A(\kappa)^2 - A(\kappa)/\kappa \end{pmatrix}. \tag{5.3.15}$$

Thus, for large n,

$$n\text{var}(\hat{\mu}) \simeq \frac{1}{\kappa A(\kappa)}, \tag{5.3.16}$$

$$n\text{var}(\hat{\kappa}) \simeq \frac{1}{1 - A(\kappa)^2 - A(\kappa)/\kappa}, \tag{5.3.17}$$

$$n\text{cov}(\hat{\mu}, \hat{\kappa}) \simeq 0, \tag{5.3.18}$$

so that $\hat{\mu}$ and $\hat{\kappa}$ are approximately independently normally distributed with means μ and κ and variances (5.3.16) and (5.3.17), respectively. Since $A(\kappa)$ increases from 0 to 1 as κ increases from 0 to ∞, it follows from (5.3.16) that μ can be estimated with much smaller precision for small κ than for large κ. This is not surprising, since the von Mises distribution $M(\mu, \kappa)$ tends to the uniform distribution as κ tends to 0. The approximation (5.3.16) can be refined by (4.8.19), which states that

$$n\text{var}(\hat{\mu}) = \frac{1}{\kappa A(\kappa)} + \frac{3\kappa(1 - A(\kappa)^2) - 5A(\kappa)}{n\kappa^2 A(\kappa)^3} + O(n^{-2}). \tag{5.3.19}$$

The sample mean vector $\bar{R}(\cos\bar{\theta}, \sin\bar{\theta})$ is an unbiased estimator of the population mean vector $A(\kappa)(\cos\mu, \sin\mu)$. Since A is a non-linear function, $\hat{\kappa}$ is a biased estimator of κ. Approximations to this bias can be obtained by substituting $p = 2$ in the general results (10.3.18)–(10.3.19) below due to Schou (1978). Thus

$$\text{E}[\hat{\kappa} - \kappa] = \frac{1}{n}\frac{A'(\kappa) - \kappa A''(\kappa)}{2\kappa A'(\kappa)^2} + O(n^{-2}), \tag{5.3.20}$$

and

$$\text{E}[\hat{\kappa} - \kappa] \simeq \frac{3\kappa}{n}, \qquad n \to \infty, \kappa \to \infty. \tag{5.3.21}$$

Application to (5.3.20) of the expansion of A given by putting $p = 2$ in (A.12) of Appendix 1 yields the approximation

$$\text{E}[\hat{\kappa} - \kappa] \simeq \frac{3\kappa}{5n} \qquad n \to \infty, \kappa \to 0. \tag{5.3.22}$$

Best & Fisher (1981) proposed the estimator

$$\begin{cases} \max(\hat{\kappa} - 2/n\hat{\kappa}, 0), & \hat{\kappa} < 2, \\ (n-1)^3\hat{\kappa}/(n^3 + n), & \hat{\kappa} \geq 2. \end{cases} \tag{5.3.23}$$

and showed by simulation that it is approximately unbiased unless both n and κ are small.

High-Concentration Asymptotics

The high-concentration approximation $2n\kappa(1-\bar{R}) \dot{\sim} \chi^2_{n-1}$ of (4.8.31), together with (5.3.9), shows that

$$n\frac{\kappa}{\hat{\kappa}} \dot{\sim} \chi^2_{n-1} \tag{5.3.24}$$

for large κ.

Restricted Maximum Likelihood Estimation

Two important classes of von Mises distributions are those in which one of the parameters μ and κ is known. If κ is known then it follows from (5.3.1) that the restricted maximum likelihood estimator of μ is equal to the unrestricted maximum likelihood estimator $\bar{\theta}$. If μ is known then (5.3.4) shows that the restricted maximum likelihood estimator $\hat{\kappa}_\mu$ of κ is

$$\hat{\kappa}_\mu = A^{-1}(\bar{C}\cos\mu + \bar{S}\sin\mu). \tag{5.3.25}$$

Approximations to the bias of $\hat{\kappa}_\mu$ can be obtained by substituting $p=2$ in the general results (10.3.20)–(10.3.22) below due to Mardia, Southworth & Taylor (1999). The analogues of (5.3.20) and (5.3.21) are

$$\mathrm{E}[\hat{\kappa}_\mu - \kappa] = -\frac{A''(\kappa)}{2nA'(\kappa)^2} + O(n^{-2}) \tag{5.3.26}$$

$$= \frac{1}{n}\frac{2\kappa^4 A'(\kappa)^3 + 3\kappa A(\kappa)^2 + 2(1-\kappa^2)A(\kappa) - \kappa}{2\left\{\kappa - A(\kappa) - \kappa A(\kappa)^2\right\}^2} + O(n^{-2}) \tag{5.3.27}$$

and

$$\mathrm{E}[\hat{\kappa}_\mu - \kappa] \simeq \frac{2\kappa}{n}, \qquad n\to\infty,\ \kappa\to\infty, \tag{5.3.28}$$

respectively.

5.3.2 *Estimation Using Marginal Likelihood of $\bar{R}$*

The fact that the maximum likelihood estimate $\hat{\kappa}$ is a function of $\bar{R}$ alone suggets that $\bar{R}$ is in some sense sufficient for κ. Although $\bar{R}$ is not sufficient for κ in the usual sense, it is G-sufficient for κ (Barndorff-Nielsen, 1978a, Section 4.4), where G is the group $SO(2)$ of rotations of $\mathbb{R}^2$. That is, (i) the distribution of $\bar{R}$ depends only on κ and not on μ, (ii) for each κ, the family of conditional distributions of $\bar{\theta}$ given $\bar{R}$ (which is $M(\mu, n\kappa\bar{R})$ by (4.5.5)) is a transformation model under G. This suggests that inference on κ should be based on the marginal distribution of $\bar{R}$.

Accordingly, Schou (1978) considered estimation of κ by $\check{\kappa}$, the maximum likelihood estimator based on the marginal distribution of $\bar{R}$. The corresponding estimate $\breve{\kappa}$ maximises the marginal likelihood, which is proportional to

$$A(\kappa)^n / A(n\kappa\bar{R}).$$

The estimate $\breve{\kappa}$ is given by

$$\begin{cases} \breve{\kappa} = 0, & \bar{R} < n^{-1/2}, \\ A(\breve{\kappa}) = A(n\breve{\kappa}\bar{R}), & \bar{R} \geq n^{-1/2}. \end{cases} \tag{5.3.29}$$

A table of $\breve{\kappa}$ as a function of $\bar{R}$ is given in Appendix 2.5. The approximation

$$\breve{\kappa} \doteq \frac{1 - 1/n}{2(1 - \bar{R})} + \frac{\bar{R} - 1/n^2}{4\bar{R}(1 - 1/n)}, \tag{5.3.30}$$

is suitable for $\bar{R} > 0.9$. Schou (1978) showed that $\breve{\kappa} \leq \hat{\kappa}$ and $\breve{\kappa} = \hat{\kappa} + O_p(n^{-1})$, and that

$$(n-1)\frac{\kappa}{\breve{\kappa}} \overset{\cdot}{\sim} \chi^2_{n-1} \tag{5.3.31}$$

for large κ.

5.4 WRAPPED CAUCHY DISTRIBUTIONS

The probability density function (3.5.68) of the wrapped Cauchy distribution $WC(\mu, \rho)$ can be expressed as

$$f(\theta; \mu, \rho) = \frac{\sqrt{1 - \|\boldsymbol{\mu}\|^2}}{2\pi(1 - \boldsymbol{\mu}^T\mathbf{x})}, \tag{5.4.1}$$

where

$$\mathbf{x} = (\cos\theta, \sin\theta)^T, \qquad \boldsymbol{\mu} = \frac{2\rho}{1+\rho^2}(\cos\mu, \sin\mu)^T, \tag{5.4.2}$$

so that $\|\boldsymbol{\mu}\| < 1$. The log-likelihood based on independent observations $\theta_1, \ldots, \theta_n$ is

$$l(\boldsymbol{\mu}; \theta_1, \ldots, \theta_n) = \frac{n}{2} - \sum_{i=1}^{n} \log(1 - \boldsymbol{\mu}^T\mathbf{x}), \tag{5.4.3}$$

where

$$\mathbf{x}_i = (\cos\theta_i, \sin\theta_i)^T. \tag{5.4.4}$$

Differentiation of (5.4.3) with respect to $\boldsymbol{\mu}$ shows that the maximum likelihood estimate $\hat{\boldsymbol{\mu}}$ of $\boldsymbol{\mu}$ satisfies

$$\sum_{i=1}^{n} w_i(\mathbf{x}_i - \hat{\boldsymbol{\mu}}) = \mathbf{0}, \tag{5.4.5}$$

where

$$w_i = 1/(1 - \boldsymbol{\mu}^T\mathbf{x}_i). \tag{5.4.6}$$

The estimate $\hat{\boldsymbol{\mu}}$ can be calculated using the following iterative reweighting algorithm (Kent & Tyler, 1988): Let $\boldsymbol{\mu}_0$ be any 2-vector with $||\boldsymbol{\mu}_0|| < 1$. Then define $\boldsymbol{\mu}_\nu : \nu = 1, \ldots$ iteratively by

$$\boldsymbol{\mu}_{\nu+1} = \frac{\sum_{i=1}^n w_{i,\nu}\mathbf{x}_i}{\sum_{i=1}^n w_{i,\nu}}, \qquad \nu = 0, 1, \ldots,$$

where the weights $w_{i,\nu}$ are defined iteratively by

$$w_{i,\nu+1} = \frac{1}{1 - \boldsymbol{\mu}_\nu^T \mathbf{x}_i}.$$

Under mild conditions, the algorithm converges to the unique maximum likelihood estimate $\hat{\boldsymbol{\mu}} = (\hat{\mu}_1, \hat{\mu}_2)^T$. The maximum likelihood estimates μ and ρ of $\hat{\mu}$ and $\hat{\rho}$ are given by

$$\hat{\mu} = \tan^{-1}\left(\frac{\hat{\mu}_2}{\hat{\mu}_1}\right), \qquad \hat{\rho}^2 = \frac{1 - \sqrt{1 - ||\hat{\boldsymbol{\mu}}||^2}}{1 + \sqrt{1 - ||\hat{\boldsymbol{\mu}}||^2}}.$$

For an accelerated version of this algorithm in a slightly more general setting, see Blake & Marinos (1990).

5.5 MIXTURES OF VON MISES DISTRIBUTIONS

Some data sets cannot be described adequately by fitting a single von Mises distribution but are fitted well by a mixture of k von Mises distributions, i.e. by a probability density function of the form

$$\frac{1}{2\pi}\sum_{i=1}^k p_i \frac{\exp\{\kappa_i \cos(\theta - \mu_i)\}}{I_0(\kappa_i)}, \tag{5.5.1}$$

where $p_i \geq 0$ and $p_1 + \ldots + p_k = 1$. For example, the turtle data in Table 1.5 can be described adequately by a mixture of two von Mises distributions. The parameters $(p_1, \mu_1, \kappa_1), \ldots, (p_k, \mu_k, \kappa_k)$ of such finite mixtures of von Mises distributions are identifiable (Fraser, Hsu & Walker, 1981; Kent, 1983a, for identifiability in a much more general class of mixtures of directional distributions). The parameters of (5.5.1) can be estimated by maximum likelihood, using the EM algorithm (Titterington, Smith & Makov, 1985, Section 4.3.2).

An important special case of (5.5.1) consists of mixtures of two von Mises distributions. An algorithm for maximum likelihood estimation in this case was given by Jones & James (1969). A convenient alternative (Spurr & Koutbeiy, 1991) to maximum likelihood estimation is estimation of the parameters $\mu_1, \kappa_1, \mu_2, \kappa_2, p$ by least-squares fitting of the trigonometric moments

$$p\frac{I_j(\kappa_1)}{I_0(\kappa_1)}\cos(j\mu_1) + (1-p)\frac{I_j(\kappa_2)}{I_0(\kappa_2)}\cos(j\mu_2),$$

$$p\frac{I_j(\kappa_1)}{I_0(\kappa_1)}\sin(j\mu_1) + (1-p)\frac{I_j(\kappa_2)}{I_0(\kappa_2)}\sin(j\mu_2)$$

to their sample analogues a_j and b_j for $j = 1, 2, 3$.

A particularly useful case of (5.5.1) consists of mixtures of two von Mises distributions with the same concentration and modes π radians apart. Then the probability density function (5.5.1) has the form

$$f(\theta; \mu, \kappa, p) = \frac{1}{2\pi I_0(\kappa)}\left\{pe^{\kappa\cos(\theta-\mu)} + (1-p)e^{-\kappa\cos(\theta-\mu)}\right\}. \tag{5.5.2}$$

Such a mixture is bimodal (with modes at μ and $\mu + \pi$) if and only if

$$1/(1 + \mathrm{e}^{2\kappa}) < p < 1/(1 + e^{-2\kappa})$$

(Mardia & Sutton, 1975). Double-wrapping of the circle onto itself as in Section 3.6.2 sends θ to $\psi = 2\theta$. Then ψ has probability density function

$$f(\psi; \mu, \kappa) = \frac{1}{2\pi I_0(\kappa)}\cosh(\kappa\cos(\psi/2 - \mu)), \tag{5.5.3}$$

which does not depend on p. The parameters μ and κ can be estimated by equating the sample and population first trigonometric moments of ψ (and using (3.5.37)), to give estimates μ^* and κ^* satisfying

$$\mu^* = \frac{1}{2}\bar{\psi}, \qquad 1 - \frac{2A(\kappa^*)}{\kappa^*} = \bar{R}_2,$$

where $\bar{\psi}$ and $\bar{R}_2$ denote the mean direction and the mean resultant length of the doubled angles $\psi_1, \ldots, \psi_n$. The mixing parameter p can be estimated by equating the sample and population first trigonometric moments of θ, to give the estimate p^* of p which satisfies

$$(2p^* - 1)A(\kappa^*) = \bar{C}\cos\mu^* + \bar{S}\sin\mu^*.$$

Some variants of this method of estimation for (5.5.2) were considered by Spurr (1981).

For the submodel of (5.5.2) in which $p = 1/2$, the probability density functions have the form

$$f(\theta; \mu, \kappa) = \frac{1}{4\pi I_0(\kappa)}\left\{e^{\kappa\cos(\theta-\mu)} + e^{\kappa\cos(\theta+\mu)}\right\}. \tag{5.5.4}$$

Maximum likelihood estimation in (5.5.4) was considered by Bartels (1984).

The methods described above are suitable for circular data. They can be applied to axial data after 'doubling the angles'.

Example 5.2

For the turtle data in Table 1.5, calculation gives $\bar{\psi} = 125°$ and $\bar{R}_2 = 0.481$. Then κ^* is a solution of $A(\kappa)/\kappa = 0.260$. Interpolation in Appendix 2.3 gives $\kappa^* = 3.15$. Also, $\bar{C} = 0.217$ and $\bar{S} = 0.447$ and so the estimate of the proportion of turtles heading towards the sea is $p^* = 0.80$. Cairns (1975) fitted a projected normal distribution to this data set. See Fraser (1979, pp. 226–231).

6

Tests of Uniformity and Tests of Goodness-of-Fit

6.1 INTRODUCTION

Because of the central role played by the uniform distribution, one of the most important hypotheses about a distribution on the circle is that of uniformity. Graphical assessment of uniformity is considered in Section 6.2. The main formal tests of uniformity are presented in Section 6.3. In Section 6.4 we show how tests of uniformity give rise to tests of goodness-of-fit.

6.2 GRAPHICAL ASSESSMENT OF UNIFORMITY

Before carrying out a formal test of uniformity it is sensible to inspect the data. Some impression of whether or not the data might reasonably come from the uniform distribution is given by a simple plot of the observations $\theta_1, \ldots, \theta_n$ on the circle. It can be more useful to construct a ***uniform probability plot.*** In such a plot, the observations are ordered as $0 \leq \theta_{(1)} \leq \cdots \leq \theta_{(n)} \leq 2\pi$ (with respect to some initial direction) and then $\theta_{(i)}/2\pi$ is plotted against $i/(n+1)$. If $\theta_1, \ldots, \theta_n$ is a random sample from the uniform distribution then the points should lie near a straight line of slope 45° passing through the origin.

Note that the plot depends on the initial direction. One way of guarding against misleading impressions arising from an unfortunate choice of initial direction is to extend the plot by adding the points $(1+i/(n+1), 1+\theta_{(i)}/2\pi)$ for (say) $1 \leq i \leq 0.2n$ and $(-1+i/(n+1), -1+\theta_{(i)}/2\pi)$ for (say) $0.8n \leq i \leq n$.

Example 6.1

In an experiment on pigeon homing (Schmidt-Koenig, 1963), the vanishing angles of 10 birds were $55°, 60°, 65°, 95°, 100°, 110°, 260°, 275°, 285°, 295°$. The uniform probability plot in Fig. 6.1 calls attention to the gap between 110° and 260°, and suggests that the directions were not selected uniformly. This impression of non-uniformity is not confirmed by the formal test which we shall apply in Example 6.2.

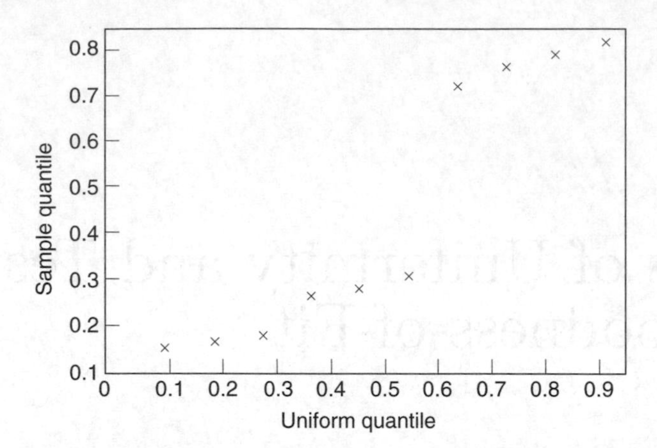

Figure 6.1 Uniform probability plot of the pigeon-homing data of Example 6.1.

6.3 TESTS OF UNIFORMITY

6.3.1 The Rayleigh Test

Perhaps the simplest test of uniformity is Rayleigh's test, which we now describe. Because $\mathrm{E}[(\cos\theta, \sin\theta)^T] = \mathbf{0}$ when θ has the uniform distribution, it is intuitively reasonable to reject uniformity when the vector sample mean $(\bar{C}, \bar{S})$ is far from $\mathbf{0}$, i.e. when $\bar{R}$ is large. As shown in the next paragraph, it is useful to take the test statistic as $2n\bar{R}^2$.

A more formal justification for the Rayleigh test is that it is the score test of uniformity within the von Mises model (3.5.17). Put $\boldsymbol{\omega} = (\kappa \cos\mu, \kappa \sin\mu)^T$. Then $\boldsymbol{\omega}$ is the natural parameter of this exponential model and the log-likelihood based on $\theta_1, \ldots, \theta_n$ is

$$l(\boldsymbol{\omega}; \theta_1, \ldots, \theta_n) = n\boldsymbol{\omega}^T \bar{\mathbf{x}} - n \log I_0(\kappa), \tag{6.3.1}$$

where

$$\bar{\mathbf{x}} = \frac{1}{n} \sum_{i=1}^{n} (\cos\theta_i, \sin\theta_i)^T$$

is the sample mean vector. The score is

$$\mathbf{U} = \frac{\partial l}{\partial \boldsymbol{\omega}^T} = n\bar{\mathbf{x}} - nA(\kappa)(\cos\mu, \sin\mu)^T. \tag{6.3.2}$$

At $\kappa = 0$, $\mathbf{U} = n\bar{\mathbf{x}}$, while it follows from (4.8.2) or from (4.7.1) and circular symmetry that

$$n\mathrm{var}(\mathbf{U}) = \tfrac{1}{2}\mathbf{I}_2.$$

Then the score statistic is

$$\mathbf{U}^T \operatorname{var}(\mathbf{U})^{-1}\mathbf{U} = 2n\bar{R}^2. \tag{6.3.3}$$

It follows from the general theory of score tests (see Section 9.3 of Cox & Hinkley, 1974) that the large-sample asymptotic distribution of $2n\bar{R}^2$ under uniformity is

$$2n\bar{R}^2 \ \dot{\sim}\ \chi_2^2, \tag{6.3.4}$$

with error of order $O(n^{-1/2})$. An alternative proof of (6.3.4) comes from the central limit theorem and the moments of $\bar{C}$ and $\bar{S}$ under uniformity, which can be obtained from (4.8.1)–(4.8.2).

It follows from general results of Cordeiro & Ferrari (1991) on correcting score tests that the modified Rayleigh statistic

$$S^* = \left(1 - \frac{1}{2n}\right) 2n\bar{R}^2 + \frac{n\bar{R}^4}{2} \tag{6.3.5}$$

has a χ_2^2 distribution with error of order $O(n^{-1})$ (see Jupp, 1999). The practical importance of this is that for all except the smallest sample sizes, there is negligible error in the significance levels if the observed S^* is compared to the usual upper quantiles of the χ_2^2 distribution.

An excellent approximation (even for small samples) to the upper tail probabilities of $n\bar{R}^2$ is given by the saddlepoint approximation (4.4.12) to the upper tail probabilities of $\bar{R}$. Approximation (4.4.12) involves explicit use of Bessel functions. These are avoided in the alternative approximations

$$\Pr(n\bar{R}^2 \geq K) = e^{-K}\left\{1 + \frac{2K - K^2}{4n} - \frac{24K - 132K^2 + 76K^3 - 9K^4}{288n^2}\right\} \tag{6.3.6}$$

(Pearson, 1906; Greenwood & Durand, 1955) and

$$\Pr(n\bar{R}^2 > K) = \exp\{[1 + 4n + 4(n^2 - nK)]^{1/2} - (1 + 2n)\} \tag{6.3.7}$$

(Wilkie, 1983).

The Rayleigh test is also the likelihood ratio test of uniformity within the von Mises family. Let w be the likelihood ratio statistic $2[(l(\hat{\omega}; \theta_1, \ldots, \theta_n) - l(\tilde{\omega}; \theta_1, \ldots, \theta_n)]$, where $\hat{\omega}$ and $\tilde{\omega}$ denote respectively the unrestricted maximum likelihood estimate and the maximum likelihood estimate under the null hypothesis. Then

$$w = 2n\{\hat{\kappa}\bar{R} - I_0(\hat{\kappa})\} = 2n\{\hat{\kappa}A(\hat{\kappa}) - I_0(\hat{\kappa})\}. \tag{6.3.8}$$

Differentiation with respect to $\hat{\kappa}$ gives

$$\begin{aligned}\frac{dw}{d\hat{\kappa}} &= 2n\{A(\hat{\kappa}) + \hat{\kappa}A'(\hat{\kappa}) - I_1(\hat{\kappa})\}\\ &= 2n\hat{\kappa}A'(\hat{\kappa}),\end{aligned}$$

and so

$$\frac{dw}{d\bar{R}} = \frac{dw}{d\hat{\kappa}} / \frac{d\bar{R}}{d\hat{\kappa}} = 2n\hat{\kappa} \geq 0.$$

Thus w is an increasing function of $\bar{R}$ and so (Mardia, 1972a, p. 134) the likelihood ratio test is equivalent to the Rayleigh test.

Example 6.2
For the data on pigeon homing in Example 6.1, is there evidence that the directions were not selected uniformly?

Here $n = 10$, $\bar{C} = 0.149$, $\bar{S} = 0.166$, so that $2n\bar{R}^2 = 0.993$ and the modified Rayleigh statistic (6.3.5) is 0.956. Comparing this with the χ_2^2 distribution gives an observed significance level of 0.62, so the directions seem to have been selected uniformly. Comparison of $2n\bar{R}^2$ with the χ_2^2 distribution gives an observed significance level of 0.61, so in this case the modification of the Rayleigh statistic makes little difference.

Example 6.3
Do the leukaemia data of Table 1.4 show evidence of seasonal effects?

If there were no seasonal variation, the corresponding angular observations (obtained by translating time of year into an angle) could be regarded as being drawn from the uniform distribution on the circle. Because months have differing lengths, it is appropriate to adjust the numbers of cases so that they correspond to 'months' of equal length. The adjusted data are shown in Table 6.1, where the first interval (0°, 30°) corresponds to the month of January, and so on. Calculation gives

$$C = -48.278, \quad S = -16.005, \quad \bar{R} = 0.101,$$

and so $2n\bar{R}^2 = 10.22$. As the 1% value of χ_2^2 is 9.21, the null hypothesis is rejected strongly. Thus there is evidence of a seasonal effect. It may be noted that if we use the grouping correction (Section 2.5), the value of $2n\bar{R}^2$ increases to 10.46.

This data set has been analysed by David & Newell (1965) using another technique.

Because the Rayleigh test is equivalent to the likelihood ratio test against von Mises alternatives, it follows from the Neyman–Pearson lemma that the Rayleigh test is most powerful against these alternatives. Also, because the statistic $\bar{R}$ is invariant under rotation and reflection of the data, the test is invariant under these operations. Indeed, the Rayleigh test is the most powerful invariant test against von Mises alternatives. To see this, note that by the Neyman–Pearson lemma, the most powerful invariant test is the likelihood ratio test based on the marginal likelihood of a maximal invariant. (see Lehmann, 1959, Chapter 6). The joint probability density function for

Table 6.1 Month of onset of cases of lymphatic leukaemia in the UK, 1946–1960 (Lee, 1963, reproduced by permission of the BMJ). Times have been converted to angles and numbers of cases have been adjusted accordingly.

Month	Angular range (in degrees)	Number of cases
January	0–30	39
Febuary	30–60	37
March	60–90	29
April	90–120	45
May	120–150	38
June	150–180	59
July	180–210	50
August	210–240	54
September	240–270	37
October	270–300	47
November	300–330	34
December	330–360	37

independent random variables $\theta_1, \ldots, \theta_n$ with the $M(\mu, \kappa)$ distribution is

$$f(\theta_1, \ldots, \theta_n; \mu, \kappa) = \frac{1}{I_0(\kappa)^n} \exp\{n\kappa \bar{R} \cos(\bar{\theta} - \mu)\}. \tag{6.3.9}$$

A maximal invariant under the group of rotations is given by $u_1, \ldots, u_{n-1}$, where

$$u_j = \theta_j - \theta_n, \quad j = 1, \ldots, n. \tag{6.3.10}$$

The joint density of $u_1, \ldots, u_{n-1}$ is

$$\begin{aligned} \check{f}(u_1, \ldots, u_{n-1}; \mu, \kappa) &= \int_0^{2\pi} f(u_1 + x, \ldots, u_n + x) dx \\ &= \frac{1}{I_0(\kappa)^n} \int_0^{2\pi} \exp\{n\kappa \bar{R} \cos(x - \bar{\theta} - \mu)\} dx \\ &= \frac{I_0(n\bar{R}\kappa)}{I_0(\kappa)^n}. \end{aligned} \tag{6.3.11}$$

The marginal likelihood $\check{L}$ of κ based on $u_1, \ldots, u_{n-1}$ is

$$\check{L}(\kappa; u_1, \ldots, u_{n-1}) = \check{f}(u_1, \ldots, u_{n-1}; \mu, \kappa).$$

It follows from (6.3.11) that the marginal likelihood ratio is

$$\frac{\check{L}(\kappa; u_1, \ldots, u_{n-1})}{\check{L}(0; u_1, \ldots, u_{n-1})} = \frac{I_0(n\bar{R}\kappa)}{I_0(\kappa)^n}. \tag{6.3.12}$$

Since $I_0'(\kappa) = I_1(\kappa) \geq 0, I_0(n\bar{R}\kappa)$ is a monotonically increasing function of $\bar{R}$ for $\kappa > 0$. Hence the test rejects uniformity for large values of $\bar{R}$, i.e. it is the Rayleigh test. Since the critical region does not depend on κ, the Rayleigh test is the uniformly most powerful invariant test.

Bhattacharyya & Johnson (1969) have shown that the Rayleigh test is locally most powerful invariant against projected normal distributions.

Note that, although the Rayleigh test is consistent against (non-uniform) von Mises alternatives, it is not consistent against alternatives with $\rho = 0$ (in particular, distributions with antipodal symmetry). Tests of uniformity which are consistent against all alternatives include Kuiper's test and Watson's U^2 test considered in Sections 6.3.2 and 6.3.3.

The Rayleigh Test When the Mean Direction Is Given

Under certain circumstances, we wish to test uniformity against an alternative in which the mean direction is specified. In this case, it is intuitively reasonable to reject uniformity for large values of

$$\bar{C} = \frac{1}{n} \sum_{i=1}^{n} \cos(\theta_i - \mu), \tag{6.3.13}$$

where μ is the hypothesised mean direction. This test can be considered as a variant of the Rayleigh test. It is the score test of uniformity ($\kappa = 0$) within the family of von Mises distributions $M(\mu, \kappa)$. The log-likelihood based on $\theta_1, \ldots, \theta_n$ is

$$l(\kappa; \theta_1, \ldots, \theta_n) = n\kappa \cos(\mathbf{x}_0 - \mu) - n \log I_0(\kappa), \tag{6.3.14}$$

and so the score is

$$\frac{\partial l}{\partial \kappa} = U = n[\cos(\bar{\theta} - \mu) - A(\kappa)]. \tag{6.3.15}$$

At $\kappa = 0$, $U = n\bar{C}$, while it follows from (4.8.2) or from (4.7.1) and circular symmetry that

$$n\text{var}(U) = \tfrac{1}{2}.$$

Thus the score statistic is

$$\frac{U^2}{\text{var}(U)} = 2n\bar{C}^2.$$

It follows from the general theory of score tests that the large-sample asymptotic distribution of $2n\bar{C}^2$ under uniformity is

$$2n\bar{C}^2 \;\dot{\sim}\; \chi_1^2. \tag{6.3.16}$$

An alternative proof of (6.3.16) comes from the central limit theorem and the moments of $\bar{C}$ under uniformity, which can be obtained from (4.8.1)–(4.8.2).

Some quantiles of $\bar{C}$ are given in Appendix 2.6. From Section 4.8.1, for large n, $(2n)^{1/2}\bar{C}$ has approximately a standard normal distribution. A better approximation to the tail probabilities is given by the Edgeworth expansion (Durand & Greenwood, 1957)

$$\Pr[(2n)^{1/2}\bar{C} \geq K] = 1 - \Phi(K)$$
$$+\phi(K)\left\{\frac{3K - K^3}{16n} + \frac{15K + 305K^3 - 125K^5 + 9K^7}{4608n^2}\right\},$$

where Φ and ϕ denote the distribution function and the probability density function of the standard normal distribution.

Example 6.4

Before the discovery of isotopes in the 1920s, it was speculated that the atomic weights of elements were integers (subject to errors). Von Mises (1918) investigated this speculation by regarding the fractional parts (converted to angles) of the atomic weights as a random sample from a distribution on the circle with mean direction 0, and testing for uniformity. The fractional parts (converted to angles) of the atomic weights (as known to von Mises in 1918) of the 24 lightest elements are given in Table 6.2.

Table 6.2 Fractional parts of the atomic weights (as known in 1918) of the the 24 lightest elements

Fractional part (in degrees)	Frequency
0	12
3.6	1
36	6
72	1
108	2
169.2	1
324	1

Here $\mu = 0$. Calculation gives $\bar{C} = 0.724$. From Appendix 2.6, the 1% value of $\bar{C}$ for $n = 24$ is 0.334, and so the null hypothesis of uniformity is rejected strongly. Since $(2n)^{1/2}\bar{C} = 5.02$, the large-sample approximation (6.3.16) yields an observed significance level of 10^{-7}.

6.3.2 Kuiper's Test

Various tests for distributions on the line are based on measuring the deviation between the empirical and hypothesised cumulative distribution functions. To construct analogues of these for distributions on the circle, we need to define the empirical distribution function in the circular case.

Once an initial direction and an orientation of the circle have been chosen, the empirical distribution function S_n is defined as follows. The ordered observations $\theta_{(1)}, \ldots, \theta_{(n)}$ are augmented by $\theta_{(0)}$ and $\theta_{(n+1)}$, which are defined by $\theta_{(0)} = 0$ and $\theta_{(n+1)} = 2\pi$. Then S_n is defined by

$$S_n(\theta) = i/n \quad \text{if } \theta_{(i)} \leq \theta < \theta_{(i+1)}, \quad i = 0, 1, \ldots, n. \tag{6.3.17}$$

Note that both $S_n(\theta)$ and $F(\theta)$ depend on the origin and orientation. It is useful to put

$$U_i = \frac{\theta_{(i)}}{2\pi}, \quad i = 0, \ldots, n+1. \tag{6.3.18}$$

Consideration of the Kolmogorov–Smirnov statistic on the line suggests that uniformity of a distribution on the circle should be rejected for large values of

$$\max(D_n^+, D_n^-), \tag{6.3.19}$$

where

$$D_n^+ = \sup_\theta \{S_n(\theta) - F(\theta)\}, \quad D_n^- = \sup_\theta \{F(\theta) - S_n(\theta)\}, \tag{6.3.20}$$

with S_n denoting the empirical distribution function given by (6.3.17) and F denoting the cumulative distribution function of the uniform distribution, given by $F(\theta) = \theta/2\pi$. However, D_n^+ and D_n^- depend on the choice of the initial direction. This led Kuiper (1960) to define

$$V_n = D_n^+ + D_n^-, \tag{6.3.21}$$

which (as we shall see below) does not depend on the choice of the initial direction. The null hypothesis of uniformity is rejected for large values of V_n. Kuiper's test is consistent against all alternatives to uniformity.

An Alternative Representation of V_n

We have

$$\begin{aligned} D_n^+ &= \max_{0 \leq i \leq n} \sup_{\theta_{(i)} \leq \theta < \theta_{(i+1)}} \left\{ \frac{i}{n} - \frac{\theta}{2\pi} \right\} \\ &= \max_{0 \leq i \leq n} \left\{ \frac{i}{n} - \inf_{\theta_{(i)} \leq \theta < \theta_{(i+1)}} \frac{\theta}{2\pi} \right\}, \end{aligned} \tag{6.3.22}$$

so that

$$D_n^+ = \max_{1 \leq i \leq n} \left\{ \frac{i}{n} - U_i \right\}, \tag{6.3.23}$$

where $U_0, \ldots U_{n+1}$ are defined in (6.3.18). Similarly,

$$D_n^- = \max_{1 \le i \le n} \left\{ U_i - \frac{i-1}{n} \right\}. \tag{6.3.24}$$

Substitution of (6.3.23) and (6.3.24) into (6.3.21) yields the following equivalent representation of V_n:

$$V_n = \max_{1 \le i \le n} \left\{ \frac{i}{n} - U_i \right\} + \max_{1 \le i \le n} \left\{ U_i - \frac{i-1}{n} \right\}. \tag{6.3.25}$$

It is sometimes useful to rewrite (6.3.25) as

$$V_n = \max_{1 \le i \le n} \left(U_i - \frac{i}{n} \right) - \min_{1 \le i \le n} \left(U_i - \frac{i}{n} \right) + \frac{1}{n}. \tag{6.3.26}$$

Rotation-Invariance of V_n

We now show that V_n does not depend on the choice of initial direction. Consider a set of points on the circle represented by angles $2\pi U_1, \ldots, 2\pi U_n$ with $0 \le U_1 \le \ldots \le U_n \le 1$. If a new initial direction is chosen making angle $2\pi c$ with the old initial direction, then there is an integer k such that $U_k \le c \le U_{k+1}$. The observations make angles $2\pi U_1', \ldots, 2\pi U_n'$ with the new initial direction, where

$$U_j' = \begin{cases} U_{k+j} - c & \text{for } j = 1, \ldots, n-k, \\ U_{k+j-n} + 1 - c & \text{for } j = n-k+1, \ldots n. \end{cases} \tag{6.3.27}$$

Consequently, for $j = 1, \ldots, n-k$, we have

$$\frac{j}{n} - U_j' = \frac{i-k}{n} - (U_i - c)$$

with $i = k + j$, whereas for $j = n - k + 1, \ldots, n$, we have

$$\frac{j}{n} - U_j' = \frac{i-k+n}{n} - (U_i + 1 - c)$$

with $i = k + j - n$. Then

$$\frac{j}{n} - U_j' = \frac{i}{n} - U_i + c - \frac{k}{n}, \quad i = k + j \ (\mathrm{mod}\, n), \ j = 1, \ldots, n. \tag{6.3.28}$$

It follows from (6.3.25) that V_n does not depend on c (or k). Hence V_n is well-defined (i.e. it does not depend on the choice of initial direction). Note that the proof shows also that V_n is invariant under rotation. A similar argument shows that V_n is invariant under change of orientation.

The Null Distribution of V_n

Under the null hypothesis of uniformity, the large-sample asymptotic tail probabilities of V_n satisfy

$$\Pr(n^{1/2}V_n \geq z) = 2\sum_{m=1}^{\infty}(4m^2z^2-1)e^{-2m^2z^2} - \frac{8z}{3n^{1/2}}\sum_{m=1}^{\infty}m^2(4m^2z^2-3)e^{-2m^2z^2} + O(n^{-1}) \quad (6.3.29)$$

(see Kuiper, 1960).

It is convenient to use the modification

$$V_n^* = n^{1/2}V_n\left(1+\frac{0.155}{\sqrt{n}}+\frac{0.24}{n}\right) \quad (6.3.30)$$

of V_n. Stephens (1970) showed that the distribution under uniformity of V_n^* varies very little with n, if $n \geq 8$. Some upper quantiles of V_n^* are given in Table 6.3.

Table 6.3 Upper quantiles of V^* (reproduced from Stephens, 1970, by permission of The Royal Statistical Society)

α	0.10	0.05	0.025	0.01
V_n^*	1.620	1.747	1.862	2.001

Some tables of upper tail probabilities of V_n^* are given by Arsham (1988). Another approximation to the null distribution of V_n is given by Maag & Dicaire (1971).

Example 6.5

The hypothesis of uniformity of the distribution giving rise to the pigeon-homing data of Example 6.1 can be tested using Kuiper's test. (The data set is also shown in Table 6.4).

Table 6.4 shows the necessary calculations. The underlined entries denote the maximum and minimum values of $U_i - i/n$. We therefore have

$$V_n = 0.053 - (-0.294) + 0.1 = 0.447,$$

so that $V_n^* = 1.517$, and the hypothesis of uniformity is accepted at the 10% significance level.

The way in which D_n^+ and D_n^- measure the difference between the empirical distribution function S_n and the cumulative distribution function F (of the uniform distribution) can be seen from Fig. 6.2.

Table 6.4 Calculations required to apply Kuiper's test to the pigeon-homing data of Example 6.1

i	1	2	3	4	5	6	7	8	9	10
$\theta_{(i)}$	55°	60°	65°	95°	100°	110°	260°	275°	285°	295°
$U_i = \theta_{(i)}/2\pi$	0.153	0.167	0.181	0.264	0.278	0.306	0.722	0.764	0.792	0.819
i/n	0.1	0.2	0.3	0.4	0.5	0.6	0.7	0.8	0.9	1.0
$U_i - i/n$	<u>0.053</u>	−0.033	−0.119	−0.136	−0.222	<u>−0.294</u>	0.022	−0.036	−0.108	−0.181

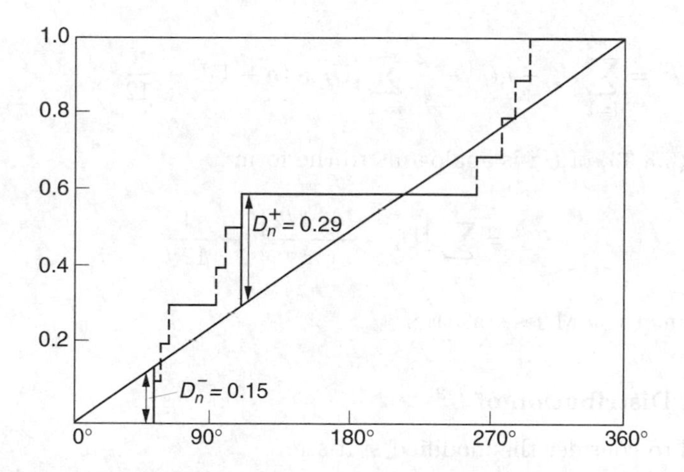

Figure 6.2 Empirical and cumulative distribution functions for the pigeon-homing data of Example 6.1.

A version of Kuiper's test for use with grouped data has been given by Freedman (1979).

6.3.3 Watson's U^2 Test

Instead of measuring the discrepancy between the empirical distribution function and the cumulative distribution function of the uniform distribution by a variant of the maximum deviation (as in Kuiper's test), we can use the (corrected) mean square deviation. This leads to Watson's (1961) statistic

$$U^2 = n \int_0^{2\pi} \left\{ S_n(\theta) - \frac{\theta}{2\pi} - \mu \right\}^2 \frac{1}{2\pi} d\theta, \tag{6.3.31}$$

where

$$\mu = \int_0^{2\pi} \left\{ S_n(\theta) - \frac{\theta}{2\pi} \right\} \frac{1}{2\pi} d\theta = \frac{1}{2} - \frac{1}{n} \sum_{i=1}^{n} \frac{\theta_{(i)}}{2\pi}. \tag{6.3.32}$$

Note that U^2 is an analogue for circular data of the Cramér–von Mises statistic W^2 for data on the real line. It follows from (6.3.31) and (6.3.32) that U^2 is well-defined and invariant under rotations and reflections. Carrying out the integration in (6.3.31) gives the explicit formula

$$U^2 = \sum_{i=1}^{n} \left[U_i - \bar{U} - \frac{i - 1/2}{n} + \frac{1}{2} \right]^2 + \frac{1}{12n}, \tag{6.3.33}$$

where $\bar{U} = (U_1 + \ldots + U_n)/n$. Further manipulation gives the alternative expression

$$U^2 = \sum_{i=1}^{n} U_i^2 - n\bar{U}^2 - \frac{2}{n} \sum_{i=1}^{n} iU_i + (n+1)\bar{U} + \frac{n}{12}. \tag{6.3.34}$$

The form (6.3.33) of U^2 is analogous to the form

$$W^2 = \sum_{i=1}^{n} \left[U_i - \frac{i - 1/2}{n} \right]^2 + \frac{1}{12n} \tag{6.3.35}$$

of the Cramér–von Mises statistic.

The Null Distribution of U^2

It is useful to consider the modified statistic

$$U^{*2} = \left(U^2 - \frac{0.1}{n} + \frac{0.1}{n^2} \right) \left(1 + \frac{0.8}{n} \right). \tag{6.3.36}$$

Stephens (1970) showed that distribution under uniformity of U^{*2} varies very little with n, for $n \geq 8$, and calculated the quantiles of U^{*2} which are given in Table 6.5. Other approximations have been discussed by Pearson & Stephens (1962) and Tiku (1965).

Table 6.5 Upper quantiles of U^{*2} (reproduced by permission of The Royal Statistical Society)

α	0.10	0.05	0.025	0.01
U^{*2}	0.152	0.187	0.221	0.267

The large-sample asymptotic distribution of U^2 under uniformity is given by

$$\lim_{n \to \infty} \Pr(U^2 > u) = 2 \sum_{m=1}^{\infty} (-1)^{m-1} e^{-2m^2\pi^2 u}. \tag{6.3.37}$$

This series converges rapidly and can be calculated easily. It follows from (6.3.58) and (6.3.72), and was obtained by Watson (1961). A more elementary proof is given in Watson (1995). Watson (1961) showed that the large-sample asymptotic distribution of U^2 is the same as that of V_n^2/π^2. He showed also that if

$$W_n(t) = n^{1/2}\{S_n(t) - F(t)\}, \tag{6.3.38}$$

where $F(t) = t/2\pi$, then under uniformity the process W_n converges in distribution as $n \to \infty$ to a Brownian bridge process.

Example 6.6

The hypothesis of uniformity of the distribution giving rise to the pigeon-homing data of Example 6.1 can be tested using Watson's test. (The data are also shown in Table 6.4).

The U_i for the data are shown in Table 6.4. Calculation yields

$$\sum_{i=1}^{n} U_i^2 = 2.728, \quad \sum_{i=1}^{n} iU_i = 31.796, \quad \bar{U} = 0.445.$$

Consequently, $U^2 = 0.116$ and $U^{*2} = 0.115$. Since the 5% value of U^{*2} in Table 6.5 is 0.187, the hypothesis of uniformity is again accepted.

Some modifications of Watson's U^2 test for use with grouped data are considered in Section 6.4.2.

As we shall show in Section 6.3.7, Watson's U^2 test is a locally most powerful invariant test against the alternatives (6.3.64) with $f(\theta) = \theta^2/2\pi^2$. An invariant test of uniformity which is most powerful against distant alternatives with probability density functions of the form $c_p f(\theta)^p$ as $p \to \infty$ is Watson's (1976) test which rejects uniformity for large values of G_n, where

$$G_n = \sqrt{n} \sup_{\theta} \left\{ S_n(\theta) - \frac{\theta}{2\pi} - \frac{1}{2} + \frac{1}{n}\sum_{i=1}^{n} \frac{\theta_{(i)}}{2\pi} \right\}.$$

The distribution function of G_n was determined and tabulated by Darling (1983).

6.3.4 Some Quick Tests

The Hodges–Ajne Test

One strategy for testing uniformity of a distribution on the circle is to reject uniformity if the observed number of points of a sample which fall in a suitable set differs greatly from the expected number. Two natural measures of the discrepancy between the observed and expected numbers are (i) the maximum difference and (ii) the mean square difference as the set varies in a suitable class. In the case where the set runs through all semicircular arcs, the former gives the Hodges–Ajne statistic and the latter gives Ajne's A_n statistic (considered in Section 6.3.5).

Let $N(\theta)$ denote the number of observations in the semicircle centred on θ, i.e. in the arc $(\theta - \pi/2, \theta + \pi/2)$. The Hodges–Ajne test rejects uniformity for large values of

$$\max_{\theta} N(\theta), \tag{6.3.39}$$

or, equivalently, for small values of

$$m = \min_{\theta} N(\theta). \tag{6.3.40}$$

This test was introduced by Ajne (1968), using $n - m$, the maximum number of observations in a semicircle, as the test statistic. Bhattacharyya & Johnson (1969) pointed out the connection with the bivariate sign test of Hodges (1955).

A combinatorial argument (following the method of Daniels, 1954) shows that

$$\Pr(m \le t) = 2^{-n+1}(n - 2t)\binom{n}{t}, \quad t < \frac{n}{3}, \tag{6.3.41}$$

a formula obtained by Hodges (1955). Tables for m are given by Hodges (1955) and Klotz (1959). For large n, it can be shown (Daniels, 1954; Ajne, 1968) that

$$\Pr\left(\frac{n - 2m}{n^{1/2}} \le t\right) \sim \frac{4t}{\sqrt{2\pi}} \sum_{k=0}^{\infty} \exp\left\{-\frac{(2k+1)^2 t^2}{2}\right\}.$$

For most purposes it is sufficient to take just the leading term, giving

$$\Pr\left(\frac{n - 2m}{n^{1/2}} \le t\right) \simeq \frac{4t}{\sqrt{2\pi}} \exp\left\{\frac{-t^2}{2}\right\}.$$

The approximate 5% and 1% values of $(n - 2m)/n^{1/2}$ are 3.023 and 3.562, respectively. Appendix 2.7 gives some quantiles of m.

Example 6.7

In a pigeon-homing experiment reported by Batschelet (1971), the vanishing angles of 15 birds were

$$115^\circ, 120^\circ, 120^\circ, 130^\circ, 135^\circ, 140^\circ, 150^\circ, 150^\circ,$$

$$150^\circ, 165^\circ, 185^\circ, 210^\circ, 235^\circ, 270^\circ, 345^\circ.$$

We test the hypothesis of uniformity using the Hodges–Ajne test. By drawing a circular plot (as in Fig. 1.1) of this data set, it can be seen readily that all the observations except that at 345° lie below the line making angle 110°

with the x-axis, and so $m = 1$. Since for $n = 15$, the 5% value of m given in Appendix 2.7 is 2, the null hypothesis is rejected at the 5% significance level.

Ajne (1968) has shown that the Hodges–Ajne test is the locally most powerful invariant test for testing uniformity against alternatives of the form

$$f(\theta) = \begin{cases} p/\pi & \text{if } \alpha \leq \theta < \pi + \alpha, \\ q/\pi & \text{if } \alpha + \pi \leq \theta < 2\pi + \alpha, \end{cases} \tag{6.3.42}$$

where $p + q = 1$ and $p \to 1$ or $p \to 0$. A locally most powerful invariant test for $p \simeq q$ will be presented in Section 6.3.5.

We now give two spacing tests, i.e. tests which are based on the sample arc lengths $T_1, \ldots, T_n$ defined by

$$T_i = \theta_{(i)} - \theta_{(i-1)}, \qquad i = 1, \ldots, n-1, \qquad T_n = 2\pi - (\theta_{(n)} - \theta_{(1)}).$$

The Range Test

The circular sample range w was defined in Section 2.3.4, as the length of the smallest arc which contains all the observations, i.e.

$$w = 2\pi - T, \quad T = \max_{1 \leq i \leq n} T_i. \tag{6.3.43}$$

Since small values of w indicate clustering of the observations, the hypothesis of uniformity is rejected for small values of w. Clearly, w is invariant under rotations.

A combinatorial argument shows that the distribution function of the circular range under the hypothesis of uniformity is

$$\Pr(w \leq r) = \sum_{k=1}^{\infty} (-1)^{k-1} \binom{n}{k} \left[1 - k\left(1 - \frac{r}{2\pi}\right)\right]^{n-1}, \tag{6.3.44}$$

where the sum is over values of k such that $1 - k(1 - r/2\pi) > 0$. This result has already appeared in various contexts (see David, 1970, Section 5.4) and was first obtained by R. A. Fisher (1929). In this context, it was given by Laubscher & Rudolph (1968) and Rao (1969). Appendix 2.8 gives some quantiles of the distribution of the circular range w.

Example 6.8

We test the hypothesis of uniformity for the roulette data in Example 1.1.

From Example 2.3, $w = 169°$. From Appendix 2.8, the 5% value of w for $n = 9$ is 88.1°, so the hypothesis of uniformity is rejected at the 5% level.

A Test of Equal Spacings

Under uniformity, $E[T_i] = 2\pi/n$. Hence it is intuitively reasonable to reject uniformity for large values of

$$L = \frac{1}{2}\sum_{i=1}^{n}\left|T_i - \frac{2\pi}{n}\right|. \tag{6.3.45}$$

Large values of L indicate clustering of the observations. This test in the above context was introduced by Rao (1969), who noted that for the linear case it was suggested by Kendall (1946) and studied by Sherman (1950). The null distribution of L is given by Rao (1976) and follows from Sherman (1950) and Darling (1953). The first and second moments of L are given in Sherman (1950).

Some quantiles of L are given in Appendix 2.9. A more extensive table of quantiles and tables of upper tail probabilities are given in Russell & Levitin (1996). Sherman (1950) has shown that

$$\frac{n^{1/2}(L - 2\pi/e)}{2\pi(2e^{-1} - 5e^{-2})^{1/2}} \dot{\sim} N(0,1)$$

for large n. A comparison of the Bahadur efficiency of this test with that of some other tests of uniformity under von Mises alternatives was given by Rao (1972a).

Example 6.9

In a pigeon-homing experiment, 13 birds were released singly in the Toggenburg Valley. Their the vanishing angles were

$$20°, 135°, 145°, 165°, 170°, 200°, 300°, 325°, 335°, 350°, 350°, 350°, 355°.$$

We test the hypothesis of uniformity using the L test. The successive values of the T_i are

$$115°, 10°, 20°, 5°, 30°, 100°, 25°, 10°, 15°, 0°, 0°, 5°, 25°.$$

Here $n = 13$, so $360°/n = 27.7°$ and $L = 162°$. Since the 5% value of L given in Appendix 2.9 is 167.8°, the hypothesis of uniformity is accepted.

6.3.5 *Ajne's A_n Test*

Let $N(\theta)$ denote the number of observations in the semicircle centred on θ, i.e. in the arc $(\theta - \pi/2, \theta + \pi/2)$. Since the expected value of $N(\theta)$ under uniformity is $n/2$, it is intuitively reasonable to reject uniformity for large values of

$$A_n = \frac{1}{2\pi n}\int_0^{2\pi}\left\{N(\theta) - \frac{n}{2}\right\}^2 d\theta, \tag{6.3.46}$$

as proposed by Ajne (1968). (Some other authors define A_n to be various multiples of (6.3.46).)

The following computational formula for A_n will be derived in Section 6.3.7:

$$A_n = \frac{n}{4} - \frac{1}{n\pi}\sum_{j=2}^{n}\sum_{i=1}^{j-1} \min\{\theta_{(j)} - \theta_{(i)}, 2\pi - [\theta_{(j)} - \theta_{(i)}]\} \quad (6.3.47)$$

$$= n\left(\frac{1}{4} - \frac{\bar{D}_0}{2\pi}\right), \quad (6.3.48)$$

where $\bar{D}_0$ is the circular mean difference given by (2.3.15).

The large-sample asymptotic null distribution of A_n under uniformity can be obtained from (6.3.72) as (Watson, 1967)

$$\lim_{n\to\infty} \Pr(A_n > a) = \frac{4}{\pi}\sum_{k=1}^{\infty}\frac{(-1)^{k-1}}{2k-1}\exp\left\{-\frac{(2k-1)^2\pi^2}{2}a\right\}. \quad (6.3.49)$$

Some upper quantiles of the large-sample asymptotic null distribution of A_n are given in Table 6.6. Upper quantiles of A_n for various values of n were given by Stephens (1969b). His table shows that for $n \geq 16$, the upper 5% quantile of A_n is within 0.01 of the value given in Table 6.6.

Table 6.6 Upper quantiles of A_n

α	0.10	0.05	0.025	0.01
A_n	0.516	0.656	0.797	0.982

It follows from a general result in Section 6.3.7 that Ajne's A_n test is locally most powerful invariant against the alternatives (6.3.42) for $p/q \to 1$.

A generalisation of A_n to

$$A_n(t) = \frac{1}{2\pi n}\int_0^{2\pi}\{N(t,\theta) - nt\}^2\, d\theta, \quad (6.3.50)$$

where $N(t,\theta)$ denotes the number of observations in the arc $(\theta - t\pi, \theta + t\pi)$ and $0 < t < 1$, was considered by Rothman (1972) and (for $1/t$ a positive integer) by Rao (1972b). Note that $A_n(1/2) = A_n$.

6.3.6 The Hermans–Rasson Test

Hermans & Rasson (1985) proposed the test which rejects uniformity for large values of

$$H_n = 2\pi A_n - 2.895\frac{1}{n}\sum_{j=2}^{n}\sum_{i=1}^{j-1}\left(|\sin(\theta_i - \theta_j)| - \frac{2}{\pi}\right). \quad (6.3.51)$$

This test is consistent against all alternatives and was constructed to be powerful against a large class of multimodal alternatives to uniformity. For $n > 10$, approximate 10%, 5% and 1% upper quantiles of H_n are 0.60, 0.75, 1.09, respectively.

6.3.7 *Beran's Class of Tests of Uniformity*

The Rayleigh test rejects uniformity for large values of the squared resultant length R^2 of the observations. The observations can be regarded as unit complex numbers $z_1, \ldots, z_n$. Since p-fold wrapping of the circle onto itself by $z \mapsto z^p$ preserves the uniform distribution, it would also be reasonable to reject uniformity for large values of the squared resultant length R_p^2 of $z_1^p, \ldots, z_n^p$, for any $p = 1, 2, \ldots$. Note that

$$R_p^2 = C_p^2 + S_p^2,$$

where

$$C_p = \sum_{j=1}^{n} \cos p\theta_j, \quad S_p = \sum_{j=1}^{n} \sin p\theta_j,$$

with $z_j = e^{i\theta_j}$. More generally, if $\rho_1, \rho_2, \ldots$ is any square-summable sequence of real numbers then we can consider

$$B_n = 2\sum_{p=1}^{\infty} \rho_p^2 R_p^2 = n\sum_{p=1}^{\infty} \rho_p^2 2n\bar{R}_p^2, \tag{6.3.52}$$

which is a weighted sum of the Rayleigh statistics $2n\bar{R}_p^2$ for different powers p of the data. It is reasonable to reject uniformity for large values of B_n. Note that B_n can be rewritten as

$$B_n = \sum_{i=1}^{n}\sum_{j=1}^{n} h(\theta_i - \theta_j), \tag{6.3.53}$$

where

$$h(\theta) = 2\sum_{p=1}^{\infty} \rho_p^2 \cos p\theta. \tag{6.3.54}$$

The tests based on B_n were introduced by Beran (1968; 1969a). These tests are sometimes called 'Sobolev tests'. (The reason for this name is indicated in Section 10.8.)

Some Special Cases

Rayleigh Test

Taking

$$\rho_p = \begin{cases} 1/\sqrt{2}, & p = 1, \\ 0, & p > 1, \end{cases} \tag{6.3.55}$$

gives

$$h(\theta) = \cos\theta, \tag{6.3.56}$$

and so

$$B_n = \sum_{i=1}^{n}\sum_{j=1}^{n} \cos(\theta_i - \theta_j) = \left(\sum_{i=1}^{n} \cos\theta_i\right)^2 + \left(\sum_{i=1}^{n} \sin\theta_i\right)^2 = R^2, \tag{6.3.57}$$

which is a multiple of the Rayleigh statistic $2n\bar{R}^2$.

Watson's U^2

Taking

$$\rho_p = \frac{1}{\pi p}, \tag{6.3.58}$$

gives

$$h(\theta) = \frac{2}{\pi^2}\sum_{p=1}^{\infty} \frac{1}{p^2} \cos p\theta = 2\left\{\frac{1}{6} - \frac{\theta}{2\pi} + \left(\frac{\theta}{2\pi}\right)^2\right\}. \tag{6.3.59}$$

Substituting (6.3.59) into (6.3.53) and writing $U_i = \theta_i/2\pi$ shows that

$$B_n = 4n\left\{\frac{1}{n}\sum_{i<j}(U_i - U_j)^2 - \frac{1}{n}\sum_{i<j}|U_i - U_j| + \frac{n}{12}\right\}.$$

Further manipulation and comparison with (6.3.34) yields

$$U^2 = \frac{B_n}{4n}. \tag{6.3.60}$$

Ajne's A_n

Taking

$$\rho_p = \begin{cases} 0, & p \text{ even}, \\ 1/\pi p, & p \text{ odd}, \end{cases} \tag{6.3.61}$$

gives

$$h(\theta) = \frac{1}{4} - \frac{|\theta|}{2\pi}, \qquad -\pi \le |\theta| \le \pi. \tag{6.3.62}$$

Calculation shows that

$$N(\theta) - \frac{n}{2} \sim 2 \sum_{p=1}^{\infty} \rho_p (C_p \cos p\theta + S_p \sin p\theta),$$

so it follows from Parseval's formula (4.2.3) that

$$A_n = \frac{B_n}{n}.$$

Substitution of $h(\theta)$ into (6.3.53) leads to the computational formula (6.3.47) for A_n.

Comparison of (6.3.58) and (6.3.61) shows that Ajne's A_n can be considered as the 'non-antipodally-symmetric part' of Watson's U^2.

$A_n(t)$ of Rothman and Rao

Taking

$$\rho_p = \frac{\sin(pt\pi)}{2pt\pi} \tag{6.3.63}$$

yields

$$A_n(t) = \frac{B_n}{n},$$

where $A_n(t)$ is the statistic of Rothman (1972) and Rao (1972b) defined in (6.3.50).

Beran's Tests as Locally Most Powerful Invariant Tests

The tests based on B_n were introduced by Beran (1968) as locally (i.e. for $\kappa \to 0$) most powerful rotation-invariant tests of uniformity against alternatives with densities of the form

$$g(\theta; \mu, \kappa) = \left\{1 + \kappa \left[f(\theta + \mu) - \frac{1}{2\pi} \right] \right\} \frac{1}{2\pi}, \quad 0 < \kappa \leq 1, \tag{6.3.64}$$

where μ is an unknown location parameter and f is related to B_n by (6.3.67) and (6.3.69).

By the Neyman–Pearson lemma, the most powerful invariant test is the likelihood ratio test based on the marginal likelihood of a maximal invariant. A slight generalisation of the argument given for the Rayleigh test in Section 6.3.1 shows that the joint density of the maximal invariant $u_1, \ldots, u_{n-1}$ (where $u_i = \theta_i - \theta_n$) is

$$\begin{aligned} \breve{g}(u_1, \ldots, u_{n-1}; \mu, \kappa) &= \frac{1}{2\pi} \int_0^{2\pi} \prod_{i=1}^{n} g(u_i + x, \ldots, u_n + x; \mu, \kappa)\, dx \\ &= \frac{1}{2\pi} \int_0^{2\pi} \prod_{i=1}^{n} g(\theta_i + \theta; \mu, \kappa)\, d\theta. \end{aligned}$$

Thus the most powerful invariant test rejects uniformity for large values of

$$I = \int_0^{2\pi} \prod_{i=1}^{n} \left[1 + \kappa \left\{ f(\theta + \theta_i) - \frac{1}{2\pi} \right\} \right] d\theta. \tag{6.3.65}$$

Expanding the integrand in a power series in κ and using the fact that $f(\theta+\theta_i)$ is a probability density function gives

$$I = 1 + \kappa^2 \int_0^{2\pi} \sum_{i=1}^{n} \sum_{j=1}^{n} \left\{ f(\theta + \theta_i) - \frac{1}{2\pi} \right\} \left\{ f(\theta + \theta_j) - \frac{1}{2\pi} \right\} d\theta + O(\kappa^3).$$

From the inversion formula (4.2.2), the Fourier expansion of $f(\theta)$ is

$$f(\theta) = \frac{1}{2\pi} \left\{ 1 + 2 \sum_{p=1}^{\infty} (\alpha_p \cos p\theta + \beta_p \sin p\theta) \right\}, \tag{6.3.66}$$

and so

$$\sum_{i=1}^{n} \left\{ f(\theta + \theta_i) - \frac{1}{2\pi} \right\}$$

$$= \frac{1}{2\pi} \left\{ \sum_{p=1}^{\infty} (\alpha_p C_p + \beta_p S_p) \cos p\theta + \sum_{p=1}^{\infty} (\beta_p C_p - \alpha_p S_p) \sin p\theta \right\}. \tag{6.3.67}$$

Then Parseval's formula (4.2.3) gives

$$\frac{1}{2\pi} \int_0^{2\pi} \left[\left\{ \sum_{i=1}^{n} f(\theta + \theta_i) \right\} - \frac{n}{2\pi} \right]^2 d\theta = 2 \sum_{p=1}^{\infty} \rho_p^2 (C_p^2 + S_p^2), \tag{6.3.68}$$

where

$$\rho_p^2 = \alpha_p^2 + \beta_p^2, \tag{6.3.69}$$

and so

$$B_n = \frac{1}{2\pi} \int_0^{2\pi} \left[\left\{ \sum_{i=1}^{n} f(\theta + \theta_i) \right\} - \frac{n}{2\pi} \right]^2 d\theta. \tag{6.3.70}$$

Then

$$I = 1 + \kappa^2 B_n - \kappa^2 \sum_{i=1}^{n} \int_0^{2\pi} \left\{ f(\theta + \theta_i) - \frac{1}{2\pi} \right\}^2 d\theta + O(\kappa^3). \tag{6.3.71}$$

Since each integral in the third term on the right-hand side of (6.3.71) does not depend on θ_i,

$$I = c + \kappa^2 B_n + O(\kappa^3)$$

for some constant c. Thus B_n is a locally most powerful invariant test.

Important special cases of this general result are as follows:

(i) the Rayleigh test is a locally most powerful invariant test against cardioid alternatives;
(ii) Watson's U^2 test is a locally most powerful invariant test against the alternatives (6.3.64) with $f(\theta) = \theta^2/2\pi^2$;
(iii) Ajne's A_n test is a locally most powerful invariant test against the alternatives (6.3.42) when p and q are nearly the same, i.e. $p/q \to 1$.

The Rayleigh test has the much stronger property (proved in Section 6.3.1) of being uniformly most powerful invariant against von Mises alternatives.

The Asymptotic Null Distribution and Consistency of B_n

Under uniformity, the large-sample asymptotic distribution of $\bar{R}^2$ is given by $2n\bar{R}^2 \mathrel{\dot\sim} \chi^2_2$. Similarly, for large n, $C_1, S_1, C_2, S_2, \ldots$ are asymptotically independently distributed as $N(0, 1/2)$, suggesting that the asymptotic characteristic function of B_n/n is

$$\prod_{p=1}^{\infty}(1 - 2\rho_p^2 it)^{-1}.$$

This argument can be made rigorous (see Beran, 1969a). When the non-vanishing ρ_p^2 are all distinct, the characteristic function may be inverted and it is found by a partial fraction expansion of the characteristic function that

$$\lim_{n\to\infty} \Pr\left(\frac{B_n}{n} > x\right) = \sum_{p=1}^{\infty} a_p \exp\left(-\frac{x}{2\rho_p^2}\right), \tag{6.3.72}$$

where $a_p = \prod_{k\neq p}\{1 - (\rho_k/\rho_p)^2\}^{-1}$. These results are due to Beran (1969a).

We now consider the consistency property of the B_n test (Beran, 1969a). Let G be the distribution function under the alternative hypothesis. From the strong law of large numbers and (6.3.53),

$$\frac{B_n}{n^2} \to \mathrm{E}\left[\frac{B_n}{n^2}\right] = \int_0^{2\pi}\int_0^{2\pi} h(x-y)dG(x)dG(y) = \sum_{p=1}^{\infty} \rho_p^2 \rho_p^{*2}, \tag{6.3.73}$$

where ρ_p^* is the mean resultant length of $(\cos p\theta, \sin p\theta)$ under the distribution function G. Hence the test B_n is consistent if there is at least one p for which both ρ_p and ρ_p^* are non-zero. Since $\rho_p^* \neq 0$ for some p whenever the alternative is not uniform, a necessary and sufficient condition for B_n to be consistent against all alternatives is that $\rho_p \neq 0$ for all $p > 0$. In particular, it follows from (6.3.58) that Watson's U^2 test is consistent against all alternatives. On the other hand, in view of (6.3.61), Ajne's A_n test is not consistent against all alternatives.

The asymptotic distribution of B_n under general alternatives was obtained by Beran (1969a). He also derived the Bahadur efficiency of the tests based on B_n.

Generalisations of B_n for testing uniformity on spheres (and other sample spaces) are given in Section 10.8.

6.3.8 *Relative Performances of Various Tests of Uniformity*

The powers of Kuiper's V_n, Watson's U^2 and Ajne's A_n tests have been compared by Stephens (1969b) in a simulation study. This suggested that Kuiper's test may be preferred for small samples. For alternatives given by (6.3.42), the A_n test (the locally most powerful invariant test) is only very slightly more powerful than V_n or U^2 (which have almost the same power). Stephens (1969b) also found that:

(i) for a class of unimodal alternatives, all the three tests are equally powerful;
(ii) for a class of bimodal alternatives, V_n and U^2 have the same power but both are much more powerful than A_n;
(iii) for a class of alternatives with four modes, V_n is more powerful than U^2, which in turn is more powerful than A_n.

The difference between the power of V_n and U^2 is less marked for moderately large samples.

Rao (1969) has shown by simulation that for small samples, the power of the L test relative to the Rayleigh test against von Mises alternatives is tolerable for large κ.

6.4 TESTS OF GOODNESS-OF-FIT

6.4.1 *The Probability Integral Transformation*

Consideration of the probability integral transformations on the line suggests the following analogue on the circle. Let F be the cumulative distribution function of a circular distribution and suppose that an orientation and initial direction have been chosen. Then the *probability integral transformation* of the distribution is the transformation of the circle which sends θ to $2\pi F(\theta)$. If F is continuous then the transformed random variable

$$U = 2\pi F(\theta) \bmod 2\pi \tag{6.4.1}$$

is distributed uniformly on the circle.

6.4.2 *Tests of Goodness-of-Fit*

By means of the probability integral transformation (6.4.1), any test of uniformity on the circle gives rise to a corresponding test of goodness-of-fit.

Consider the hypothesis that a distribution on the circle is a given distribution, with continuous cumulative distribution function F_0. Given observations $\theta_1, \ldots, \theta_n$, this hypothesis can be tested by applying any test of uniformity to $U_1, \ldots, U_n$, where $U_i = 2\pi F_0(\theta_i)$.

A more common problem is to test the hypothesis that the distribution generating the data belongs to some given family. After fitting a distribution in this family, we are interested in in assessing how well the fitted distribution fits the data. Goodness-of-fit of a fitted distribution with cumulative distribution function $\hat{F}$ to data $\theta_1, \ldots, \theta_n$ is judged by assessing whether or not the uniform order statistics of $2\pi\hat{F}(\theta_1), \ldots, 2\pi\hat{F}(\theta_n)$ could reasonably be a random sample from the uniform distribution. More precisely, a test of uniformity is applied to $2\pi\hat{F}(\theta_1), \ldots, 2\pi\hat{F}(\theta_n)$. Usually the alternative is completely general and so an omnibus test of uniformity, such as Kuiper's test or Watson's U^2 test is appropriate. Because parameters have been estimated, the null distribution of such goodness-of-fit tests will *not* be the same as the null distribution of the corresponding test of uniformity, although the difference will be small for large samples. For the goodness-of-fit test of von Mises distributions based on Watson's U^2, quantiles have been calculated by Lockhart & Stephens (1985), and are given in Appendix 2.10.

Some specific tests of 'von Misesness' are given in Section 7.5.

Example 6.10

Does a von Mises distribution provide a good fit to the pigeon-homing data set of Example 6.7?

The maximum likelihood estimates of the parameters are $\hat{\mu} = 155.8°$ and $\hat{\kappa} = 1.63$. Using the goodness-of-fit test based on Watson's U^2 gives $U^2 = 0.238$. From Appendix 2.10, the hypothesis that the data come from a von Mises distribution is rejected at the 0.5% level.

Watson's U^2 for Grouped Data

Because the probability integral transformation (6.4.1) transforms a given distribution to the uniform distribution only when the given distribution is continuous, goodness-of-fit tests based on tests of uniformity have to be modified before they are used on grouped data. We now describe two such modifications of Watson's U^2 test.

Grouping of circular data puts observations into ordered cells. Suppose that there are k cells and that the null hypothesis specifies the probability of an observation falling in the jth cell as p_j, for $j = 1, \ldots, k$. For n independent observations, denote the observed and expected numbers of observations in the jth cell by O_j and E_j, respectively. Then $E_j = np_j$.

Define

$$S_j = \sum_{i=1}^{j}(O_i - E_i), \quad j = 1, \ldots, k, \tag{6.4.2}$$

and

$$\bar{S} = \sum_{i=1}^{k} p_j S_j. \tag{6.4.3}$$

Then

$$U_G^2 = \sum_{i=1}^{k} (S_j - \bar{S})^2 p_j \tag{6.4.4}$$

is an analogue of U^2 which is suitable for grouped data. It was introduced by Choulakian, Lockhart & Stephens (1994). The statistic U_G^2 is invariant under cyclic permutations and order-reversing permutations of the cells, so it is appropriate for grouped data on the circle.. In the case $p_1 = \ldots = p_k$, (6.4.4) reduces to

$$U_G^2 = \frac{1}{nk} \sum_{j=1}^{k} \left(S_j - \frac{1}{k} \sum_{i=1}^{k} S_i \right)^2, \tag{6.4.5}$$

which is Freedman's (1981) grouped version of U^2 and can be obtained by replacing U_i and $(i - 1/2)/n$ in (6.3.33) by O_i/n and p_i, respectively. For the case $p_1 = \ldots = p_k$ and various values of k, some quantiles of the large-sample asymptotic distribution of U_G^2 are given in Choulakian, Lockhart & Stephens (1994). In particular, for $n \geq 8$, these 10%, 5% and 1% quantiles are within 0.01 of the corresponding quantiles of U^{2*} given in Table 6.5.

An alternative grouped version of U^2 was obtained by Brown (1994) from careful consideration of the effect of grouping on the process W_n of (6.3.38). First define

$$Y_j = \sum_{i=1}^{j-1} (O_i - E_i) + \tfrac{1}{2}(O_j - E_j). \tag{6.4.6}$$

Then Brown's grouped version of U^2 is

$$\begin{aligned} U_d^2 &= \frac{1}{n} \left\{ \sum_{j=1}^{k} p_j Y_j^2 - \left(\sum_{j=1}^{k} p_j Y_j \right)^2 \right\} \\ &\quad + \frac{1}{6} \sum_{j=1}^{k} p_j^2 \left(1 - \frac{p_j}{2} \right) + \frac{1}{12n} \sum_{j=1}^{k} p_j (O_j - E_j)^2. \end{aligned} \tag{6.4.7}$$

The first term in (6.4.7) is a grouped analogue of the usual U^2 statistic (6.3.33) and is similar to U_G^2, while the other two terms are grouping corrections. The statistic U_d^2 is invariant under cyclic permutations and order-reversing permutations of the cells. The null distribution of U_d^2 is close to that of U^2.

Example 6.11

Does a von Mises distribution provide a good fit to the leukaemia data of Example 6.3?

The maximum likelihood estimates of the parameters are $\hat{\mu} = 198.3°$ and $\hat{\kappa} = 0.202$. Brown's statistic (6.4.7) is $U_d^2 = 0.041$. Since the upper 10% quantile of U^{*2} given in Table 6.5 is 0.152, we conclude that the fitted von Mises distribution is a good fit to the data. Comparison of U_d^2 with the upper 15% quantiles given in Appendix 2.10 for $\kappa = 0$ and $\kappa = 0.5$ leads to the same conclusion.

If it is accepted that the data come from a von Mises distribution, then the tests in the next chapter can be applied.

7

Tests on von Mises Distributions

7.1 INTRODUCTION

Inference on distributions on the circle has been most highly developed for the von Mises distributions. This development is due to their elegant structure as an exponential transformation model under rotations. In this chapter we discuss inference on von Mises distributions. Section 7.2 treats tests and confidence intervals based on a single sample, analogous to those of standard normal theory. Two-sample inference is considered in Section 7.3, and this is extended to the multi-sample case in Section 7.4. A test of von Misesness is described in Section 7.5.

7.2 SINGLE-SAMPLE TESTS

In this section we consider inference on a von Mises distribution $M(\mu, \kappa)$ on the basis of a random sample $\theta_1, \ldots, \theta_n$. We shall assume that $\kappa > 0$, so that the distribution is not uniform.

When a mean direction μ_0 is specified, it will be useful to modify the notation of (2.2.1) by writing

$$\bar{C} = \frac{1}{n}\sum_{j=1}^{n}\cos(\theta_j - \mu_0), \qquad \bar{S} = \frac{1}{n}\sum_{j=1}^{n}\sin(\theta_j - \mu_0). \tag{7.2.1}$$

7.2.1 Tests for the Mean Direction

Concentration Parameter Known

The Likelihood Ratio Test

Suppose that we wish to test

$$H_0 : \mu = \mu_0 \quad \text{against} \quad H_1 : \mu \neq \mu_0. \tag{7.2.2}$$

The likelihood ratio test rejects H_0 for large values of

$$w = 2n\kappa(\bar{R} - \bar{C}). \tag{7.2.3}$$

Under H_0 the large-sample approximation

$$2n\kappa(\bar{R} - \bar{C}) \;\dot{\sim}\; \chi^2_1$$

holds. For moderate sample sizes, this approximation can be improved by replacing w by the Bartlett-corrected version

$$w^* = \left\{1 - \frac{1}{4n\kappa A(\kappa)}\right\} w$$

(this is the case $p = 2$ of (10.4.14)). An alternative improvement is Stephens's approximation (4.8.29):

$$2n\gamma(\bar{R} - \bar{C}) \;\dot{\sim}\; \chi^2_1, \tag{7.2.4}$$

where

$$\frac{1}{\gamma} = \frac{1}{\kappa} + \frac{3}{8\kappa^2}. \tag{7.2.5}$$

This approximation is adequate for $\kappa > 2$.

A Conditional Test

The likelihood function can be expressed as

$$L(\mu, \kappa; \theta_1, \ldots, \theta_n) = g_1(\bar{\theta}|R; \mu, \kappa)\breve{g}(R; \kappa)h(\theta), \tag{7.2.6}$$

where g_1 denotes the conditional probability density function of $\bar{\theta}$ given R, $\breve{g}$ denotes the marginal probability density function of R, and h is a function of $\theta_1, \ldots, \theta_n$ only. Since $\breve{g}(R; \kappa)$ does not depend on μ, no inferences about μ are possible if only R is given, while the observed value of R can be regarded as determining the precision with which inferences about μ can be made. A more detailed consideration of (7.2.6) shows that R is G-ancillary for μ (Barndorff-Nielsen, 1978a, Section 4.4), where $G = SO(2)$ is the rotation group of the circle. By the ancillarity principle (Fisher, 1959, Section IV4; Stuart & Ord, 1991, pp. 1202–1203), it is appropriate to test (7.2.2) using only the distribution of $\bar{\theta}|R$. From (4.5.5), $\bar{\theta}|R \sim M(\mu, \kappa R)$. Since the likelihood ratio statistic is given by (7.2.3), the conditional likelihood ratio test rejects H_0 for large values of $R - C$, i.e. for small values of $\cos\bar{\theta}$. Hence the critical region of size α is given by (see Fig. 7.1)

$$\pi - \delta < \bar{\theta} - \mu_0 < \pi + \delta, \tag{7.2.7}$$

where δ is determined by

$$\frac{1}{2\pi I_0(\kappa R)} \int_{\pi}^{\pi+\delta} \exp\{\kappa R \cos\theta\} d\theta = \frac{\alpha}{2} \tag{7.2.8}$$

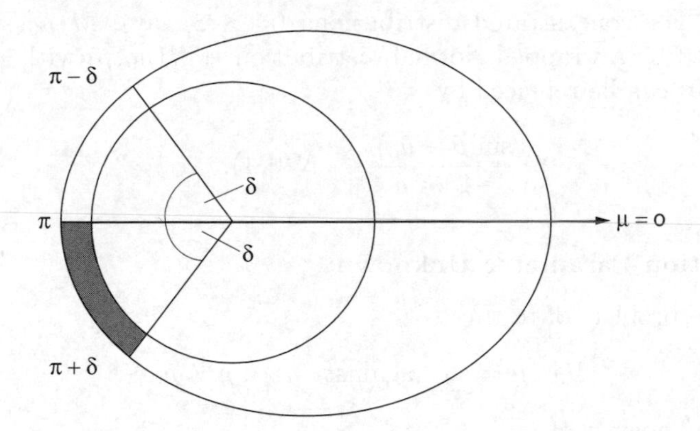

Figure 7.1 Density of $\bar{\theta}|R$ for $\mu = 0$ and $\kappa R = 1$. Shaded area $= \frac{1}{2}\alpha$. The arc $(\pi - \delta, \pi + \delta)$ is the critical region of size α.

and can be calculated from Appendix 2.2. For $\kappa R > 10$, the wrapped normal approximation $M(\mu, \kappa R) \simeq N(\mu, 1/(\kappa R))$ can be used.

It can be shown that this test is the only unbiased test against H_1 among all the tests of size α having critical regions containing π. (See Stuart & Ord, 1991, pp. 833–834, Example 22.7 for the normal case; Mardia, 1972a, pp. 140–141 for this case.)

Example 7.1
From a large-scale survey, it is known that the dip directions of cross-beds of a section of a river have mean direction $\mu = 342°$ and concentration $\kappa = 0.8$. In a pilot survey of a neighbouring section of the river, 10 observations gave $\bar{\theta} = 278°$ and $\bar{R} = 0.35$. Can the mean direction for the neighbouring section be taken as 342°?

An appropriate null hypothesis is $\mu = 342°$. We have $\kappa R = 2.8$. From Appendix 2.2, for $\kappa = 2.8$ and $\mu = 0$, the critical region of size 0.05 defined by (7.2.7) is $(180° \pm 101.3°)$. Hence the critical region when $\mu_0 = 342°$ is $(342° - 180° \pm 101.3°)$, i.e. $(60.7°, 263.3°)$. Consequently, we accept the null hypothesis at the 5% significance level.

The Score Test

From (5.3.1) the score at μ_0 is

$$\left.\frac{\partial l(\mu, \kappa; \theta_1, \ldots, \theta_n)}{\partial \mu}\right|_{\mu=\mu_0} = -n\kappa\bar{S} = -n\kappa\cos(\bar{\theta} - \mu_0), \tag{7.2.9}$$

so the score test rejects H_0 for large values of $|S|$. Under H_0,

$$\frac{n\kappa}{A(\kappa)}\bar{S}^2 \mathrel{\dot{\sim}} \chi_1^2, \tag{7.2.10}$$

for n large. For concentrated distributions ($\kappa > 2$, say), $M(\mu_0, \kappa)$ can be approximated by a wrapped Normal distribution $WN(\mu_0, \rho)$ with $A(\kappa) = \rho$. Then (7.2.10) can be replaced by

$$\frac{\sin(\bar{\theta} - \mu_0)}{-2\log\rho} \overset{\cdot}{\sim} N(0, 1). \tag{7.2.11}$$

Concentration Parameter Unknown

Consider the problem of testing

$$H_0 : \mu = \mu_0 \quad \text{against} \quad H_1 : \mu \neq \mu_0 \tag{7.2.12}$$

when κ is unknown.

The Likelihood Ratio Test and Variants

The likelihood ratio statistic for (7.2.12) is

$$w = 2n\left(\hat{\kappa}\bar{R} - \tilde{\kappa}\bar{C} - \log I_0(\hat{\kappa}) + \log I_0(\tilde{\kappa})\right), \tag{7.2.13}$$

where $\tilde{\kappa} = A^{-1}(\bar{C})$ denotes the maximum likelihood estimate of κ under H_0. The likelihood ratio test rejects H_0 for large values of w. A simpler statistic than (7.2.13) is

$$2n\hat{\kappa}(\bar{R} - \bar{C}), \tag{7.2.14}$$

obtained by replacing κ in (7.2.13) by its maximum likelihood estimate given by (5.3.6). It follows from Wilks's theorem that, under H_0, (7.2.14) is distributed approximately as χ_1^2 for large n.

More refined approximations by Upton (1973) to the likelihood ratio statistic w of (7.2.13) lead to

$$\frac{4n(\bar{R}^2 - \bar{C}^2)}{2 - \bar{C}^2} \overset{\cdot}{\sim} \chi_1^2, \tag{7.2.15}$$

which is appropriate for $n \geq 5$ and $\bar{C} \leq 2/3$, and to

$$\frac{2n^3}{n^2 + C^2 + 3n} \log \frac{1 - \bar{C}^2}{1 - \bar{R}^2} \overset{\cdot}{\sim} \chi_1^2, \tag{7.2.16}$$

which is appropriate for $n \geq 5$ and $\bar{C} > 2/3$.

Another useful approximation to to the likelihood ratio test is the test (Yamamoto & Yanagimoto, 1995) which rejects H_0 for large values of

$$2n\check{\kappa}(\bar{R} - \bar{C}), \tag{7.2.17}$$

where $\check{\kappa}$ is the marginal maximum likelihood estimate of κ given in (5.3.29). An advantage of this test is that, under H_0,

$$2n\check{\kappa}(\bar{R} - \bar{C}) \overset{\cdot}{\sim} F_{1,n-1} \tag{7.2.18}$$

if either n or κ is large. This follows from (7.2.22) and (7.2.14), respectively.

The Score Test

From (7.2.9), the score at $(\mu_0, \hat{\kappa})$ is

$$\left.\frac{\partial l(\hat{\kappa}, \mu; \theta_1, \ldots, \theta_n)}{\partial \mu}\right|_{(\mu,\kappa)=(\mu_0,\hat{\kappa})} = -n\hat{\kappa}\bar{S}. \tag{7.2.19}$$

Under H_0, we have

$$\frac{n\hat{\kappa}}{A(\hat{\kappa})}\bar{S}^2 \;\dot{\sim}\; \chi_1^2, \tag{7.2.20}$$

for n large. For most practical purposes, we can replace (7.2.20) by

$$(n\bar{R}\hat{\kappa})^{1/2} \sin\bar{\theta} \;\dot{\sim}\; N(0,1). \tag{7.2.21}$$

This approximation is satisfactory if $\hat{\kappa} \geq 2$ or if $\hat{\kappa} \geq 0.4$ and $n \geq 10$.

Some Other Tests

Another test of (7.2.12) is based on the approximation

$$\frac{\bar{R} - \bar{C}}{(1 - \bar{R})/(n-1)} \;\dot{\sim}\; F_{1,n-1}, \tag{7.2.22}$$

for large κ. This approximation follows from the high-concentration asymptotic distribution of $(2n\kappa(\bar{R} - \bar{C}), 2n\kappa(1 - \bar{C}))$ given in Section 4.8.2 and was first suggested by Watson & Williams (1956). It is suitable when $\bar{C} \geq 5/6$.

Approximation (7.2.22) can be related as follows to the usual t test of the mean of a normal distribution. Recall that the decomposition

$$2\kappa(n - C) = 2\kappa(n - R) + 2\kappa(n - R). \tag{7.2.23}$$

given in (4.8.30) is analogous to the decomposition

$$\frac{1}{\sigma^2}\sum_{i=1}^{n}(x_i - \mu)^2 = \frac{1}{\sigma^2}\sum_{i=1}^{n}(x_i - \bar{x})^2 + \frac{n}{\sigma^2}(\bar{x} - \mu)^2 \tag{7.2.24}$$

given in (4.8.34) for a random sample $x_1, \ldots, x_n$ from $N(\mu, \sigma^2)$. As discussed in Section 4.8.2, if κ is large then the terms in (7.2.23) have χ^2-distributions analogous to those of the corresponding terms in (7.2.24). Then (7.2.22) is analogous to

$$\frac{n(\bar{x} - \mu)^2}{s^2} \sim t_{n-1}^2 = F_{1,n-1}.$$

A test which is suitable when κ is small is based on the approximation

$$2n(\bar{R}^2 - \bar{C}^2) \;\dot{\sim}\; \chi_1^2. \tag{7.2.25}$$

This approximation follows from the large-sample asymptotic distribution of $\bar{S}$ under uniformity (see Section 4.8.1). It is suitable when $\bar{C} \leq 1/3$. Related tests which are appropriate when $1/3 < \bar{C} < 5/6$ are given in Stephens (1962a).

Confidence Intervals for μ

Confidence intervals for μ can be obtained in the usual way from tests of (7.2.2).

An approximate $100(1-\alpha)\%$ confidence interval for μ based on (7.2.14) is

$$\bar{\theta} \pm \cos^{-1}\left(\frac{1-\chi^2_{1;\alpha}}{2\hat{\kappa}R}\right), \tag{7.2.26}$$

where $\chi^2_{1;\alpha}$ denotes the upper α quantile of the χ^2_1 distribution. An approximate $100(1-\alpha)\%$ confidence interval for μ with coverage probability closer to $1-\alpha$ is

$$\bar{\theta} \pm \cos^{-1}\left(\left\{\frac{2n(2R^2 - n\chi^2_{1;\alpha})}{R^2(4n - \chi^2_{1;\alpha})}\right\}^{1/2}\right), \tag{7.2.27}$$

which is based on (7.2.15). This is appropriate when $\bar{R} \leq 2/3$. Similarly, (7.2.16) leads to the approximate $100(1-\alpha)\%$ confidence interval

$$\bar{\theta} \pm \cos^{-1}\left(\frac{\{n^2 - (n^2 - R^2)\exp(\chi^2_{1;\alpha}/n)\}^{1/2}}{R}\right) \tag{7.2.28}$$

for μ. This is appropriate when $\bar{R} > 2/3$.

Using (7.2.21) an approximate $100(1-\alpha)\%$ confidence interval for μ is

$$\bar{\theta} \pm \sin^{-1}\left(\frac{z_{\alpha/2}}{n\bar{R}\hat{\kappa}^{1/2}}\right), \tag{7.2.29}$$

where $z_{\alpha/2}$ denotes the upper $\alpha/2$ quantile of the $N(0,1)$ distribution. This is appropriate if $\hat{\kappa} \geq 2$ or if $\hat{\kappa} \geq 0.4$ and $n \geq 10$.

Confidence intervals for μ can also be based on the conditional distribution of $\bar{\theta}$ given R, as used in the likelihood ratio test given by (7.2.7). Replacing κ by its maximum likelihood estimate $\hat{\kappa}$ gives

$$\bar{\theta} \pm \delta \tag{7.2.30}$$

as an approximate $100(1-\alpha)\%$ confidence interval for μ (see Fig. 7.2), where δ is determined by

$$\frac{1}{2\pi I_0(\hat{\kappa}R)} \int_{\pi}^{\pi+\delta} \exp\{\hat{\kappa}R\cos\theta\}d\theta = \frac{\alpha}{2} \tag{7.2.31}$$

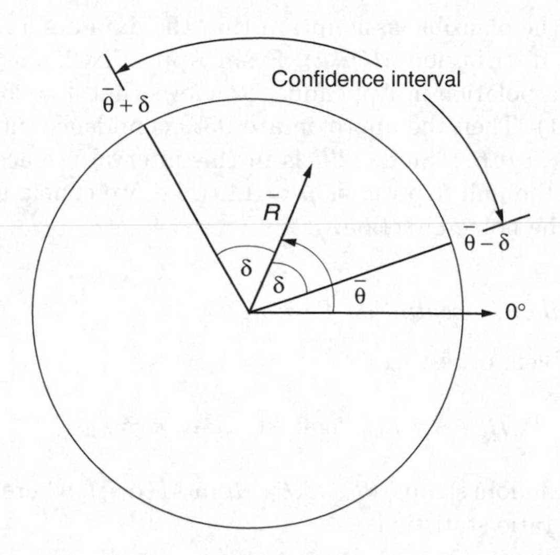

Figure 7.2 Confidence interval $(\bar{\theta}+\delta, \bar{\theta}-\delta)$ for the mean direction μ.

and can be found from Appendix 2.2.

If $\bar{R} \geq 0.4$ and $n \geq 30$ then (7.2.30) can be approximated by

$$\bar{\theta} \pm \frac{z_{\alpha/2}}{\sqrt{\hat{\kappa} R}}. \tag{7.2.32}$$

Because the marginal maximum likelihood estimate of κ defined by (5.3.29) is less biased than $\hat{\kappa}$, the confidence interval

$$\bar{\theta} \pm \frac{z_{\alpha/2}}{\sqrt{\breve{\kappa} R}} \tag{7.2.33}$$

(provided that $\breve{\kappa} > 0$) may have better coverage than (7.2.32).

Example 7.2

In an experiment on homing pigeons (Schmidt-Koenig, 1963), the vanishing angles of 15 birds were

$$85°, 135°, 135°, 140°, 145°, 150°, 150°, 150°,$$

$$160°, 185°, 200°, 210°, 220°, 225°, 270°.$$

The home direction was 149°. Was there any preference for the home direction?

Here $n = 15$ and calculation gives $\bar{\theta} = 168.5°$, $\bar{R} = 0.74$. First we test the hypothesis of uniformity using the Rayleigh test. Since $2n\bar{R}^2 = 16.43 > 10.591 = \chi^2_{2;0.001}$, we reject the hypothesis of uniformity, i.e. there is a preferred direction.

Next we make the plausible assumption that the data are a random sample from a von Mises distribution $M(\mu, \kappa)$. From Appendix 2.4, $\hat{\kappa} = 2.29$, and so $n\bar{R}\hat{\kappa} = 25.42$. Interpolation in Appendix 2.2 shows that $\delta = 23.1°$, where δ is defined by (7.2.31). Then the approximate 95% confidence interval given by (7.2.30) is $(145.4°, 191.6°)$. Since 149° is in this interval, we accept at the 5% significance level the null hypothesis $\mu = 149°$, i.e. we conclude that there is a preference for the home direction.

7.2.2 *Tests for the Concentration Parameter*

Consider the problem of testing

$$H_0 : \kappa = \kappa_0 \quad \text{against} \quad H_1 : \kappa \neq \kappa_0 \tag{7.2.34}$$

on the basis of a random sample $\theta_1, \ldots, \theta_n$ from $M(\mu, \kappa)$, where μ is unknown.

The likelihood ratio statistic is

$$2n\{(\hat{\kappa} - \kappa_0)\bar{R} - \log I_0(\hat{\kappa})/I_0(\kappa_0)\}. \tag{7.2.35}$$

Thus the likelihood ratio test of $\kappa = \kappa_0$ against $\kappa > \kappa_0$ rejects H_0 for large values of R, while the test of $\kappa = \kappa_0$ against $\kappa < \kappa_0$ rejects H_0 for small values of R.

We now show that these one-sided tests are uniformly most powerful invariant tests under rotations of the circle. From (5.3.1), $\bar{\theta}$ and R are sufficient statistics for μ and κ. Since R is a maximal invariant after this sufficient reduction, invariant tests of $\kappa = \kappa_0$ depend only on R. From (4.5.4), the marginal probability density function of R is

$$\check{f}(R; \kappa) = \{I_0(\kappa)\}^{-n} I_0(\kappa R) h_n(R), \tag{7.2.36}$$

where $h_n(R)$ does not depend on κ. Differentiation gives

$$\frac{\partial^2 \log \check{f}(R; \kappa)}{\partial R \partial \kappa} = A(\kappa R) + \kappa R A'(\kappa R). \tag{7.2.37}$$

Application of (3.5.4) to the $(1, 1)$ exponential model $M(0, \kappa R)$ shows that if $\theta \sim M(0, \kappa R)$ then $\operatorname{var}(\cos\theta) = A'(\kappa R)$. Thus $A'(\kappa R) \geq 0$. Since $A(\kappa R) \geq 0$ the right-hand side of (7.2.37) is non-negative. Hence the marginal density $\check{f}$ has a monotone likelihood ratio in κ. Applying the Neyman–Pearson lemma to (7.2.36) gives the result.

Appendix 2.11 gives 90% and 98% equal-tailed confidence intervals for κ.

For $\kappa > 2$, an approximate $(1 - \alpha)$ confidence interval for κ is

$$\left(\frac{1 + (1 + 3a)^{1/2}}{4a}, \frac{1 + (1 + 3b)^{1/2}}{4b}\right), \tag{7.2.38}$$

where

$$a = (n - R)/\chi^2_{n-1;1-\alpha/2}, \tag{7.2.39}$$
$$b = (n - R)/\chi^2_{n-1;\alpha/2}. \tag{7.2.40}$$

To see this, recall Stephens's approximation (4.8.32):

$$2\gamma(n - R) \stackrel{\cdot}{\sim} \chi^2_{n-1}, \tag{7.2.41}$$

where γ is defined in (7.2.5). From (7.2.41), we have

$$\Pr(a^{-1} < 2\gamma < b^{-1}) \simeq 1 - \alpha.$$

The inequality $2\gamma < b^{-1}$ implies that $16b\kappa^2 - 8\kappa - 3 < 0$. The roots of the quadratic equation $16b\kappa^2 - 8\kappa - 3 = 0$ are of opposite signs and if β is the positive root then $\kappa < \beta$, which is the right end-point of the interval (7.2.38). Similarly, the inequality $2\gamma > a^{-1}$ leads to the left end-point of the interval.

Tests of $\kappa = \kappa_0$ against one-sided alternatives can be obtained similarly. (For large n, the normal approximation for $\bar{R}$ with mean and variance given by (4.8.17) and (4.8.18) is satisfactory.) Stephens (1969a) has given quantiles of $\bar{R}$ for $\alpha = 0.01, 0.05, 0.95, 0.99$, which are based on approximations by Pearson curves.

For $\kappa_0 \leq 2$, the approximations of Section 4.8.3 can be used.

Example 7.3

In the pigeon homing experiment used in Example 6.7, the vanishing angles for 15 birds were

$$115°, 120°, 120°, 130°, 135°, 140°, 150°, 150°, 150°,$$

$$165°, 185°, 210°, 235°, 270°, 345°.$$

Obtain a 90% confidence interval for the concentration parameter.

We have $n = 15$ and $\bar{R} = 0.626$. Let κ_l and κ_u be the lower and the upper 90% confidence limits, respectively. From the lower curves in Appendix 2.11a at $\bar{R} = 0.626$, we find that $\kappa_l = 0.44$ for $n = 10$ and $\kappa_l = 0.88$ for $n = 20$. Similarly, from the upper curves at $\bar{R} = 0.626$, $\kappa_u = 2.62$ for $n = 10$ and $\kappa_u = 2.40$ for $n = 20$. For $n = 15$, interpolation gives $\kappa_l = 0.66$ and $\kappa_u = 2.51$. Hence a 90% confidence interval for κ is $(0.66, 2.51)$.

For illustrative purposes, we also give the 90% confidence interval for κ from (7.2.38). We have

$$n - R = 5.604, \quad \chi^2_{14;0.05} = 23.68 \quad \text{and} \quad \chi^2_{14;095} = 6.57.$$

Consequently, $a = 0.853$ and $b = 0.237$, so the approximate 90% confidence interval for κ given by (7.2.38) is $(0.85, 2.44)$. Note that this interval is not necessarily an approximate 95% confidence interval, because $\hat{\kappa}$ is less than 2.

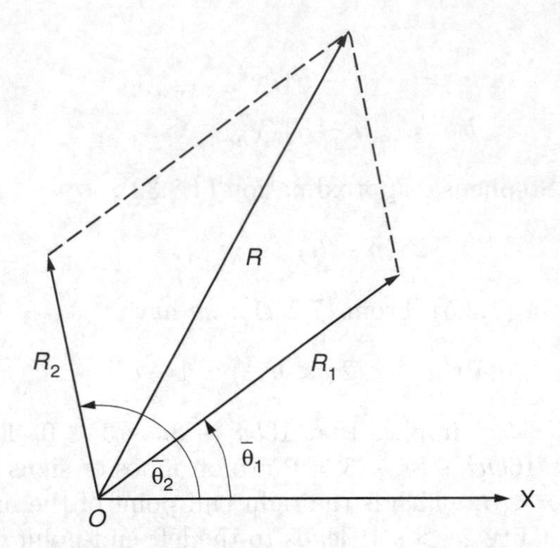

Figure 7.3 Geometry of resultant vectors.

7.3 TWO-SAMPLE TESTS

Suppose that $\theta_{11}, \ldots, \theta_{1n_1}$ and $\theta_{21}, \ldots, \theta_{2n_2}$ are independent random samples of sizes n_1, n_2 from $M(\mu_1, \kappa_1)$ and $M(\mu_2, \kappa_2)$, respectively. Let the corresponding sample mean directions be $\bar{\theta}_1, \bar{\theta}_2$ and the resultant lengths be R_1, R_2. Suppose that the mean direction and the resultant length of the combined sample are $\bar{\theta}$ and R, respectively. Let

$$C_i = \sum_{j=1}^{n_i} \cos\theta_{ij}, \quad S_i = \sum_{j=1}^{n_i} \sin\theta_{ij}, \quad i = 1, 2.$$

7.3.1 Tests for Equality of Mean Directions

We are interested in testing

$$H_0 : \mu_1 = \mu_2 \quad \text{against} \quad H_1 : \mu_1 \neq \mu_2, \tag{7.3.1}$$

where the concentration parameters κ_1 and κ_2 are equal (to κ, say) and unknown. One way of constructing suitable tests is suggested by the geometry (see Fig. 7.3) of the resultant vectors $R_1(\cos\bar{\theta}_1, \sin\bar{\theta}_1)$, $R_2(\cos\bar{\theta}_2, \sin\bar{\theta}_2)$ of the two samples and $R(\cos\bar{\theta}, \sin\bar{\theta})$ of the combined sample.

The cosine rule gives

$$R^2 = R_1^2 + R_2^2 + 2R_1R_2\cos(\bar{\theta}_2 - \bar{\theta}_1),$$

so that $R_1 + R_2 \geq R$. If H_0 is true then $\bar{\theta}_1 \simeq \bar{\theta}_2$, and so $R_1 + R_2 \simeq R$. If H_0 is false then $R_1 + R_2 - R$ will tend to be large. Thus it is reasonable to reject

H_0 for large values of

$$R_1 + R_2 - R. \tag{7.3.2}$$

When κ is known, the likelihood ratio statistic for testing H_0 against H_1 of (7.3.1) is

$$w = 2\kappa(R_1 + R_2 - R), \tag{7.3.3}$$

and so the test based on (7.3.2) is equivalent to the likelihood ratio test. However, when κ is unknown, the distribution of (7.3.2) depends on the nuisance parameter κ. We next give two ways (due to Watson & Williams, 1956) of eliminating κ: (i) by conditioning on R; (ii) by 'cancelling' κ in F tests based on normal approximations.

The Two-Sample Watson–Williams Test

The Test

We have shown in Section 4.6.3 that, when H_0 is true, the conditional distribution of (R_1, R_2) given R (with probability density (4.6.22)) does not depend on κ. Hence an appropriate test rejects H_0 for large values of $R_1 + R_2$ given R. For given R, this test is a similar test, since under H_0 the value of K such that

$$\Pr(R_1 + R_2 > K|R) = \alpha \tag{7.3.4}$$

does not depend on κ.

The Null Distribution

On transforming the variables R_1 and R_2 to

$$u = R_1 + R_2, \quad v = R_1 - R_2$$

in the probability density function of (R_1, R_2) given R in (4.6.22), we obtain the probability density function of (u, v) given R. On integrating this probability density function with respect to v, we find that, under H_0, the probability density function of u given R is

$$\frac{R}{\pi h_n(R)(u^2 - R^2)^{1/2}} \int_{-R}^{R} \frac{h_{n_1}\{(u+v)/2\}h_{n_2}\{(u-v)/2\}}{(R^2 - v^2)^{1/2}} dv, \tag{7.3.5}$$

where $R < u < n$ and h_n is the probability density function of R for the uniform case, which is given by (4.4.5).

Let $n_1 \le n_2$. Put $r = n_1/n$ and $\bar{R}' = (R_1 + R_2)/n$. Appendices 2.12a–2.12b give the 5% quantiles for the test when $r = 1/2$ and $r = 1/3$, respectively, i.e. for $n_1 = n_2$ and $n_2 = 2n_1$. For $0 < \bar{R} < 0.4$ and $1/3 < r < 1/2$, we read the values of $\bar{R}'$ from Appendices 2.12a and 2.12b for $r = 1/2$ and $1/3$ respectively and then obtain the value of $\bar{R}'$ for given r by interpolation. For $\bar{R} > 0.4$ and any moderate value of r, Appendix 2.12a can be used.

A High-Concentration F Test

For large κ, we have from approximation (4.8.36) due to Watson & Williams (1956) that

$$2\kappa(R_1 + R_2 - R) \overset{\cdot}{\sim} \chi_1^2 \tag{7.3.6}$$

and

$$2\kappa(n - R_1 - R_2) \overset{\cdot}{\sim} \chi_{n-2}^2. \tag{7.3.7}$$

Further, for large κ, these statistics are approximately independently distributed. Hence, under the null hypothesis,

$$\frac{R_1 + R_2 - R}{(n - R_1 - R_2)/(n-2)} \overset{\cdot}{\sim} F_{1,n-2} \tag{7.3.8}$$

asymptotically for large κ. (Note that, from general asymptotic theory of likelihood ratio tests applied to (7.3.3), under H_0 (7.3.6) holds also for large n_1 and n_2 and for all values of κ.) The approximation (7.3.8) can be refined using Stephens's approximation (4.8.32)

$$2\gamma(n_i - R_i) \overset{\cdot}{\sim} \chi_{n_i-1}^2, \tag{7.3.9}$$

where

$$\frac{1}{\gamma} = \frac{1}{\kappa} + \frac{3}{8\kappa^2}. \tag{7.3.10}$$

Using (7.3.6) and (7.3.9), and replacing the unknown κ by its maximum likelihood estimate $\hat{\kappa}$, gives

$$\left(1 + \frac{3}{8\hat{\kappa}}\right) \frac{(n-2)(R_1 + R_2 - R)}{n - R_1 - R_2} \overset{\cdot}{\sim} F_{1,n-2}. \tag{7.3.11}$$

Simulations (Stephens, 1972) support this approximation for $\hat{\kappa} > 2$, i.e. for $\bar{R} > 0.7$. The approximation (7.3.11) differs negligibly from (7.3.8) for $\hat{\kappa} > 10$, i.e. $\bar{R} > 0.95$.

An Approximate Confidence Interval

We now give a method of obtaining approximate confidence intervals for the difference

$$\delta = \mu_1 - \mu_2.$$

From (4.5.5), $\bar{\theta}_1 | R_1$ and $\bar{\theta}_2 | R_2$ are distributed as $M(\mu_1, \kappa R_1)$ and $M(\mu_2, \kappa R_2)$, respectively. Hence, from (3.5.44) the distribution conditional on (R_1, R_2) of the difference

$$d = \bar{\theta}_1 - \bar{\theta}_2$$

is approximately $M(\delta, \kappa^*)$, where κ^* is given by

$$A(\kappa^*) = A(\kappa R_1) A(\kappa R_2). \tag{7.3.12}$$

In practice, we may replace κ by its maximum likelihood estimate $\hat{\kappa}$ under the null hypothesis, which is given by

$$A(\hat{\kappa}) = \frac{R_1 + R_2}{n}. \tag{7.3.13}$$

Then (Mardia, 1972a, p. 156) the distribution of d can be approximated by $M(\delta, \hat{\kappa}^*)$, where $\hat{\kappa}^*$ is defined by

$$A(\hat{\kappa}^*) = A(\hat{\kappa}R_1)A(\hat{\kappa}R_2). \tag{7.3.14}$$

For given α, define ν by

$$\Pr(\pi - \nu < \theta < \pi + \nu) = \alpha, \tag{7.3.15}$$

where $\theta \sim M(0, \hat{\kappa}^*)$. Then an approximate $100(1-\alpha)\%$ confidence interval for δ is the arc

$$(d + \pi - \nu, d + \pi + \nu). \tag{7.3.16}$$

This method is illustrated in Example 7.4.

Example 7.4

In an experiment on pigeon-homing (Schmidt-Koenig, 1958), the 'internal clocks' of 10 birds were reset by 6 hours clockwise, while the clocks of 9 birds were left unaltered. It is predicted from sun-azimuth compass theory that the mean direction of the vanishing angles in the experimental group should deviate by about 90° in the anticlockwise direction with respect to the mean direction of the angles of the birds in the control group. The vanishing angles (measured in the clockwise sense) of the birds for this experiment are as follows:

$$\text{Control group } (\theta_{i1}) : 75°, 75°, 80°, 80°, 80°, 95°, 130°, 170°, 210°.$$

$$\text{Experimental group } (\theta_{i2}) : 10°, 50°, 55°, 55°, 65°, 90°, 285°, 285°, 325°, 355°.$$

Do the data support sun-azimuth compass theory?

We may assume that the two samples come from von Mises distributions $M(\mu_1, \kappa_1)$ and $M(\mu_2, \kappa_2)$ with $\kappa_1 = \kappa_2$ (see Example 7.5). An appropriate null hypothesis is $\mu_1 = \mu_2 + 90°$ (where the circle is oriented *clockwise*). To test this we rotate the angles for the experimental group clockwise by 90°, transforming θ_{i2} to $\theta_{i2} + 90°$, and then apply a test for equality of the mean directions. The data for the control group and the rotated data for the experimental group give

$$\begin{aligned} n_1 = 9, \quad C_1 = -1.542, \quad & S_1 = 6.332, \quad R_1 = 6.507 \\ n_2 = 10, \quad C_2 = 1.892, \quad & S_2 = -5.530, \quad R_2 = 5.845 \\ R = 12.340, \quad \bar{R} = \; & 0.650, \quad \bar{R}' = (R_1 + R_2)/n = 0.650. \end{aligned}$$

We use the Watson–Williams test. Since $\bar{R} > 0.4$, we use Appendix 2.12a. For $\bar{R} = 0.65$, the 5% values of $\bar{R}'$ for $n = 16$ and $n = 20$ are 0.72 and 0.71, respectively. Thus, for $n = 19$, interpolation gives the 5% value of $\bar{R}'$ as 0.71. Since the observed value of $\bar{R}'$ is 0.650, the null hypothesis is accepted at the 5% significance level, and we conclude that the data support the sun-azimuth compass theory.

For this example, we now obtain an approximate 95% confidence interval for δ using (7.3.16). We have

$$\bar{\theta}_1 = 103.7° \quad \text{and} \quad \bar{\theta}_2 = 18.9°, \quad \text{so that} \quad d = 84.8°.$$

By using (7.3.13) and Appendix 2.4, we find that $\hat{\kappa} = 1.739$. Appendix 2.3 gives $A(\hat{\kappa}R_1) = 0.955$ and $A(\hat{\kappa}R_2) = 0.950$. Hence, from (7.3.14), $A(\hat{\kappa}^*) = 0.901$, which gives $\hat{\kappa}^* = 5.67$ from Appendix 2.4. Using Appendix 2.2, we find that ν satisfying (7.3.15) is 49.7°. Hence, from (7.3.16), an approximate 95% confidence interval for δ is $(35.1°, 134.5°)$. Note that 90° lies in this arc (in agreement with the above test) but 0° does not.

The Likelihood Ratio Test

The likelihood ratio statistic for testing $H_0 : \mu_1 = \mu_2$ against $H_1 : \mu_1 \neq \mu_2$ is

$$w = 2\left\{\hat{\kappa}_{12}(R_1 + R_2) - \hat{\kappa}R - \log I_0(\hat{\kappa}_{12}) + \log I_0(\hat{\kappa})\right\}, \tag{7.3.17}$$

where $\hat{\kappa}_{12}$ and $\hat{\kappa}$ denote the maximum likelihood estimates of κ under H_1 and H_0, respectively. For large κ, it can be shown by using

$$A(\kappa) \simeq 1 - \frac{1}{2\kappa}, \quad I_0(\kappa) \simeq (2\pi\kappa)^{-1/2}e^{\kappa}, \tag{7.3.18}$$

that

$$w \simeq \log \frac{n - R}{n - (R_1 + R_2)}, \tag{7.3.19}$$

which is an increasing function of the F statistic given in (7.3.8). For small κ, on using the approximations

$$A(\kappa) \simeq \frac{\kappa}{2}, \quad I_0(\kappa) \simeq 1 + \frac{\kappa^2}{4}, \tag{7.3.20}$$

we find that

$$w \simeq \frac{2}{n}\{(R_1 + R_2)^2 - R^2\}. \tag{7.3.21}$$

Thus, in these two extreme cases, the likelihood ratio statistic w is approximately an increasing function of $R_1 + R_2$ for given R, and so the test based on w given R is close to the two-sample Watson–Williams test.

7.3.2 Tests of Equality of Concentration Parameters

Consider the problem of testing

$$H_0 : \kappa_1 = \kappa_2 \quad \text{against} \quad H_1 : \kappa_1 \neq \kappa_2, \tag{7.3.22}$$

where the mean directions μ_1, μ_2 and the concentrations κ_1, κ_2 are unknown.

We now obtain tests suitable for various ranges of $\bar{R}$. For $\bar{R} \leq 0.7$, these tests are based on the variance-stabilising transformations of Section 4.8.3. For $\bar{R} > 0.7$, the test is based on the high-concentration approximations of Section 4.8.2.

Case I. $\bar{R} < 0.45$. In this case, we can use the statistic

$$\frac{2}{\sqrt{3}} \frac{g_1(2\bar{R}_1) - g_1(2\bar{R}_2)}{\{1/(n_1 - 4) + 1/(n_2 - 4)\}^{1/2}}, \tag{7.3.23}$$

where g_1 is defined by (4.8.40). Under H_0, (7.3.23) is distributed approximately as $N(0, 1)$. The critical region consists of both tails. If $n_1 = n_2$, the mean of (7.3.23) is of order $O(n^{-1})$, but if $n_1 \neq n_2$ then it follows from (4.8.43) that the bias in $g_1(2\bar{R}_1) - g_1(2\bar{R}_2)$ is

$$\left(\frac{1}{n_1} - \frac{1}{n_2}\right) a^3\kappa + o\left(\frac{1}{n}\right).$$

This bias is negligible if either κ is small or $n_1 \simeq n_2$.

Case II. $0.45 \leq \bar{R} \leq 0.70$. For this case, we take our test statistic as

$$\frac{g_2(\bar{R}_1) - g_2(\bar{R}_2)}{c_3\{1/(n_1 - 3) + 1/(n_2 - 3)\}^{1/2}}, \tag{7.3.24}$$

where the function g_2 is defined by (4.8.46) and $c_3 = 0.893$. Under H_0, this statistic is distributed approximately as $N(0, 1)$. For $n_1 \neq n_2$, the bias to order n^{-1} can be obtained from (4.8.48). Example 7.5 illustrates this test.

Case III. $\bar{R} > 0.70$. In this case, it follows from (7.3.9) that, under H_0,

$$\frac{(n_1 - R_1)/(n_1 - 1)}{(n_2 - R_2)/(n_2 - 1)} \stackrel{.}{\sim} F_{n_1-1, n_2-1}. \tag{7.3.25}$$

The critical region of the test consists of both tails of the F_{n_1-1,n_2-1} distribution. Simulations have shown that this approximation is adequate for $\bar{R} > 0.70$.

The likelihood ratio statistic w for testing $\kappa_1 = \kappa_2$ is again complicated. However, on following the method used to obtain (7.3.19) and (7.3.21), we find that, for large κ, w is approximately an increasing function of the F statistic in (7.3.25). For small κ

$$w \simeq \frac{2n_1 n_2}{n}(\bar{R}_1 - \bar{R}_2)^2. \tag{7.3.26}$$

Example 7.5
Using the data in Example 7.4, we test the equality of the concentration parameters of the corresponding populations.

We have $\bar{R} = 0.48$, $\bar{R}_1 = 0.723$, $\bar{R}_2 = 0.585$, $n_1 = 9$, $n_2 = 10$. Since $\bar{R} > 0.45$, we use the test based on (7.3.24). It is found that

$$g_2(\bar{R}_1) = -1.150, \quad g_2(\bar{R}_2) = -1.425.$$

The value of the denominator in (7.3.24) is 0.497. Consequently, the value of the test statistic is 0.553, which is less than 1.96, and so the null hypothesis is accepted at the 5% significance level.

7.4 MULTI-SAMPLE TESTS

Suppose that (for $i = 1, \ldots, q$) $\theta_{i1}, \ldots, \theta_{in_i}$ are q independent random samples of sizes n_i from $M(\mu_i, \kappa_i)$. Let $\bar{\theta}_i$ and R_i denote the mean direction and the resultant length of the ith sample, and $\bar{\theta}$ and R denote the mean direction and the resultant length of the combined sample. Then

$$R_i^2 = C_i^2 + S_i^2, \quad R^2 = C^2 + S^2, \tag{7.4.1}$$

where

$$C_i = \sum_{j=1}^{n_i} \cos\theta_{ij}, \qquad S_i = \sum_{j=1}^{n_i} \sin\theta_{ij}, \quad i = 1, \ldots, q,$$

$$C = \sum_{j=1}^{q} C_j, \qquad S = \sum_{j=1}^{q} S_j,$$

7.4.1 One-Way Classification

We wish to test

$$H_0 : \mu_1 = \ldots = \mu_q \tag{7.4.2}$$

against the alternative that at least one of the equalities does not hold. In this subsection we shall assume that $\kappa_1 = \ldots = \kappa_q$, where the common concentration κ is unknown. Tests can be constructed using the approaches employed in the two-sample case.

The Multi-Sample Watson–Williams Test

The multi-sample analogue of the two-sample Watson–Williams test is the conditional test which rejects H_0 for large values of

$$R_1 + \ldots + R_q \tag{7.4.3}$$

given R. From (4.6.17), the probability density function of $R_1, \ldots, R_q$ given R does not depend on κ. Thus, for given R, this test is a similar test.

A High-Concentration F Test

On following the same argument as for the modified F-approximation (7.3.8), we find that under H_0

$$F_{WW} \mathrel{\dot{\sim}} F_{q-1,n-q}, \tag{7.4.4}$$

where

$$F_{WW} = \frac{(\sum_{i=1}^{q} R_i - R)/(q-1)}{(n - \sum_{i=1}^{q} R_i)/(n-q)}. \tag{7.4.5}$$

The approximation (7.4.5) can be refined as in (7.3.11) to

$$\left(1 + \frac{3}{8\hat{\kappa}}\right) F_{WW} \mathrel{\dot{\sim}} F_{q-1,n-q}, \tag{7.4.6}$$

where $\hat{\kappa}$ is the maximum likelihood estimate of κ based on $\bar{R}$, and is given by (5.3.6). It is found from simulation (Stephens, 1972) that approximation (7.4.6) is adequate for $\kappa \geq 1$, i.e. $\bar{R} \geq 0.45$. For $\hat{\kappa} > 10$, the factor in $\hat{\kappa}$ in (7.4.6) is negligible. The calculations can be displayed in an analysis of variance (ANOVA) table (see Table 7.1). The last column of this table can be modified to incorporate the correction factor in (7.4.6).

Table 7.1 ANOVA table

Source	d.f.	SS	Mean Square	F
Between Samples	$q-1$	$\sum_{i=1}^{q} R_i - R$	$(\sum_{i=1}^{q} R_i - R)/(q-1)$ $= MS_B$	MS_B/MS_W
Within Samples	$n-q$	$n - \sum_{i=1}^{q} R_i$	$(n - \sum_{i=1}^{q} R_i)/(n-q)$ $= MS_W$	
Total	$n-1$	$n-R$		

Confidence intervals for the simple contrasts $\mu_i - \mu_j$ can be obtained by the method used to derive (7.3.16).

For large κ, Rao & Sengupta (1970) investigated the problem of choice of sample size in attaining a desired precision in estimating the mean direction. They also considered the problem of optimum allocation of resources under this model.

The Likelihood Ratio Test

When κ is known, the likelihood ratio statistic is

$$2\kappa \sum_{i=1}^{q} R_i \{1 - \cos(\bar{\theta}_i - \bar{\theta})\} = \kappa \sum_{i=1}^{q} R_i \|\bar{\mathbf{u}}_{i.} - \bar{\mathbf{u}}_{..}\|^2, \tag{7.4.7}$$

where

$$\bar{\mathbf{u}}_{i.} = \frac{\bar{\mathbf{x}}_{i.}}{\|\bar{\mathbf{x}}_{i.}\|}, \qquad \bar{\mathbf{u}}_{..} = \frac{\bar{\mathbf{x}}_{..}}{\|\bar{\mathbf{x}}_{..}\|}$$

are the unit vectors corresponding to the mean directions $\bar{\theta}_i$ and $\bar{\theta}$ of the ith sample and the combined sample, respectively. Under the null hypothesis,

$$2\kappa \sum_{i=1}^{q} R_i(1 - \cos(\bar{\theta}_i - \bar{\theta})) \ \dot{\sim}\ \chi^2_{q-1}. \tag{7.4.8}$$

A better approximation (Cordeiro, Paula & Botter, 1994, Section 4.4) is

$$\left\{1 - \frac{1}{4\kappa} A(\kappa) \left(\sum_{i=1}^{q} \frac{1}{n_i} - \frac{1}{n} \right) \right\} 2\kappa \sum_{i=1}^{q} R_i(1 - \cos(\bar{\theta}_i - \bar{\theta})) \ \dot{\sim}\ \chi^2_{q-1}.$$

In the case where κ is unknown, Anderson & Wu (1995) suggested replacing κ in (7.4.7) by its maximum likelihood estimate $\hat{\kappa}$ to obtain the test which rejects H_0 for large values of

$$\begin{aligned} w &= 2\hat{\kappa} \sum_{i=1}^{q} R_i(1 - \cos(\bar{\theta}_i - \bar{\theta})) \\ &= \hat{\kappa} \sum_{i=1}^{q} R_i \|\bar{\mathbf{u}}_{i.} - \bar{\mathbf{u}}_{..}\|^2. \end{aligned} \tag{7.4.9}$$

Under the null hypothesis, the large-sample asymptotic distribution of (7.4.9) is χ^2_{q-1}.

For data from a concentrated von Mises distribution with $n_1 = \ldots = n_q$,

$$w \simeq \left(1 + \frac{1}{4\hat{\kappa}}\right) \frac{n(q-1)}{n-q} F_{WW},$$

where F_{WW} is defined in (7.4.5).

For small κ, the approximations (7.3.20) show that the likelihood ratio statistic w of (7.4.9) is approximately equal to U, where

$$U = \frac{2}{n}\left\{\left(\sum_{i=1}^{q} R_i\right)^2 - R^2\right\}. \tag{7.4.10}$$

When H_0 is true, $U \overset{\cdot}{\sim} \chi^2_{q-1}$ for large n. This approximation can be improved as follows, using a multiplicative correction to bring the mean closer to $q-1$. From (4.7.3) and (4.8.17), we have

$$\mathrm{E}[R_i] = n_i A(\kappa) + \frac{1}{2\kappa} + O\left(\frac{1}{n_i}\right), \quad \mathrm{E}[R_i^2] = n_i + n_i(n_i-1)A(\kappa)^2.$$

Substituting these into $E[U]$ gives

$$\mathrm{E}[U] \simeq (q-1)\left\{\frac{2A(\kappa)}{\kappa} + \frac{q}{2n\kappa^2}\right\}.$$

Using $A(\kappa)/\kappa \simeq \frac{1}{2}(1-\frac{1}{8}\kappa^2) + O(\kappa^4)$, we have

$$cU \overset{\cdot}{\sim} \chi^2_{q-1}, \tag{7.4.11}$$

where

$$\frac{1}{c} = 1 - \frac{1}{8}\kappa^2 + \frac{q}{2n\kappa^2}.$$

This approximation (Mardia, 1972a, p. 164) is found to be satisfactory for moderately small values of n, provided that κ is not near 0 or 1. In practice, we replace κ by its maximum likelihood estimate and we reject H_0 for large values of cU.

Example 7.6

Table 7.2 gives wind directions in degrees at Gorleston, England, between 11 a.m. and 12 noon on Sundays in 1968, classified by season. For this data set, $q = 4$, $n_1 = n_2 = n_4 = 12$, $n_3 = 13$, $n = 49$. (Three readings are missing because there was no wind on the corresponding days.) Do the data indicate that the mean wind directions for the four seasons are different?

Table 7.2 Wind directions in degrees at Gorleston on Sundays in 1968, classified by season.

Season	Wind directions (in degrees)												
Winter	50	120	190	210	220	250	260	290	290	320	320	340	
Spring	0	20	40	60	160	170	200	220	270	290	340	350	
Summer	10	10	20	20	30	30	40	150	150	150	170	190	290
Autumn	30	70	110	170	180	190	240	250	260	260	290	350	

Table 7.3 Calculations of various statistics for the wind direction data in Table 7.2.

Season	C_i	S_i	R_i	$\hat{\mu}_i$	$\hat{\kappa}_i$
Winter	0.166	−5.116	5.119	272°	0.94
Spring	1.842	−1.074	2.132	330°	0.36
Summer	2.121	3.234	3.868	57°	0.62
Autumn	−1.966	−2.509	3.188	232°	0.55
Combined sample	2.163	−5.464	5.877	292°	0.24

We assume that the concentration parameters are equal. (This assumption will be justified in Example 7.7.) Table 7.3 gives the values of the relevant statistics for this data set. Since $\hat{\kappa} = 0.24$, we use (7.4.11). All the $\hat{\kappa}_i$ are less than 1. We have $c = 0.59$, $\sum_{i=1}^{q} R_i = 14.306$, and $R = 5.877$, giving $cU = 4.10$. Since the 5% value of χ_3^2 is 7.81, we accept that the mean wind directions for the four seasons are the same.

In passing, we note from (7.3.20) that the likelihood ratio statistic for testing the null hypothesis $\mu_1 = \ldots = \mu_q$ and $\kappa_1 = \ldots = \kappa_q$ against the general alternative is approximately

$$2\left(\sum_{i=1}^{q} \frac{R_i^2}{n_i} - \frac{R^2}{n}\right), \tag{7.4.12}$$

when $\kappa_1, \ldots, \kappa_q$ are small. Under the null hypothesis, the asymptotic large-sample distribution of (7.4.12) is $\chi^2_{2(q-1)}$.

ANOVA Based on the Embedding Approach

Another approach to analysis of variance for circular data was proposed by Harrison, Kanji & Gadsden (1986) and Harrison & Kanji (1988). Their idea is to take the embedding approach and so to consider the observations as unit vectors in the plane. Recall that for a single sample $\mathbf{x}_1, \ldots, \mathbf{x}_n$ on the circle we have

$$\sum_{i=1}^{n} ||\mathbf{x}_i - \bar{\mathbf{x}}|| = n(1 - \bar{R}^2).$$

Similarly, for q samples $\mathbf{x}_{11}, \ldots, \mathbf{x}_{1n_1}; \ldots; \mathbf{x}_{q1}, \ldots, \mathbf{x}_{qn_q}$, the basic ANOVA decomposition

$$\sum_{i=1}^{q}\sum_{j=1}^{n_i} ||\mathbf{x}_{ij} - \bar{\mathbf{x}}_{..}||^2 = \sum_{i=1}^{q}\sum_{j=1}^{n_i} ||\mathbf{x}_{ij} - \bar{\mathbf{x}}_{i.}||^2 + \sum_{i=1}^{q} n_i ||\bar{\mathbf{x}}_{i.} - \bar{\mathbf{x}}_{..}||^2 \tag{7.4.13}$$

can be rewritten as

$$n(1-\bar{R}^2) = \left(\sum_{i=1}^{q} n_j\bar{R}_j^2 - n\bar{R}^2\right) + \left(n - \sum_{i=1}^{q} n_j\bar{R}_j^2\right), \qquad (7.4.14)$$

which is reminiscent of (9.6.25). In (7.4.14) the terms represent respectively the total variation, the variation between samples, and the variation within samples. This leads to the test which rejects H_0 for large values of

$$\frac{(\sum_{i=1}^{q} n_i\bar{R}_i^2 - \bar{R}^2)/(q-1)}{(n-\sum_{i=1}^{q} n_i\bar{R}_i^2)/(n-q)}. \qquad (7.4.15)$$

Under H_0, the high-concentration asymptotic distribution of (7.4.15) is $F_{q-1,n-q}$. Harrison, Kanji & Gadsden (1986) give also the more refined approximation

$$\left(1-\frac{1}{5\hat{\kappa}}-\frac{1}{10\hat{\kappa}^2}\right)\frac{(\sum_{i=1}^{q} n_i\bar{R}_i^2 - \bar{R}^2)/(q-1)}{(n-\sum_{i=1}^{q} n_i\bar{R}_i^2)/(n-q)} \stackrel{\sim}{\sim} F_{q-1,n-q}. \qquad (7.4.16)$$

ANOVA Based on a Tangential Approach

One way of measuring the deviation between observations θ and ϕ is by $\sin(\theta-\phi)$. This can be regarded as a 'tangential approach', since $\sin(\theta-\phi)$ can be thought of as the length of a tangent to the circle at θ. This approach leads to the test which rejects H_0 for large values of

$$\frac{(n-q)\sum_{i=1}^{q} n_i(\bar{d}_i-\bar{d})^2}{(q-1)\sum_{i=1}^{q}\sum_{j=1}^{n_i}(d_{ij}-\bar{d}_i)^2}, \qquad (7.4.17)$$

where

$$d_{ij} = |\sin(\theta_{ij}-\bar{\theta}_i)|, \qquad j=1,\ldots,n_i, \qquad (7.4.18)$$

$$\bar{d}_i = \frac{1}{n_i}\sum_{i=1}^{q} d_{ij}, \qquad \bar{d} = \frac{1}{n}\sum_{i=1}^{q} n_i\bar{d}_i. \qquad (7.4.19)$$

Under the null hypothesis, the asymptotic large-sample distribution of (7.4.17) is $F_{q-1,n-q}$. Fisher (1993, Section 5.4.4) recommends this test on the grounds of its robustness against both outliers and departure from the von Mises assumption.

7.4.2 Tests for the Homogeneity of Concentration Parameters

Because the tests of Section 7.4.1 are based on the assumption of equal concentrations, it is often necessary to test the hypothesis

$$H_0 : \kappa_1 = \ldots = \kappa_q, \qquad (7.4.20)$$

where $\mu_1, \ldots, \mu_q$ and the common concentration κ are not specified. Following Section 4.8.3, we divide our test procedure into three parts.

Case I. $\bar{R} < 0.45$. Applying the $\sin^{-1}$ transformation g_1 given by (4.8.40) to $2\bar{R}_i$, and using (4.8.42), shows that, under H_0, $g_1(2\bar{R}_1), \ldots, g_1(2\bar{R}_n)$ are approximately distributed as independent $N(g_1(\kappa), \sigma_i^2)$ variables, where

$$\sigma_i^2 = \frac{3}{4(n_i - 4)} = \frac{1}{w_i}, \quad \text{say.} \tag{7.4.21}$$

Standard normal theory shows that the weighted least-squares estimate of $g_1(\kappa)$ based on $g_1(2\bar{R}_1), \ldots, g_1(2\bar{R}_n)$ is

$$\hat{g}_1(\kappa) = \frac{\sum_{i=1}^q w_i g_1(2\bar{R}_i)}{\sum_{i=1}^q w_i} \tag{7.4.22}$$

and that, under H_0,

$$U_1 = \sum_{i=1}^q w_i g_1(2\bar{R}_i)^2 - \frac{\{\sum_{i=1}^q w_i g_1(2\bar{R}_i)\}^2}{\sum_{i=1}^q w_i} \mathrel{\dot{\sim}} \chi^2_{q-1}. \tag{7.4.23}$$

Example 7.7 illustrates this procedure.

Case II. $0.45 \leq \bar{R} \leq 0.70$. On proceeding as in case I but applying the $\sinh^{-1}$ transformation g_2 defined by (4.8.46), we find that an appropriate statistic is

$$U_2 = \sum_{i=1}^q w_i g_2^2(\bar{R}_i) - \frac{\{\sum_{i=1}^q w_i g_2(\bar{R}_i)\}^2}{\sum_{i=1}^q w_i} \mathrel{\dot{\sim}} \chi^2_{q-1}, \tag{7.4.24}$$

where now

$$\frac{1}{w_i} = \frac{0.798}{n_i - 3}.$$

Case III. $\bar{R} > 0.70$. For this case, the high-concentration approximation $M(\mu_i, \kappa) \simeq N(\mu_i, 1/\kappa)$ gives the sample variance of $\theta_{i1}, \ldots, \theta_{in_i}$ as approximately $2\kappa(n_i - R_i)$. Then it is appropriate to use Bartlett's test of homogeneity (Stuart & Ord, 1991, pp. 875–876), and so to reject H_0 for large values of

$$U_3 = \frac{1}{1+d}\left\{\nu \log\left(\frac{n - \sum_{i=1}^q R_i}{\nu}\right) - \sum_{i=1}^q \nu_i \log\left(\frac{n_i - R_i}{\nu_i}\right)\right\}, \tag{7.4.25}$$

where

$$\nu_i = n_i - 1, \quad \nu = n - q, \quad d = \frac{1}{3(q-1)}\left\{\sum_{i=1}^q \frac{1}{\nu_i} - \frac{1}{\nu}\right\}.$$

Under H_0 and for large κ, $U_3 \mathrel{\dot{\sim}} \chi^2_{q-1}$.

Table 7.4 Calculations for testing homogeneity of the concentrations for the wind direction data in Table 7.2.

n_i	$2a\bar{R}_i$	$g_1(2a\bar{R}_i)$	w_i	$w_i g_1(2a\bar{R}_i)$	$w_i g_1(2a\bar{R}_i)^2$
12	0.522	0.550	10.667	5.861	3.221
12	0.218	0.219	10.667	2.340	0.514
13	0.364	0.373	12.000	4.475	1.669
12	0.325	0.332	10.667	3.538	1.174
Total			44.000	16.215	6.577

Example 7.7
We test the homogeneity of the concentration parameters for the wind direction data in Example 7.6.

Since $\bar{R} = 0.199$, we use U_1 of (7.4.23). Table 7.4 shows the relevant calculations where $2a\bar{R}_i = 1.225\bar{R}_i$, $w_i = 4(n_i - 4)/3$. Using the totals in the last three columns in (7.4.23), we obtain

$$U = 6.577 - (16.215)^2/44.000 = 0.601.$$

Since the 5% value of χ_3^2 is 7.81, we accept that the concentration parameters for the wind directions are the same for all four seasons.

Various suggestions for two-way and multi-way ANOVA for circular data are given by Stephens (1982), Harrison, Kanji & Gadsden (1986), Harrison & Kanji (1988) and Anderson & Wu (1995).

7.4.3 The Heterogeneous Case

If it is not reasonable to make the homogeneity assumption $\kappa_1 = \ldots = \kappa_q$ then it is not appropriate to use the methods of Section 7.4.1 to test

$$H_0 : \mu_1 = \ldots = \mu_q \qquad (7.4.26)$$

against the alternative that at least one of the equalities does not hold. We now consider tests of (7.4.26) in the heterogeneous case.

If $\kappa_1, \ldots, \kappa_q$ are known then the likelihood ratio statistic for (7.4.26) is

$$2\left(\sum_{i=1}^{q} \kappa_i R_i - R_W\right), \qquad (7.4.27)$$

where

$$R_W = \left\{\left(\sum_{i=1}^{q} \kappa_i R_i \cos\bar{\theta}_i\right)^2 + \left(\sum_{i=1}^{q} \kappa_i R_i \sin\bar{\theta}_i\right)^2\right\}^{1/2}.$$

Under (7.4.26), the large-sample asymptotic distribution of (7.4.27) is χ^2_{q-1}. The statistic (7.4.27) generalises (7.4.7).

If $\kappa_1, \ldots, \kappa_q$ are not known then they can be replaced by their maximum likelihood estimates to give

$$2\left(\sum_{i=1}^{q} \hat{\kappa}_i R_i - R_W\right), \tag{7.4.28}$$

where now

$$R_W = \left\{\left(\sum_{i=1}^{q} \hat{\kappa}_i R_i \cos\bar{\theta}_i\right)^2 + \left(\sum_{i=1}^{q} \hat{\kappa}_i R_i \sin\bar{\theta}_i\right)^2\right\}^{1/2}.$$

Under (7.4.26), the large-sample asymptotic distribution of (7.4.28) is χ^2_{q-1}. The test which rejects (7.4.26) for large values of (7.4.28) was introduced by Watson (1983b).

7.5 TESTING VON MISESNESS

Most of the inferential procedures described in this chapter have been based on the assumption that the underlying distribution is a von Mises distribution. Thus is it is useful to be able to assess whether or not this assumption is reasonable. One way of testing that a given distribution is a von Mises distribution is to test for von Misesness within some larger model. In particular, we can extend the von Mises model to the exponential model (3.5.46) with probability density functions proportional to

$$\exp\{\kappa\cos(\theta - \mu) + \psi_1 \cos(2\theta) + \psi_2 \sin(2\theta)\}.$$

The von Mises distributions are those members of this model for which

$$\psi_1 = \psi_2 = 0. \tag{7.5.1}$$

Define

$$U = \sum_{i=1}^{n} (\cos\theta_i, \sin\theta_i) \qquad V = \sum_{i=1}^{n} (\cos(2\theta_i), \sin(2\theta_i)). \tag{7.5.2}$$

Then a suitable test of (7.5.1) is the score test based on the conditional distribution of V given U. Using a large-sample approximation to this conditional distribution leads to the statistic

$$S = \frac{s_c^2}{v_c(\hat{\kappa})} + \frac{s_s^2}{v_s(\hat{\kappa})}, \tag{7.5.3}$$

where s_c and s_s are the components

$$s_c = \sum_{i=1}^{n} \cos 2(\theta_i - \hat{\mu}) - nA(\hat{\kappa}), \qquad s_s = \sum_{i=1}^{n} \sin 2(\theta_i - \hat{\mu})$$

of the conditional score,

$$\begin{aligned} v_c(\kappa) &= \frac{I_0(\kappa)^2 + I_0(\kappa)I_4(\kappa) - 2I_2(\kappa)^2}{2I_0(\kappa)^2} \\ &\quad - \frac{(I_0(\kappa)I_3(\kappa) + I_0(\kappa)I_1(\kappa) - 2I_1(\kappa)I_2(\kappa))^2}{2I_0(\kappa)^2(I_0(\kappa)^2 + I_0(\kappa)I_1(\kappa) - 2I_1(\kappa)^2)} \\ v_s(\kappa) &= \frac{(I_0(\kappa) - I_4(\kappa))(I_0(\kappa) - I_2(\kappa)) - (I_0(\kappa) - I_3(\kappa))^2}{2I_0(\kappa)(I_0(\kappa) - I_2(\kappa))}, \end{aligned}$$

are approximations to their variances, and $\hat{\mu}$, $\hat{\kappa}$ denote the maximum likelihood estimates under the von Mises model. The null hypothesis (7.5.1) of von Misesness is rejected for large values of S. Under the null hypothesis, the large-sample asymptotic distribution of S is χ_2^2. This test was proposed by Cox (1975). Further details can be found in Barndorff-Nielsen & Cox (1989, pp. 227–228).

8

Non-parametric Methods

8.1 INTRODUCTION

If preliminary investigation of a data set on the circle indicates that it is unlikely to have been generated by some von Mises distribution (or a distribution in some other standard parametric family) then inference has to be carried out either by resampling methods such as bootstrapping or by non-parametric methods. Bootstrapping is considered in Section 12.5. Non-parametric methods on the circle are considered in this chapter. These are based on cumulative distribution functions and ranks, as on the line. Section 8.2 considers tests of symmetry. Two-sample tests are considered in Section 8.3. In particular, we discuss two-sample versions of Kuiper's test and Watson's U^2 test, and show how any rotation-invariant test for uniformity gives rise to a two-sample test. Extensions to the multi-sample case are given in Section 8.4. For a survey of non-parametric methods in directional statistics up to 1984, the reader is referred to Jammalamadaka (1984).

8.2 TESTS OF SYMMETRY

8.2.1 Tests of Symmetry about a Given Axis

To test the hypothesis that a specified direction μ_0 is an axis of symmetry against the alternative that it is not, we can use the usual tests of symmetry on the line. After a suitable rotation of the circle, we may assume that $\mu_0 = 0$. We describe briefly two standard procedures adapted for the circular case: (i) The sign test rejects the null hypothesis if the number of observations on the upper semicircle is either too large or too small. (ii) We can also use the one-sample Wilcoxon test as follows. Let $\theta_{(1)}, \ldots, \theta_{(n)}$ be the ordered observations. Suppose that s of these observations lie in the upper semicircle. Put

$$\theta_i^* = \theta_{(i)} - \pi, \quad i = s+1, \ldots, n.$$

Let $r_1, \ldots, r_s$ be the ranks of these s observations in the sequence $\theta_{(1)}, \ldots, \theta_{(s)}, \theta_{s+1}^*, \ldots, \theta_n^*$. The null hypothesis is rejected if the sum $r_1 + \ldots + r_s$ is either too large or too small.

Various parametric competitors of these tests have been given in Section 7.2.1. Schach (1969a) has shown that the asymptotic relative efficiency of the Wilcoxon test relative to the locally most powerful test for von Mises distributions tends to $6/\pi^2$ as $\kappa \to 0$ and $3/\pi$ as $\kappa \to \infty$. For the sign test, the asymptotic relative efficiency tends to $8/\pi^2$ as $\kappa \to 0$ and $2/\pi$ as $\kappa \to \infty$. Further, in each case, the asymptotic relative efficiency is an increasing function of κ. As we would expect, for $\kappa \to \infty$ the asymptotic relative efficiencies of both non-parametric tests coincide with the corresponding asymptotic relative efficiencies for detecting shifts of a normal distribution. These comparisons indicate that for small κ and large n, the sign test may be preferred to the Wilcoxon test.

Example 8.1

We test the hypothesis of symmetry about $\theta = 149°$ for the pigeon-homing data of Example 7.2 by applying the sign test.

The number of observations in the interval $(149°, 329°)$ is 10. Since the probability of $n' \geq 11$ for $n = 15$ is 0.118 (see Siegel, 1956, p. 250), we accept at the 10% level the null hypothesis that the population circular median is 149°.

8.2.2 Tests of l-fold Symmetry

Consider the hypothesis that the distribution has l-fold symmetry, i.e.

$$H_0 : F(\theta_2) - F(\theta_1) = F(\theta_2 + 2\pi/l) - F(\theta_1 + 2\pi/l), \tag{8.2.1}$$

so that F is invariant under rotation by $2\pi/l$. If $l = 2$ then this hypothesis is that of antipodal (or central) symmetry, i.e. the probability density function satisfies $f(x + \pi) = f(x)$, so that the distribution is suitable for modelling axial data. One way of testing (8.2.1) is to extend the given sample $\theta_1, \ldots, \theta_n$ to a 'symmetrised sample' $\phi_1, \ldots, \phi_{nl}$ consisting of

$$\left\{\theta_i + j\frac{2\pi}{l} : 1 \leq i \leq n, 1 \leq j \leq l\right\}.$$

If $\phi_1, \ldots, \phi_{nl}$ are moved around the circle until they are equally spaced then $\theta_1, \ldots, \theta_n$ are transformed to uniform scores, as described in Section 8.3.1. Then (8.2.1) may be tested (Jupp & Spurr, 1983) by applying any test of uniformity to these uniform scores.

8.3 TWO-SAMPLE TESTS

In many circumstances we may wish to test the equality of two distributions. More precisely, we wish to test

$$H_0 : F_1 = F_2 \quad \text{against} \quad H_1 : F_1 \neq F_2, \tag{8.3.1}$$

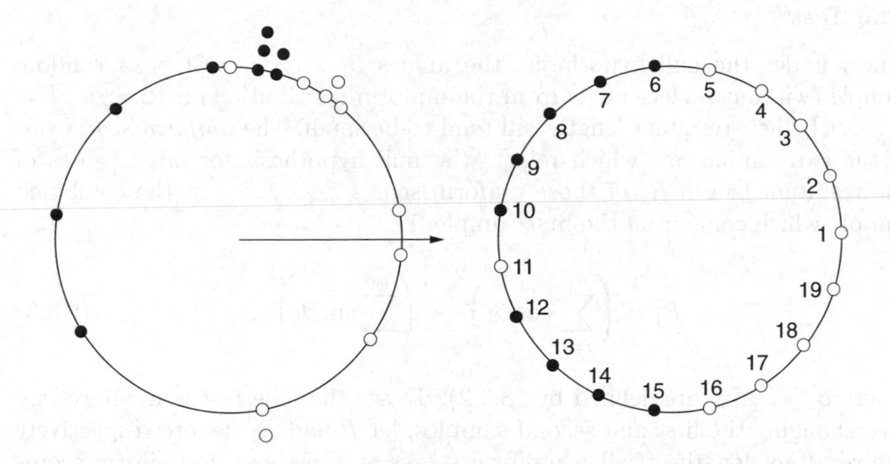

Figure 8.1 Generation of equidistant sample points and the combined ranks for the two samples in Example 8.3.

where F_1 and F_2 denote the distribution functions, on the basis of independent random samples $\theta_{11}, \ldots, \theta_{1n_1}$ from the first distribution and $\theta_{21}, \ldots, \theta_{2n_2}$ from the second distribution. The combined sample has size $n = n_1 + n_2$.

8.3.1 The Uniform-Scores Test

Suppose that the elements $\theta_{11}, \ldots, \theta_{1n_1}, \theta_{21}, \ldots, \theta_{2n_2}$ of the combined sample are distinct (as happens with probability 1 if the distributions are continuous). We then move these points continuously around the unit circle (keeping them distinct) until the spaces between successive points are of length $2\pi/n$ (see Fig. 8.1). That is, we replace the angular observations in the combined sample by the *uniform scores* $2\pi k/n, k = 1, \ldots, n$, on the circle. Then the n_1 observations in the first sample are transformed to

$$\beta_i = \frac{2\pi r_i}{n}, \quad i = 1, \ldots, n_1, \tag{8.3.2}$$

where $r_1, \ldots, r_{n_1}$ denote the linear ranks in the combined sample of the observations in the first sample. For the data in Fig. 8.1, the ranks r_i are 6, 7, 8, 9, 10, 12, 13, 14, 15. Under the null hypothesis of equality of the two populations, the angles $\beta_1, \ldots, \beta_{n_1}$ will be a random sample (without replacement) from the uniform distribution on $\{2\pi k/n : k = 1, \ldots, n\}$. However, if the null hypothesis is false, the angles β_i will tend to cluster in some proper subset of $\{2\pi k/n : k = 1, \ldots, n\}$.

The Test

Since, under the null hypothesis, the angles $\beta_1, \ldots, \beta_{n_1}$ will be a random sample (without replacement) from the uniform distribution on $\{2\pi k/n : k = 1, \ldots, n\}$, their resultant length will tend to be small. The *uniform-scores test* is the two-sample test which rejects the null hypothesis for large values of the resultant length R_1 of those uniform scores $\beta_1, \ldots, \beta_{n_1}$ in the combined sample which come from the first sample, i.e.

$$R_1^2 = \left(\sum_{i=1}^{n_1} \cos\beta_i\right)^2 + \left(\sum_{i=1}^{n_1} \sin\beta_i\right)^2, \tag{8.3.3}$$

where $\beta_1, \ldots, \beta_{n_1}$ are defined by (8.3.2). To see that the test is unaltered by interchanging the first and second samples, let R and R_2 denote respectively the resultant lengths of all n uniform scores and of those n_2 uniform scores which come from the second sample. Since

$$\sum_{k=1}^{n} \cos\left(\frac{2\pi k}{n}\right) = \sum_{k=1}^{n} \sin\left(\frac{2\pi k}{n}\right) = 0,$$

it follows that $R = 0$ and $R_1 = R_2$.

This test was proposed by Wheeler & Watson (1964), following a suggestion of J. L. Hodges Jr. As shown by Mardia (1969b), the test is a particular case of a test for the bivariate location problem given independently by Mardia (1967; 1968).

Note that the uniform-scores test is the Rayleigh test applied to the uniform scores $\beta_1, \ldots, \beta_{n_1}$ of the first sample. This idea will be generalised in Section 8.3.5.

Null Distribution and Power

Some selected quantiles of R_1^2 are given in Appendix 2.13. For $n > 20$, we can use the result (Mardia, 1967; 1969a) that under H_0 the large-sample $(n_1 \to \infty, n_2 \to \infty, 0 < \lim n_1/n < 1)$ asymptotic distribution of

$$R^* = \frac{2(n-1)R_1^2}{n_1 n_2} \tag{8.3.4}$$

is

$$R^* \;\dot{\approx}\; \chi_2^2.$$

For $\alpha < 0.025$, greater accuracy can be achieved by using the approximation (Mardia, 1967)

$$\frac{R^*}{n-1-R^*} \;\dot{\approx}\; F_{d,(n-3)d},$$

where

$$d = 1 + \frac{n(n+1) - 6n_1 n_2}{n(n_1 - 1)(n_2 - 1)}.$$

This approximation is obtained by fitting a beta distribution to the null distribution of R_1^2 with the help of the first two moments.

The asymptotic power of this test is given in Mardia (1969b) and Schach (1969b). The test is consistent against those alternatives (F_1, F_2) where F_1 and F_2 are unimodal and symmetrical about their mean directions (Mardia, 1969b). For shift-type alternatives, it is shown in Mardia (1969b) and Schach (1969b) that the asymptotic relative efficiency of this test relative to its locally most powerful competitor tends to 1 when the underlying population is von Mises and κ tends to zero.

Example 8.2

In an experiment on pigeon homing, similar to that described in Example 7.4, it was predicted from sun-azimuth compass theory that the angles of the birds in the experimental group should deviate about 90° in the anticlockwise direction relative to the angles of the birds in the control group. The vanishing angles of the birds for this experiment with $n_1 = 12$ and $n_2 = 14$ are given below in degrees (Schmidt-Koenig's data cited in Watson, 1962).

Control group (θ_{i1}): 50, 290, 300, 300, 305, 320, 330, 330, 335, 340, 340, 355.

Experimental group (θ_{i2}): 70, 155, 190, 195, 215, 235, 235, 240, 255, 260, 290, 300, 300, 300.

Let us first test the null hypothesis that the two samples are drawn from the same population. We break the ties in the combined sample by replacing the observations 290, 300, 300 of the control group by 285, 295, 295, so that all ties between samples are broken in favour of H_0. Since R_1 is invariant under rotations, we may rank the observations by taking the first angle in the combined sample as 50° and proceed in the anticlockwise direction. The ranks r_i for the control group are $1, 2, 3, 4, 5, 6, 7, 8, 9, 13, 14, 16$. Calculation gives the resultant vector of the uniform scores as $(C_1, S_1)^T = (-0.115, 5.966)^T$. Then, from (8.3.3) and (8.3.4), we have

$$R^* = 10.60.$$

As the 1% value of χ_2^2 is 9.21, H_0 is rejected strongly.

Since the sun-azimuth theory predicts a shift in location of 90°, we now test

$$F_2(\theta) = F_1(\theta + 90°).$$

We do this by applying the uniform-scores test to the data set consisting of the control group and the *rotated* experimental group (rotated clockwise by 90°, as in Example 7.4). Breaking the ties by randomisation, we find that the data support the theory. For this example the conclusion is the same, irrespective of how the ties are broken.

We now give an example in which $n < 20$.

Example 8.3
We test the hypothesis of equality of the two populations for the following pigeon homing data (in degrees) of Example 7.4. (The data and the combined ranks are shown in Fig. 8.1.)

Sample 1: 75, 75, 80, 80, 80, 95, 130, 170, 210.
Sample 2: 10, 50, 55, 55, 65, 90, 285, 285, 325, 355.

On taking the r_i as 1, 2, 3, 4, 5, 7, 8, 9, 10, we have

$$n_1 = 9, \quad n_2 = 10, \quad C_1 = -1.085, \quad S_1 = 4.954, \quad R_1^2 = 25.72.$$

From Appendix 2.13, the 1% value of R_1^2 is 21.07. Hence the null hypothesis is rejected strongly.

8.3.2 Kuiper's Two-Sample Test

Since the sample versions of the distribution functions F_1 and F_2 are the empirical distribution functions S_1 and S_2 of the two samples, it is appropriate to reject the null hypothesis (8.3.1) when S_1 and S_2 are 'far apart'. Measuring the distance between S_1 and S_2 by the maximum deviation leads to Kuiper's (1960) two-sample test, which rejects H_0 for large values of

$$V_{n_1,n_2} = \sup_\theta\{S_1(\theta) - S_2(\theta)\} - \inf_\theta\{S_1(\theta) - S_2(\theta)\}. \tag{8.3.5}$$

This test is the two-sample analogue of Kuiper's one-sample test V_n considered in Section 6.3.2, and is a modification for the circle of the two-sample Kolmogorov–Smirnov test. Kuiper (1960) has shown that the large-sample asymptotic null distribution of V_{n_1,n_2} is the same as that of V_n. Steck (1969) has given a method of evaluating its null distribution but the quantiles have not yet been tabulated. For an example and further details, the reader is referred to Batschelet (1981, Section 6.5). In the linear case Abrahamson (1967) has compared the asymptotic relative behaviours of the Kolmogorov–Smirnov and Kuiper's tests.

8.3.3 Watson's Two-Sample U^2 test

Instead of measuring the discrepancy between the empirical distribution functions S_1 and S_2 of the two samples by a variant of the maximum deviation (as in Kuiper's two-sample test), we can use a form of (corrected) mean square deviation. This leads to Watson's (1962) two-sample test, which rejects H_0 for large values of

$$U^2_{n_1,n_2} = \frac{n_1 n_2}{n} \int_0^\infty \left[S_1(\theta) - S_2(\theta) - \int_0^\infty \{S_1(\xi) - S_2(\xi)\} dS(\xi) \right]^2 dS(\theta), \tag{8.3.6}$$

where S is the distribution function of the combined sample, given by

$$S(\theta) = \frac{n_1}{n} S_1(\theta) + \frac{n_2}{n} S_2(\theta).$$

This test is the two-sample analogue of Watson's one-sample U^2 test considered in Section 6.3.3, and is a modification for the circle of the two-sample Cramér–von Mises test.

To obtain a more explicit form of $U^2_{n_1,n_2}$, consider the ordered combined sample. Suppose that a_i is the number of observations from the first sample among the first i order statistics of the combined sample and b_i is the corresponding number of observations from the second sample, so that $a_i + b_i = i$. From (8.3.6) we have

$$U^2_{n_1,n_2} = \frac{n_1 n_2}{n^2} \sum_{k=1}^{n} (d_k - \bar{d})^2, \tag{8.3.7}$$

where

$$d_k = \frac{b_k}{n_2} - \frac{a_k}{n_1}, \quad \bar{d} = \sum \frac{d_k}{n}.$$

Since $U^2_{n_1,n_2}$ depends only on the relative ranks of the two samples, it is invariant under rotations.

Following Burr (1964), we now express (8.3.7) in terms of the linear ranks $r_1, \ldots, r_{n_1}$ of the first sample. This alternative form is more useful for calculations. Let s_i be the number of observations of the second sample which precede the ith observation of the first sample in the combined sample. Then $s_i = r_i - i$ and

$$d_1 = \frac{1}{n_2}, \ldots, d_{r_1} = \frac{r_1}{n_2}, \qquad d_{r_1+1} = \frac{r_1+1}{n_2} - \frac{1}{n_1}, \ldots,$$

$$d_{r_2} = \frac{r_2}{n_2} - \frac{1}{n_1}, \qquad d_{r_2+1} = \frac{r_2}{n_2} - \frac{2}{n_1}, \ldots, d_n = \frac{n_2}{n_2} - \frac{n_1}{n_1}.$$

Hence, from (8.3.7), we obtain

$$U^2_{n_1,n_2} = \frac{1}{n_1 n_2} \sum_{i=1}^{n_1} \left[(r_i - \bar{r}) - \frac{n(2_i - 1) - n_1}{2n_1} \right]^2 + \frac{n + n_1}{12 n n_1}. \tag{8.3.8}$$

By following the same procedure as for the Cramér–von Mises statistic, we have from (8.3.6), under H_0,

$$\mathrm{E}[U^2_{n_1,n_2}] = \frac{n+1}{12n}, \qquad \mathrm{var}(U^2_{n_1,n_2}) = \frac{(n_1-1)(n_2-1)(n+1)}{360 n n_1 n_2}.$$

Stephens (1965) obtained the first four moments of $U^2_{n_1,n_2}$ and fitted a Pearson curve to its null distribution. It follows from the general results of Beran (1969b) (and was shown by Watson, 1962) that, under H_0, the large-sample asymptotic distribution of $U^2_{n_1,n_2}$ is the same as that of Watson's one-sample U^2 statistic. Burr (1964) has given the exact tail probabilities of $U^2_{n_1,n_2}$ for $n \leq 17$, and some selected quantiles are given in Appendix 2.14. For $n > 17$ and $n_1 < n_2$ with n_1/n_2 not near zero, we can use the values of $U^2_{\infty,\infty}$ given in Appendix 2.14. For $n > 17$, the quantiles from the Pearson curve approximation are given in Stephens (1965).

Beran (1969b) has shown that the test is consistent against all alternatives. However, no small-sample comparisons with other tests are yet available.

Persson (1979) has shown that

$$U^2_{n_1,n_2} = \frac{1}{nn_1n_2}\left\{\frac{n_1n_2(n_1n_2+2)}{12} - T\right\},$$

where

$$T = \sum_{i<j}\sum_{k<l} I_{ijkl}$$

with

$$I_{ijkl} = \begin{cases} 1 & \text{if } \theta_{1i} \text{ and } \theta_{1j} \text{ lie on opposite sides of the chord from } \theta_{2k} \text{ to } \theta_{2l}, \\ 0 & \text{otherwise.} \end{cases}$$

Thus Watson's two-sample U^2 test can be regarded as an analogue on the circle of the Mann–Whitney test on the line.

Example 8.4
For the bird migration data in Example 8.2, we test the hypothesis that the two populations are the same.

From Example 8.2, the ranks for the control group are 1, 2, 3, 4, 5, 6, 7, 8, 9, 13, 14, 16. Then (8.3.8) gives $U^2_{12,14} = 0.320$. From Appendix 2.14, the 1% value of $U^2_{\infty,\infty}$ is 0.268. Hence H_0 is rejected (as it was in Example 8.2). For the data set consisting of the control group and the *rotated* experimental group (rotated clockwise by 90°, as in Example 7.4), this test gives the same conclusion as was reached as in Example 8.2.

A version of $U^2_{n_1,n_2}$ which is suitable for grouped data can be obtained by putting $q = 2$ in (8.4.7) given below.

8.3.4 The Runs Test

Consider the combined sample plotted on the unit circle. As in the linear case, a *run* is an uninterrupted sequence of points belonging to one of the samples. Let r be the total number of runs in the two samples on the circle. The hypothesis H_0 is rejected if r is small, since a small number of runs indicates a separation of the two samples. On the circle, the number of runs is always even, as the number of runs for the first sample is same as for the second sample.

Barton & David (1958) and David & Barton (1962, pp. 94–95, 132–136) have given a method of enumerating the null distribution of runs. Asano (1965) has tabulated the distribution of runs for $n \leq 40$ and Appendix 2.15 gives some selected quantiles for the test obtained from these tables. For $n \geq 40$, we can use the result that

$$\frac{r-\mu}{\sigma} \dot{\sim} N(0,1),$$

where

$$\mu = \frac{2n_1n_2}{n}, \qquad \sigma^2 = \frac{2n_1n_2}{n(n-1)}\left(\frac{2n_1n_2}{n} - 1\right). \tag{8.3.9}$$

The normal approximation (8.3.9) is derived as follows, by using a relationship between runs on the circle and runs on the line obtained by cutting the circle. Let $P_c(r)$ and $P_l(r)$ be the probabilities of r runs on the circle and on the line, respectively. If there are r runs on the circle then there are $r+1$ or r runs on the line, depending on whether or not the cut-point occurs within a run. Thus (Mardia, 1972a, Section 7.4.4)

$$P_c(r) = P_l(r) + P_l(r+1), \tag{8.3.10}$$

and so

$$P_c(r \leq 2h) = P_l(r \leq 2h+1). \tag{8.3.11}$$

Hence the lower tail of the null distribution for the circular case can be obtained from the corresponding linear case. The approximation (8.3.9) now follows by using the well-known normal approximation to the distribution of runs for the linear case.

This test provides a quick method of testing the equality of two populations, but we should expect it to be less powerful than the tests discussed earlier. Its consistency properties are the same as in the linear case (Rao, 1969). One test which is asymptotically more efficient than the runs test is the test which rejects equality of the two populations for large values of

$$\frac{1}{n_1}\sum_{i=1}^{n_1}(r_{i+1} - r_i - 1)^2,$$

where $r_1, \ldots, r_{n_1}$ are the ranks in the combined sample of the observations in the first sample (Rao, 1976).

Example 8.5

For the pigeon-homing data of Example 8.3, we test the hypothesis of equality of the two populations.

We order the combined sample starting from the first observation of 75°. If we write '0' for an observation from the first sample and '1' for an observation from the second sample, the combined sample reduces to

$$0,0,0,0,0,1,0,0,0,0,1,1,1,1,1,1,1,1,1.$$

(For obvious reasons, the cut-point is selected so that it is a starting point of a new run.) Hence $r = 4$. We have $n_1 = 9$ and $n_2 = 10$. From Appendix 2.15, the 5% value of r is 6 and the 1% value is 4. Hence the null hypothesis is rejected, as it was in Example 8.2.

8.3.5 Derivations of Two-Sample Tests from Tests of Uniformity

The uniform-scores test of Section 8.3.1 can be obtained by applying the Rayleigh test to the uniform scores $\beta_1, \ldots, \beta_{n_1}$ of the first sample. More generally, any test of uniformity gives rise to a corresponding two-sample test, as follows (Beran, 1969b). The test of uniformity is applied to the uniform scores $\beta_1, \ldots, \beta_{n_1}$ of the first sample. If the test of uniformity is invariant under rotation and reversal of orientation then the corresponding two-sample test is invariant under homeomorphisms (continuous transformations with continuous inverses) of the circle, and so is distribution-free.

In particular, we can obtain two-sample tests from tests of uniformity in Beran's class considered in Section 6.3.7. Replacing θ_i by β_i in (6.3.53) gives the corresponding two-sample statistic

$$B_n^* = \sum_{i=1}^{n_1}\sum_{j=1}^{n_1} h\left(\frac{2\pi[r_i - r_j]}{n}\right), \tag{8.3.12}$$

where the function h is defined by (6.3.54). An alternative form of B_n^* can be obtained from (6.3.70). A generalisation of B_n^* in which h can depend on n was considered by Schach (1967). On using the special forms of h given in Section 6.3.7 for some important tests of uniformity, we can now obtain the corresponding two-sample tests. For example, on substituting $\theta_i = \beta_i$ in the statistic R^2 used in the Rayleigh test (see Section 6.3.1), we obtain the uniform-scores test statistic R_1^2 given by (8.3.3). Using $U_i = r_i/n_1$ in Watson's U^2 statistic (6.3.33) gives

$$U^2 = \frac{n_1 + n_2 - 1}{n_1 + n_2} U^2_{n_1,n_2} + \frac{1}{12}\left(\frac{n_1 + n_2 - 1}{n_1 + n_2} - \frac{1}{n_1} - \frac{1}{n_1 + n_2}\right),$$

where $U^2_{n_1,n_2}$ is Watson's two-sample U^2 statistic given by (8.3.8). From Ajne's A_n statistic (6.3.47), we obtain the corresponding two-sample statistic

$$A_n^* = \frac{\pi n_1}{4} - \frac{4\pi}{n_1} \sum_{i<j} d\left(\frac{r_i}{n}, \frac{r_j}{n}\right), \tag{8.3.13}$$

where

$$d(x, y) = \min(|x - y|, 1 - |x - y|) \tag{8.3.14}$$

is the length of the smaller arc between x and y on a circle of unit circumference. The large-sample ($n_1, n_2 \to \infty, n_1/n$ bounded away from 0 and 1) asymptotic null distribution of

$$\frac{(n-1)B_n^*}{n_1 n_2} \tag{8.3.15}$$

is the same as the asymptotic null distribution of B_n (Beran, 1969b). The asymptotic null distribution of Schach's (1967) generalisation of B_n^* is similar (Schach, 1969b).

Consistency of the B_n^* Test

Let S_1 and S be the empirical distribution functions of the first sample and the combined sample, respectively. As in the linear case, the empirical distribution function of the transformed observations of the first sample after ranking and spacing out the combined sample is $S_1 \circ S^{-1}$, where $S_1 \circ S^{-1}(\theta) = S_1\{S^{-1}(\theta)\}$. Suppose that $n_1, n_2 \to \infty$ in such a way that

$$\lim \frac{n_1}{n} = \lambda,$$

where $0 < \lambda < 1$. Then as $n \to \infty$, $S_1 \circ S^{-1}$ tends to the distribution function G, where

$$G(\theta) = F_1(H^{-1}(\theta)), \tag{8.3.16}$$

with

$$H = \lambda F_1 + (1 - \lambda)F_2.$$

If $F_1 = F_2$ then $G(\theta) = \theta$. Hence, if B_n generates a test of uniformity consistent against the alternative G, B_n^* gives a two-sample test consistent against alternatives (F_1, F_2) which satisfy (8.3.16). Therefore, from the consistency properties of B_n given in Section 6.3.7, we deduce that Watson's two-sample U^2 test is consistent against all alternatives (F_1, F_2), whereas the A_n^* test is consistent at least against those alternatives (F_1, F_2) for which G has a symmetric unimodal density.

By the same argument, it follows that the uniform-scores test is consistent only for the alternatives for which the distribution of G has non-zero resultant length (Beran, 1969b). By varying λ, Mardia (1969b) has shown by a different method that the test is consistent when F_1 and F_2 correspond to unimodal densities.

It is expected that the B_n^* test will perform favourably against the alternatives of (6.3.64), since for these alternatives the B_n test is a locally most powerful invariant test.

8.3.6 Tests Obtained from Two-Sample Rank Tests on the Line

Distribution-free two-sample tests on the circle can also be derived from ordinary two-sample rank tests on the line. Given a rank test T on the line, the corresponding test statistic for circular data is $\max T(r_1, \ldots, r_n)$, the maximum being over all choices of origin and orientation of the circle. The circular test obtained in this way from the Mann–Whitney test was introduced by Batschelet (1965, p. 37) and studied by Eplett (1979; 1982).

8.4 MULTI-SAMPLE TESTS

Suppose that we wish to test

$$H_0 : F_1 = \ldots = F_q, \tag{8.4.1}$$

where $F_1, \ldots F_q$ are the distribution functions of q populations, against the alternative that $F_1, \ldots F_q$ are not all the same. We shall asume that $F_1, \ldots F_q$ are continuous. The data will consist of q independent random samples $\theta_{ij}, j = 1, \ldots, n_i, i = 1, \ldots, q$, of sizes $n_1, \ldots, n_q$ from the q populations. The combined sample has size $n = n_1 + \ldots + n_q$.

Any test of uniformity gives rise to a corresponding q-sample test, by the following generalisation of the construction of Section 8.3.5. If the test of uniformity is invariant under rotation and reversal of orientation then the corresponding q-sample test is invariant under homeomorphisms of the circle, and so is distribution-free.

Let $r_{ij}, j = 1, \ldots, n_i$, be the ranks of the angles in the ith sample, for $i = 1, \ldots, q$. We replace the n angles by their uniform scores

$$\beta_{ij} = \frac{2\pi r_{ij}}{n}, \quad j = 1, \ldots, n_i, \quad i = 1, \ldots, q.$$

The corresponding q-sample test rejects the null hypothesis for large values of W, where

$$W = \sum_{i=1}^{q} T_i \tag{8.4.2}$$

and $T_i = T(\beta_{i1}, \ldots \beta_{in_i})$ is the test statistic for uniformity applied to the uniform scores of the ith sample.

In particular, if the test of uniformity is the Rayleigh test then the corrresponding q-sample test is the *q-sample uniform-scores test*, which rejects the null hypothesis for large values of W, where

$$W = 2 \sum_{i=1}^{q} \frac{R_i^2}{n_i}, \tag{8.4.3}$$

in which R_i is the resultant length of those uniform scores in the combined sample which come from the ith sample, i.e.

$$R_i^2 = \left(\sum_{j=1}^{n_i} \cos\beta_{ij}\right)^2 + \left(\sum_{j=1}^{n_i} \sin\beta_{ij}\right)^2.$$

Appendix 2.16 gives some quantiles of W for $q = 3$. For other values of n and q, we can use the large-sample result (Mardia, 1970)

$$W \stackrel{.}{\sim} \chi^2_{2(q-1)}.$$

For $q = 2$, $R_1 = R_2$, and so

$$W = \frac{2nR_1^2}{n_1 n_2}.$$

Mardia (1972b) has shown that the Bahadur efficiency of the q-sample uniform-scores test relative to its parametric competitor (7.4.12) for von Mises distributions tends to 1 as $\kappa \to 0$.

Example 8.6

We test equality of the populations for the wind direction data of Example 7.6. (The data are given in Table 7.2.)

Table 8.1 shows the ranks r_{ij} for the data obtained with 0 as the starting point. The ties were broken by a process of randomisation. On using the totals C_i and S_i from Table 8.1 in (8.4.3), it is found that $W = 12.81$. We have here $q = 4$ and $n = 49$. Since $\Pr(\chi^2_6 > 12.81) = 0.046$, the null hypothesis is (just) rejected at the 5% level. The conclusion is not the same as in Example 7.6. This difference in behaviour of the two tests may be due to the influence of the ties on the q-sample uniform-scores test. However, in general, the q-sample uniform-scores test detects not only changes of location but also changes of dispersion.

Table 8.1 Calculations required for an application of the uniform-scores test to the wind direction data of Example 8.6.

Season	Ranks r_{ij}	C_i	S_i
Winter	12, 16, 27, 29, 30, 33, 37, 40, 41, 44, 45, 46	0.113	−4.743
Spring	4, 10, 13, 20, 22, 28, 31, 38, 43, 47, 48	1.528	0.195
Summer	2, 3, 5, 6, 8, 9, 11, 17, 18, 19, 21, 25, 42	1.216	6.476
Autumn	7, 14, 15, 23, 24, 26, 32, 34, 35, 36, 39, 49	−2.857	−1.928

Watson's two-sample $U^2_{n_1,n_2}$ statistic has been generalised to the multi-sample case by Maag (1966) as

$$U^2_{(q-1)} = \frac{1}{2\pi}\int_0^{2\pi} \sum_{j=1}^{q} n_j \left[S_j(x) - S(x) - \frac{1}{2\pi}\int_0^{2\pi} \{S_j(y) - S(y)\}\, dy\right]^2 dS(x), \tag{8.4.4}$$

where S_j and S denote the empirical distribution functions of the jth sample and of the combined sample, respectively. Carrying out the integration in (8.4.4) gives the explicit formula

$$U^2_{(q-1)} = \sum_{j=1}^{q} \frac{n_j}{n} \left(\sum_{i=1}^{n} \{S_j(x_i) - S(x_i)\}^2 - \frac{1}{n}\left[\sum_{i=1}^{n} \{S_j(x_i) - S(x_i)\}\right]^2 \right). \tag{8.4.5}$$

The large-sample asymptotic null distribution of $U^2_{(q-1)}$ is given in Maag (1966). Its quantiles for small samples have not yet been tabulated.

For grouped data, the following variant of (8.4.4) was given by Brown (1994). Suppose that the data are grouped into m cells. Let O_{ji} be the number of observations from the jth sample in the ith cell. Put

$$T_i = O_{1i} + \ldots + O_{qi}, \qquad p_i = \frac{n_i}{n}, \qquad E_{ji} = \frac{n_j T_i}{n}.$$

Then T_i and E_{ji} are the total number of observations in the ith cell and the expected number of observations from the jth sample in the ith cell, respectively. Define

$$Y_{ji} = \sum_{k=1}^{i-1} (O_{jk} - E_{jk}) + \tfrac{1}{2}(O_{ji} - E_{ji}). \tag{8.4.6}$$

Then Brown's grouped version of $U^2_{(q-1)}$ is

$$\begin{aligned} U^2_{(q-1),d} &= (q-1)\sum_{i=1}^{m} \frac{p_j^2}{6}\left(1 - \frac{p_j}{2}\right) + \frac{1}{12}\sum_{j=1}^{q}\frac{1}{n_j}\sum_{i=1}^{m}(O_{ji} - E_{ji})^2 \\ &\quad + \sum_{j=1}^{q}\frac{1}{n_j}\left\{\sum_{i=1}^{m} p_i Y_{ji}^2 - \left(\sum_{i=1}^{m} p_i Y_{ji}\right)\right\} \\ &\quad - n\left\{\sum_{i=1}^{m}\left(\sum_{j=1}^{q} Y_{ji}\right)^2 - \left(\sum_{i=1}^{m}\sum_{j=1}^{q} Y_{ji}\right)^2\right\}. \end{aligned} \tag{8.4.7}$$

The runs test can easily be extended to the multi-sample case. For $q = 3$ and 4, quantiles of the runs test statistic can be obtained on using the enumeration of the null distribution of runs given by Barton & David (1958).

9

Distributions on Spheres

9.1 SPHERICAL DATA

There are various practical situations in which observations are made on directions in three dimensions. Examples of such data include:

(i) directions of palaeomagnetism in rock,
(ii) directions from the earth to stars,
(iii) directions of optical axes in quartz crystals.

In many cases, as in (i) and (ii) above, the directions arise from ordinary multivariate (vectorial) data when the magnitudes of the observed vectors are unknown or irrelevant. Sometimes, as in (iii) above, the direction is determined only up to sign, i.e. the observation is of an *axis.* Data which are directions may be represented as points on a sphere. Data which are axes may be represented as antipodal pairs of points on a sphere. In either case, the observations are of spherical data. In order to handle such data, the techniques of spherical statistics are required.

Much of the theory of spherical statistics is analogous to that for circular statistics. Further, one can consider directions in p dimensions, i.e. unit vectors in p-dimensional Euclidean space $\mathbb{R}^p$. In this chapter and the next we shall consider the general p-dimensional case because

(i) this gives a unified treatment of the circular ($p = 2$) and spherical ($p = 3$) cases,
(ii) from many points of view the general p-dimensional case is just as easy as the three-dimensional case.

Furthermore, one of the important tools of the theory of directional statistics can be considered as embedding the sample space S^{p-1} into an infinite-dimensional sphere (see Section 10.8). Of course, most applications are to two or three dimensions. However,

(i) because S^3 is related to the group $SO(3)$ of rotations of three-dimensional space (see (13.2.2) below), the case $p = 4$ provides a way of dealing with rotational data (see Section 13.2 below),

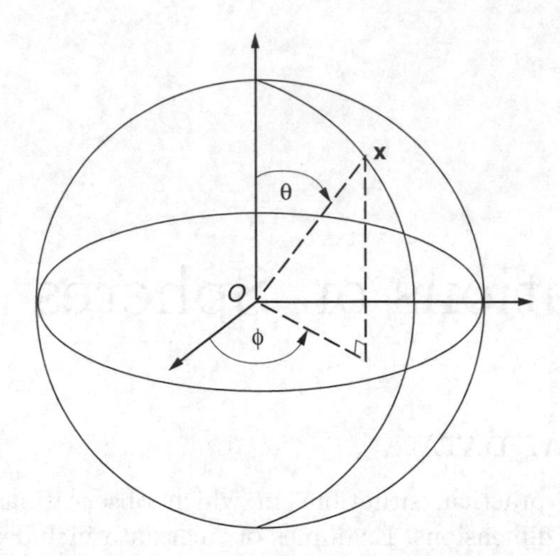

Figure 9.1 Spherical polar coordinates: θ = colatitude, ϕ = longitude.

(ii) compositional data on the $(p-1)$-simplex, i.e. observations of the form

$$(y_1, \ldots, y_p), \qquad y_i \geq 0, \quad i = 1, \ldots, p, \qquad y_1 + \ldots + y_p = 1,$$

can be transformed to $(p-1)$-dimensional spherical data by

$$(y_1, \ldots, y_p) \mapsto (\sqrt{y}_1, \ldots, \sqrt{y}_p).$$

Directions in p dimensions can be represented as unit vectors $\mathbf{x}$, i.e. as points on $S^{p-1} = \{\mathbf{x} : \mathbf{x}^T\mathbf{x} = 1\}$, the $(p-1)$-dimensional sphere with unit radius and centre at the origin.

When $p = 3$ we can use spherical polar coordinates (θ, ϕ) defined by

$$\mathbf{x} = (\cos\theta, \sin\theta\cos\phi, \sin\theta\sin\phi)^T,$$

which are indicated in Fig. 9.1. In geographical terminology, θ denotes colatitude (so that $\pi/2 - \theta$ denotes latitude) and ϕ denotes longitude. If $\mathbf{x}$ is a random unit vector which is distributed uniformly on S^2 (see Section 9.3.1) then (θ, ϕ) has probability density

$$\frac{1}{4\pi}\sin\theta, \quad 0 \leq \theta \leq \pi, \ 0 \leq \phi \leq 2\pi.$$

In order to visualise data on the unit sphere in three-dimensional space $\mathbb{R}^3$, it is useful to project it onto a plane using an equal-area projection. For many purposes, the most useful such projection is *Lambert's equal-area projection*

$$(\cos\theta, \sin\theta\cos\phi, \sin\theta\sin\phi)^T \mapsto 2\sin\left(\frac{\theta}{2}\right)(\cos\phi, \sin\phi)^T, \qquad (9.1.1)$$

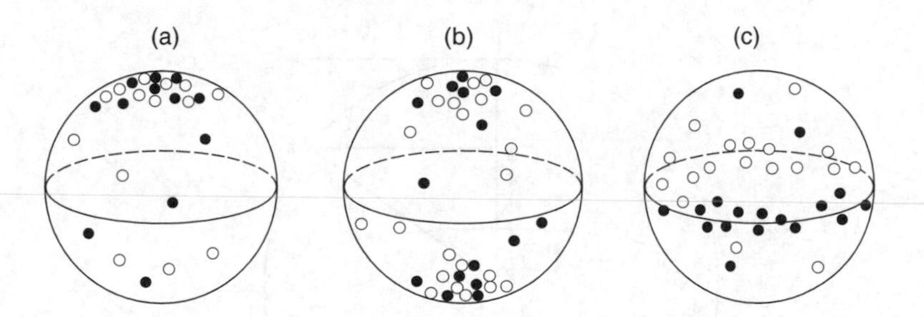

Figure 9.2 Configurations of observations on the sphere from (a) unimodal, (b) bimodal, (c) girdle distributions. An open circle indicates an observation on the far side of the sphere.

which maps the unit sphere in $\mathbb{R}^3$ to the disc of radius 2. In the geological literature, Lambert's equal-area projection is known as the *Schmidt projection* and the image in the disc of a net of curves of constant latitude or longitude is known as a *Schmidt net.* Calculation shows that if

$$(y_1, y_2) = 2 \sin \left(\frac{\theta}{2} \right) (\cos \phi, \sin \phi)$$

then $dy_1 \, dy_2 = \sin \theta d\theta d\phi$, and so (9.1.1) is area-preserving, i.e. it sends the uniform distribution on the unit sphere to the uniform distribution on the disc of radius 2. Detailed reference books on map projections include Maling (1992) and Richardus & Adler (1972).

The projection (9.1.1) has the minor disadvantage that it identifies the points with spherical polar coordinates (θ, ϕ) and $(\pi - \theta, \phi)$. If the data are spread over both the upper and lower hemispheres then it is helpful to use Lambert's equal-area projection to project the two hemispheres onto separate discs in the plane.

Some common configurations of spherical data are shown in Fig. 9.2.

A key idea in distribution theory and data analysis on spheres is the 'tangent-normal' decomposition: any unit vector $\mathbf{x}$ can be decomposed as

$$\mathbf{x} = t\boldsymbol{\mu} + (1 - t^2)^{\frac{1}{2}} \boldsymbol{\xi} \tag{9.1.2}$$

with $\boldsymbol{\xi}$ a unit tangent to S^{p-1} at $\boldsymbol{\mu}$. Figure 9.3 shows this decomposition.

From the theoretical point of view, there are three basic approaches to directional statistics, which may be termed the embedding, wrapping and intrinsic approaches:

(a) in the *embedding approach*, the sphere S^{p-1} is regarded as a subset of $\mathbb{R}^p$;

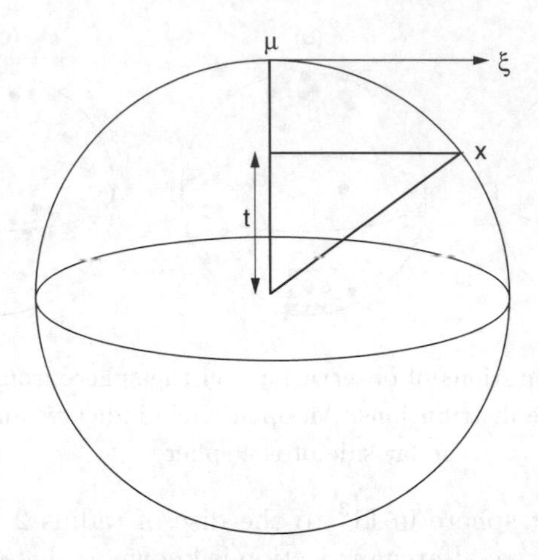

Figure 9.3 Tangent-normal decomposition $\mathbf{x} = t\boldsymbol{\mu} + (1 - t^2)^{\frac{1}{2}}\boldsymbol{\xi}$.

(b) in the *wrapping approach*, tangent vectors $\mathbf{x}$ to the sphere at $\boldsymbol{\mu}$ are wrapped onto the sphere by

$$\mathbf{x} \mapsto (\sin \|\mathbf{x}\|)\boldsymbol{\mu} + (\cos \|\mathbf{x}\|)\mathbf{x}, \tag{9.1.3}$$

where $\mathbf{x}^T\boldsymbol{\mu} = 0$ (in the circular case, $p = 2$ and (9.1.3) is essentially addition mod 2π);

(c) in the *intrinsic approach* the sphere is regarded just as a manifold in its own right without reference to any embedding.

We have already used these approaches in the circular case, for example:

(a) the embedding approach was used in the construction of projected normal distributions in Section 3.5.6;
(b) the wrapping approach was used in the construction of wrapped distributions in Section 3.5.7;
(c) the intrinsic approach was used when obtaining von Mises distributions from diffusions on the circle (Section 3.5.4).

Further examples of these approaches are as follows:

(a) the embedding approach is used in the construction in Section 9.2 of descriptive measures for spherical data;
(b) the wrapping approach is used in Section 9.3.3 to construct the Brownian motion distributions;
(c) the intrinsic approach is useful for constructing splines on spheres (Section 12.8.1) and for inference on rotations (Section 13.2).

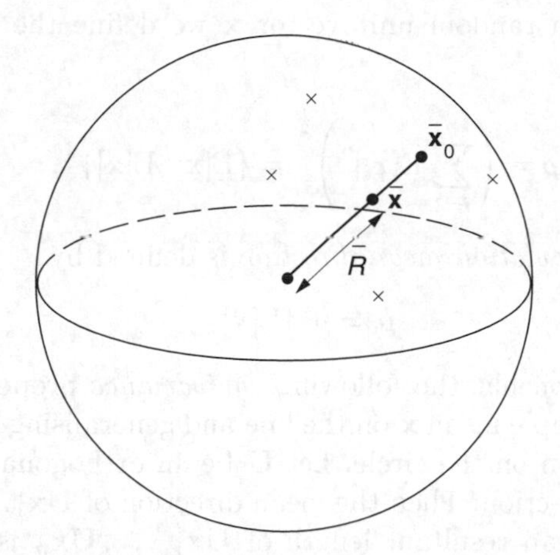

Figure 9.4 Descriptive statistics of spherical data: $\bar{\mathbf{x}}_0$ = mean direction, $\bar{R}$ = mean resultant length.

9.2 DESCRIPTIVE MEASURES

9.2.1 The Mean Direction and the Mean Resultant Length

Let $\mathbf{x}_1, \ldots, \mathbf{x}_n$ be points on S^{p-1}. Then the location of these points can be summarised by their sample mean vector in $\mathbb{R}^p$, which is

$$\bar{\mathbf{x}} = \frac{1}{n}\sum_{i=1}^{n} \mathbf{x}_i. \tag{9.2.1}$$

As in the circular case, it is useful to express the vector $\bar{\mathbf{x}}$ in polar form as

$$\bar{\mathbf{x}} = \bar{R}\bar{\mathbf{x}}_0, \tag{9.2.2}$$

where $\bar{\mathbf{x}}_0$ is a unit vector and $\bar{R} \geq 0$, so that

$$\bar{R} = \|\bar{\mathbf{x}}\|$$

and

$$\bar{\mathbf{x}}_0 = \|\bar{\mathbf{x}}\|^{-1}\bar{\mathbf{x}}.$$

The unit vector $\bar{\mathbf{x}}_0$ is called the *mean direction* of the sample and $\bar{R}$ is called the *mean resultant length*. If the points $\mathbf{x}_1, \ldots, \mathbf{x}_n$ are considered as having equal mass then their centre of mass is $\bar{\mathbf{x}}$, which has direction $\mathbf{x}_0$ and distance $\bar{R}$ from the origin. Figure 9.4 shows these quantities. In the case $p = 2$, (9.2.2) is equivalent to (2.2.2)–(2.2.3) and $\bar{\mathbf{x}}_0$ is the unit vector with angle $\bar{\theta}$.

Analogously, for a random unit vector $\mathbf{x}$ we define the *population mean resultant length* ρ by

$$\rho = \left(\sum_{i=1}^{p} \mathrm{E}[x_i]^2\right)^{\frac{1}{2}} = \{\mathrm{E}[\mathbf{x}]^T \mathrm{E}[\mathbf{x}]\}^{\frac{1}{2}}.$$

When $\rho > 0$, the *population mean direction* is defined by

$$\boldsymbol{\mu} = \rho^{-1}\mathrm{E}[\mathbf{x}].$$

The mean direction has the following *equivariance* property analogous to that of the usual sample mean $\bar{\mathbf{x}}$ on the line and generalising property (2.2.11) of the mean direction on the circle. Let $\mathbf{U}$ be an orthogonal transformation, i.e. a rotation or reflection. Then the mean direction of $\mathbf{U}\mathbf{x}_1, \ldots, \mathbf{U}\mathbf{x}_n$ is $\mathbf{U}\bar{\mathbf{x}}_0$. By contrast, the mean resultant length of $\mathbf{U}\mathbf{x}_1, \ldots, \mathbf{U}\mathbf{x}_n$ is $\bar{R}$, so that $\bar{R}$ is invariant under rotation and reflection.

The mean resultant length has the following minimisation property, which was given for the circle in Section 2.3.2. For $\mathbf{a}$ in S^{p-1}, let $S(\mathbf{a})$ be the arithmetic mean of the squared Euclidean distances between $\mathbf{x}_i$ and $\mathbf{a}$. Then

$$\begin{aligned} S(\mathbf{a}) &= \frac{1}{n}\sum_{i=1}^{n} \|\mathbf{x}_i - \mathbf{a}\|^2 \\ &= 2(1 - \bar{\mathbf{x}}^T \mathbf{a}) \\ &= \bar{R}^2 - 2\bar{R}\bar{\mathbf{x}}_0'\mathbf{a} + 1. \end{aligned} \tag{9.2.3}$$

From the middle expression in (9.2.3), it follows that $S(\mathbf{a})$ is minimised (subject to the constraint $\mathbf{a}^T\mathbf{a} = 1$) when $\mathbf{a} = \bar{\mathbf{x}}_0$ and that

$$\min_{\mathbf{a}} S(\mathbf{a}) = 2(1 - \bar{R}). \tag{9.2.4}$$

Because of (9.2.4) and (9.2.8) below, the quantity $2(1-\bar{R})$ is sometimes called the sample *spherical variance.* However, in view of (9.2.11) below, the total variation $1 - \bar{R}^2$ may be a more appropriate analogue of the sample variance. Note that $\bar{R} \simeq 0$ when $\mathbf{x}_1, \ldots, \mathbf{x}_n$ are widely dispersed and that $\bar{R} \simeq 1$ when $\mathbf{x}_1, \ldots, \mathbf{x}_n$ are heavily concentrated. Thus $\bar{R}$ measures clustering around the mean direction.

Let $\boldsymbol{\mu}$ be any unit vector and put

$$\bar{C} = \frac{1}{n}\sum_{i=1}^{n} \mathbf{x}_i^T \boldsymbol{\mu}, \tag{9.2.5}$$

so that $\bar{C}$ is the sample mean of the components of $\mathbf{x}_1, \ldots, \mathbf{x}_n$ along $\boldsymbol{\mu}$. Since

$$2n(1 - \bar{C}) = \sum_{i=1}^{n} \|\mathbf{x}_i - \boldsymbol{\mu}\|^2,$$

$2n(1-\bar{C})$ can be regarded as the total variation of $\mathbf{x}_1, \ldots, \mathbf{x}_n$ about $\boldsymbol{\mu}$. As in the circular case (see (4.8.30)), this total variation can be decomposed as

$$2n(1-\bar{C}) = 2n(1-\bar{R}) + 2n(\bar{R}-\bar{C}). \tag{9.2.6}$$

To relate (9.2.6) to a more familiar decomposition, note that (by the tangent-normal decomposition (9.1.2)) $\mathbf{x}_i$ may be written as

$$\mathbf{x}_i = (1 - \|\mathbf{y}_i\|^2)^{\frac{1}{2}} \boldsymbol{\mu} + \mathbf{y}_i,$$

with $\mathbf{y}_i$ normal to $\boldsymbol{\mu}$. If $\mathbf{x}_1, \ldots, \mathbf{x}_n$ are close to $\boldsymbol{\mu}$ (as happens, for example, if $\mathbf{x}_1, \ldots, \mathbf{x}_n$ are a sample from a concentrated distribution with mean direction $\boldsymbol{\mu}$) then $\|\mathbf{y}_i\|$ is small and a little algebra yields

$$2n(1-\bar{C}) = \sum_{i=1}^{n} \|\mathbf{y}_i\|^2 + O(\|\mathbf{y}_i\|^4), \tag{9.2.7}$$

$$2n(1-\bar{R}) = \sum_{i=1}^{n} \|\mathbf{y}_i - \bar{\mathbf{y}}\|^2 + O(\|\mathbf{y}_i\|^4), \tag{9.2.8}$$

$$2n(\bar{R}-\bar{C}) = n\|\bar{\mathbf{y}}\|^2 + O\left(\sum_{i=1}^{n} \|\mathbf{y}_i\|^4\right). \tag{9.2.9}$$

Thus the decomposition (9.2.6) is approximately the tangent-space part of the familiar 'Euclidean' decomposition

$$\sum_{i=1}^{n} \|\mathbf{x}_i - \boldsymbol{\mu}\|^2 = \sum_{i=1}^{n} \|\mathbf{x}_i - \bar{\mathbf{x}}\|^2 + n\|\bar{\mathbf{x}} - \boldsymbol{\mu}\|^2.$$

In (9.2.6) the component $2n(1-\bar{R})$ is an approximate measure of the variation of $\mathbf{x}_1, \ldots, \mathbf{x}_n$ about $\bar{\mathbf{x}}$, whereas $2n(\bar{R}-\bar{C})$ is an approximate measure of the deviation of $\bar{\mathbf{x}}$ from $\boldsymbol{\mu}$.

9.2.2 The Moment of Inertia

An important measure of dispersion is the scatter matrix $\bar{\mathbf{T}}$ about the origin, defined by

$$\bar{\mathbf{T}} = \frac{1}{n} \sum_{i=1}^{n} \mathbf{x}_i \mathbf{x}_i^T. \tag{9.2.10}$$

Readers familiar with mechanics may find it helpful to interpret $\bar{\mathbf{T}}$ as the inertia tensor about the origin of particles of weight n^{-1} at each of the points $\mathbf{x}_1, \ldots, \mathbf{x}_n$. The use of $\bar{\mathbf{T}}$ in exploratory data analysis will be considered in Section 10.2.

Let $\mathbf{S}$ denote the sample variance matrix given by

$$\mathbf{S} = \frac{1}{n-1}\sum_{i=1}^{n}(\mathbf{x}_i - \bar{\mathbf{x}}_0)(\mathbf{x}_i - \bar{\mathbf{x}}_0)^T.$$

Then

$$\bar{\mathbf{T}} = \frac{n-1}{n}\mathbf{S} + \bar{\mathbf{x}}_0\bar{\mathbf{x}}_0^T.$$

Note that the restriction $\mathbf{x}_i^T\mathbf{x}_i = 1$ leads to

$$\operatorname{tr}\bar{\mathbf{T}} = 1$$

and so

$$\frac{n-1}{n}\operatorname{tr}\mathbf{S} + \bar{R}^2 = 1, \qquad (9.2.11)$$

generalising (4.7.2).

Similarly, the restriction that a random vector $\mathbf{x}$ lies on S^{p-1} leads to the connection

$$\operatorname{tr}\mathbf{\Sigma} + \rho^2 = 1 \qquad (9.2.12)$$

between ρ and the variance matrix $\mathbf{\Sigma}$ of $\mathbf{x}$. Thus the singular nature of the sample space for directional data means that (in contrast to the usual unrestricted multivariate case) there is a connection between the mean $\mathrm{E}[\mathbf{x}]$ and the variance matrix $\mathbf{\Sigma}$ of the random vector $\mathbf{x}$. Since $\mathbf{S}$ is an unbiased estimator of $\mathbf{\Sigma}$, taking the expectation of (9.2.11) yields

$$\mathrm{E}[\bar{R}^2] = \rho^2 + \frac{1}{n}(1-\rho^2), \qquad (9.2.13)$$

as in (4.7.3).

Example 9.1

The directions of remnant magnetisation in nine specimens of Icelandic lava flows of 1947–1948 considered by R. A. Fisher (1953) are given in Table 9.1.

Table 9.1 Directions of magnetised lava flows in 1947–1948 in Iceland. (Hospers's data cited in R. A. Fisher, 1953).

North	West	Up
0.388	0.117	−0.914
0.171	−0.321	−0.932
0.272	−0.204	−0.940
0.123	−0.062	−0.991
0.182	0.003	−0.983
0.291	−0.029	−0.956
0.225	−0.272	−0.935
0.518	0.022	−0.855
0.449	−0.433	−0.782

Calculation shows that the sample mean vector of these nine directions is

$$\bar{\mathbf{x}} = (0.291, -0.131, -0.921)^T,$$

and so the sample mean direction is

$$\bar{\mathbf{x}}_0 = (0.299, -0.135, -0.945)^T$$

and the sample mean resultant length is $\bar{R} = 0.975$. Since $\bar{R} \simeq 1$, the sample is highly concentrated around its mean direction. The scatter matrix $\bar{\mathbf{T}}$ is

$$\bar{\mathbf{T}} = \begin{pmatrix} 0.101 & -0.036 & -0.261 \\ -0.036 & 0.047 & 0.116 \\ -0.261 & 0.116 & 0.852 \end{pmatrix}.$$

The eigenvalues $\bar{t}_1, \bar{t}_2, \bar{t}_3$ of $\bar{\mathbf{T}}$ are 0.950, 0.031 and 0.019. The facts that $\bar{t}_1 \simeq 1$ and $\bar{R} \simeq 1$ suggest that the data come from a concentrated unimodal distribution (see Table 10.1 below).

9.3 MODELS FOR SPHERICAL DATA

9.3.1 *The Uniform Distribution*

The most basic probability distribution on S^{p-1} is the *uniform distribution*, in which the probability of a set is proportional to its $(p-1)$-dimensional area. The uniform distribution is the unique distribution which is invariant under rotation and reflection. It follows that $\mathrm{E}[\mathbf{x}] = \mathbf{0}$ for this distribution, and so $\rho = 0$ and the mean direction $\boldsymbol{\mu}$ is not defined.

Note that the intersection of S^{p-1} with the hyperplane through $t\boldsymbol{\mu}$ and normal to $\boldsymbol{\mu}$ is a $(p-2)$-dimensional sphere of radius $\sqrt{1-t^2}$. It follows that the density of t is

$$B\left(\frac{p-1}{2}, \frac{1}{2}\right)^{-1} (1-t^2)^{\frac{(p-3)}{2}}, \qquad -1 \leq t \leq 1. \tag{9.3.1}$$

Some moments of the uniform distribution are considered in Section 9.6.1.

9.3.2 *Von Mises–Fisher Distributions*

From (3.5.17), the log-density of the von Mises distribution $M(\mu, \kappa)$ can be written as

$$\log f(\theta; \mu, \kappa) = \kappa \cos(\theta - \mu) - \log I_0(\kappa) - \log 2\pi = \kappa \boldsymbol{\mu}^T \mathbf{x} - \log I_0(\kappa) - \log 2\pi, \tag{9.3.2}$$

where

$$\mathbf{x} = (\cos\theta, \sin\theta)^T, \qquad \boldsymbol{\mu} = (\cos\mu, \sin\mu)^T.$$

Thus an appropriate generalisation to S^{p-1} consists of the distributions with log-densities which are linear in $\mathbf{x}$, i.e. which have densities $f(\cdot;\boldsymbol{\mu},\kappa)$ satisfying

$$\log f(\mathbf{x};\boldsymbol{\mu},\kappa) = \kappa\boldsymbol{\mu}^T\mathbf{x} + \text{constant}. \tag{9.3.3}$$

These are the von Mises–Fisher distributions.

Definition

A unit random vector $\mathbf{x}$ has the $(p-1)$-dimensional *von Mises–Fisher* (or *Langevin*, in G. S. Watson's terminology) distribution $M_p(\boldsymbol{\mu},\kappa)$ if its probability density function with respect to the uniform distribution is

$$f(\mathbf{x};\boldsymbol{\mu},\kappa) = \left(\frac{\kappa}{2}\right)^{p/2-1} \frac{1}{\Gamma(p/2)I_{p/2-1}(\kappa)} \exp\{\kappa\boldsymbol{\mu}^T\mathbf{x}\}, \tag{9.3.4}$$

where $\kappa \geq 0$, $\|\boldsymbol{\mu}\| = 1$, and I_ν denotes the modified Bessel function of the first kind and order ν, defined in (3.5.27) and (A.1) of Appendix 1. For reasons given below, the parameters $\boldsymbol{\mu}$ and κ are called the *mean direction* and the *concentration parameter*, respectively.

In the case $p = 3$, the von Mises–Fisher distributions are called Fisher distributions, because R. A. Fisher (1953) studied them in detail. We shall write $F(\boldsymbol{\mu},\kappa)$ for $M_3(\boldsymbol{\mu},\kappa)$. In this case, the normalising constant simplifies and the density of the Fisher distribution $F(\boldsymbol{\mu},\kappa)$ with respect to the uniform distribution is

$$\frac{\kappa}{2\sinh\kappa}\exp\{\kappa\boldsymbol{\mu}^T\mathbf{x}\}.$$

If we write $\mathbf{x}$ and $\boldsymbol{\mu}$ in spherical polar coordinates as

$$\begin{aligned} \mathbf{x} &= (\cos\theta, \sin\theta\cos\phi, \sin\theta\sin\phi)^T, \\ \boldsymbol{\mu} &= (\cos\alpha, \sin\alpha\cos\beta, \sin\alpha\sin\beta)^T, \end{aligned}$$

then the density of (θ,ϕ) is

$$\frac{\kappa}{4\pi\sinh\kappa}\exp\{\kappa[\cos\theta\cos\alpha + \sin\theta\sin\alpha\cos(\phi-\beta)]\}\sin\theta.$$

To verify that the density (9.3.4) integrates to 1, note that the intersection of S^{p-1} with the hyperplane through $t\boldsymbol{\mu}$ and normal to $\boldsymbol{\mu}$ is a $(p-2)$-dimensional sphere of radius $\sqrt{1-t^2}$. Since I_ν satisfies

$$I_\nu(\kappa) = \frac{(\kappa/2)^\nu}{\Gamma(\nu+\frac{1}{2})\Gamma(\frac{1}{2})}\int_{-1}^{1} e^{\kappa t}(1-t^2)^{\nu-\frac{1}{2}}\,dt \tag{9.3.5}$$

(see (A.3) of Appendix 1), it follows that

$$\begin{aligned} \int_{S^{p-1}} \exp\{\kappa\boldsymbol{\mu}^T\mathbf{x}\}d\mathbf{x} &= B\left(\frac{p-1}{2},\frac{1}{2}\right)^{-1}\int_{-1}^{1} e^{\kappa t}(1-t^2)^{(p-3)/2}dt \\ &= \Gamma\left(\frac{p}{2}\right)\left(\frac{\kappa}{2}\right)^{1-p/2} I_{p/2-1}(\kappa), \end{aligned} \tag{9.3.6}$$

as required.

Mean Direction

Because (9.3.4) is symmetrical about $\boldsymbol{\mu}$, the mean direction of $\mathbf{x}$ is $\boldsymbol{\mu}$. Also,

$$\mathrm{E}[\mathbf{x}] = \rho\boldsymbol{\mu}, \tag{9.3.7}$$

where

$$\begin{aligned}\rho = A_p(\kappa) &= \frac{\int_{-1}^{1} t^2 e^{\kappa t}(1-t^2)^{(p-3)/2}dt}{\int_{-1}^{1} e^{\kappa t}(1-t^2)^{(p-3)/2}dt} \\ &= \frac{I_{p/2}(\kappa)}{I_{p/2-1}(\kappa)}.\end{aligned} \tag{9.3.8}$$

In the case $p = 3$, $A_p(\kappa)$ has the simple form

$$A_3(\kappa) = \coth\kappa - \frac{1}{\kappa}. \tag{9.3.9}$$

As $\mathbf{x}$ runs through S^{p-1}, $\boldsymbol{\mu}^T\mathbf{x}$ is maximised at $\boldsymbol{\mu}$ and minimised at $-\boldsymbol{\mu}$. Thus, (provided that $\kappa > 0$) the density (9.3.4) has a mode at $\boldsymbol{\mu}$ and an antimode at $-\boldsymbol{\mu}$.

Transformation Property

If $\mathbf{U}$ is an orthogonal transformation then the density (9.3.4) satisfies

$$f(\mathbf{Ux}; \mathbf{U}\boldsymbol{\mu}, \kappa) = f(\mathbf{x}; \boldsymbol{\mu}, \kappa),$$

so that if $\mathbf{x}$ is distributed as $M_p(\boldsymbol{\mu}, \kappa)$ then $\mathbf{Ux}$ is distributed as $M_p(\mathbf{U}\boldsymbol{\mu}, \kappa)$. Thus for fixed κ, the set of $M_p(\boldsymbol{\mu}, \kappa)$ distributions is a transformation model (see Section 3.5.1) under the group $O(p)$ of orthogonal transformations of $\mathbb{R}^p$.

Shape of the Distribution

For $\kappa > 0$, the distribution has a mode at the mean direction $\boldsymbol{\mu}$, whereas when $\kappa = 0$ the distribution is uniform. The larger the value of κ, the greater is the clustering around the mean direction. This behaviour explains why κ is called the concentration parameter. Since the density (9.3.4) depends on $\mathbf{x}$ only through $\mathbf{x}^T\boldsymbol{\mu}$, the Fisher distribution $M_p(\boldsymbol{\mu}, \kappa)$ is rotationally symmetric about $\boldsymbol{\mu}$. Further, in the tangent-normal decomposition

$$\mathbf{x} = t\boldsymbol{\mu} + (1-t^2)^{\frac{1}{2}}\boldsymbol{\xi},$$

t is invariant under rotation about $\boldsymbol{\mu}$ while $\boldsymbol{\xi}$ is equivariant (any such rotation $\mathbf{U}$ takes $\boldsymbol{\xi}$ to $\mathbf{U}\boldsymbol{\xi}$). Thus the conditional distribution of $\boldsymbol{\xi}|t$ is uniform on S^{p-2}. It follows that

$$\boldsymbol{\xi} \text{ and } t \text{ are independent,} \tag{9.3.10}$$

and

$$\boldsymbol{\xi} \text{ is uniform on } S^{p-2}. \tag{9.3.11}$$

Further, the calculation leading to (9.3.6) shows that the marginal density of t is

$$\left(\frac{\kappa}{2}\right)^{p/2-1}\left\{\Gamma\left(\frac{p-1}{2}\right)\Gamma\left(\frac{1}{2}\right)I_{(p-1)/2}(\kappa)\right\}^{-1} e^{\kappa t}(1-t^2)^{(p-3)/2} \qquad (9.3.12)$$

on $[-1, 1]$.

In particular, when $p = 3$, if we use spherical polar coordinates (θ, ϕ) with $\boldsymbol{\mu}$ as the pole then θ and ϕ are independent, the probability density function of θ is

$$f(\theta; \kappa) = \frac{\kappa}{2\sinh\kappa} e^{\kappa\cos\theta} \sin\theta \qquad (9.3.13)$$

on $[0, \pi]$ and ϕ is uniform on the unit circle with probability density function

$$h(\phi) = \frac{1}{2\pi}. \qquad (9.3.14)$$

Figure 9.5 shows the density (9.3.13) for some selected values of κ. It illustrates the way in which the probability mass becomes more concentrated about $\theta = 0$ as κ increases. Selected quantiles of θ are given in Appendix 3.1.

History

Fisher distributions (with $p = 3$) first appeared in statistical mechanics in the following context (Langevin, 1905). Consider a collection of weakly interacting dipoles of moments $\mathbf{m}$ which are subject to an external electric field $\mathbf{H}$. The energies U of the dipoles have a Maxwell–Boltzmann distribution with probability density function

$$\frac{1}{kT}\exp\left(-\frac{U}{kT}\right),$$

where k is Boltzmann's constant and T is the absolute temperature. Since

$$U = -\mathbf{H}^T\mathbf{m},$$

the directions $\|\mathbf{m}\|^{-1}\mathbf{m}$ of the dipoles have a Fisher distribution. Arnold (1941) considered maximum likelihood estimation in Fisher distributions and gave a corresponding characterisation. Kuhn & Grün (1942) found Fisher distributions as approximate solutions to a problem associated with paths and chains of random segments in three dimensions. Significant advances in statistical applications were made by R. A. Fisher (1953), who used these distributions to investigate certain statistical problems in palaeomagnetism. The extension to $p > 3$ is due to Watson & Williams (1956).

Characterisations

The characterisations of the von Mises distribution given in Section 3.5.4 extend to von Mises–Fisher distributions on spheres of arbitrary dimensions.

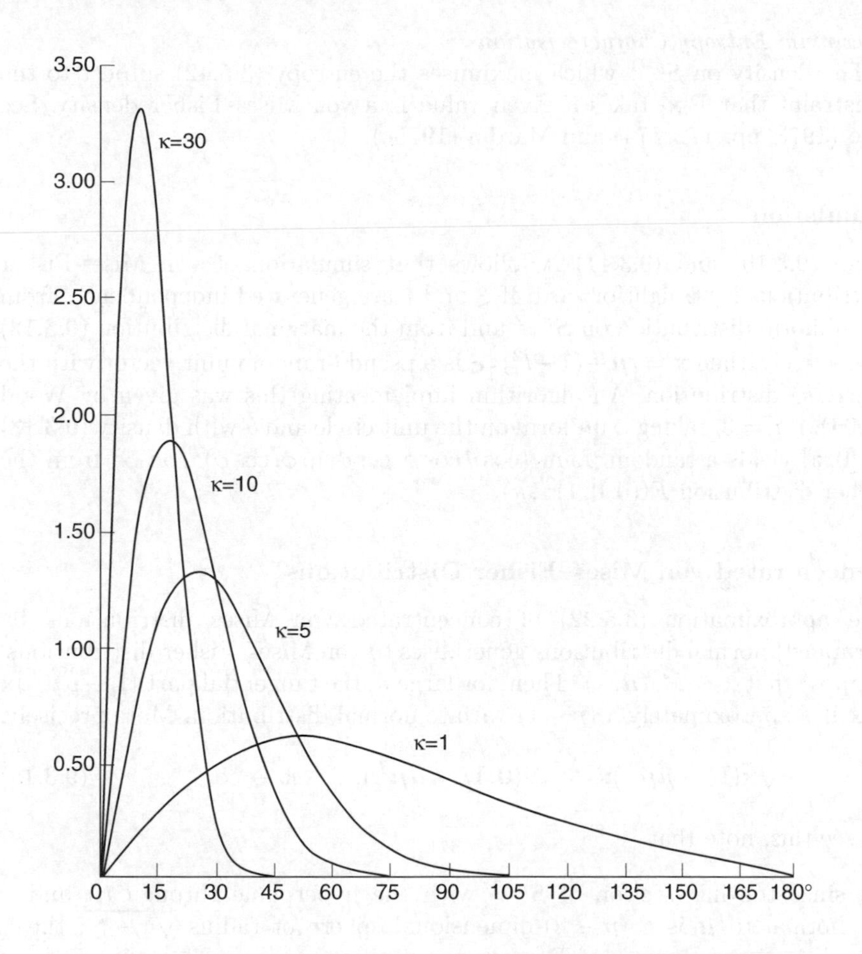

Figure 9.5 Density of colatitude θ for the Fisher distribution for $p = 3$ and $\kappa = 1, 5, 10, 30$.

Maximum Likelihood Characterisation

Let $f(\mathbf{x}; \boldsymbol{\mu})$ be a probability density function on S^{p-1} with mean direction $\boldsymbol{\mu}$ and $\rho > 0$. If

(i) for all random samples $\bar{\mathbf{x}}_0$ is a maximum likelihood estimate of $\boldsymbol{\mu}$,
(ii) $f(\mathbf{x}; \boldsymbol{\mu}) = g(\mathbf{x}^T \boldsymbol{\mu})$ for all $\mathbf{x}$ in S^{p-1}, where the function g is lower semi-continuous from the left at 1,

then $f(\mathbf{x}; \boldsymbol{\mu})$ is a von Mises–Fisher density.

In the case $p = 3$ this result is due to Arnold (1941) and was proved by Breitenberger (1963) by a simpler method. The general result was proved by Bingham & Mardia (1975).

Maximum Entropy Characterisation

The density on S^{p-1} which maximises the entropy (3.5.42) subject to the constraint that E[$\mathbf{x}$] takes a given value is a von Mises–Fisher density. See Rao (1973, pp. 172–174) and Mardia (1975c).

Simulation

From (9.3.10) and (9.3.11) it follows that simulation of von Mises-Fisher distributions is straightforward. If $\boldsymbol{\xi}$ and t are generated independently from the uniform distribution on S^{p-2} and from the marginal distribution (9.3.12) respectively, then $\mathbf{x} = t\boldsymbol{\mu} + (1-t^2)^{\frac{1}{2}}\boldsymbol{\xi}$ is a pseudo-random unit vector with the $M_p(\boldsymbol{\mu}, \kappa)$ distribution. An algorithm implementing this was given by Wood (1994). If $p = 3$, taking ϕ uniform on the unit circle and θ with density (9.3.13) on $[0, \pi]$ yields a random point $(\cos\theta\cos\phi, \cos\theta\sin\phi, \cos\phi)^T$ on S^2 from the Fisher distribution $F((0,0,1)^T, \kappa)$.

Concentrated von Mises–Fisher Distributions

The approximation (3.5.22) of concentrated von Mises distributions by (wrapped) normal distributions generalises to von Mises–Fisher distributions. Suppose that $\mathbf{x} \sim M_p(\boldsymbol{\mu}, \kappa)$. Then, for large κ, the tangential part $(\mathbf{I}_p - \boldsymbol{\mu}\boldsymbol{\mu}^T)\mathbf{x}$ of $\mathbf{x}$ has approximately a $(p-1)$-variate normal distribution. More precisely,

$$\sqrt{\kappa}(\mathbf{I}_p - \boldsymbol{\mu}\boldsymbol{\mu}^T)\mathbf{x} \stackrel{\cdot}{\sim} N(\mathbf{0}, \mathbf{I}_p - \boldsymbol{\mu}\boldsymbol{\mu}^T), \qquad \kappa \to \infty. \tag{9.3.15}$$

To see this, note that

(a) since the intersection of S^{p-1} with the hyperplane through $t\boldsymbol{\mu}$ and normal to $\boldsymbol{\mu}$ is a $(p-2)$-dimensional sphere of radius $\sqrt{1-t^2}$, the projection $\mathbf{I}_p - \boldsymbol{\mu}\boldsymbol{\mu}^T$ of $\mathbb{R}^p$ onto the tangent space to S^{p-1} at $\boldsymbol{\mu}$ maps the uniform distribution on S^{p-1} to the distribution on the unit disc $\{\mathbf{y} : \|\mathbf{y}\| < 1\}$ which has density $(1 - \|\mathbf{y}\|^2)^{\frac{1}{2}}$ with respect to Lebesgue measure;

(b)
$$2(1 - \mathbf{x}^T\boldsymbol{\mu}) = \|\mathbf{x} - \boldsymbol{\mu}\|^2 = \|\mathbf{y}\|^2 + O(\|\mathbf{y}\|^4),$$

where $\mathbf{y} = (\mathbf{I}_p - \boldsymbol{\mu}\boldsymbol{\mu}^T)\mathbf{x}$ is the tangential part of $\mathbf{x}$.

It follows that, for large κ, the density (9.3.4) is approximately proportional to

$$\exp\left\{-\frac{\kappa}{2}\|\mathbf{y}\|^2\right\}.$$

Further, from (9.3.15), we find that

$$2\kappa(1 - \mathbf{x}^T\boldsymbol{\mu}) \stackrel{\cdot}{\sim} \chi^2_{p-1}, \qquad \kappa \to \infty, \tag{9.3.16}$$

generalising the approximation (4.8.22) for concentrated von Mises distributions.

When $p = 3$, we may write $\mathbf{x} = (\cos\theta, \sin\theta\cos\phi, \sin\theta\sin\phi)^T$, where (θ, ϕ) are spherical polar coordinates with $\theta = 0$ at $\boldsymbol{\mu}$. Then (9.3.15) and the approximations $\sin\theta \simeq \theta$ and $2(1 - \cos\theta) \simeq \theta^2$ for $\theta \simeq 0$ give

$$\sqrt{\kappa}(\theta\cos\phi, \theta\sin\phi) \quad \dot{\sim} \quad N(\mathbf{0}, \mathbf{I}_2), \qquad \kappa \to \infty \tag{9.3.17}$$

$$\kappa\theta^2 \quad \dot{\sim} \quad \chi^2_2, \qquad \kappa \to \infty. \tag{9.3.18}$$

Further, (9.3.16) gives

$$1 - \mathbf{x}^T\boldsymbol{\mu} \;\dot{\sim}\; \mathcal{E}(\kappa) \qquad \kappa \to \infty, \tag{9.3.19}$$

where $\mathcal{E}(\kappa)$ denotes the exponential distribution with mean $1/\kappa$, namely the distribution with probability density function

$$\kappa e^{-\kappa x}, \qquad x \geq 0.$$

Genesis through Conditioning

It was shown in Section 3.5.4 that von Mises distributions can be obtained by conditioning suitable bivariate normal distributions. This generalises to von Mises–Fisher distributions. Let $\mathbf{x}$ have an $N(\boldsymbol{\mu}, \kappa^{-1}\mathbf{I}_p)$ distribution with $||\boldsymbol{\mu}|| = 1$. Then the conditional distribution of $\mathbf{x}$ given that $||\mathbf{x}|| = 1$ is $M_p(\boldsymbol{\mu}, \kappa)$. Downs (1966) and Downs & Gould (1967) have investigated this relationship much further.

Let $\mathbf{x}$ have an $M_p(\boldsymbol{\mu}, \kappa)$ distribution. Put $\mathbf{x} = (\mathbf{x}_1^T, \mathbf{x}_2^T)^T$ and $\boldsymbol{\mu} = (\boldsymbol{\mu}_1^T, \boldsymbol{\mu}_2^T)^T$, with $\mathbf{x}_1$ and $\boldsymbol{\mu}_1$ in $\mathbb{R}^q$. Then it follows from the above that the conditional distribution of $\mathbf{x}_1$ given that $\mathbf{x}_2 = \mathbf{0}$ is $M_p(\boldsymbol{\mu}_2, \kappa)$.

Infinite Divisibility

Hartman & Watson (1974) showed that the von Mises–Fisher distribution $M_p(\boldsymbol{\mu}, \kappa)$ is a mixture over λ of Brownian motion distributions $BM_p(\boldsymbol{\mu}, \lambda)$ (defined below in Section 9.3.3). By showing the infinite divisibility of the mixing distribution, Kent (1977) proved that $M_p(\boldsymbol{\mu}, \kappa)$ is infinitely divisible. The 'mth roots' of $M_p(\boldsymbol{\mu}, \kappa)$ can also be expressed as a mixture over λ of $BM_p(\boldsymbol{\mu}, \lambda)$.

9.3.3 Other Distributions

Brownian Motion Distributions

The Brownian motion distribution $BM_p(\boldsymbol{\mu}, \kappa)$ on S^{p-1} is the distribution at time κ^{-1} of a random point which starts at $\boldsymbol{\mu}$ and moves on S^{p-1} under an isotropic diffusion with infinitesimal variance $\mathbf{I}_{p-1}$ (the identity matrix on the appropriate tangent space of S^{p-1}). In the case $p = 2$, $BM_2(\boldsymbol{\mu}, \kappa)$ is the wrapped normal distribution $WN(\mu, A(\kappa))$ considered in Section 3.5.7 with $\boldsymbol{\mu} = (\cos\mu, \sin\mu)^T$.

In the case $p = 3$, the probability density function of $BM_3(\boldsymbol{\mu}, \kappa)$ with respect to the uniform distribution is

$$f(\theta, \phi) = \sum_{k=1}^{\infty} (2k+1) e^{-k(k+1)/4\kappa} P_k(\cos\theta), \tag{9.3.20}$$

where (θ, ϕ) are spherical polar coordinates with $\theta = 0$ at $\boldsymbol{\mu}$ (Perrin, 1928). Here P_k denotes the Legendre polynomial of order k. It can be shown that the mean resultant length ρ is

$$\rho = \exp\{-(2\kappa)^{-1}\}. \tag{9.3.21}$$

Roberts & Ursell (1960) showed that if the distributions $BM_3(\boldsymbol{\mu}, \kappa)$ and $F(\boldsymbol{\mu}, A_3^{-1}(\rho))$ have the same mean resultant length (i.e. ρ is given by (9.3.21)) then they are very close, as in the approximation (3.5.23) of $M(\mu, \kappa)$ by $WN(\mu, A(\kappa))$. Hence the methodology for Fisher distributions will be approximately applicable to Brownian motion distributions. Roberts & Ursell (1960) looked also at random walks on spheres and more general compact Riemannian manifolds.

The result (3.5.24) on the closeness of the densities of $M(\mu, \kappa)$ and $WN(\mu, A(\kappa))$ extends to the corresponding distributions on S^{p-1}. Kent (1978) showed that

$$f_{VM}(\mathbf{x}; \boldsymbol{\mu}, \kappa) - f_{BM}(\mathbf{x}; \boldsymbol{\mu}, \kappa) = O(\kappa^{(p-3)/2}), \qquad \kappa \to \infty,$$

where $f_{VM}(\cdot; \boldsymbol{\mu}, \kappa)$ and $f_{BM}(\cdot; \boldsymbol{\mu}, \kappa)$ denote the densities of the von Mises–Fisher distribution $M_p(\boldsymbol{\mu}, \kappa)$ and the Brownian motion distribution $BM_p(\boldsymbol{\mu}, \kappa)$, respectively.

Fisher–Bingham Distributions

The most important feature of the Fisher distributions (9.3.4) is that their log-densities are linear in the observation $\mathbf{x}$. A natural generalisation (Beran, 1979) is to replace $\mathbf{x}$ in the exponent in (9.3.4) by higher polynomials $\mathbf{t}(\mathbf{x})$ in $\mathbf{x}$. In particular, the use of general quadratics in $\mathbf{x}$ yields the Fisher–Bingham model (Mardia, 1975a) with densities

$$f(\mathbf{x}; \boldsymbol{\mu}, \kappa, \mathbf{A}) = \frac{1}{a(\kappa, \mathbf{A})} \exp\{\kappa \boldsymbol{\mu}^T \mathbf{x} + \mathbf{x}^T \mathbf{A} \mathbf{x}\}, \tag{9.3.22}$$

where $\mathbf{A}$ is a symmetric $p \times p$ matrix and the constraint $\mathbf{x}^T \mathbf{x} = 1$ allows us to assume without loss of generality that $\operatorname{tr} \mathbf{A} = 0$. These distributions can also be obtained by conditioning p-variate normal distributions on $\|\mathbf{x}\| = 1$. Let $\mathbf{x}$ have an $N(-\frac{1}{2}\kappa(\mathbf{A} + c\mathbf{I}_p)^{-1}\boldsymbol{\mu}, -\frac{1}{2}(\mathbf{A} + c\mathbf{I}_p)^{-1})$ distribution, where c is such that $\mathbf{A} + c\mathbf{I}_p$ is negative definite. Then the conditional density of $\mathbf{x}$ given that $\|\mathbf{x}\| = 1$ is (9.3.22).

If $\mathbf{A} = \mathbf{0}$ then (9.3.22) reduces to the von Mises–Fisher density (9.3.4), while if $\kappa = 0$ then (9.3.22) reduces to the Bingham density given below in (9.4.3). When $p = 2$, the Fisher–Bingham densities (9.3.22) have the form (3.5.46). Further models with interesting geometrical properties appropriate for modelling phenomena from various fields can be obtained by suitable restriction of the parameters of the Fisher–Bingham model (9.3.22).

The relationships between the various exponential models discussed in this section are shown in Table 9.2.

Table 9.2 Some exponential families of distributions on S^{p-1}, showing name, equation number in text, and dimension. Arrows denote inclusion.

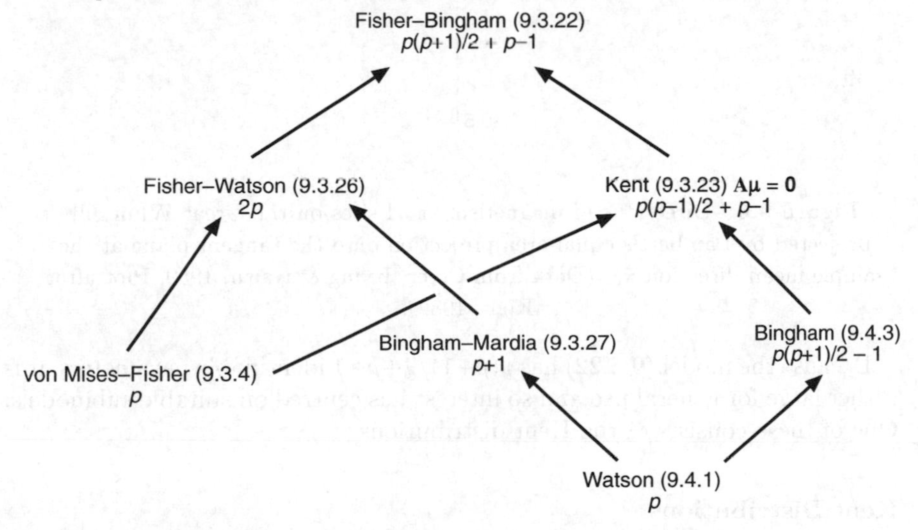

Although exponential models have many pleasant inferential properties, the need to evaluate the normalising constant (or at least the first derivative of its logarithm) can be a practical difficulty. For directional distributions, these normalising constants are often given in terms of special functions such as Bessel functions. Some simplifications have been effected by de Waal (1979) and Wood (1988).

In the general context of exponential models on spheres, Beran (1979) has developed a regression-based estimator (using non-parametric density estimation) which bypasses the need for calculation of the normalising constant.

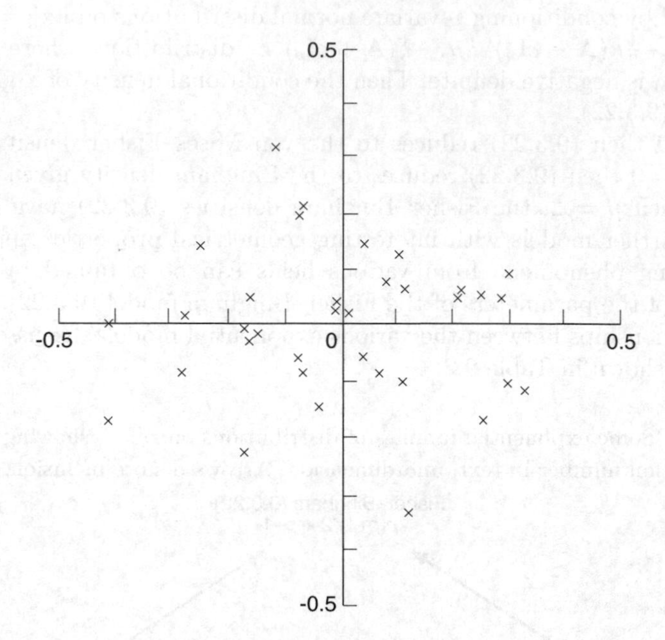

Figure 9.6 Directions of magnetism at 34 sites on the Great Whin Sill projected by Lambert's equal-area projection onto the tangent plane at the sample mean direction $\bar{\mathbf{x}}_0$. (Data from Creer, Irving & Nairn, 1959. Plot after Kent, 1982).

Because the model (9.3.22) has $p(p+1)/2+p-1$ identifiable parameters, it is rather large for general use and so interest has centred on suitable submodels. One of these consists of the Kent distributions.

Kent Distributions

Some data sets (such as that plotted in Fig. 9.6) are not fitted adequately by Fisher distributions and appear to arise from distributions with oval density contours. In order to model these, Kent (1982) introduced the model with densities

$$f(\mathbf{x}; \boldsymbol{\mu}, \kappa, \mathbf{A}) = \frac{1}{a(\kappa, \mathbf{A})} \exp\{\kappa \boldsymbol{\mu}^T \mathbf{x} + \mathbf{x}^T \mathbf{A} \mathbf{x}\}, \tag{9.3.23}$$

where $\mathbf{A}$ is a symmetric $p \times p$ matrix with $\operatorname{tr} \mathbf{A} = 0$ and $\mathbf{A}\boldsymbol{\mu} = \mathbf{0}$. Putting $\mathbf{z} = (1-t^2)^{1/2}\boldsymbol{\xi}$ in (9.1.2) and using the approximation $t \simeq ||\mathbf{z}||^2$ gives the approximation

$$f(\mathbf{x}; \boldsymbol{\mu}, \kappa, \mathbf{A}) \simeq \frac{1}{a(\kappa, \mathbf{A})} \exp\left\{ -\frac{1}{2}\mathbf{z}^T \left(\mathbf{I}_p - \boldsymbol{\mu}\boldsymbol{\mu}^T\right)(\kappa \mathbf{I}_p - 2\mathbf{A})\left(\mathbf{I}_p - \boldsymbol{\mu}\boldsymbol{\mu}^T\right)\mathbf{z} \right\}, \tag{9.3.24}$$

for large κ. Thus, for large κ, the distribution of the tangential part $\mathbf{z}$ of $\mathbf{x}$ is approximately $(p-1)$ variate normal. In particular, if κ is large enough, the density (9.3.23) has a mode at $\boldsymbol{\mu}$ and density contours which are approximately ellipsoidal. This provides the main motivation for the Kent distributions.

When $p = 3$, the parameter matrix $\mathbf{A}$ can be written as

$$\mathbf{A} = \beta\left(\boldsymbol{\xi}_1\boldsymbol{\xi}_1^T - \boldsymbol{\xi}_2\boldsymbol{\xi}_2^T\right),$$

where $\beta \geq 0$ and $\boldsymbol{\xi}_1$, $\boldsymbol{\xi}_2$, $\boldsymbol{\mu}$ are orthogonal. Then (9.3.23) takes the form

$$f(\mathbf{x}; \kappa, \beta, \boldsymbol{\Gamma}) = \frac{1}{c(\beta, \kappa)} \exp\{\kappa x_1 + \beta(x_2^2 - x_3^2)\}, \tag{9.3.25}$$

where $\boldsymbol{\Gamma}$ is the orthogonal matrix $\boldsymbol{\Gamma} = (\boldsymbol{\mu}, \boldsymbol{\xi}_1, \boldsymbol{\xi}_2)$ and $(x_1, x_2, x_3) = (\boldsymbol{\mu}^T\mathbf{x}, \boldsymbol{\xi}_1^T\mathbf{x}, \boldsymbol{\xi}_2^T\mathbf{x})$.

The Kent distributions (9.3.23) can be obtained from the Fisher–Bingham distributions (9.3.22) by imposing the restriction $\mathbf{A}\boldsymbol{\mu} = \mathbf{0}$. Relaxing the restriction $\mathbf{A}\boldsymbol{\mu} = \mathbf{0}$ on the parameter of (9.3.23) to the condition that $\boldsymbol{\mu}$ be an eigenvector of $\mathbf{A}$, yields the one-parameter extension of the Kent model suggested by Rivest (1984). This extension is a (d, d) exponential model with $d = p(p+1)/2$ (see Section 3.5.1).

Fisher–Watson Distributions

A smaller useful model is the $2p$-parameter Fisher–Watson model introduced by Wood (1988), which is obtained from (9.3.22) by the restriction that the matrix $\mathbf{A}$ has rank 1 (instead of $\operatorname{tr}\mathbf{A} = 0$), so that the densities have the form

$$f(\mathbf{x}; \boldsymbol{\mu}, \boldsymbol{\mu}_0, \kappa, \kappa_0) = \frac{1}{a(\kappa_0, \kappa, \boldsymbol{\mu}_0^T\boldsymbol{\mu})} \exp\{\kappa_0\boldsymbol{\mu}_0^T\mathbf{x} + \kappa(\boldsymbol{\mu}^T\mathbf{x})^2\}. \tag{9.3.26}$$

Bingham–Mardia Distributions

Certain problems in the earth sciences require rotationally-symmetric spherical distributions which have a 'modal ridge' along a small circle rather than a mode at a single point. One such problem is that of modelling plumes as a tectonic plate rotates over a volcanic hot spot. Suitable models include the 'small circle' model of C. Bingham & Mardia (1978) which has probability density functions

$$f(\mathbf{x}; \boldsymbol{\mu}, \kappa, \nu) = \frac{1}{a(\kappa)} \exp\{\kappa(\boldsymbol{\mu}^T\mathbf{x} - \nu)^2\}. \tag{9.3.27}$$

This is a $(p+1)$-parameter curved exponential submodel of the Fisher–Watson model (9.3.26). Another model consisting of 'small circle' distributions on the sphere was introduced by Mardia & Gadsden (1977) and has densities proportional to

$$\exp\left\{\alpha\boldsymbol{\mu}^T\mathbf{x}+\beta\left[1-(\boldsymbol{\mu}^T\mathbf{x})^2\right]^{\frac{1}{2}}\right\}.$$

Wood Distributions

In order to model bimodal data on S^{p-1}, Wood (1982) modified the Fisher distributions by 'doubling the longitude'. The density functions are

$$\begin{aligned} f(\mathbf{x};\boldsymbol{\mu}_1,\boldsymbol{\mu}_2,\kappa,\alpha) &= \frac{1}{a(\boldsymbol{\mu}_1,\boldsymbol{\mu}_2,\kappa,\alpha)} \\ &\quad\times\exp\left\{\kappa\cos(\alpha\mathbf{x}^T\boldsymbol{\mu}_1)+\kappa\sin\left(\alpha\frac{[2(\mathbf{x}^T\boldsymbol{\mu}_2)^2-1]}{[1-(\mathbf{x}^T\boldsymbol{\mu}_1)^2]^{1/2}}\right)\right\}, \end{aligned} \tag{9.3.28}$$

where $\boldsymbol{\mu}_1,\boldsymbol{\mu}_2$ are orthogonal unit vectors, $\kappa \geq 0$ and α is real. These distributions have mean $\boldsymbol{\mu}_1$ and two modes of equal strength at $\cos\alpha\,\boldsymbol{\mu}_1 \pm \sin\alpha\,\boldsymbol{\mu}_2$. Note that (9.3.28) is not an exponential model.

Projected Distributions

Distributions on S^{p-1} can be obtained by radial projection of distributions on $\mathbb{R}^p$. Given a a random vector $\mathbf{x}$ in $\mathbb{R}^p$ such that $\Pr(\mathbf{x}=\mathbf{0})=0$, the corresponding *projected* (or *offset*) *distribution* on S^{p-1} is that of $||\mathbf{x}||^{-1}\mathbf{x}$. If $\mathbf{x}$ has the p-variate normal distribution $N_p(\boldsymbol{\mu},\boldsymbol{\Sigma})$ then the corresponding projected distribution on S^{p-1} is called the *projected normal* (or *offset normal*) distribution $PN_p(\boldsymbol{\mu},\boldsymbol{\Sigma})$. The case $p=2$ was considered in Section 3.5.6. The density of $PN_p(\boldsymbol{\mu},\boldsymbol{\Sigma})$ was given in an infinite series by Bingham (see Watson, 1983a, pp. 226–231) and in simpler form by Pukkila & Rao (1988).

Another interesting class of projected distributions consists of the *offset* (or *decentred*) *directional distributions*, which are obtained by displacing a distribution on S^{p-1} by a constant vector $\mathbf{c}$ and then projecting radially back onto S^{p-1}, so that $\mathbf{y}$ on S^{p-1} is sent to

$$\mathbf{x}=\frac{\mathbf{y}+\mathbf{c}}{||\mathbf{y}+\mathbf{c}||}.$$

One way in which offset directional distributions arise is through directions being detected by an observer at $\mathbf{c}$, instead of at the origin. For example, meteorites hitting the earth's atmosphere are usually recorded by an observer on the earth's surface. The *offset uniform distributions* arise by applying this procedure to the uniform distribution on S^{p-1}. They have densities

$$f(\mathbf{x};\lambda,\boldsymbol{\mu})=A(\mathbf{x},\mathbf{c})^{-1}\left[\lambda\mathbf{x}^T\boldsymbol{\mu}+A(\mathbf{x},\mathbf{c})\right]^{p-1} \tag{9.3.29}$$

on S^{p-1} for $\lambda \leq 1$, and

$$f(\mathbf{x};\lambda,\boldsymbol{\mu}) = A(\mathbf{x},\mathbf{c})^{-1}\left\{\left[\lambda\mathbf{x}^T\boldsymbol{\mu} + A(\mathbf{x},\mathbf{c})\right]^{p-1} + \left[\lambda\mathbf{x}^T\boldsymbol{\mu} - A(\mathbf{x},\mathbf{c})\right]^{p-1}\right\} \tag{9.3.30}$$

on $\{\mathbf{x} : \mathbf{x}^T\boldsymbol{\mu} \leq (\lambda^2-1)^{1/2}/\lambda\}$ for $\lambda > 1$. Here $\boldsymbol{\mu} \in S^{p-1}$, $\lambda \geq 0$ and

$$A(\mathbf{x},\mathbf{c}) = \left\{1 - \lambda^2[1-(\mathbf{x}^T\boldsymbol{\mu})^2]\right\}^{1/2},$$

with $\mathbf{c} = \lambda\boldsymbol{\mu}$. Formula (9.3.30) was derived by Kent & Mardia (1997), who also give a complete derivation for all λ. The density (9.3.29) in the case $\lambda \leq 1$ was found by Boulerice & Ducharme (1994). Formulae (9.3.29)–(9.2.30) show that the offset uniform distributions are rotationally symmetric about $\boldsymbol{\mu}$ and (if $\lambda > 0$) have mode and mean direction $\boldsymbol{\mu}$.

Models with Rotational Symmetry

A very important property of the von Mises–Fisher distributions is that they are rotationally symmetric about their modal directions. Saw (1978; 1983; 1984) and Watson (1983a, pp. 92, 136–186) have abstracted this property by considering general distributions with such symmetry (see also Bingham & Mardia, 1975). Among such distributions, those which are continuous have probability density functions of the form

$$f(\mathbf{x}) = g(\boldsymbol{\mu}^T\mathbf{x}), \tag{9.3.31}$$

where g is a known function. In the tangent-normal decomposition

$$\mathbf{x} = t\boldsymbol{\mu} + (1-t^2)^{\frac{1}{2}}\boldsymbol{\xi} \tag{9.3.32}$$

(where $\boldsymbol{\mu}^T\boldsymbol{\xi} = 0$), symmetry implies that the unit tangent $\boldsymbol{\xi}$ at $\boldsymbol{\mu}$ to S^{p-1} is uniformly distributed on S^{p-2} and is independent of t.

For distributions with rotational symmetry about a direction $\boldsymbol{\mu}$,

$$\mathrm{E}[\mathbf{x}] = \mathrm{E}[t]\boldsymbol{\mu} \tag{9.3.33}$$

and

$$\mathrm{var}(\mathbf{x}) = \mathrm{var}(t)\boldsymbol{\mu}\boldsymbol{\mu}^T + \frac{1-\mathrm{E}[t^2]}{p-1}(\mathbf{I}_p - \boldsymbol{\mu}\boldsymbol{\mu}^T), \tag{9.3.34}$$

where $t = \mathbf{x}^T\boldsymbol{\mu}$.

Among the families of distributions of the form (9.3.31) considered by Saw (1983) is a p-parameter model with the property that if $\mathbf{x} = (\mathbf{x}_1^T, \mathbf{x}_2^T)^T$ has a distribution in this model then $\|\mathbf{x}_1\|^{-1}\mathbf{x}_1$ and $\|\mathbf{x}_2\|^{-1}\mathbf{x}_2$ are independent with distributions in the corresponding models on S^{q-1} and S^{p-q-1}. Watson (1983a, Chapter 4) showed that many models of the form (9.3.31) have the same asymptotic (large-sample or high-concentration) behaviour as the von

Mises-Fisher model. For example, the large-sample asymptotic distribution of $\bar{\mathbf{x}}$ is given by (9.6.5)–(9.6.6) below. Distributions with symmetry of the form (9.3.31) can arise through (i) symmetrisation performed by the observation process, e.g. the rotation of the earth as in the motivating problem for the 'rotating spherical cap' distribution of Mardia & Edwards (1982), or (ii) the inability of the observation procedure to distinguish between $\mathbf{x}$ and $\mathbf{Ux}$, where $\mathbf{U}$ is a rotation about $\boldsymbol{\mu}$, so that only the colatitude (or equivalently $\mathbf{x}^T\boldsymbol{\mu}$) is observed, as in Clark's (1983) motivation for the 'marginal Fisher' distributions introduced by Mardia & Edwards (1982). In the case $p = 3$, the probability density function of the colatitude θ is

$$f(\theta; \theta_0, \kappa) = \frac{\kappa \sin\theta}{2\pi \sinh\kappa} \exp\{\kappa \cos\theta \cos\theta_0\} I_0(\kappa \sin\theta \sin\theta_0), \tag{9.3.35}$$

where $0 \leq \theta_0 \leq \pi/2$.

Another useful model of the form (9.3.31) has densities

$$f(\mathbf{x}; \boldsymbol{\mu}, \kappa) = c(\kappa) \exp\{\kappa \cos^{-1}(\mathbf{x}^T\boldsymbol{\mu})\}.$$

It was introduced by Purkayastha (1991), who showed that it is characterised by the property that the maximum likelihood estimate of $\boldsymbol{\mu}$ is the sample median direction. For large κ, the use of this distribution in data analysis was considered by Cabrera & Watson (1990).

The rotational symmetry about the axis $\boldsymbol{\mu}$ possessed by densities of the form (9.3.31) can be generalised to rotational symmetry about some subspace V of $\mathbb{R}^p$. The corresponding distributions have probability density functions of the form

$$f(\mathbf{x}) = g(\mathbf{x}_V), \tag{9.3.36}$$

where $\mathbf{x}_V$ is the orthogonal projection of $\mathbf{x}$ onto V (see Watson, 1983a, pp. 94–95; 1983c).

9.4 MODELS FOR AXIAL DATA

9.4.1 Introduction

Sometimes the observations are not *directions* but *axes*; that is, the unit vectors $\mathbf{x}$ and $-\mathbf{x}$ are indistinguishable, so that it is $\pm\mathbf{x}$ which is observed. In this context it is appropriate to consider probability density functions for $\mathbf{x}$ on S^{p-1} which are antipodally symmetric, i.e.

$$f(-\mathbf{x}) = f(\mathbf{x}).$$

In such cases the observations $\pm\mathbf{x}_1, \ldots, \pm\mathbf{x}_n$ can be regarded as being on the projective space $\mathbb{R}P^{p-1}$, which is obtained by identifying opposite points on the sphere S^{p-1}.

9.4.2 Watson Distributions

One of the simplest models for axial data is the (Dimroth–Scheidegger–)Watson model, which has densities

$$f(\pm\mathbf{x};\boldsymbol{\mu},\kappa) = M\left(\frac{1}{2},\frac{p}{2},\kappa\right)^{-1} \exp\{\kappa(\boldsymbol{\mu}^T\mathbf{x})^2\}, \tag{9.4.1}$$

where $M(1/2, p/2, \cdot)$ denotes the Kummer function

$$M\left(\frac{1}{2},\frac{p}{2},\kappa\right) = B\left(\frac{p-1}{2},\frac{1}{2}\right)^{-1} \int_{-1}^{1} e^{\kappa t^2}(1-t^2)^{(p-3)/2}dt \tag{9.4.2}$$

obtained by taking $a = 1/2$ and $b = p/2$ in (A.17) of Appendix 1.

The distribution $W(\boldsymbol{\mu},\kappa)$ with density (9.4.1) is rotationally symmetric about $\boldsymbol{\mu}$. For $\kappa > 0$, the density has maxima at $\pm\boldsymbol{\mu}$, and so the distribution is bipolar. As κ increases, the distribution becomes more concentrated about $\pm\boldsymbol{\mu}$. Thus the parameter κ measures concentration. For $\kappa < 0$, the distribution is concentrated around the great circle orthogonal to $\boldsymbol{\mu}$, and so the distribution is a symmetric girdle distribution. These distributions were studied and introduced independently by Dimroth (1962; 1963) and Watson (1965).

9.4.3 Bingham Distributions

Because some axial data sets do not show the rotational symmetry of the Watson distributions, it is useful to consider the Bingham distributions. These have densities

$$f(\pm\mathbf{x};\mathbf{A}) = {}_1F_1\left(\frac{1}{2},\frac{p}{2},\mathbf{A}\right)^{-1} \exp\{\mathbf{x}^T\mathbf{A}\mathbf{x}\}, \tag{9.4.3}$$

where the hypergeometric function ${}_1F_1(1/2, p/2, \cdot)$ of matrix argument is defined in (A.26) of Appendix 1. Because of the constraint $\mathbf{x}^T\mathbf{x} = 1$, the matrices $\mathbf{A}$ and $\mathbf{A} + c\mathbf{I}_p$ give the same distribution. Thus we may assume without loss of generality that $\operatorname{tr}\mathbf{A} = 0$. In applied work it is sometimes useful to assume instead that the smallest eigenvalue of $\mathbf{A}$ is zero.

The densities (9.4.3) form a transformation model under the group $O(p)$ of rotations and reflections of $\mathbb{R}^p$. For $\mathbf{U}$ in $O(p)$ we have

$$f(\mathbf{U}\mathbf{x};\mathbf{U}\mathbf{A}\mathbf{U}^T) = f(\mathbf{x};\mathbf{A}). \tag{9.4.4}$$

It follows that

$${}_1F_1\left(\frac{1}{2},\frac{p}{2},\mathbf{U}\mathbf{A}\mathbf{U}^T\right) = {}_1F_1\left(\frac{1}{2},\frac{p}{2},\mathbf{A}\right). \tag{9.4.5}$$

In particular, if the spectral decomposition of $\mathbf{A}$ is $\mathbf{A} = \mathbf{U}^T\mathbf{K}\mathbf{U}$ with $\mathbf{U}$ orthogonal and $\mathbf{K}$ diagonal then

$${}_1F_1\left(\frac{1}{2},\frac{p}{2},\mathbf{A}\right) = {}_1F_1\left(\frac{1}{2},\frac{p}{2},\mathbf{K}\right). \tag{9.4.6}$$

In general, the larger the elements of $\mathbf{K}$, the greater the clustering around the principal axes of $\mathbf{A}$ (i.e. the axes given by the columns of $\mathbf{U}^T$). Varying the values of $\mathbf{K}$ in (9.4.3) gives a wide range of distributional shapes, including the uniform distribution ($\mathbf{K} = \kappa \mathbf{I}_p$), symmetric and asymmetric girdle distributions (in which the probability mass is clustered around a great circle) and bimodal distributions.

For $p = 2$, the distributions with density (9.4.3) reduce to the 2-wrapped von Mises distributions (considered in Section 3.6.1) obtained by 'doubling the angles'. There is no simple analogue of this process for $p \geq 3$.

In the case $p = 3$, the distributions with densities (9.4.3) were proposed by Bingham (1964; 1974), who investigated their statistical properties extensively.

The Watson distributions are precisely those Bingham distributions which have rotational symmetry about some axis.

Bingham distributions can be obtained by conditioning p-variate normal distributions on $\|\mathbf{x}\| = 1$. Let $\mathbf{x}$ have an $N_p(\mathbf{0}, \boldsymbol{\Sigma})$ distribution. Then the conditional distribution of $\mathbf{x}$ given that $\|\mathbf{x}\| = 1$ is Bingham with parameter matrix $-\frac{1}{2}\boldsymbol{\Sigma}^{-1}$.

Asymptotic expansions for the normalising constant in various types of concentrated Bingham distribution have been developed by Kent (1987).

An interesting submodel of the Bingham model (9.4.3) consists of the *complex Bingham distributions* (Kent, 1994), which are considered in Section 14.6 and provide useful models for two-dimensional shapes.

9.4.4 Angular Central Gaussian Distributions

Because the normalising constant of the Bingham distribution is rather complicated, maximum likelihood estimation in this model can be tedious. This led Tyler (1987) to introduce the *angular central Gaussian distributions* $ACG(\mathbf{A})$ on S^{p-1} with probability density functions

$$f(\mathbf{x}; \mathbf{A}) = |\mathbf{A}|^{-1/2}(\mathbf{x}^T \mathbf{A}^{-1}\mathbf{x})^{-p/2}. \tag{9.4.7}$$

Here $\mathbf{A}$ is a symmetric $p \times p$ positive definite parameter matrix and is identifiable up to multiplication by a positive scalar. Note that

$$f(-\mathbf{x}; \mathbf{A}) = f(\mathbf{x}; \mathbf{A})$$

and so (9.4.7) defines a distribution on $\mathbb{R}P^{p-1}$. The reason for the name 'angular central Gaussian' is that if $\mathbf{x} \sim N_p(\mathbf{0}, \mathbf{A})$ then $\|\mathbf{x}\|^{-1}\mathbf{x} \sim ACG(\mathbf{A})$. For $p = 2$, the distribution $ACG(\mathbf{A})$ is the angular central Gaussian distribution $PN_2(\mathbf{0}, \mathbf{A})$ considered in Section 3.5.6. The transformation property (3.5.53) of the angular central Gaussian distributions on the circle generalises to the angular central Gaussian distributions on spheres. Each invertible linear transformation $\mathbf{A}$ of $\mathbb{R}^p$ gives rise to an invertible transformation $\varphi_{\mathbf{A}}$ of S^{p-1} by

$$\varphi_{\mathbf{A}}(\mathbf{x}) = \frac{1}{\|\mathbf{A}\mathbf{x}\|}\mathbf{A}\mathbf{x}. \tag{9.4.8}$$

Since $\mathbf{A} \mapsto \varphi_{\mathbf{A}}$ is scale-invariant, i.e. $\varphi_{c\mathbf{A}} = \varphi_{\mathbf{A}}$ for all non-zero c, we may assume without loss of generality that $|\mathbf{A}| = 1$, i.e. that $\mathbf{A}$ is a unimodular matrix. A simple calculation shows that

$$\mathbf{x} \sim ACG(\mathbf{\Sigma}) \quad \Rightarrow \quad \mathbf{A}\mathbf{x} \sim ACG(\mathbf{A}\mathbf{\Sigma}\mathbf{A}^T). \tag{9.4.9}$$

Thus the angular central Gaussian distributions on S^{p-1} form a transformation model under the action of the unimodular group $SL_p(\mathbb{R})$ on $\mathbb{R}^p$.

9.4.5 Other Axial Distributions

Girdle distributions on S^2 with probability density functions proportional to

$$\exp(-\kappa|\cos\theta|)$$

and

$$\exp(\kappa \sin\theta),$$

where $\cos\theta = \mathbf{x}^T\boldsymbol{\mu}$, have been investigated by Arnold (1941) and Selby (1964).

A one-parameter extension of the Bingham model was introduced by Kelker & Langenberg (1982) in order to model axes concentrated asymmetrically near a small circle. It has density functions

$$f(\mathbf{x}; \boldsymbol{\mu}_1, \boldsymbol{\mu}_2, \kappa_1, \kappa_2, \gamma) = \frac{1}{a(\kappa_0, \kappa, \gamma)} \exp\{-\kappa_1[\sin(\theta-\gamma)]^2 + \kappa_2 \cos(\theta-\gamma)[\sin\phi]^2\}, \tag{9.4.10}$$

where $\boldsymbol{\mu}_1, \boldsymbol{\mu}_2$ are orthogonal unit vectors, $\kappa_1, \kappa_2 > 0$, $\gamma \in [0, 2\pi]$ and (θ, ϕ) is defined by $\cos\theta = \boldsymbol{\mu}_1^T\mathbf{x}$ and $\sin\theta\cos\phi = \boldsymbol{\mu}_2^T\mathbf{x}$ with $\theta \in [0, \pi]$. This model is a composite transformation model under the orthogonal group $O(p)$.

9.5 DISTRIBUTION THEORY

We now generalise the principal distributional results of Sections 4.4 and 4.5 to various distributions on spheres. In particular, we obtain the distributions of the resultant length, the mean direction, and of some related statistics for random samples from the uniform and von Mises–Fisher distributions. In Chapter 10 these results will be applied to inference on spherical distributions.

9.5.1 The Uniform Distribution

Let $\bar{\mathbf{x}}_0$ and $\bar{R}$ be the mean direction and mean resultant length of a random sample from the uniform distribution on S^{p-1}.

Because $\bar{R}$ is invariant under rotations of $\mathbb{R}^p$ while $\bar{\mathbf{x}}_0$ is equivariant, the conditional distribution of $\bar{\mathbf{x}}_0|\bar{R}$ is uniform on S^{p-1}. It follows that

$$\bar{\mathbf{x}}_0 \text{ is distributed uniformly on } S^{p-1} \tag{9.5.1}$$

and

$$\bar{\mathbf{x}}_0 \text{ and } \bar{R} \text{ are independent,} \tag{9.5.2}$$

generalising (4.4.1)–(4.4.2). Kent, Mardia & Rao (1979) showed that property (9.5.2) characterises the uniform distribution on S^{p-1} (among distributions with positive continuous density).

The marginal distribution of $\bar{R}$ can be found by inverting its characteristic function (see Mardia, 1975a, Section 3.2; Mardia, Kent & Bibby, 1979, Section 15.4.1). Denote the resultant length $n\bar{R}$ by R. The density of R is

$$h_n(R) = 2^{(1-p/2)(n-1)}\Gamma\left(\frac{p}{2}\right)^{n-1} R\int_0^\infty \rho^{n-p(n-1)/2} J_{p/2-1}(\rho R) J_{p/2-1}(\rho)^n d\rho, \tag{9.5.3}$$

where $0 \le R \le n$ and J_ν denotes the Bessel function of the first kind and order ν (which is defined in (A.15) of Appendix 1).

For $p = 3$ and $n = 2, 3$, this simplifies to

$$h_2(R) = R/2, \quad 0 \le R \le 2, \tag{9.5.4}$$

and

$$h_3(R) = \begin{cases} R^2/2, & 0 \le R \le 1, \\ R(3-R)/4, & 1 \le R \le 3. \end{cases} \tag{9.5.5}$$

The large-sample asymptotic distribution of $\bar{R}$ will be given in Section 9.6.2.

9.5.2 Von Mises–Fisher Distributions

Single-Sample Case

The joint distribution of $(\bar{R}, \bar{\mathbf{x}}_0)$ for samples from a von Mises–Fisher distribution can be obtained by straightforward extension of the arguments of Sections 4.5.1–4.5.2. Suppose that $\mathbf{x}_1, \ldots, \mathbf{x}_n$ is a random sample from $M_p(\boldsymbol{\mu}, \kappa)$. The joint density of $(\mathbf{x}_1, \ldots, \mathbf{x}_n)$ is

$$f(\mathbf{x}_1, \ldots, \mathbf{x}_n) = c(\kappa)^n \exp\{n\kappa\bar{R}\bar{\mathbf{x}}_0^T\boldsymbol{\mu}\}, \tag{9.5.6}$$

where $c(\kappa)$ is the normalising constant of the von Mises–Fisher distribution (9.3.4). Integration over

$$\{(\mathbf{x}_1, \ldots, \mathbf{x}_n)|\bar{\mathbf{x}} = \bar{R}\bar{\mathbf{x}}_0\}, \tag{9.5.7}$$

gives the joint density of $(\bar{R}, \bar{\mathbf{x}}_0)$ as

$$\begin{aligned} g(\bar{R}, \bar{\mathbf{x}}_0; \boldsymbol{\mu}, \kappa) &= c(\kappa)^n \exp\{n\kappa\bar{R}\bar{\mathbf{x}}_0^T\boldsymbol{\mu}\} g(\bar{R}, \bar{\mathbf{x}}_0; \boldsymbol{\mu}, 0) \\ &= c(\kappa)^n \exp\{n\kappa\bar{R}\bar{\mathbf{x}}_0^T\boldsymbol{\mu}\} h_n(n\bar{R}), \end{aligned} \tag{9.5.8}$$

where h_n is the marginal density (9.5.3) of R in the uniform case. Integration over $\bar{\mathbf{x}}_0$ yields the marginal density of $\bar{R}$ as

$$f(\bar{R};\kappa) = c(\kappa)^n c(n\kappa\bar{R})^{-1} h_n(n\bar{R}). \tag{9.5.9}$$

It follows that the conditional density of $\bar{\mathbf{x}}_0|R$ is proportional to

$$\exp\{\bar{\mathbf{x}}_0^T(n\kappa\bar{R}\boldsymbol{\mu}_0)\},$$

and so

$$\bar{\mathbf{x}}_0|\bar{R} \sim M_p(\boldsymbol{\mu}, n\kappa\bar{R}), \tag{9.5.10}$$

generalising (4.5.5). As we shall see in Section 10.4.2, this result is very useful for inference on von Mises–Fisher distributions.

Integrating (9.5.8) over

$$\{(\mathbf{x}_1,\ldots,\mathbf{x}_n) : \|\mathbf{x}_i\| = 1, \bar{\mathbf{x}}_0^T\boldsymbol{\mu} = \bar{C}\}$$

gives the joint density of $(\bar{R},\bar{C})$ as

$$g^*(\bar{R},\bar{C};\boldsymbol{\mu},\kappa) = c(\kappa)^n \exp\{n\kappa\bar{C}\} g^*(\bar{R},\bar{C};\boldsymbol{\mu},0).$$

Thus the conditional density of $\bar{R}|\bar{C}$ is

$$\frac{g^*(\bar{R},\bar{C};\boldsymbol{\mu},0)}{h_n(n\bar{R})}, \tag{9.5.11}$$

which does not depend on κ. This result is due to Watson & Williams (1956) and is the basis of a test considered in Section 10.4.2. See also Stephens (1962b) and Mardia (1975b).

Multi-sample Case

Let $\mathbf{x}_{11},\ldots,\mathbf{x}_{1n_1};\ldots;\mathbf{x}_{q1},\ldots,\mathbf{x}_{qn_q}$ be q independent random samples of sizes $n_1,\ldots,n_q$, from $M_p(\boldsymbol{\mu},\kappa)$. Then the combined sample has size n, where $n = n_1 + \ldots + n_q$. Let R_i be the resultant length of the ith sample and R be the resultant length of the combined sample. Put $\bar{C} = \bar{\mathbf{x}}^T\boldsymbol{\mu}$, as in (9.2.5). An argument analogous to that used to derive (9.5.8) shows that the joint density of $(\bar{R},\bar{C},\bar{R}_1,\ldots,\bar{R}_q)$ is

$$f(\bar{R},\bar{C},\bar{R}_1,\ldots,\bar{R}_q;\boldsymbol{\mu},\kappa) = c(\kappa)^n \exp\{n\kappa\bar{C}\} f(\bar{R},\bar{C},\bar{R}_1,\ldots,\bar{R}_q;\boldsymbol{\mu},0). \tag{9.5.12}$$

Integrating out $\bar{C}$ gives the joint density of $(\bar{R}_1,\ldots,\bar{R}_q)$ as

$$f(\bar{R},\bar{R}_1,\ldots,\bar{R}_q;\boldsymbol{\mu},\kappa) = c(\kappa)^n c(n\kappa\bar{R}) f(\bar{R},\bar{R}_1,\ldots,\bar{R}_q;\boldsymbol{\mu},0). \tag{9.5.13}$$

Dividing (9.5.13) by the marginal density (9.5.9) shows that the conditional density of $(\bar{R}_1, \ldots, \bar{R}_q)|\bar{R}$ is

$$f(\bar{R}_1, \ldots, \bar{R}_q|\bar{R}; \kappa) = \frac{f(\bar{R}_1, \ldots, \bar{R}_q)}{h_n(n\bar{R})}, \tag{9.5.14}$$

which does not depend on κ. For $p = 3$ and $q = 2$, an expression for $f(\bar{R}_1, \bar{R}_2|\bar{R}; \kappa)$ was given by R.A. Fisher (1953).

9.6 MOMENTS AND LIMITING DISTRIBUTIONS

Because distribution theory can be complicated, various asymptotic approximations are useful. Asymptotic properties of directional distributions fall into three categories, according to whether the sample size n, the concentration κ, or the dimension $p-1$ tends to infinity. First, we calculate the moments of the uniform distribution on S^{p-1}. These are useful in calculating the large-sample asymptotic distributions of various tests of uniformity.

9.6.1 *Moments of the Uniform Distribution*

If $\mathbf{X} = (X_1, \ldots, X_p)^T$ is uniformly distributed on S^{p-1} then $-\mathbf{X}$ has the same distribution as $\mathbf{X}$ and so all odd moments of $\mathbf{X}$ are zero. Also, symmetry considerations and (9.2.12) give

$$\mathrm{E}[X_i X_j] = \begin{cases} 0 & \text{if } i \neq j, \\ p^{-1} & \text{if } i = j. \end{cases} \tag{9.6.1}$$

Further, from

$$\mathrm{E}\left[\left(\sum_{i=1}^{n} X_i^2\right)\left(\sum_{j=1}^{n} X_j^2\right)\right] = 1$$

and

$$\mathrm{E}[X_i^4] = \mathrm{E}\left[\left(\frac{X_i + X_j}{\sqrt{2}}\right)^4\right]$$

it follows that

$$\mathrm{E}[X_i^2 X_j^2] = \begin{cases} \{p(p+2)\}^{-1} & \text{if } i \neq j, \\ 3\{p(p+2)\}^{-1} & \text{if } i = j. \end{cases} \tag{9.6.2}$$

Higher-order moments $\mathrm{E}[X_{i_1} \ldots X_{i_s}]$ of the uniform distribution on S^{p-1} can be obtained from the following simple relationship (9.6.3) between $\mathrm{E}[X_{i_1} \ldots X_{i_s}]$ and the corresponding moment $\mu_{i_1 \ldots i_s}$ of the $N_p(\mathbf{0}, \mathbf{I}_p)$ distribution. If $\mathbf{Z} \sim N_p(\mathbf{0}, \mathbf{I}_p)$ then for any $\mathbf{U}$ in $O(p)$, $\mathbf{UZ} \sim N_p(\mathbf{0}, \mathbf{I}_p)$, and so, for any positive r, the conditional distribution of $\|\mathbf{Z}\|^{-1}\mathbf{Z}$ given $\|\mathbf{Z}\| = r$ is uniform on S^{p-1}. It follows that

$$\begin{aligned}
\mu_{i_1 \ldots i_s} &= \frac{1}{(2\pi)^{p/2}} \int_{\mathbb{R}^p} z_{i_1} \ldots z_{i_s}\, e^{-\mathbf{z}^T\mathbf{z}/2}\, d\mathbf{z} \\
&= \frac{1}{(2\pi)^{p/2}} \int_0^\infty r^s \mathrm{E}[X_{i_1} \ldots X_{i_s}]\, e^{-r^2/2} \frac{2\pi^{p/2}}{\Gamma(p/2)} r^{p-1}\, dr \\
&= \left(\frac{2^{s/2}}{\Gamma(p/2)} \int_0^\infty e^{-t} t^{(s+p)/2-1} dt \right) \mathrm{E}[X_{i_1} \ldots X_{i_s}],
\end{aligned}$$

where $t = r^2/2$, and so

$$\begin{aligned}
\mathrm{E}[X_{i_1} \ldots X_{i_s}] &= \frac{\Gamma(p/2)}{2^{s/2}\Gamma((s+p)/2)} \mu_{i_1 \ldots i_s} \\
&= \frac{\Gamma(p/2)}{2^{s/2}\Gamma((s+p)/2)} \sum \delta_{k_1 k_2} \delta_{k_3 k_4} \ldots \delta_{k_{s-1} k_s}, \qquad (9.6.3)
\end{aligned}$$

where the sum is over all partitions $\{k_1, k_2\}, \ldots, \{k_{s-1}, k_s\}$ of $\{i_1, \ldots, i_s\}$ into subsets of size 2, and δ is the Kronecker delta ($\delta_{jk} = 1$ if $j = k$, $\delta_{jk} = 0$ if $j \neq k$). Alternatively, these moments can be obtained by transforming to spherical polar coordinates and then integrating over the radius.

9.6.2 Large-Sample Asymptotics

Large-sample asymptotic results can be obtained by using the embedding approach and applying the central limit theorem. First, we consider results for axially symmetric densities. Let $\mathbf{x}_1, \ldots, \mathbf{x}_n$ be independent random variables on S^{p-1} with an axially symmetric density of the form (9.3.31). The tangent-normal decomposition (9.1.2) can be rewritten as

$$\mathbf{x} = \mathbf{x}_{\boldsymbol{\mu}} + \mathbf{x}_{\perp} \qquad (9.6.4)$$

where

$$\mathbf{x}_{\boldsymbol{\mu}} = (\mathbf{x}^T\boldsymbol{\mu})\boldsymbol{\mu} = t\boldsymbol{\mu}, \qquad \mathbf{x}_{\perp} = (\mathbf{I}_p - \boldsymbol{\mu}\boldsymbol{\mu}^T)\mathbf{x} = (1-t^2)^{\frac{1}{2}}\mathbf{x},$$

so that $\mathbf{x}_{\boldsymbol{\mu}}$ and $\mathbf{x}_{\perp}$ are the components of $\mathbf{x}$ parallel to $\boldsymbol{\mu}$ and orthogonal to it, respectively. Similar notation will be used for the analogous decomposition of the vector sample mean $\bar{\mathbf{x}}$. Put $\bar{C} = \bar{\mathbf{x}}^T\boldsymbol{\mu}$, as in (9.2.5). Then it follows (Watson, 1983a, p. 137; 1983b; 1983c) from the central limit theorem, symmetry of the distribution, and (9.3.33)–(9.3.34) that, for large n,

$$\begin{aligned}
n^{\frac{1}{2}}\bar{\mathbf{x}}_{\perp} &\stackrel{\cdot}{\sim} N\left(\mathbf{0}, \frac{1-\mathrm{E}[t^2]}{p-1}(\mathbf{I}_p - \boldsymbol{\mu}\boldsymbol{\mu}^T)\right), \qquad (9.6.5) \\
n^{\frac{1}{2}}(\bar{C} - \mathrm{E}[t]) &\stackrel{\cdot}{\sim} N(0, \mathrm{var}(t)), \qquad (9.6.6)
\end{aligned}$$

and that $\bar{\mathbf{x}}_\perp$ and $\bar{C}$ are asymptotically independent. If $\mathrm{E}[t] \neq 0$, it follows from (9.6.5)–(9.6.6) and the approximations

$$2(1-\cos\theta) \simeq \theta^2, \qquad \sin\theta \simeq \theta \qquad \text{for}\, \theta \simeq 0$$

(with $\sin\theta = \|\bar{\mathbf{x}}_\perp\|/\|\bar{\mathbf{x}}\|$), that

$$2n(p-1)\frac{\mathrm{E}[t]^2}{1-\mathrm{E}[t^2]}(1-\bar{\mathbf{x}}_0^T\boldsymbol{\mu}) \;\dot{\sim}\; \chi^2_{p-1}, \qquad n \to \infty. \tag{9.6.7}$$

The special case of von Mises–Fisher distributions is considered in Watson (1983b) and extensions to distributions symmetric about a subspace are given in Watson (1983c). The distribution of $n\|\bar{\mathbf{x}}_\perp\|^2$ under local alternatives to the hypothesis of given mean direction is a multiple of a non-central χ^2_{p-1} distribution. Details are given in Watson (1983a, pp. 140–143).

Large-sample asymptotic results on the distribution of the sample mean direction $\bar{\mathbf{x}}_0$ of almost any distribution on S^{p-1} can be obtained by Taylor expansion of the radial projection map $\mathbf{v} \mapsto \mathbf{v}/\|\mathbf{v}\|$ from $\mathbb{R}^p \setminus \{\mathbf{0}\}$ to S^{p-1}. Let $\bar{\mathbf{x}}_0$ be the sample mean direction of a random sample from a distribution on S^{p-1} with mean direction $\boldsymbol{\mu}$, mean resultant length ρ, and variance matrix $\boldsymbol{\Sigma}$. Then, if $\rho > 0$,

$$\mathrm{E}[\bar{\mathbf{x}}_0] = \boldsymbol{\mu} - \frac{1}{n}\boldsymbol{\nu} + \frac{1}{n}\boldsymbol{\tau} + O(n^{-3/2}),$$

where the normal component $\boldsymbol{\nu}$ and the tangential component $\boldsymbol{\tau}$ of the asymptotic bias are given by

$$\boldsymbol{\nu} = \frac{1}{2\rho^2}\left(\mathrm{tr}\boldsymbol{\Sigma} - \boldsymbol{\mu}^T\boldsymbol{\Sigma}\boldsymbol{\mu}\right)\boldsymbol{\mu},$$

$$\boldsymbol{\tau} = \frac{-1}{\rho^2}\left(\mathbf{I}_p - \boldsymbol{\mu}\boldsymbol{\mu}^T\right)\boldsymbol{\Sigma}\boldsymbol{\mu}.$$

If $\rho > 0$ and $\boldsymbol{\Sigma}$ is non-singular then

$$\sqrt{n}\left(\bar{\mathbf{x}}_0 - \boldsymbol{\mu}\right) \;\dot{\sim}\; N\left(\mathbf{0}, \frac{1}{\rho^2}\left(\mathbf{I}_p - \boldsymbol{\mu}\boldsymbol{\mu}^T\right)\boldsymbol{\Sigma}\left(\mathbf{I}_p - \boldsymbol{\mu}\boldsymbol{\mu}^T\right)\right) \qquad n \to \infty,$$

and so

$$2n\left(1 - \bar{\mathbf{x}}_0^T\boldsymbol{\mu}\right) \;\dot{\sim}\; \sum_{i=1}^{p-1}\lambda_i Z_i^2, \qquad n \to \infty, \tag{9.6.8}$$

where $Z_1, \ldots, Z_{p-1}$ are independent standard normal random variables and $\lambda_1, \ldots, \lambda_{p-1}$ are the non-zero eigenvalues of $\rho^{-2}\left(\mathbf{I}_p - \boldsymbol{\mu}\boldsymbol{\mu}^T\right)\boldsymbol{\Sigma}\left(\mathbf{I}_p - \boldsymbol{\mu}\boldsymbol{\mu}^T\right)$. See Hendriks, Landsman & Ruymgaart (1996). If the distribution on S^{p-1} has rotational symmetry about $\boldsymbol{\mu}$ then (9.6.8) reduces to (9.6.7).

The Uniform Distribution

If $\mathbf{x}$ is uniformly distributed on S^{p-1} then so is $\mathbf{Ux}$, for any orthogonal matrix $\mathbf{U}$. It follows that $\mathrm{E}[\mathbf{x}] = \mathbf{0}$ and (by (9.2.12)) that $\mathrm{var}(\mathbf{x}) = p^{-1}\mathbf{I}_p$. Alternatively, the variance can be obtained from (9.6.1). By the central limit theorem,

$$np\bar{R}^2 \;\dot{\sim}\; \chi^2_p. \tag{9.6.9}$$

The case $p = 2$ was given in (4.8.14). The result (9.6.9) was first proved by Rayleigh (1919) and forms the basis of the Rayleigh test for uniformity considered in Section 10.4.1.

Von Mises–Fisher Distributions

Now suppose that $\mathbf{x} \sim M_p(\boldsymbol{\mu}, \kappa)$. Then from (9.3.7),

$$\mathrm{E}[\bar{C}] = A_p(\kappa). \tag{9.6.10}$$

It follows that (9.6.5)–(9.6.6) specialise to

$$n^{\frac{1}{2}}\bar{\mathbf{x}}_\perp \;\dot{\sim}\; N\left(\mathbf{0}, \frac{1 - A_p(\kappa)^2 - A_p'(\kappa)}{p-1}(\mathbf{I}_p - \boldsymbol{\mu}\boldsymbol{\mu}^T)\right), \tag{9.6.11}$$

$$n^{\frac{1}{2}}(\bar{C} - A_p(\kappa)) \;\dot{\sim}\; N(0, A_p'(\kappa)), \tag{9.6.12}$$

and that $\bar{\mathbf{x}}_\perp$ and $\bar{C}$ are asymptotically independent.

9.6.3 High-Concentration Asymptotics

As the concentration of the distribution of a random unit vector $\mathbf{x}$ increases, $\mathbf{x}$ is 'distributed on a smaller portion of S^{p-1}' and the embedding, intrinsic and wrapped approaches become indistinguishable. In many cases, high-concentration asymptotic results can be obtained either by straightforward expansion of the density about the mean direction or as examples of the general results of Jørgensen (1987) on dispersion models.

Von Mises–Fisher Distributions

For von Mises–Fisher distributions, there are the following high-concentration asymptotic results which extend the high-concentration results on von Mises distributions which were considered in Section 4.8.2.

From (9.3.16), it follows immediately that

$$2n\kappa(1 - \bar{C}) \;\dot{\sim}\; \chi^2_{n(p-1)}, \tag{9.6.13}$$

extending (4.8.23). For large κ, $\bar{R} \simeq \bar{C} \simeq 1$, so that

$$2n\kappa(\bar{R} - \bar{C}) \simeq n\kappa(\bar{R}^2 - \bar{C}^2) = n\kappa\|(\mathbf{I}_p - \boldsymbol{\mu}\boldsymbol{\mu}^T)\bar{\mathbf{x}}\|^2. \tag{9.6.14}$$

Combining this with (9.3.15) gives

$$2n\kappa(\bar{R}-\bar{C}) \mathrel{\dot{\sim}} \chi^2_{p-1}, \tag{9.6.15}$$

extending (4.8.28).

From (9.6.13), (9.6.15) and the 'analysis of variance' decomposition

$$2n\kappa(1-\bar{C}) = 2n\kappa(1-\bar{R}) + 2n\kappa(\bar{R}-\bar{C}), \tag{9.6.16}$$

we deduce that

$$2n\kappa(1-\bar{R}) \mathrel{\dot{\sim}} \chi^2_{(n-1)(p-1)} \tag{9.6.17}$$

and that $2n\kappa(1-\bar{R})$ and $2n\kappa(\bar{R}-\bar{C})$ are approximately independently distributed for large κ. Hence, the two components (9.6.15) and (9.6.17) of (9.6.13) behave in the same way as the components in the usual ANOVA decomposition. These results are due to Watson (1956a; 1983a). In the case $p=3$, Stephens (1967) has shown that these approximations are tolerable for κ as low as 3 and are accurate for $\kappa \geq 5$. These results form the basis of various F tests which will be considered in Sections 10.4–10.6. Watson (1983a, pp. 157–165; 1984) has found the non-null distributions of many of these F statistics.

Another consequence of (9.6.16) is that, for large κ and for t and $n-t$ moderately small compared with n,

$$\frac{(n-t-1)(t+R_{n-t}-R_n)}{t(n-t-R_{n-t})} \mathrel{\dot{\sim}} F_{(p-1)t,(p-1)(n-t-1)}, \tag{9.6.18}$$

where $R_m = \|\mathbf{x}_1 \ldots + \mathbf{x}_m\|$. Fisher & Willcox (1978) derived (9.6.18) and showed by simulation that (for $p=2,3$ and $t=1$) this approximation to an F distribution is reasonable for $\kappa \geq 3$ and $n \geq 5$.

The approximation (9.6.13) can be improved by multiplicative correction, i.e.

$$2n\gamma(1-\bar{C}) \mathrel{\dot{\sim}} \chi^2_{n(p-1)} \tag{9.6.19}$$

to a higher order of approximation, where γ is chosen to make the mean of $2n\gamma(1-\bar{C})$ close to the mean of $\chi^2_{n(p-1)}$. Taking

$$\frac{1}{\gamma} = \frac{2}{p-1}\left[1-A_p(\kappa)\right] \tag{9.6.20}$$

ensures that

$$\mathrm{E}[2n\gamma(1-\bar{C})] = 2n(p-1).$$

Unless κ is very large, it is sensible to replace κ by γ in (9.6.13)–(9.6.17).

Applying the asymptotic expansion for $I_\nu(\kappa)$ given in (A.4) of Appendix 1 to (9.6.20) and performing some manipulation gives (Stephens, 1992)

$$\frac{1}{\gamma} = \frac{1}{\kappa} - \frac{p-3}{4\kappa^2} - \frac{p-3}{4\kappa^3} + O(\kappa^{-4}). \tag{9.6.21}$$

Thus no correction of order higher than $1/\kappa$ is available when $p = 3$. In the case $p = 3$, a good choice of γ is given by

$$\frac{1}{\gamma} = \frac{1}{\kappa} - \frac{1}{5\kappa^3}. \tag{9.6.22}$$

It is found that this approximation is quite satisfactory for $\kappa \geq 1.5$ and is acceptable for $\kappa \geq 1$. The choice (9.6.22) was obtained initially by studying the behaviour of the residuals after fitting the chi-squared approximation (9.6.17) to the distribution of $\bar{R}$.

Von Mises–Fisher Distributions: The Multi-sample Case

As in Section 4.8.2, the high-concentration approximations can be applied in the multi-sample case. Let $\mathbf{x}_{11}, \ldots, \mathbf{x}_{1n_1}; \ldots; \mathbf{x}_{q1}, \ldots, \mathbf{x}_{qn_q}$ be q independent random samples of sizes $n_1, \ldots, n_q$ from $M_p(\boldsymbol{\mu}, \kappa)$ with mean resultant lengths $\bar{R}_1, \ldots, \bar{R}_q$, respectively. Let $\bar{R}$ denote the mean resultant length of the combined sample of length $n = n_1 + \ldots + n_q$. It follows from (9.6.17) that, if κ is large,

$$2n\kappa(1 - \bar{R}) \quad \dot{\approx} \quad \chi^2_{(n-1)(p-1)}, \tag{9.6.23}$$

$$2n_j\kappa(1 - \bar{R}_j) \quad \dot{\approx} \quad \chi^2_{(n_j-1)(p-1)}, \quad j = 1, \ldots, q. \tag{9.6.24}$$

Together with the decomposition

$$2\kappa(n - R) = 2\kappa\left(n - \sum_{j=1}^{q} R_j\right) + 2\kappa\left(\sum_{j=1}^{q} R_j - R\right), \tag{9.6.25}$$

this implies that

$$2\kappa\left(n - \sum_{j=1}^{q} R_j\right) \quad \dot{\approx} \quad \chi^2_{(n-q)(p-1)}, \tag{9.6.26}$$

$$2\kappa\left(\sum_{j=1}^{q} R_j - R\right) \quad \dot{\approx} \quad \chi^2_{(q-1)(p-1)} \tag{9.6.27}$$

and that $2\kappa(n - \sum_{j=1}^{q} R_j)$ and $2\kappa(\sum_{j=1}^{q} R_j - R)$ are approximately independent. This result is due to Watson (1956a) and extends (4.8.36). Unless κ is very large, it is sensible to replace κ by γ in (9.6.23)–(9.6.27).

Note that in the decomposition (9.6.25), the terms again represent respectively (κ times) the variation of the combined sample, the variation within samples, and the variation between samples.

Bingham distributions

If $\mathbf{x}$ has the Bingham distribution (9.4.3) with parameter matrix $\mathbf{A}$, then

$$2\left(\kappa_1 - \mathbf{x}^T\mathbf{A}\mathbf{x}\right) \;\dot{\sim}\; \chi^2_{p-1}, \qquad \kappa_1 - \kappa_2 \to \infty, \tag{9.6.28}$$

where $\kappa_1, \ldots \kappa_p$ (with $\kappa_1 \geq \ldots \geq \kappa_p$) are the eigenvalues of $\mathbf{A}$. This follows from general high-concentration results on exponential dispersion models (see Jørgensen, 1987, Section 4) or from

$$\mathbf{x}^T\mathbf{A}\mathbf{x} = \mathrm{tr}\;\mathbf{A} - \sum_{i=2}^{p}(\kappa_1 - \kappa_i)y_i^2,$$

where $\mathbf{U}^T\mathbf{x} = (y_1, \ldots, y_p)$ with

$$\mathbf{A} = \mathbf{U}\;\mathrm{diag}\left(\kappa_1, \ldots, \kappa_p\right)\mathbf{U}^T.$$

An improvement of the approximation (9.6.28) was given by Bingham, Chang & Richards (1992).

9.6.4 High-Dimensional Asymptotics

Asymptotic results are also available when the dimension $p-1$ of the sphere tends to infinity. So far, these results have been almost entirely of theoretical interest, although a possible application to large-sample asymptotics of certain permutation distributions was given in Watson (1988). Let $\mathbf{x}$ be a random vector in S^{p-1} having a distribution which is symmetrical about some q-dimensional subspace V of $\mathbb{R}^p$, so that its density is of the form $f(\mathbf{x}) = c_p^{-1}g(\mathbf{x}_V)$, where $\mathbf{x}_V$ denotes the orthogonal projection of $\mathbf{x}$ onto V, g is a given function on V and c_p is a normalising constant. Watson (1983d) proved that, as $p \to \infty$ with V fixed,

$$p^{\frac{1}{2}}\mathbf{x}_V \;\dot{\sim}\; N(\mathbf{0}, \mathbf{I}_q). \tag{9.6.29}$$

Further, if $\mathbf{x}_1, \ldots, \mathbf{x}_r$ are independent and uniformly distributed on S^{p-1} then the distribution of $\{p^{\frac{1}{2}}\mathbf{x}_i^T\mathbf{x}_j : 1 \leq i \leq j \leq r\}$ tends to $N(\mathbf{0}, \mathbf{I}_{r(r+1)/2})$ as $p \to \infty$ (Stam, 1982). Generalisations of (9.6.29) for von Mises–Fisher and Watson distributions and for uniform distributions on Stiefel manifolds (see Section 13.2.1) were given by Watson (1983d; 1988). In particular, he showed that, if $\mathbf{x}$ has a von Mises–Fisher distribution on S^{p-1} with mean direction $\boldsymbol{\mu}$ in V and with concentration $p^{\frac{1}{2}}\kappa$, then for large p,

$$\begin{aligned} p^{\frac{1}{2}}\boldsymbol{\mu}^T\mathbf{x} &\;\dot{\sim}\; N(\kappa, 1), \\ p^{\frac{1}{2}}\{\mathbf{x}_V - (\mathbf{x}^T\boldsymbol{\mu})\boldsymbol{\mu}\} &\;\dot{\sim}\; N(\mathbf{0}, \mathbf{I}_q - \boldsymbol{\mu}\boldsymbol{\mu}^T) \end{aligned}$$

and $\mathbf{x}^T\boldsymbol{\mu}$ and $\mathbf{x}_V - (\boldsymbol{\mu}^T\mathbf{x})\boldsymbol{\mu}$ are asymptotically independent.

10

Inference on Spheres

10.1 INTRODUCTION

In this chapter we consider inference for distributions on spheres of arbitrary dimension. Exploratory data analysis on S^2 is considered in Section 10.2. In Section 10.3 we treat estimation for the von Mises–Fisher, Watson, Bingham, Kent and angular central Gaussian distributions. The next three sections deal mainly with hypothesis testing for von Mises–Fisher distributions, although extensions to other distributions with rotational symmetry are considered. Single-sample, two-sample and multi-sample tests are discussed in Sections 10.4, 10.5 and 10.6, respectively. Many of the tests are analogous to those for the circular case. However, on spheres other than the circle inference has a special geometrical flavour and that is emphasised here. Non-parametric techniques are almost non-existent on higher-dimensional spheres. Tests for axial distributions are considered in Section 10.7. A general framework for testing uniformity is given in Section 10.8.

10.2 EXPLORATORY DATA ANALYSIS

Before undertaking formal tests, it is advisable to carry out exploratory data analysis. Because it is the case of greatest practical importance, we restrict the discussion here to data on the usual sphere S^2.

When inspecting spherical data, it is usually necessary to rotate the sphere. Let $\mathbf{a}$ and $\mathbf{b}$ be points on the sphere with $\mathbf{a} \neq \mathbf{b}$. Then a convenient rotation which takes $\mathbf{a}$ to $\mathbf{b}$ is $\mathbf{H}(\mathbf{a}, \mathbf{b})$, defined by

$$\mathbf{H}(\mathbf{a}, \mathbf{b}) = \frac{(\mathbf{a}+\mathbf{b})(\mathbf{a}+\mathbf{b})^T}{1+\mathbf{a}^T\mathbf{b}} - \mathbf{I}_p. \tag{10.2.1}$$

This is the rotation by π about the axis through $\mathbf{a}+\mathbf{b}$. It interchanges $\mathbf{a}$ and $\mathbf{b}$.

A useful impression of a spherical data set can usually be obtained by plotting suitable projections of the raw data (e.g. by Lambert's equal-area projection which was given in (9.1.1)) or of contour plots or shade plots of estimated densities. Contour plots can be obtained from the computer program of Diggle & Fisher (1985). A more detailed discussion of methods of displaying spherical data is given in Fisher, Lewis & Embleton (1987, Section 3.3).

Further useful information can be obtained from the mean resultant length $\bar{R}$, the sample mean direction $\bar{\mathbf{x}}_0$, and the scatter matrix $\bar{\mathbf{T}}$. In particular, the eigenvalues $\bar{t}_1, \bar{t}_2, \bar{t}_3$ of $\bar{\mathbf{T}}$ give an indication of the general shape of the data set. A guide to such interpretation of these eigenvalues is given in Table 10.1.

Table 10.1 Descriptive interpretation of the shapes of spherical distributions in terms of the eigenvalues $\bar{t}_1, \bar{t}_2, \bar{t}_3$ of $\bar{\mathbf{T}}$ and the mean resultant length $\bar{R}$

Relative magnitudes of eigenvalues	Type of distribution	Other features
$\bar{t}_1 \simeq \bar{t}_2 \simeq \bar{t}_3$	uniform	
$\bar{t}_1$ large; $\bar{t}_2, \bar{t}_3$ small		
(i) $\bar{t}_2 \neq \bar{t}_3$	unimodal if $\bar{R} \simeq 1$,	concentrated at one end of $\mathbf{t}_1$
	bimodal otherwise	concentrated at both ends of $\mathbf{t}_1$
(ii) $\bar{t}_2 \simeq \bar{t}_3$	unimodal if $\bar{R} \simeq 1$,	rotational symmetry about $\mathbf{t}_1$
	bipolar otherwise	
$\bar{t}_3$ small; $\bar{t}_1$, $\bar{t}_2$ large		
(i) $\bar{t}_1 \neq \bar{t}_2$	girdle	concentrated about great circle in plane of $\mathbf{t}_1, \mathbf{t}_2$
(ii) $\bar{t}_1 \simeq \bar{t}_2$	symmetric girdle	rotational symmetry about $\mathbf{t}_3$

Probability plots for spherical data are particularly useful tools in exploratory data analysis. Consider first observations $\mathbf{x}_1, \ldots, \mathbf{x}_n$ on S^2 which might have come from a Fisher distribution. Let (θ'_i, ϕ'_i) denote the spherical polar coordinates of $\mathbf{x}_i$ in some coordinate system in which the sample mean direction is the pole $\theta' = 0$. A convenient choice of (θ'_i, ϕ'_i) is as the spherical polar coordinates of $\mathbf{H}(\bar{\mathbf{x}}_0, \mathbf{n})\mathbf{x}_i$, where $\mathbf{H}(\bar{\mathbf{x}}_0, \mathbf{n})$ is the rotation given by (10.2.1) which takes the sample mean direction $\bar{\mathbf{x}}_0$ to the north pole $\mathbf{n} = (0, 0, 1)^T$. Similarly, let (θ''_i, ϕ''_i) denote the spherical polar coordinates of $\mathbf{x}_i$ in the coordinate system in which the sample mean direction $\bar{\mathbf{x}}_0$ has spherical polar coordinates $(\theta'', \phi'') = (\pi/2, 0)$. More precisely, define the rotation matrix $\mathbf{A}$ by

$$\mathbf{A} = \begin{pmatrix} \sin\hat{\alpha}\cos\hat{\beta} & \sin\hat{\alpha}\sin\hat{\beta} & \cos\hat{\alpha} \\ \sin\hat{\beta} & -\cos\hat{\beta} & 0 \\ \cos\hat{\alpha}\cos\hat{\beta} & \cos\hat{\alpha}\sin\hat{\beta} & -\sin\hat{\alpha} \end{pmatrix},$$

where

$$\bar{\mathbf{x}}_0 = \begin{pmatrix} \sin\hat{\alpha}\cos\hat{\beta} \\ \sin\hat{\alpha}\sin\hat{\beta} \\ \cos\hat{\alpha} \end{pmatrix}.$$

The spherical polar coordinates (θ_i'', ϕ_i'') are defined by

$$\mathbf{A}\mathbf{x}_i = \begin{pmatrix} \sin\theta_i''\cos\phi_i'' \\ \sin\theta_i''\sin\phi_i'' \\ \cos\theta_i'' \end{pmatrix}, \tag{10.2.2}$$

with ϕ_i'' in the range $(-\pi, \pi]$. The probability plots for Fisher distributions are constructed as follows.

(i) The *colatitude plot* is the plot of the order statistics $1 - \cos\theta'_{(i)}$ against $-\log\{1-(i-\frac{1}{2})/n\}$. It follows from (9.3.19) that, if κ is not too small (say $\kappa \geq 2$), this plot should be close to a straight line through the origin with slope κ^{-1}.

(ii) The *longitude plot* is the plot of the order statistics $\phi'_{(i)}/2\pi$ against $(i-\frac{1}{2})/n$. It follows from the symmetry of $F(\boldsymbol{\mu}, \kappa)$ about $\boldsymbol{\mu}$ that this plot should be close to a straight line through the origin with unit slope. Note that this plot depends on the choice of zero for the longitude ϕ'.

(iii) The *two-variable plot* is the plot of the ordered values of $\phi_i''(\sin\theta_i'')^{1/2}$ against standard normal quantiles. It follows from the approximation

$$\kappa(1,0,0)\mathbf{x} = \kappa\sin\theta''\cos\phi'' \simeq \kappa\sin\theta'' - \tfrac{1}{2}\kappa\left(\phi''\sqrt{\sin\theta''}\right)^2$$

that, if κ is not too small (say $\kappa \geq 2$), this plot should be close to a straight line through the origin with slope $\kappa^{-1/2}$.

These plots provide graphical tests of goodness-of-fit to a Fisher distribution, quick estimates of the concentration parameter κ, and a method of detecting outliers. Formal tests of goodness-of-fit based on these probability plots will be considered in Section 12.3. The above probability plots were introduced by Lewis & Fisher (1982).

Example 10.1

Probability plots for the remnant-magnetisation data set of Example 9.1 are given in Fig. 10.1.

All three plots are reasonably linear, so there is no suggestion that the data may not be from a Fisher distribution. This impression will be confirmed in Example 12.1 by formal tests. The colatitude and two-variable plots give rough estimates of 45 and 27, respectively, for the concentration parameter κ.

Probability plots are available also for axial data. For observations $\pm\mathbf{x}_1, \ldots, \pm\mathbf{x}_n$ on $\mathbb{R}P^2$ which might have come from a Watson distribution, colatitude plots and longitude plots are constructed as follows.

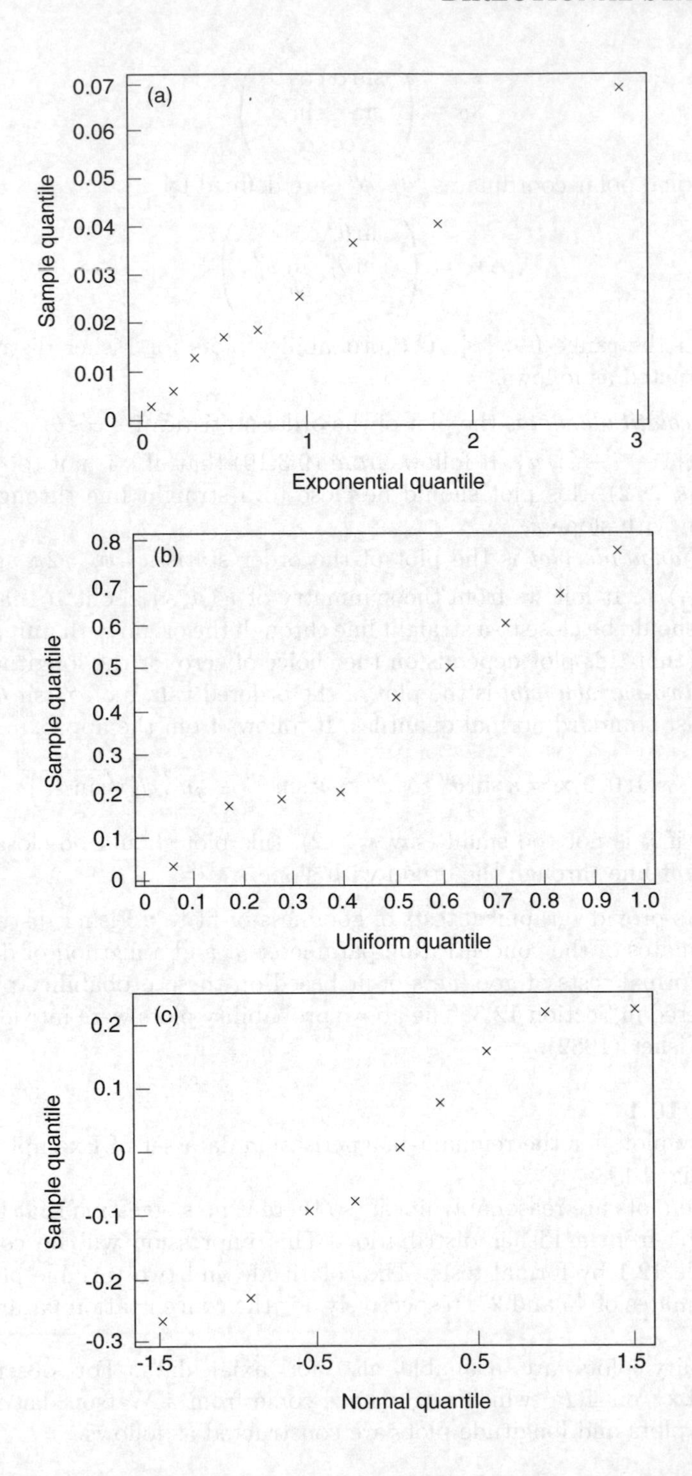

Figure 10.1 Probability plots for the data of Example 9.1: (a) colatitude plot; (b) longitude plot; (c) two-variable plot.

Bipolar Case

Let (θ'_i, ϕ'_i) be spherical polar coordinates of $\mathbf{x}_i$ about the dominant eigenvector $\mathbf{t}_1$ of $\bar{\mathbf{T}}$.

(i) The *colatitude plot* is the plot of the ordered values of $1-\cos^2\theta'_i$ against $-\log\{1-(i-\frac{1}{2})/n\}$. It follows from (10.7.25) that if $-\kappa$ is not too small (say $-\kappa \geq 5$) then this plot should be close to a straight line through the origin with slope $|\kappa|^{-1}$.

(ii) The *longitude plot* is the plot of the order statistics $\phi'_{(i)}/\pi$ against $(i-\frac{1}{2})/n$. This plot should be close to a straight line through the origin with unit slope. Note that this plot depends on the choice of zero for the longitude ϕ'.

Girdle Case

Let (θ'_i, ϕ'_i) be spherical polar coordinates of $\mathbf{x}_i$ about the eigenvector $\mathbf{t}_3$ of $\bar{\mathbf{T}}$ corresponding to the smallest eigenvalue.

(i) The *colatitude plot* is the plot of the ordered values of $\cos^2\theta'_i$ against $F^{-1}(i-\frac{1}{2})/n)$, where F denotes the distribution function of the χ^2_1 distribution. It follows from (10.7.23) that if κ is not too small (say $\kappa \geq 5$) then this plot should be close to a straight line through the origin with slope κ^{-1}.

(ii) The *longitude plot* is the plot of the order statistics $\phi'_{(i)}/\pi$ against $(i-\frac{1}{2})/n$. This plot should be close to a straight line through the origin with unit slope. Note that this plot depends on the choice of zero for the longitude ϕ'.

These probability plots for Watson distributions were introduced by Best & Fisher (1986).

10.3 POINT ESTIMATION

10.3.1 Von Mises–Fisher Distributions

Let $\mathbf{x}_1, \ldots, \mathbf{x}_n$ be a random sample from the von Mises–Fisher distribution with probability density function

$$\left(\frac{\kappa}{2}\right)^{p/2-1} \frac{1}{\Gamma(p/2) I_{p/2-1}(\kappa)} \exp\{\kappa \boldsymbol{\mu}^T \mathbf{x}\} \qquad (10.3.1)$$

with respect to the uniform distribution. The vector mean of $\mathbf{x}_1, \ldots, \mathbf{x}_n$ is

$$\bar{\mathbf{x}} = \frac{1}{n}\sum_{i=1}^{n} \mathbf{x}_i.$$

It follows from the general theory of exponential models (see (3.5.6)) that $\bar{\mathbf{x}}$ is sufficient for κ and $\boldsymbol{\mu}$.

Maximum Likelihood Estimation

Because the model with densities (10.3.1) is a regular exponential model with canonical statistic $\mathbf{x}$, the maximum likelihood estimate of the canonical parameter $\kappa\boldsymbol{\mu}$ is obtained by putting the population mean equal to the sample mean, i.e.

$$\hat{\rho}\hat{\boldsymbol{\mu}} = \bar{\mathbf{x}}$$

(see (3.5.7)). It follows that

$$\hat{\rho} = \bar{R}, \tag{10.3.2}$$

$$\hat{\boldsymbol{\mu}} = \bar{\mathbf{x}}_0, \tag{10.3.3}$$

and so

$$A_p(\hat{\kappa}) = \bar{R}, \quad \text{i.e.} \quad \hat{\kappa} = A_p^{-1}(\bar{R}), \tag{10.3.4}$$

where A_p is defined by (9.3.8). The function A_3^{-1} is tabulated in Appendix 3.2. For large κ, the asymptotic formula

$$\begin{aligned} I_\nu(\kappa) \quad = \quad & (2\pi\kappa)^{-\frac{1}{2}} e^\kappa \left\{ 1 - \frac{4\nu^2 - 1}{8\kappa} - \frac{(4\nu^2-1)(4\nu^2-9)}{2(8\kappa)^2} \right\} \\ & + O(\kappa^{-3}) \end{aligned} \tag{10.3.5}$$

gives

$$A_p(\kappa) = 1 - \frac{p-1}{2\kappa} + \frac{(p-1)(p-3)}{8\kappa^2} + O(\kappa^{-3}). \tag{10.3.6}$$

Hence,

$$\hat{\kappa} = \frac{p-1}{2(1-\bar{R})} + \frac{p-3}{4} + O(n(1-\bar{R})), \quad \bar{R} \to 1.$$

Thus, for $\bar{R} \simeq 1$,

$$\hat{\kappa} \simeq \frac{p-1}{2(1-\bar{R})}. \tag{10.3.7}$$

When $p = 3$, this approximation is satisfactory for $\bar{R} \geq 0.9$.

From the series expansion

$$I_p(\kappa) = \sum_{r=0}^{\infty} \frac{1}{\Gamma(p+r+1)\Gamma(r+1)} \left(\frac{\kappa}{2}\right)^{2r+p}, \tag{10.3.8}$$

it follows that, for small κ,

$$A_p(\kappa) = \frac{\kappa}{p} \left\{ 1 - \frac{1}{p(p+2)}\kappa^2 + \frac{1}{p^2(p+2)(p+4)}\kappa^4 + O(\kappa^6) \right\} \tag{10.3.9}$$

so that, for small $\bar{R}$,

$$\hat{\kappa} = p\bar{R} \left\{ 1 + \frac{p}{p+2}\bar{R}^2 + \frac{p^2(p+8)}{(p+2)^2(p+4)}\bar{R}^4 + O(\bar{R}^6) \right\}. \tag{10.3.10}$$

When $p = 3$, taking just the first term in (10.3.10) gives a satisfactory approximation for $\bar{R} < 0.05$.

Because (10.3.1) is a regular exponential model with canonical parameter $\boldsymbol{\theta} = \kappa\boldsymbol{\mu}$, it follows from general theory (see (3.5.4) and (ii) after (3.5.4)) that the Fisher information is

$$\frac{\partial}{\partial\boldsymbol{\theta}^T} A_p(\kappa)\boldsymbol{\mu}.$$

A calculation using

$$\frac{\partial||\boldsymbol{\theta}||}{\partial\boldsymbol{\theta}^T} = \frac{\boldsymbol{\theta}}{||\boldsymbol{\theta}||} \tag{10.3.11}$$

and

$$\frac{\partial^2||\boldsymbol{\theta}||}{\partial\boldsymbol{\theta}\partial\boldsymbol{\theta}^T} = \frac{1}{||\boldsymbol{\theta}||}\left(\mathbf{I}_p - \frac{\boldsymbol{\theta}}{||\boldsymbol{\theta}||}\frac{\boldsymbol{\theta}^T}{||\boldsymbol{\theta}||}\right) \tag{10.3.12}$$

shows that the Fisher information matrix is

$$\begin{pmatrix} I_{\kappa\kappa} & I_{\kappa\boldsymbol{\mu}} \\ I_{\boldsymbol{\mu}\kappa} & I_{\boldsymbol{\mu}\boldsymbol{\mu}} \end{pmatrix} = \begin{pmatrix} A_p'(\kappa) & 0 \\ 0 & \kappa A_p(\kappa)(\mathbf{I}_p - \boldsymbol{\mu}\boldsymbol{\mu}^T) \end{pmatrix}. \tag{10.3.13}$$

Hence, for large n (provided that $\kappa > 0$), $\hat{\kappa}$ and $\hat{\boldsymbol{\mu}}$ are asymptotically independently normally distributed with means κ and $\boldsymbol{\mu}$. Also,

$$n\text{var}(\hat{\kappa}) \simeq A_p'(\kappa)^{-1}, \quad n\text{var}(\hat{\boldsymbol{\mu}}) \simeq \frac{1}{\kappa A_p(\kappa)}(\mathbf{I}_p - \boldsymbol{\mu}\boldsymbol{\mu}^T). \tag{10.3.14}$$

Note that the limit of $n\text{var}(\hat{\boldsymbol{\mu}})$ is singular, since $\boldsymbol{\mu}^T\boldsymbol{\mu} = 1$. If $\mathbf{x} \sim M_p(\boldsymbol{\mu}, \kappa)$ then the score based on $\mathbf{x}$ is

$$\frac{\partial l(\boldsymbol{\theta}; \mathbf{x})}{\partial\boldsymbol{\theta}^T} = \mathbf{x} - A_p(\kappa) \tag{10.3.15}$$

(where $l(\boldsymbol{\theta}; \mathbf{x})$ denotes the log-likelihood), and so it follows from (10.3.13) that the variance matrix $\boldsymbol{\Sigma}$ of $\mathbf{x}$ is

$$\boldsymbol{\Sigma} = \begin{pmatrix} A_p'(\kappa) & 0 \\ 0 & \dfrac{A_p(\kappa)}{\kappa}(\mathbf{I}_p - \boldsymbol{\mu}\boldsymbol{\mu}^T) \end{pmatrix}. \tag{10.3.16}$$

Applying (9.2.12) gives the differential equation ((A.14) of Appendix 1)

$$A_p'(\kappa) = 1 - A_p(\kappa)^2 - \frac{p-1}{\kappa}A_p(\kappa) \tag{10.3.17}$$

for A_p.

Taylor expansion of $A_p^{-1}(\bar{R})$, together with (9.3.7)–(9.3.8), the formula

$$\left.\frac{\partial^2 A_p^{-1}(||\mathbf{x}||)}{\partial\mathbf{x}\partial\mathbf{x}^T}\right|_{\mathbf{x}=A_p(\kappa)(1,0,\ldots,0)^T} =$$

$$\text{diag}\left(-\frac{A_p''(\kappa)}{A_p'(\kappa)^3}, \frac{1}{A_p(\kappa)A_p'(\kappa)}, \ldots, \frac{1}{A_p(\kappa)A_p'(\kappa)}\right)$$

and (10.3.16), shows that

$$\mathrm{E}[\hat{\kappa}-\kappa] = \frac{1}{n}\frac{(p-1)A_p'(\kappa)-\kappa A_p''(\kappa)}{2\kappa A_p'(\kappa)^2} + O(n^{-2}), \tag{10.3.18}$$

for large n (Schou, 1978). Substituting (10.3.6) in (10.3.18) gives the approximation

$$\mathrm{E}[\hat{\kappa}-\kappa] \simeq \frac{(p+1)\kappa}{n(p-1)}, \qquad n\to\infty,\ \frac{\kappa}{p}\to\infty. \tag{10.3.19}$$

For the model $M_p(\boldsymbol{\mu},\kappa)$ in which $\boldsymbol{\mu}$ is known, the analogues of (10.3.18) and (10.3.19) are

$$\mathrm{E}[\hat{\kappa}_{\boldsymbol{\mu}}-\kappa] = -\frac{A_p''(\kappa)}{2nA_p'(\kappa)^2} + O(n^{-2}) \tag{10.3.20}$$

$$= \frac{1}{n}\frac{2\kappa^4 A_p'(\kappa)^3 + 3(p-1)\kappa A_p(\kappa)^2 + (p^2-p-2\kappa^2)A_p(\kappa) - (p-1)\kappa}{2\left\{\kappa-(p-1)A_p(\kappa)-(p-1)\kappa A_p(\kappa)^2\right\}^2} + O(n^{-2}) \tag{10.3.21}$$

and

$$\mathrm{E}[\hat{\kappa}_{\boldsymbol{\mu}}-\kappa] \simeq \frac{2\kappa}{n(p-1)}, \qquad n\to\infty,\ \frac{\kappa}{p}\to\infty, \tag{10.3.22}$$

respectively (Mardia, Southworth & Taylor, 1999).

When κ is large, (9.6.17) gives

$$2n\kappa(1-\bar{R}) \stackrel{\cdot}{\sim} \chi^2_{(n-1)(p-1)}. \tag{10.3.23}$$

Combining this with (10.3.7) gives

$$n(p-1)\frac{\kappa}{\hat{\kappa}} \stackrel{\cdot}{\sim} \chi^2_{(n-1)(p-1)} \tag{10.3.24}$$

for large κ.

The above expressions for the bias of $\hat{\kappa}$ can be used to produce bias-corrected estimators of κ. For example, since $\mathrm{E}[1/\chi^2_f] = 1/(f-2)$ for $f>2$, it follows from (10.3.23) that the estimator κ^* given by

$$\kappa^* = \frac{(n-1)(p-1)-2}{2n(1-\bar{R})} \tag{10.3.25}$$

is approximately unbiased. In the case $p=3$, Best & Fisher (1981) calculated the approximate bias of the estimator given by the right-hand side of (10.3.7) and showed by simulation that the estimator

$$\left(1-\frac{1}{n}\right)^2 \frac{p-1}{2(1-\bar{R})}$$

is approximately unbiased unless both n and κ are small.

Example 10.2
Assuming that the Icelandic remnant-magnetisation data of Example 9.1 came from a Fisher distribution $F(\boldsymbol{\mu}, \kappa)$, we estimate the parameters $\boldsymbol{\mu}$ and κ.
From Example 9.1,

$$\bar{\mathbf{x}}_0 = (0.298, -0.135, -0.945)^T, \qquad \bar{R} = 0.975.$$

From Appendix 3.2, $A_3^{-1}(0.975) = 39.53$. Thus

$$\hat{\boldsymbol{\mu}} = (0.298, -0.135, -0.945)^T, \qquad \hat{\kappa} = 39.53.$$

The estimate given by (10.3.25) is $\kappa^* = 30.75$. The values of both $\hat{\kappa}$ and κ^* indicate high concentration.

Marginal Maximum Likelihood Estimation

As in the case $p = 2$ considered in Section 5.3.2, $\bar{R}$ is G-sufficient for κ, where $G = O(p)$, the group of orthogonal transformations of $\mathbb{R}^p$. Accordingly, Schou (1978) considered estimation of κ by $\check{\kappa}$, the maximum likelihood estimator based on the marginal distribution of $\bar{R}$. The corresponding estimate maximises the marginal likelihood

$$\check{L}(\kappa) = \frac{A_p(\kappa)^n}{A_p(n\kappa\bar{R})}$$

and is given by

$$\begin{aligned} \check{\kappa} &= 0, & \bar{R} < n^{-\frac{1}{2}} \\ A_p(\check{\kappa}) &= \bar{R}A_p(n\check{\kappa}\bar{R}), & \bar{R} \geq n^{-\frac{1}{2}}. \end{aligned} \qquad (10.3.26)$$

For the case $p = 3$, a table giving $\check{\kappa}$ in terms of $\bar{R}$ is given in Appendix 3.3.
The relationship between $\hat{\kappa}$ and $\check{\kappa}$ is that

$$\check{\kappa} \leq \hat{\kappa} \quad \text{and} \quad \check{\kappa} = \hat{\kappa} + O_p(n^{-1}).$$

Expansion of (10.3.26) for $\bar{R} \simeq 1$ yields

$$\check{\kappa} = \frac{p-1}{2}\frac{1-1/n}{1-\bar{R}} - \frac{p-3}{4}\frac{1-1/n^2\bar{R}}{1-1/n} + O(n(1-\bar{R})), \quad \bar{R} \to 1.$$

The large-sample and high-concentration asymptotic distributions of $\check{\kappa}$ are

$$\sqrt{n}(\check{\kappa} - \kappa) \stackrel{\cdot}{\sim} N(0, A_p(\kappa)^{-1}), \qquad \kappa \to \infty, \qquad (10.3.27)$$

and

$$(n-1)(p-1)\frac{\kappa}{\check{\kappa}} \stackrel{\cdot}{\sim} \chi^2_{(n-1)(p-1)}, \qquad \kappa \to \infty. \qquad (10.3.28)$$

In the case $p = 3$, simulations by Schou (1978) indicate that $\check{\kappa}$ tends to have a smaller bias than $\hat{\kappa}$.

10.3.2 Watson Distributions

Let $\pm\mathbf{x}_1, \ldots, \pm\mathbf{x}_n$ be a random sample from the Watson distribution (9.4.1) with probability density function

$$f(\pm\mathbf{x}; \boldsymbol{\mu}, \kappa) = M\left(\frac{1}{2}, \frac{p}{2}, \kappa\right)^{-1} \exp\{\kappa(\mathbf{x}^T\boldsymbol{\mu})^2\}, \qquad (10.3.29)$$

where $M(1/2, p/2, \cdot)$ denotes a Kummer function (see (A.17) of Appendix 1). The log-likelihood function is

$$\begin{aligned} l(\boldsymbol{\mu}, \kappa; \pm\mathbf{x}_1, \ldots, \pm\mathbf{x}_n) &= \kappa \sum_{i=1}^{n} (\mathbf{x}_i^T\boldsymbol{\mu})^2 - n \log M\left(\frac{1}{2}, \frac{p}{2}, \kappa\right) \\ &= n\left\{\kappa\boldsymbol{\mu}^T\bar{\mathbf{T}}\boldsymbol{\mu} - \log M\left(\frac{1}{2}, \frac{p}{2}, \kappa\right)\right\}, \qquad (10.3.30) \end{aligned}$$

where $\bar{\mathbf{T}}$ is the scatter matrix of $\mathbf{x}_1, \ldots, \mathbf{x}_n$ given by (9.2.10). Differentiation of (10.3.30) with respect to κ, together with (A.17) and (A.20) of Appendix 1, gives

$$D_p(\hat{\kappa}) = \hat{\boldsymbol{\mu}}^T\bar{\mathbf{T}}\hat{\boldsymbol{\mu}}, \qquad (10.3.31)$$

where

$$D_p(\kappa) = \frac{\int_0^1 t^2 e^{\kappa t^2}(1-t^2)^{(p-3)/2} dt}{\int_0^1 e^{\kappa t^2}(1-t^2)^{(p-3)/2} dt} = \frac{M(3/2, p/2+1, \kappa)}{pM(1/2, p/2, \kappa)}. \qquad (10.3.32)$$

Let $\bar{t}_1, \ldots, \bar{t}_p$ be the eigenvalues of $\bar{\mathbf{T}}$ with

$$\bar{t}_1 \geq \ldots \geq \bar{t}_p,$$

and let $\pm\mathbf{t}_1, \ldots, \pm\mathbf{t}_p$ be the corresponding unit eigenvectors. Since $\boldsymbol{\mu}^T\boldsymbol{\mu} = 1$, it follows from (10.3.30) that

$$\pm\,\hat{\boldsymbol{\mu}} = \begin{cases} \pm\mathbf{t}_1, & \hat{\kappa} > 0, \\ \pm\mathbf{t}_p, & \hat{\kappa} < 0. \end{cases} \qquad (10.3.33)$$

The expansions (A.23)–(A.25) of D_p in Appendix 1 yield the approximations

$$D_p(\kappa) = \begin{cases} 1 - (p-1)/2\kappa + O(\kappa^{-2}), & \kappa \to \infty, \\ p^{-1}(1 + 2(p-1)\kappa/p(p+2)) + O(\kappa^2), & \kappa \simeq 0, \\ 1/(2|\kappa|) + O(\kappa^{-2}), & \kappa \to -\infty. \end{cases} \qquad (10.3.34)$$

For $p = 3$, these approximations are adequate for $\kappa \geq 10$, $|\kappa| \leq 0.2$ and $\kappa \leq -10$, respectively. The function D_3^{-1} is tabulated in Appendices 3.4 and 3.5.

Example 10.3
For 30 measured directions on the c-axis of calcite grains from the Taconic mountains of New York (Bingham, 1964), the scatter matrix is

$$\bar{\mathbf{T}} = \begin{pmatrix} 0.471 & 0.000 & 0.085 \\ 0.000 & 0.353 & 0.057 \\ 0.085 & 0.057 & 0.176 \end{pmatrix}.$$

We fit a Watson distribution to this data set.

The eigenvalues and unit eigenvectors of $\bar{\mathbf{T}}$ are

$$\begin{aligned} \bar{t}_1 &= 0.495, & \pm\mathbf{t}_1 &= (0.954, 0.114, 0.276)^T \\ \bar{t}_2 &= 0.365, & \pm\mathbf{t}_2 &= (-0.176, 0.961, 0.212)^T \\ \bar{t}_3 &= 0.139, & \pm\mathbf{t}_3 &= (-0.241, -0.251, 0.938)^T. \end{aligned}$$

Since $\bar{t}_1 \simeq \bar{t}_2 > \bar{t}_3$, it is appropriate to fit a girdle Watson distribution with rotational symmetry about $\mathbf{t}_3$. The assumption of such symmetry will be justified by a formal test in Example 10.13.

The maximum likelihood estimate $\pm\hat{\boldsymbol{\mu}}$ of $\pm\boldsymbol{\mu}$ is

$$\pm\hat{\boldsymbol{\mu}} = \pm\mathbf{t}_3 = \pm(-0.241, -0.251, 0.938)^T.$$

From Appendix 3.4,

$$\hat{\kappa} = D_3^{-1}(0.139) = -3.33,$$

indicating that $|\kappa|$ is moderately large.

10.3.3 Bingham Distributions

Let $\pm\mathbf{x}_1, \ldots, \pm\mathbf{x}_n$ be a random sample from the Bingham distribution with probability density function

$$f(\pm\mathbf{x}; \mathbf{A}) = {}_1F_1\left(\frac{1}{2}, \frac{p}{2}, \mathbf{A}\right)^{-1} \exp\{\mathbf{x}^T\mathbf{A}\mathbf{x}\}, \qquad (10.3.35)$$

where $\mathbf{A}$ is a symmetric matrix. The log-likelihood function is

$$l(\mathbf{A}; \pm\mathbf{x}_1, \ldots, \pm\mathbf{x}_n) = n\left\{\log\mathrm{tr}(\mathbf{A}\bar{\mathbf{T}}) - \log{}_1F_1\left(\frac{1}{2}, \frac{p}{2}, \mathbf{A}\right)\right\}.$$

Write $\mathbf{A}$ and $\bar{\mathbf{T}}$ in polar form as

$$\begin{aligned} \mathbf{A} &= \mathbf{U}\mathbf{K}\mathbf{U}^T, \\ \bar{\mathbf{T}} &= \mathbf{V}\boldsymbol{\Lambda}\mathbf{V}^T, \end{aligned}$$

with $\mathbf{U}$ and $\mathbf{V}$ orthogonal, $\mathbf{K} = \mathrm{diag}(\kappa_1, \ldots, \kappa_p)$ and $\boldsymbol{\Lambda} = (\bar{t}_1, \ldots, \bar{t}_p)$, where $\kappa_1 \geq \ldots \geq \kappa_p$ and $\bar{t}_1 \geq \ldots \geq \bar{t}_p$. Then it follows (e.g. from a result of

Theobald, 1975) that the maximum of $\operatorname{tr} \mathbf{A}\bar{\mathbf{T}}$ with respect of $\mathbf{A}$ occurs when $\mathbf{U} = \mathbf{V}$. Thus

$$\hat{\mathbf{U}} = \mathbf{V}. \tag{10.3.36}$$

Since the Bingham distributions with densities (10.3.35) form a regular exponential model with canonical statistic $\mathbf{x}\mathbf{x}^T$, general results on exponential models (see (3.5.7) and (3.5.3)) show that

$$\bar{t}_i = \left. \frac{\partial \log {}_1F_1(1/2, p/2, \mathbf{K})}{\partial \kappa_i} \right|_{\mathbf{K}=\hat{\mathbf{K}}}, \quad i = 1, \ldots, p. \tag{10.3.37}$$

Note that $\kappa_1, \ldots, \kappa_p$ are estimable only up to an additive constant, because $\mathbf{A}$ and $\mathbf{A} + c\mathbf{I}_p$ define the same distribution, for any real c.

For $p = 3$, solution of (10.3.37) can often be expedited by using Kent's (1987) high-concentration asymptotic expansions for $1/{}_1F_1(\frac{1}{2}, \frac{3}{2}, \mathbf{K})$. For $p =$ 3, the $\hat{\kappa}_i$ can be obtained from the tables of Mardia & Zemroch (1977).

If $p = 3$ and the eigenvalues $\bar{t}_1, \ldots, \bar{t}_3$ of $\bar{\mathbf{T}}$ indicate that the distribution is nearly rotationally symmetric, i.e. either bipolar or symmetric girdle, the following approximation (suggested by C. Bingham) can be used. In the bipolar case, define

$$d = \bar{t}_2 - \bar{2}_3, \qquad s = \bar{t}_1 + \bar{t}_2, \qquad \hat{\kappa}_0 = -D_3^{-1}(\bar{t}_1). \tag{10.3.38}$$

Then

$$\hat{\kappa}_1 \simeq 0, \qquad \hat{\kappa}_2 \simeq \hat{\kappa}_0 + \delta, \qquad \hat{\kappa}_3 \simeq \hat{\kappa}_0 - \delta, \tag{10.3.39}$$

where

$$\delta = \frac{2d\hat{\kappa}_0}{s(\hat{\kappa}_0 - 1.5) + 1}. \tag{10.3.40}$$

In the girdle case, define

$$d = \bar{t}_1 - \bar{t}_2, \qquad s = \bar{t}_1 + \bar{t}_2, \qquad \hat{\kappa}_0 = D_3^{-1}(\bar{t}_3). \tag{10.3.41}$$

Then

$$\hat{\kappa}_1 \simeq 0, \qquad \hat{\kappa}_2 \simeq -2\delta, \qquad \hat{\kappa}_3 \simeq \hat{\kappa}_0 - \delta, \tag{10.3.42}$$

where δ is given by (10.3.40).

Example 10.4

Assuming that the calcite grain data set of Example 10.3 came from a Bingham distribution, let us estimate the parameter matrix $\mathbf{A}$.

From the unit eigenvectors of $\bar{\mathbf{T}}$ given in Example 10.3, the maximum likelihood estimate $\hat{\mathbf{U}}$ of the orthogonal matrix $\mathbf{U}$ in the parameter matrix $\mathbf{A} = \mathbf{U}\mathbf{K}\mathbf{U}^T$ is

$$\begin{pmatrix} 0.954 & -0.176 & -0.241 \\ 0.114 & 0.961 & -0.251 \\ 0.276 & 0.212 & 0.938 \end{pmatrix}.$$

From Example 10.3, the eigenvalues of $\bar{\mathbf{T}}$ are

$$\bar{t}_1 = 0.495, \qquad \bar{t}_2 = 0.365, \qquad \bar{t}_3 = 0.139.$$

Numerical solution of the likelihood equations (10.3.37) gives

$$\hat{\kappa}_1 = 0, \quad \hat{\kappa}_2 = -0.68, \quad \hat{\kappa}_3 = -3.62. \tag{10.3.43}$$

Since Example 10.3 suggests that the calcite grain distribution is approximately girdle in form, we could avoid solving (10.3.37) numerically by using instead the approximation (10.3.42). Simple calculations give

$$d = 0.130, \qquad s = 0.860, \qquad \hat{\kappa}_0 = D_3^{-1}(0.139) = -3.33, \qquad \delta = 0.336,$$

and so

$$\hat{\kappa}_1 = 0, \qquad \hat{\kappa}_2 = -0.67, \qquad \hat{\kappa}_3 = -3.67. \tag{10.3.44}$$

These approximations to $\hat{\kappa}_1$, $\hat{\kappa}_2$, and $\hat{\kappa}_3$ are very close to the estimates in (10.3.43).

10.3.4 Kent Distributions

In the case $p = 3$, convenient estimators of the parameters of a Kent distribution are the following moment estimators, which were proposed by Kent (1982). Let $\mathbf{H}$ be any rotation (such as $\mathbf{H}(\bar{\mathbf{x}}_0, \mathbf{n})$, given by (10.2.1)) which takes the sample mean direction $\bar{\mathbf{x}}_0$ to the north pole $\mathbf{n} = (0,0,1)^T$. Put

$$\begin{aligned}
\mathbf{B} &= \begin{pmatrix} b_{11} & b_{12} & b_{13} \\ b_{21} & b_{22} & b_{23} \\ b_{31} & b_{32} & b_{33} \end{pmatrix} = \mathbf{H}^T \bar{\mathbf{T}} \mathbf{H}, \\
\psi &= \frac{1}{2} \tan^{-1}\left(\frac{2 b_{12}}{b_{11} - b_{22}} \right), \\
\mathbf{K} &= \begin{pmatrix} \cos\psi & -\sin\psi & 0 \\ \sin\psi & \cos\psi & 0 \\ 0 & 0 & 1 \end{pmatrix}.
\end{aligned}$$

Then the moment estimate of the parameter matrix $\mathbf{\Gamma}$ of (9.3.25) is $\tilde{\mathbf{\Gamma}}$, where

$$\tilde{\mathbf{\Gamma}} = \mathbf{HK}.$$

Let l_1 and l_2 be the eigenvalues of

$$\begin{pmatrix} b_{11} & b_{12} \\ b_{21} & b_{22} \end{pmatrix}$$

and put

$$Q = l_1 - l_2.$$

The moment estimates $\tilde{\kappa}$ and $\tilde{\beta}$ of κ and β are given implicitly by

$$\left.\frac{\partial c}{\partial \kappa}\right|_{(\kappa,\beta)=(\tilde{\kappa},\tilde{\beta})} = \bar{R}, \qquad \left.\frac{\partial c}{\partial \beta}\right|_{(\kappa,\beta)=(\tilde{\kappa},\tilde{\beta})} = Q.$$

When κ is large, the limiting bivariate normal approximation (9.3.24) gives the high-concentration approximations

$$\tilde{\kappa} \simeq \frac{1}{2(1-\bar{R})-Q} + \frac{1}{2(1-\bar{R})+Q},$$
$$\tilde{\beta} \simeq \frac{1}{2}\left\{\frac{1}{2(1-\bar{R})-Q} - \frac{1}{2(1-\bar{R})+Q}\right\}.$$

Kent (1982) showed that the moment estimators are consistent and are reasonably efficient compared with maximum likelihood estimators.

10.3.5 Angular Central Gaussian Distributions

The maximum likelihood estimate $\hat{\mathbf{A}}$ of the parameter matrix $\mathbf{A}$ in the angular central Gaussian distributions with probability density functions

$$f(\mathbf{x};\mathbf{A}) = |\mathbf{A}|^{-1/2}(\mathbf{x}^T\mathbf{A}^{-1}\mathbf{x})^{-p/2} \tag{10.3.45}$$

cannot be found explicitly. However, $\hat{\mathbf{A}}$ (with $\operatorname{tr}\hat{\mathbf{A}} = p$) can be found iteratively from the equation

$$\hat{\mathbf{A}} = p\left\{\sum_{i=1}^{n}\frac{1}{\mathbf{x}_i^T\hat{\mathbf{A}}^{-1}\mathbf{x}_i}\right\}^{-1}\sum_{i=1}^{n}\frac{\mathbf{x}_i\mathbf{x}_i^T}{\mathbf{x}_i^T\hat{\mathbf{A}}^{-1}\mathbf{x}_i}. \tag{10.3.46}$$

Further details can be found in Tyler (1987).

10.4 SINGLE-SAMPLE TESTS

10.4.1 Tests of Uniformity

Because of the central role played by the uniform distribution, one of the most important hypotheses about a distribution on a sphere is that of uniformity, as in the circular case.

The Rayleigh Test

Perhaps the simplest test of uniformity is Rayleigh's test. This generalises the Rayleigh test on the circle, which was considered in Section 6.3.1. Because $\mathrm{E}[\mathbf{x}] = \mathbf{0}$ when $\mathbf{x}$ has the uniform distribution on S^{p-1}, it is intuitively reasonable to reject uniformity when the vector sample mean $n^{-1}(\mathbf{x}_1+\ldots+\mathbf{x}_n)$

is far from $\mathbf{0}$, i.e. when $\bar{R}$ is large. For $p = 3$ this test is due to Rayleigh (1919) and was first formulated explicitly by Watson (1956b). As shown in the next paragraph, it is useful to take the test statistic as

$$pn\bar{R}^2.$$

A more formal justification for the Rayleigh test is that it is the score test of uniformity within the von Mises–Fisher model (10.3.1). Let $\boldsymbol{\theta} = \kappa\boldsymbol{\mu}$ be the canonical parameter of the exponential model (10.3.1). Then the log-likelihood based on $\mathbf{x}_1, \ldots, \mathbf{x}_n$ is

$$l(\boldsymbol{\theta}; \mathbf{x}_1, \ldots, \mathbf{x}_n) = n\left\{\boldsymbol{\theta}^T\bar{\mathbf{x}} - a_p(\kappa)\right\}, \tag{10.4.1}$$

where

$$a_p(\kappa) = \log\left(\left(\frac{\kappa}{2}\right)^{1-p/2}\Gamma\left(\frac{p}{2}\right)I_{p/2-1}(\kappa)\right). \tag{10.4.2}$$

The score is

$$\mathbf{U} = n\bar{\mathbf{x}} - nA_p(\kappa), \tag{10.4.3}$$

where A_p is defined by (9.3.8).

At $\kappa = 0$, $\mathbf{U} = n\bar{\mathbf{x}}$, while (9.6.1) gives

$$n\mathrm{var}(\bar{\mathbf{x}}) = p^{-1}\mathbf{I}_p.$$

Then the score statistic is

$$S = pn\bar{R}^2. \tag{10.4.4}$$

Under uniformity, the asymptotic large-sample distribution of $pn\bar{R}^2$ is

$$pn\bar{R}^2 \overset{\cdot}{\sim} \chi^2_p. \tag{10.4.5}$$

The error in the approximation (10.4.5) is of order $O(n^{-1/2})$. The modified Rayleigh statistic S^* given by

$$S^* = \left(1 - \frac{1}{2n}\right)S + \frac{1}{2n(p+2)}S^2 \tag{10.4.6}$$

has a χ^2_p distribution with error of order $O(n^{-1})$ (Jupp, 1999). The practical importance of this is that, for all except the smallest sample sizes, there is negligible error in the significance levels if the observed S^* is compared to the usual upper quantiles of the χ^2_p distribution. A more refined approximation to the tail probability is given by the saddlepoint approximation

$$\begin{aligned}\Pr(\bar{R} > z) &= \frac{n^{p/2-1}}{\Gamma(p/2)2^{p/2-1}}e^{-nt^2/2}z^{p-2}\left\{\left(\frac{z}{\hat{\kappa}}\right)^{p-3}\left(1 - z^2 - (p-1)\frac{z}{\hat{\kappa}}\right)\right\}^{-1/2} \\ &\quad + \delta_{p3}\frac{2n^{1/2}}{\sqrt{3}}\{1 - \Phi(\sqrt{n}t)\} + O\left(\frac{1}{n}\exp\left\{-\frac{nt^2}{2}\right\}\right)\end{aligned} \tag{10.4.7}$$

(cf. Jensen, 1995, pp. 162–165), where

$$\hat{\kappa} = A_p^{-1}(z), \qquad t = \{2[\hat{\kappa}z - a_p(\hat{\kappa})]\}^{1/2},$$

$\delta_{p3} = 1$ if $p = 3$ and is zero otherwise, and Φ denotes the cumulative distribution function of the $N(0,1)$ distribution. The approximation (10.4.7) holds uniformly in κ.

The Rayleigh test is also the likelihood ratio test of uniformity within the von Mises–Fisher model. Let w be the likelihood ratio statistic. Then

$$\begin{aligned} w &= 2n(\hat{\kappa}\bar{R} - a_p(\hat{\kappa})) \\ &= 2n(\hat{\kappa}A_p(\hat{\kappa}) - a_p(\hat{\kappa})), \end{aligned}$$

where a_p is defined in (10.4.2). Differentiation with respect to $\hat{\kappa}$ gives

$$\begin{aligned} \frac{dw}{d\hat{\kappa}} &= 2n[A_p(\hat{\kappa}) + \hat{\kappa}A_p'(\hat{\kappa}) - A_p(\hat{\kappa})] \\ &= 2n\hat{\kappa}A_p'(\hat{\kappa}), \end{aligned}$$

and so

$$\frac{dw}{d\bar{R}} = \frac{dw}{d\hat{\kappa}} \bigg/ \frac{d\bar{R}}{d\hat{\kappa}} = 2n\hat{\kappa} \geq 0.$$

Thus w is an increasing function of $\bar{R}$, and so the likelihood ratio test is equivalent to the Rayleigh test.

Since w can be written as a function of $\bar{R}$, it is clear that w is invariant under rotations and reflections of S^{p-1}. From the Neyman–Pearson lemma, the Rayleigh test is also the (uniformly) most powerful invariant test for testing uniformity against the alternative of a von Mises–Fisher distribution.

Example 10.5

Bernoulli (1735) discussed the question whether the close coincidence of the orbital planes of the seven planets (then known) could have arisen 'by chance'. Although there are conceptual problems in considering a population from which the planets might be a random sample, Bernoulli's question provides a nice illustration of the Rayleigh test (Watson, 1970). Each planetary orbit determines a direction – the directed unit normal to the orbital plane. Each orbit is specified by (i, Ω), where i is the *inclination* of the orbital plane to the ecliptic, and Ω is the *longitude of the ascending node*. The corresponding directed unit normal is

$$\mathbf{x} = (\sin\Omega \sin i, -\cos\Omega \sin i, \cos i)^T.$$

The data (for the nine planets now known) are shown in Table 10.2. Do the data look like a random sample from the uniform distribution on S^2?

Calculation shows that $\bar{R} = 0.996$. Then $3n\bar{R}^2 = 26.77$ and $S^* = 33.24$. Since $\chi^2_{3;0.01} = 11.345$, the hypothesis of uniformity is rejected strongly.

Table 10.2 Orbits of the nine planets (Watson, 1970, reproduced by permission of Uppsala Universitet)

Planet	Inclination i	Longitude Ω
Mercury	$7°0'$	$47°08'$
Venus	$3°23'$	$75°47'$
Earth	$0°$	$0°$
Mars	$1°51'$	$48°47'$
Jupiter	$1°19'$	$99°26'$
Saturn	$2°30'$	$112°47'$
Uranus	$0°46'$	$73°29'$
Neptune	$1°47'$	$130°41'$
Pluto	$17°10'$	$109°0'$

Some Other Tests of Uniformity

A disadvantage of the Rayleigh test is that it is not consistent against alternatives with zero mean resultant length. A test of uniformity on S^2 which is consistent against all alternatives is Giné's (1975) F_n test, which rejects uniformity for large values of

$$F_n = \frac{3n}{2} - \frac{4}{n\pi} \sum_{i<j} (\Psi_{ij} + \sin \Psi_{ij}), \tag{10.4.8}$$

where

$$\Psi_{ij} = \cos^{-1}(\mathbf{x}_i^T \mathbf{x}_j) \tag{10.4.9}$$

is the smaller of the two angles between $\mathbf{x}_i$ and $\mathbf{x}_j$. In the large-sample limiting distribution of F_n, the 10%, 5% and 1% quantiles are 2.355, 2.748 and 3.633, respectively (Keilson *et al.*, 1983).

Ajne's A_n statistic (6.3.46) for testing uniformity on the circle generalises (Beran, 1968) to S^{p-1} as

$$A_n = \frac{n}{4} - \frac{1}{n\pi} \sum_{i<j} \Psi_{ij}, \tag{10.4.10}$$

where Ψ_{ij} is given by (10.4.9). For $p = 3$, selected quantiles of $4A_n$ are given by Keilson *et al.* (1983). In particular, the limiting large-sample 10%, 5% and 1% quantiles of $4A_n$ are 1.816, 2.207 and 3.090, respectively. For $p = 3$, Prentice (1978) derived the approximation

$$\Pr\left(A_n > \frac{x}{16}\right) \simeq 1.652 \left[\Pr(\chi_3^2 > x) - 0.516 \Pr(\chi_1^2 > x)\right], \tag{10.4.11}$$

for large n and x. The test which rejects uniformity for large values of A_n is locally most powerful invariant against the alternatives with probability density function f on S^{p-1} defined by

$$f(\mathbf{x}) = \begin{cases} c & \text{if } \mathbf{x}^T\boldsymbol{\mu} > 0, \\ 0 & \text{otherwise,} \end{cases} \tag{10.4.12}$$

where $\boldsymbol{\mu}$ is an unknown unit vector and $0 < c < 1$.

Other useful tests of uniformity on S^{p-1} include those which reject uniformity for large values of Bingham's statistic (10.7.1) or of Giné's G_n statistic (10.7.5) for the axes $\pm\mathbf{x}_1, \ldots, \pm\mathbf{x}_n$ determined by the observed directions $\mathbf{x}_1, \ldots, \mathbf{x}_n$.

General machinery for constructing tests of uniformity on S^{p-1} (and more general manifolds) is considered in Section 10.8.

A comparative power study of the Rayleigh test, Giné's (1975) F_n test (10.4.8) and a third test (based on the number of pairs of data points less than a certain distance apart) was carried out by Diggle, Fisher & Lee (1985). They found that, against equal mixtures of Fisher distributions with the modes not too far apart, the Rayleigh test was most powerful in small samples but was dominated by the F_n test in larger samples. Against mixtures of Fisher distributions in which the modes were 180° apart, the third test was most powerful.

10.4.2 Tests for the Mean Direction

A common problem is that of testing

$$H_0 : \boldsymbol{\mu} = \boldsymbol{\mu}_0 \text{ against } H_1 : \boldsymbol{\mu} \neq \boldsymbol{\mu}_0. \tag{10.4.13}$$

The main tests for this are the likelihood ratio and score tests. We give these (and some variants), together with modifications which bring the (small-sample or low-concentration) null distributions close to their asymptotic chi-squared distributions. Geometrical interpretations of these tests are indicated in Fig. 10.2.

Except where otherwise stated, we shall assume in the remainder of Section 10.4 that $\mathbf{x}_1, \ldots, \mathbf{x}_n$ is a random sample of size n from the von Mises–Fisher distribution with probability density function (10.3.1).

Concentration Parameter Known

The Likelihood Ratio Test

The likelihood ratio statistic is

$$w = 2n\kappa\{\|\bar{\mathbf{x}}\| - \boldsymbol{\mu}_0^T\bar{\mathbf{x}}\} = 2n\kappa(\bar{R} - \bar{C}),$$

where $\bar{C} = \bar{\mathbf{x}}^T\boldsymbol{\mu}_0$ denotes the component along $\boldsymbol{\mu}_0$ of the vector mean $\bar{\mathbf{x}}$. H_0 is rejected for large values of w. Under H_0, the asymptotic (large-sample or

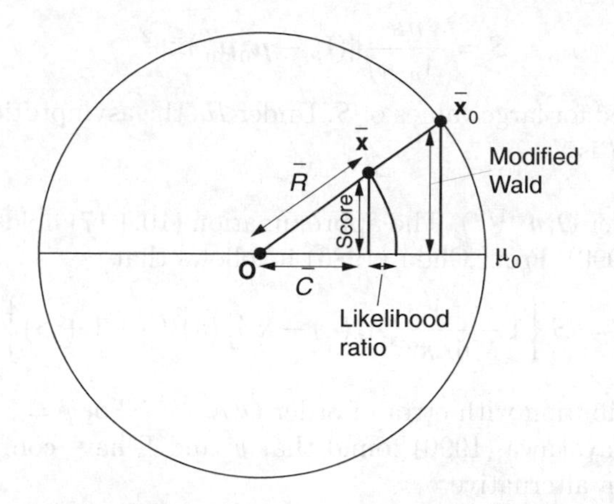

Figure 10.2 Geometrical interpretations of the likelihood ratio, score and modified Wald tests of $H_0 : \boldsymbol{\mu} = \boldsymbol{\mu}_0$ (after Yamamoto & Yanagimoto, 1995). The tests are based on the lengths of the vectors shown, all of which are in the plane of $\boldsymbol{\mu}_0$ and $\bar{\mathbf{x}}_0$.

high-concentration) distribution of w is χ_p^2 with error of order $O(n^{-1})$. The Bartlett-corrected version

$$w^* = \left\{1 + \frac{1}{n}\frac{p-3}{4\kappa A_p(\kappa)}\right\} w \tag{10.4.14}$$

of w derived by Hayakawa (1990) has a χ_p^2 distribution with error of order $O(n^{-2})$.

A Conditional Test

As in the circular case considered in Section 7.2.1, $\bar{R}$ is G-ancillary for $\boldsymbol{\mu}$, μ (Barndorff-Nielsen, 1978a, Section 4.4), where $G = O(p)$ is the orthogonal group of $\mathbb{R}^p$. By the ancillarity principle, it is appropriate to test (10.4.13) using the conditional distribution of $\bar{\mathbf{x}}_0|\bar{R}$, as first suggested by Mardia (1972a, p. 259). From (9.5.10), under H_0, $\bar{\mathbf{x}}_0|\bar{R}$ has an $M_p(\boldsymbol{\mu}_0, \kappa\bar{R})$ distribution. In the case $p = 3$, some selected quantiles of $\cos^{-1}(\bar{\mathbf{x}}_0^T\boldsymbol{\mu}_0)$ can be obtained from Appendix 3.1.

The Score Test

The score at $\boldsymbol{\mu}_0$ is

$$\left.\frac{\partial l}{\partial \boldsymbol{\mu}^T}(\boldsymbol{\mu}, \kappa; \mathbf{x}_1, \ldots, \mathbf{x}_n)\right|_{\boldsymbol{\mu}=\boldsymbol{\mu}_0} = n(\mathbf{I}_p - \boldsymbol{\mu}_0\boldsymbol{\mu}_0^T)\bar{\mathbf{x}}, \tag{10.4.15}$$

so that the geometrical interpretation of the score test is that H_0 is rejected when the tangential part $(\mathbf{I}_p - \boldsymbol{\mu}_0\boldsymbol{\mu}_0^T)\bar{\mathbf{x}}$ of $\bar{\mathbf{x}} - \boldsymbol{\mu}_0$ is large. The score test statistic is

$$S = \frac{n\kappa}{A_p(\kappa)} \|(\mathbf{I}_p - \boldsymbol{\mu}_0\boldsymbol{\mu}_0^T)\bar{\mathbf{x}}\|^2 \tag{10.4.16}$$

and H_0 is rejected for large values of S. Under H_0 the asymptotic large-sample distribution of S is

$$S \dot{\sim} \chi^2_{p-1} \tag{10.4.17}$$

with error of order $O(n^{-1/2})$. The approximation (10.4.17) holds also for large κ (Hayakawa, 1990). From Chou (1986) it follows that

$$S^* = S\left\{1 - \frac{1}{4n\kappa^3}(A_p(\kappa) - \kappa A_p'(\kappa))(p + 1 + S)\right\}$$

has a χ^2_{p-1} distribution with error of order $O(n^{-3/2})$. For $p = 3$ and $n = 20$, a simulation by Hayakawa (1990) found that w and S have comparable power against a Pitman alternative.

Concentration Parameter Unknown

The Likelihood Ratio Test The likelihood ratio statistic is

$$w = n\{\hat{\kappa}\|\bar{\mathbf{x}}\| - \tilde{\kappa}\boldsymbol{\mu}_0^T\bar{\mathbf{x}} - a_p(\hat{\kappa}) + a_p(\tilde{\kappa})\}.$$

Under H_0, $w \dot{\sim} \chi^2_p$ with error of order $O(n^{-1})$. A Bartlett-corrected version of w, given by Hayakawa (1990), is

$$w^* = w\left\{1 + \frac{1}{4n\kappa}\left[(p-3)\left(\frac{1}{A_p(\kappa)} - \frac{1}{\kappa A_p'(\kappa)}\right) + \frac{2A_p''(\kappa)}{A_p'(\kappa)^2}\right]\right\}.$$

This has a χ^2_p distribution with error of order $O(n^{-2})$.

The Score Test and Variants

The score test statistic is

$$S = \frac{n\tilde{\kappa}}{A_p(\tilde{\kappa})} \|(\mathbf{I}_p - \boldsymbol{\mu}_0\boldsymbol{\mu}_0^T)\bar{\mathbf{x}}\|^2, \tag{10.4.18}$$

where $\tilde{\kappa}$ denotes the restricted maximum likelihood estimator calculated under H_0.

From Hayakawa (1990) it follows that, under H_0,

$$S^* = S\left\{1 + \frac{1}{4n\kappa}\left[\left(\frac{1}{A_p(\kappa)} - \frac{1}{\kappa A_p'(\kappa)}\right)(3(p-1) + S) + \frac{2A_p''(\kappa)}{A_p'(\kappa)^2}\right]\right\}$$

has a χ^2_{p-1} distribution with error of order $O(n^{-3/2})$.

Watson (1983a) modified the score test by replacing $\tilde{\kappa}$ in (10.4.18) by $\hat{\kappa}$ to obtain

$$W = \frac{n\hat{\kappa}}{A_p(\hat{\kappa})} \|(\mathbf{I}_p - \boldsymbol{\mu}_0\boldsymbol{\mu}_0^T)\bar{\mathbf{x}}\|^2.$$

Under H_0 the asymptotic large-sample distribution of W is χ^2_{p-1} with error of order $O(n^{-1/2})$. Chou (1986) showed that

$$\begin{aligned} W^* &= W\left\{1 - \frac{1}{4n\kappa}\left\{\left[\frac{(p+1)\kappa A_p'(\kappa)}{A_p(\kappa)^2} - \frac{5p+9}{A_p(\kappa)} + \frac{4(p+2)}{\kappa A_p'(\kappa)} + \frac{2A_p''(\kappa)}{A_p'(\kappa)^2}\right.\right.\right. \\ &\qquad \left. + \frac{2(p+1)\kappa}{A_p(\kappa)}\left(\frac{1}{A_p(\kappa)} - \frac{1}{\kappa A_p'(\kappa)}\right)\right] \\ &\qquad \left.\left. + \left(\frac{1}{A_p(\kappa)} - \frac{1}{\kappa A_p'(\kappa)}\right)\left(4 - \frac{3\kappa A_p'(\kappa)}{A_p(\kappa)}\right)W\right\}\right\} \end{aligned} \tag{10.4.19}$$

has a χ^2_{p-1} distribution with error of order $O(n^{-3/2})$.

Whereas the score test and modified score test measure the discrepancy between the sample and the hypothesised distribution by the difference $(\mathbf{I}_p - \boldsymbol{\mu}_0\boldsymbol{\mu}_0^T)\bar{\mathbf{x}}$ in means, the Wald test uses the difference $\bar{\mathbf{x}}_0 - \boldsymbol{\mu}_0$ in maximum likelihood estimates. It is convenient to use the modified version

$$W_{\mathrm{MW}} = \frac{n\tilde{\kappa}}{A_p(\tilde{\kappa})}\|(\mathbf{I}_p - \boldsymbol{\mu}_0\boldsymbol{\mu}_0^T)\bar{\mathbf{x}}_0\|^2$$

of the Wald test introduced by Hayakawa & Puri (1985). H_0 is rejected for large values of W_{MW}. Hayakawa (1990) showed that

$$\begin{aligned} W^*_{\mathrm{MW}} &= W_{\mathrm{MW}}\left\{1 + \frac{1}{4n\kappa}\left(\left[\frac{5p+9}{A_p(\kappa)} + \frac{3p-1}{\kappa A_p'(\kappa)} - \frac{2A_p''(\kappa)}{A_p'(\kappa)^2}\right]\right.\right. \\ &\qquad \left.\left. + \left[\frac{3}{A_p(\kappa)} + \frac{1}{\kappa A_p'(\kappa)}\right]W_{\mathrm{MW}}\right)\right\} \end{aligned}$$

has a χ^2_{p-1} distribution with error of order $O(n^{-3/2})$.

For $p = 3$ and $n = 20$, a simulation by Hayakawa (1990) found that w, S and W_{MW} had comparable power against a Pitman alternative.

For large κ and under H_0, the approximations

$$w \;\dot{\sim}\; n(p-1)\log\left(1 + \frac{1}{n-1}F\right), \tag{10.4.20}$$

$$S \;\dot{\sim}\; n(p-1)\frac{F}{n-1+F}, \tag{10.4.21}$$

$$W \;\dot{\sim}\; \frac{n(p-1)}{n-1}F \tag{10.4.22}$$

can be used (Hayakawa, 1990), where $F \sim F_{n-1,(n-1)(p-1)}$.

Some Other Tests

Under H_0, $\bar{R} \simeq \bar{C}$, where $\bar{C} = \bar{\mathbf{x}}^T\boldsymbol{\mu}$. This is the intuitive basis of three useful tests which reject H_0 for large values of $\bar{R}$ compared with $\bar{C}$.

(i) The test proposed by Watson & Williams (1956) uses the result (9.5.11) that the distribution of $\bar{R}$ given $\bar{C}$ does not depend on κ. The hypothesis H_0 is rejected for large values of $\bar{R}$. For $p = 3$, nomograms were given by Stephens (1962b).

(ii) For concentrated von Mises–Fisher distributions, a suitable test is the one which rejects H_0 for large values of $(n-1)(\bar{R}-\bar{C})/(1-\bar{R})$. The high-concentration approximations (9.6.15) and (9.6.17) give

$$\frac{(\bar{R}-\bar{C})/(p-1)}{(1-\bar{R})/(n-1)(p-1)} \mathrel{\dot\sim} F_{p-1,(n-1)(p-1)}, \qquad \kappa \to \infty. \tag{10.4.23}$$

When $p = 2$, (10.4.23) reduces to (7.2.22). When $p = 3$, (10.4.23) is exact in the region $\bar{R} > 1 - 2/n$, both conditionally on $\bar{C}$ and unconditionally. When $p = 3$, the approximation (10.4.23) is reasonable for $\bar{R} > 0.6$.

(iii) For von Mises–Fisher distributions with small concentrations, a suitable test is the one which rejects H_0 for large values of $pn(\bar{R}^2 - \bar{C}^2)$. It follows from (9.6.1) and (9.6.5) that

$$pn(\bar{R}^2 - \bar{C}^2) \mathrel{\dot\sim} \chi^2_{p-1}, \qquad \kappa \to 0. \tag{10.4.24}$$

When $p = 2$, (10.4.24) reduces to (7.2.25). When $p = 3$, approximation (10.4.24) is reasonable for $\bar{R} < 0.25$.

Confidence Regions for the Mean Direction

Confidence regions for the mean direction $\boldsymbol{\mu}$ can obtained from the usual connection betweeen tests and confidence regions.

A Confidence Region Obtained from a Conditional Distribution

From (9.5.10), $\bar{\mathbf{x}}_0|\bar{R} \sim M_p(\boldsymbol{\mu}, n\kappa\bar{R})$. Then an approximate $100(1-\alpha)\%$ confidence region for $\boldsymbol{\mu}$ is

$$\{\boldsymbol{\mu} : \boldsymbol{\mu}^T\bar{\mathbf{x}}_0 \geq \cos\delta\}, \tag{10.4.25}$$

where δ is defined by

$$\Pr(\mathbf{x}^T\boldsymbol{\mu} \geq \delta) = \alpha, \tag{10.4.26}$$

with $\mathbf{x} \sim M_p(\boldsymbol{\mu}, n\hat{\kappa}\bar{R})$. When $p = 3$, values of δ can be found from Appendix 3.1. The confidence region (10.4.25) is the intersection of the unit sphere S^{p-1} with the cone having vertex at the origin, axis the sample mean direction $\bar{\mathbf{x}}_0$ and semi-vertical angle δ (see Fig. 10.3).

A Large-Sample Confidence Region

It follows from (9.6.10)–(9.6.11), the large-sample approximations

$$2n(1-\bar{\mathbf{x}}_0^T\boldsymbol{\mu}) \simeq n\frac{\|\bar{\mathbf{x}}_\perp\|^2}{\|\bar{\mathbf{x}}\|^2} \simeq n\frac{\|\bar{\mathbf{x}}_\perp\|^2}{\bar{R}^2}, \tag{10.4.27}$$

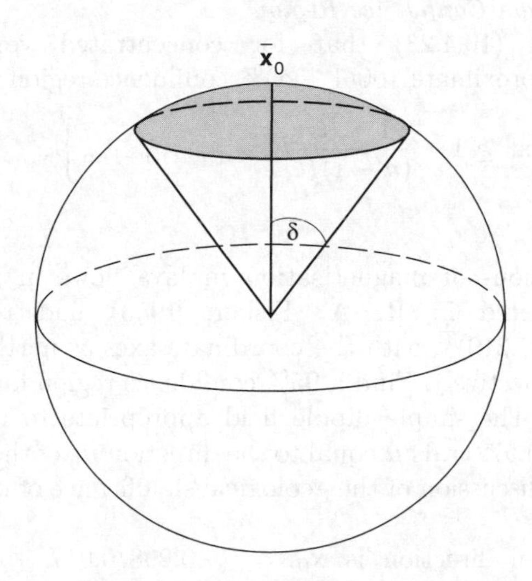

Figure 10.3 Confidence region (shaded) for the mean direction.

and (10.3.4) that, for large n, an approximate $100(1-\alpha)\%$ confidence region for $\boldsymbol{\mu}$ is

$$\left\{\boldsymbol{\mu} : 2n(p-1)\frac{A_p(\hat{\kappa})^2}{1-A_p'(\hat{\kappa})-A_p(\hat{\kappa})^2}(1-\boldsymbol{\mu}^T\bar{\mathbf{x}}_0) < \chi^2_{p-1;\alpha}\right\}. \qquad (10.4.28)$$

More generally, it follows from (9.6.7) that, for large n and any distribution which is symmetrical about $\boldsymbol{\mu}$, an approximate $100(1-\alpha)\%$ confidence region for $\boldsymbol{\mu}$ is given by

$$\left\{\boldsymbol{\mu} : 2n(p-1)\frac{\hat{\mathrm{E}}[t]^2}{1-\hat{\mathrm{E}}[t^2]}(1-\boldsymbol{\mu}^T\bar{\mathbf{x}}_0) < \chi^2_{p-1;\alpha}\right\}, \qquad (10.4.29)$$

where $\hat{\mathrm{E}}[t]$ and $\hat{\mathrm{E}}[t^2]$ are any consistent estimates of $\mathrm{E}[t]$ and $\mathrm{E}[t^2]$ with $t = \mathbf{x}^T\boldsymbol{\mu}$. (See Watson, 1983a, pp. 137–138; 1983b, 1983c.) When $p = 3$, a useful form of (10.4.29) is

$$\{\boldsymbol{\mu} : \boldsymbol{\mu}^T\bar{\mathbf{x}}_0 > \cos\delta\}, \qquad (10.4.30)$$

where

$$\cos\delta = \sqrt{\frac{(-\log\alpha)(1-\bar{\mathbf{x}}_0^T\bar{\mathbf{T}}\bar{\mathbf{x}}_0)}{n\bar{R}^2}} \qquad (10.4.31)$$

(see Fisher & Lewis, 1983; Fisher, Lewis & Embleton, 1987, pp. 115–116).

A High-Concentration Confidence Region

It follows from (10.4.23) that for concentrated von Mises–Fisher distributions an approximate $100(1-\alpha)\%$ confidence region for $\boldsymbol{\mu}$ is

$$\left\{\boldsymbol{\mu} : \boldsymbol{\mu}^T\bar{\mathbf{x}}_0 \geq 1 - \frac{1-\bar{R}}{(n-1)\bar{R}}F_{p-1,(n-1)(p-1);\alpha}\right\}. \tag{10.4.32}$$

Example 10.6

A set of 45 directions of magnetisation in lava flows in western Iceland (Hospers's data cited in R. A. Fisher, 1953) had resultant vector $(-11.612, 0.671, -37.219)^T$, with the coordinate axes as northward, eastward and downward, respectively. Find a 95% confidence region for the population mean direction $\boldsymbol{\mu}$. The simple dipole field appropriate to the location has direction $(0.233, 0, 0.972)^T$. Is $\boldsymbol{\mu}$ equal to the direction $\boldsymbol{\mu}_0$ of the *reversed* dipole field? (For further discussion of the geological significance of this problem, see Hospers, 1955.)

The sample mean direction is $\bar{\mathbf{x}}_0 = (-0.298, 0.017, -0.955)^T$ and the mean resultant length is $\bar{R} = 0.867$. From Appendix 3.2, $\hat{\kappa} = 7.51$. (The approximation $\hat{\kappa} \simeq 1/(1-\bar{R})$ gives $\hat{\kappa} \simeq 7.3$.) Since $\hat{\kappa}$ is large, we may use (10.4.32). Thus an approximate 95% confidence region for $\boldsymbol{\mu}$ is $\{\boldsymbol{\mu} : \boldsymbol{\mu}^T\bar{\mathbf{x}}_0 \geq 0.989\}$. This is the spherical cap with axis $\bar{\mathbf{x}}_0$ and semi-vertical angle $\cos^{-1}(0.989) = 8.5°$.

Since $\boldsymbol{\mu}_0^T\bar{\mathbf{x}}_0 = 0.998$, where $\boldsymbol{\mu}_0 = (-0.233, 0, -0.972)^T$, $\boldsymbol{\mu}_0$ lies in the confidence region and the null hypothesis is accepted at the 5% significance level.

10.4.3 Mean Direction in a Given Subspace

Sometimes the null hypothesis specifies not the mean direction $\boldsymbol{\mu}$ itself but a subspace V in which $\boldsymbol{\mu}$ lies. Thus we need to test

$$H_0 : \boldsymbol{\mu} \in V \text{ against } H_1 : \boldsymbol{\mu} \notin V.$$

Most of the tests of Section 10.4.2 generalise to this context, by replacing $\bar{C}$ by $\mathbf{P}_V\bar{\mathbf{x}}$, where $\mathbf{P}_V$ denotes the orthogonal projection of $\mathbb{R}^p$ onto V. Let s denote the dimension of V. Note that $\mathbf{P}_V$ is idempotent of rank s and that $\mathbf{I}_p - \mathbf{P}_V$ is the orthogonal projection of $\mathbb{R}^p$ onto the orthogonal complement of V. In matrix terms, $\mathbf{P}_V = \mathbf{V}^T(\mathbf{V}\mathbf{V}^T)^{-1}\mathbf{V}$, where $\mathbf{V}^T$ is any $p \times s$ matrix with columns that span V.

Concentration Parameter Known

The Likelihood Ratio Test

The likelihood ratio statistic is

$$w = 2n\kappa\{\|\bar{\mathbf{x}}\| - \|\mathbf{P}_V\bar{\mathbf{x}}\|\}.$$

Fujikoshi & Watamori (1992) showed that, for large κ, both w^* and the slightly simpler statistic

$$w\left\{1+\frac{p+s-4}{4n\kappa}\right\}$$

have χ^2_{p-s} distributions with error $O(\kappa^{-2})$.

The Score Test

The score test statistic is

$$S=\frac{n\kappa}{A_p(\kappa)}\|(\mathbf{I}_p-\mathbf{P}_V)\bar{\mathbf{x}}\|^2,$$

which generalises (10.4.16). The intuitive idea is that H_0 is rejected for large values of $(\mathbf{I}_p-\mathbf{P}_V)\bar{\mathbf{x}}$, the component of $\bar{\mathbf{x}}$ normal to V. Under H_0 the asymptotic large-sample distribution of S is χ^2_{p-s}. Watamori (1992) showed that

$$S^*=S\left\{1+\frac{1}{4n}\frac{\kappa A_p'(\kappa)-A_p(\kappa)}{\kappa A_p(\kappa)^2}(p-s+2-S)\right\} \tag{10.4.33}$$

has a χ^2_{p-s} distribution with error $O(n^{-3/2})$.

For large κ it is useful to consider the approximation

$$T=n\kappa\|(\mathbf{I}_p-\mathbf{P}_V)\bar{\mathbf{x}}\|^2$$

to S. Fujikoshi & Watamori (1992) showed that

$$T\left\{1+\frac{1}{4\kappa}\left(2[p-1]-\frac{p-s+2}{n}\right)\left(1+\frac{T}{p-s+2}\right)\right\}$$

has a χ^2_{p-s} distribution with error $O(\kappa^{-2})$.

Concentration Parameter Unknown

The Likelihood Ratio Test

The likelihood ratio statistic is

$$w=2n\{\hat{\kappa}\|\bar{\mathbf{x}}\|-\tilde{\kappa}\|\mathbf{P}_V\bar{\mathbf{x}}\|-a_p(\hat{\kappa})+a_p(\tilde{\kappa})\}.$$

Watamori (1992) calculated the Bartlett-corrected statistic as

$$w^*=w\left\{1+\frac{1}{4n\kappa}\left[(p+s-4)\left(\frac{1}{A_p(\kappa)}-\frac{1}{\kappa A_p'(\kappa)}+\frac{2A_p''(\kappa)}{A_p'(\kappa)^2}\right)\right]\right\},$$

which has a χ^2_{p-s} distribution with error $O(n^{-2})$.

The Score Test and Variants

The score test statistic is

$$S = \frac{n\tilde{\kappa}}{A_p(\tilde{\kappa})}\|(\mathbf{I}_p - \mathbf{P}_V)\bar{\mathbf{x}}\|^2. \tag{10.4.34}$$

Under H_0, the asymptotic large-sample distribution of S is χ^2_{p-s}. Watamori (1992) showed that

$$S^* = S\left\{1 + \frac{1}{4n\kappa}\left[\frac{A_p''(\kappa)}{A_p'(\kappa)^2} + \frac{\kappa A_p'(\kappa) - A_p(\kappa)}{A_p(\kappa)^2}(p + 3s - 6 + S)\right]\right\}$$

has a χ^2_{p-s} distribution with error $O(n^{-3/2})$.

Watson (1983a) modified the score test by replacing $\tilde{\kappa}$ in (10.4.34) by $\hat{\kappa}$ to get

$$W = \frac{n\hat{\kappa}}{A_p(\hat{\kappa})}\|(\mathbf{I}_p - \mathbf{P}_V)\bar{\mathbf{x}}\|^2.$$

Under H_0 the asymptotic large-sample distribution of W is χ^2_{p-s} with error of order $O(n^{-1/2})$.

For large κ and to order $O(\kappa^{-1})$ the score test is more powerful against local alternatives than the likelihood ratio test, which is more powerful than Watson's test.

10.4.4 A Test for the Concentration Parameter

Consider the problem of testing

$$H_0 : \kappa = \kappa_0 \quad \text{against} \quad H_1 : \kappa \neq \kappa_0 \tag{10.4.35}$$

on the basis of a random sample $\mathbf{x}_1, \ldots, \mathbf{x}_n$ from $M_p(\boldsymbol{\mu}, \kappa)$, where $\boldsymbol{\mu}$ is unknown.

As in the circular case, discussed in Section 7.2.2, the likelihood ratio test of $\kappa = \kappa_0$ against $\kappa > \kappa_0$ rejects H_0 for large values of R, while the test of $\kappa = \kappa_0$ against $\kappa < \kappa_0$ rejects H_0 for small values of R. It follows from (9.5.6) that $\bar{\mathbf{x}}_0$ and $\bar{R}$ are sufficient statistics for $\boldsymbol{\mu}$ and κ. Since R is a maximal invariant (under rotation) after this sufficient reduction, invariant tests of $\kappa = \kappa_0$ depend only on R. As in Section 7.2.2, consideration of the marginal probability density function (9.5.9) shows that the above one-sided tests are uniformly most powerful invariant tests under rotations.

For $p = 3$, some selected quantiles of $\bar{R}$ are given in Appendix 3.6. For large κ, we can use the approximation (9.6.17)

$$2n\kappa_0(1 - \bar{R}) \mathrel{\dot{\sim}} \chi^2_{(n-1)(p-1)} \tag{10.4.36}$$

or the refinement

$$2n\gamma_0(1 - \bar{R}) \mathrel{\dot{\sim}} \chi^2_{(n-1)(p-1)}, \tag{10.4.37}$$

where γ_0 is given by replacing κ by κ_0 in (9.6.21). For $p = 3$, a suitable choice of γ_0 is obtained by replacing κ by κ_0 in (9.6.22).

10.5 TWO-SAMPLE TESTS

Suppose that $\mathbf{x}_{11}, \ldots, \mathbf{x}_{1n_1}$ and $\mathbf{x}_{21}, \ldots, \mathbf{x}_{2n_2}$ are two independent random samples of sizes n_1 and n_2 from $M_p(\boldsymbol{\mu}_1, \kappa_1)$ and $M_p(\boldsymbol{\mu}_2, \kappa_2)$, respectively. Let R_1, R_2 and R be the lengths of the resultants of the first sample, the second sample and the combined sample, respectively.

10.5.1 Tests for Equality of Mean Directions

We assume that $\kappa_1 = \kappa_2$. We wish to test $H_0 : \boldsymbol{\mu}_1 = \boldsymbol{\mu}_2$ against $H_1 : \boldsymbol{\mu}_1 \neq \boldsymbol{\mu}_2$.

The Two-Sample Watson–Williams Test

It follows from (9.5.14) that the conditional distribution of R_1, R_2 given R does not depend on κ when H_0 is true. Hence, as in the circular case, an appropriate test rejects H_0 for large values of $R_1 + R_2$ given R (Watson & Williams, 1956). For the case $p = 3$, selected quantiles of $(R_1 + R_2)/n$ given $\bar{R}$ are tabulated in Appendices 3.7a–3.7b.

A High-Concentration F Test

The high-concentration approximations (9.6.26)–(9.6.27) give the F-approximation

$$\frac{(R_1 + R_2 - R)/(p-1)}{(n - R_1 - R_2)/(n-2)(p-1)} \stackrel{\cdot}{\sim} F_{p-1,(p-1)(n-2)} \tag{10.5.1}$$

for large κ. The null hypothesis is rejected for large values of this statistic.

Example 10.7

Table 10.3 shows the directions of needle-shaped crystals in two exposures from the Yukspor Mountain, Kola Peninsula (Vistelius, 1966, p. 92). Do the two populations have the same mean direction?

Here

$$\begin{aligned} n_1 &= 10, & \bar{\mathbf{x}}_1 &= (-0.723, -0.140, 0.247)^T, & R_1 &= 7.76, \\ n_2 &= 10, & \bar{\mathbf{x}}_2 &= (-0.642, -0.191, 0.207)^T, & R_2 &= 7.01. \end{aligned}$$

The resultant length R of the combined sample is 14.75. These values give the observed value of (10.5.1) as 0.07. Since the 5% value of $F_{2,36}$ is 3.26, the null hypothesis is accepted. The hypothesis of equality of concentration parameters will be tested in Example 10.8.

Table 10.3 Directions of crystals in two exposures from the Yukspor Mountain, Kola Peninsula (Vistelius, 1966).

Exposure I		Exposure II	
Azimuth (degrees)	Dip (degrees)	Azimuth (degrees)	Dip (degrees)
18	26	5	4
34	22	31	10
64	10	145	20
265	6	279	10
314	8	309	8
334	6	325	20
340	24	342	8
342	20	344	10
345	14	350	12
355	8	359	18

Confidence Intervals for the Angle between Two Mean Directions

Let $\bar{\mathbf{x}}_{01}$ and $\bar{\mathbf{x}}_{02}$ be the mean directions of samples of sizes n_1 and n_2 from distributions with mean directions $\boldsymbol{\mu}_1$ and $\boldsymbol{\mu}_2$. An intuitively appealing estimate of $\boldsymbol{\mu}_1^T\boldsymbol{\mu}_2$ is $\bar{\mathbf{x}}_{01}^T\bar{\mathbf{x}}_{02}$. In the case where $p = 3$ and the distributions are rotationally symmetric (so that the density functions have the form (9.3.31)) large-sample approximate confidence intervals for $\boldsymbol{\mu}_1^T\boldsymbol{\mu}_2$ were obtained by Lewis & Fisher (1995). They showed that, for large samples, the distribution of $1 - \boldsymbol{\mu}_1^T\boldsymbol{\mu}_2$ is approximately scaled beta. They showed also that if $\boldsymbol{\mu}_1 \simeq \boldsymbol{\mu}_2$, then the approximate large-sample distribution of $1 - \boldsymbol{\mu}_1^T\boldsymbol{\mu}_2$ is exponential with mean

$$\frac{1 - \mathrm{E}[\cos 2\theta_1]}{4n_1\rho_1^2} + \frac{1 - \mathrm{E}[\cos 2\theta_2]}{4n_2\rho_2^2},$$

where $\cos\theta_i = \mathbf{x}_i^T\boldsymbol{\mu}_i$ and ρ_i denotes the mean resultant length of the ith distribution for $i = 1, 2$.

In the cases $p = 2$ and $p = 3$, nonparametric confidence regions for rotations taking $\boldsymbol{\mu}_1$ to $\boldsymbol{\mu}_2$ were introduced by Beran & Fisher (1998).

10.5.2 Tests for Equality of Concentration Parameters

Consider the problem of testing

$$H_0 : \kappa_1 = \kappa_2 \qquad \text{against} \qquad H_1 : \kappa_1 \neq \kappa_2, \tag{10.5.2}$$

where the mean directions $\boldsymbol{\mu}_1$ and $\boldsymbol{\mu}_2$ are unknown.

A High-Concentration F Test

If H_0 is true then $\bar{R}_1 \simeq \bar{R}_2$, so it is intuitively reasonable to reject H_0 if $(1-\bar{R}_1)/(1-\bar{R}_2)$ is not close to 1. It is more appropriate to use the statistic

$$F = \frac{(n_1 - R_1)/(n_1-1)(p-1)}{(n_2 - R_2)/(n_2-1)(p-1)} \tag{10.5.3}$$

in this two-tailed test. It follows from (9.6.17) that, under H_0,

$$F \mathrel{\dot\sim} F_{(n_1-1)(p-1),(n_2-1)(p-1)}, \qquad \kappa \to \infty. \tag{10.5.4}$$

For $p = 3$, approximation (10.5.4) is reasonable for $\bar{R} \geq 0.67$.

Example 10.8
We test the hypothesis of equality of concentration parameters for the crystal direction data of Example 10.7.

From Example 10.7, $n_1 = n_2 = 10$, $\bar{R}_1 = 0.776$, $\bar{R}_2 = 0.701$, $\bar{R} = 0.74$, so it is appropriate to use (10.5.3). Here $F = 1.33$. Since $F_{18,18;0.025} = 2.63$, the null hypothesis of equal concentrations is accepted at the 5% level.

Tests Based on Variance-Stabilising Transformations

When the distributions are not concentrated, tests can be constructed using suitable variance-stabilising transformations of the mean resultant length. We now give appropriate transformations for $p = 3$.

Case I. $\bar{R} < 0.44$. Put

$$U_1 = \sqrt{\frac{5}{3}}\, \frac{g_1(3\bar{R}_1) - g_1(3\bar{R}_2)}{1/(n_1-5) + 1/(n_2-5)}, \tag{10.5.5}$$

where

$$g_1(r) = \sin^{-1}\left(\frac{r}{\sqrt{5}}\right). \tag{10.5.6}$$

Then, under H_0,

$$U_1 \mathrel{\dot\sim} N(0,1), \tag{10.5.7}$$

for n_1, n_2 large.

Case II. $0.44 \leq \bar{R} \leq 0.67$. Put

$$U_2 = \frac{g_2(\bar{R}_1) - g_2(\bar{R}_2)}{0.627\{1/(n_1-4) + 1/(n_2-4)\}}, \tag{10.5.8}$$

where

$$g_2(r) = \sin^{-1}\left(\frac{r + 0.176}{1.029}\right). \tag{10.5.9}$$

Then, under H_0,

$$U_2 \stackrel{.}{\sim} N(0,1), \tag{10.5.10}$$

for n_1, n_2 large.

10.6 MULTI-SAMPLE TESTS

Suppose that $\mathbf{x}_{i1}, \dots, \mathbf{x}_{in_i}$ $(i = 1, \dots, q)$ are q independent random samples of sizes $n_1, \dots, n_q$ from $M_p(\boldsymbol{\mu}_i, \kappa_i)$, for $i = 1, \dots, q$. Put $n = n_1 + \dots + n_q$. Let R_i and R denote the resultant length of the ith sample and of the combined sample, respectively.

10.6.1 One-Way Classification

We wish to test

$$H_0 : \boldsymbol{\mu}_1 = \dots = \boldsymbol{\mu}_q \tag{10.6.1}$$

against the alternative that at least one of the equalities is not satisfied. The tests considered in Section 7.4.1 for the circular case extend readily to the spherical case. In this subsection we shall assume that $\kappa_1 = \dots = \kappa_q$, where the common concentration κ is unknown.

The Multi-sample Watson–Williams Test

The multi-sample analogue of the two-sample Watson–Williams test is the conditional test which rejects H_0 for large values of

$$R_1 + \dots + R_q \tag{10.6.2}$$

given R. From (9.5.14), the probability density function of $R_1, \dots, R_q$ given R does not depend on κ. Thus, for given R, this test is a similar test. However, its quantiles are not yet available (except in the two-sample case which was considered in Section 10.5.1). As in the circular case, we shall consider some alternative procedures which are equivalent to the above test for large n or large κ.

The Watson–Williams test and the tests given by (10.6.4) and (10.6.10) have the following geometrical interpretation. Under (10.6.1), the vector sample means $\bar{\mathbf{x}}_1, \dots \bar{\mathbf{x}}_q$ are almost parallel, so that $R_1 + \dots + R_q \simeq R$. (See Fig. 7.3.) Thus it is intuitively reasonable to reject (10.6.1) for large values of $R_1 + \dots + R_q - R$.

A High-Concentration F Test

It follows fron the high-concentration approximations (9.6.27) and (9.6.26) to the distributions of

$$2\kappa\left(\sum_{i=1}^{q} R_i - R\right) \quad \text{and} \quad 2\kappa\left(n - \sum_{i=1}^{q} R_i\right) \tag{10.6.3}$$

that, under H_0,

$$\frac{(\sum_{i=1}^{q} R_i - R)/(q-1)(p-1)}{(n - \sum_{i=1}^{q} R_i)/(n-q)(p-1)} \mathrel{\dot{\sim}} F_{(q-1)(p-1),(n-q)(p-1)}, \tag{10.6.4}$$

for large κ. An improved approximation is

$$\frac{\hat{\kappa}}{\hat{\gamma}} \frac{(\sum_{i=1}^{q} R_i - R)/(q-1)(p-1)}{(n - \sum_{i=1}^{q} R_i)/(n-q)(p-1)} \mathrel{\dot{\sim}} F_{(q-1)(p-1),(n-q)(p-1)}, \tag{10.6.5}$$

where $\hat{\kappa} = A_p^{-1}(\bar{R})$ and $\hat{\gamma}$ is obtained by replacing κ by $\hat{\kappa}$ in (9.6.21). For $p = 3$, a good choice of $\hat{\gamma}$ is obtained by replacing κ by $\hat{\kappa}$ in (9.6.22). In this case, approximation (10.6.5) is found to be adequate for $\hat{\kappa} \geq 1$, and approximation (10.6.4) adequate for $\bar{R} \geq 0.67$. The relevant calculations can be displayed in an analysis of variance table as was done in the circular case in Section 7.4.1. This procedure is illustrated in Example 10.9.

Example 10.9
Directions at three sites in the Torridonian Sandstone Series are summarised in Table 10.4 (E. Irving's data analysed in Watson, 1956a). Are the population mean directions the same at all three sites?

Table 10.4 Directions from three sites in the Torridonian Sandstone Series (E. Irving's data analysed in Watson 1956a, reproduced by permission of Blackwell Science Ltd)

Sample number	Size	$\mathbf{\bar{x}}_0$	R
1	10	$(0.637, -0.269, 0.101)^T$	6.990
2	11	$(0.649, -0.154, 0.336)^T$	8.212
3	15	$(0.795, -0.158, 0.059)^T$	12.194

We assume that the three samples come from populations with the same concentration parameter (see Example 10.10). For the combined sample of 36 observations, the resultant length R is 26.902. Since $\bar{R} = 0.747$, we can use the F test given by (10.6.4). Further calculations are displayed in Table 10.5.

Since the 5% value of $F_{4,66}$ is 2.52, we accept the null hypothesis that the mean directions for the three sites are the same.

Table 10.5 Analysis of variance for Example 10.9.

Source	d.f.	SS	Mean square	F
Between sites	4	0.494	0.123	0.95
Within sites	66	8.604	0.130	
Total	70	9.098		

The Likelihood Ratio Test

The likelihood ratio statistic for (10.6.1) is

$$w = 2\left\{\hat{\kappa}\sum_{i=1}^{q} R_i - \tilde{\kappa}R - na_p(\hat{\kappa}) + na_p(\tilde{\kappa})\right\}, \tag{10.6.6}$$

where $\tilde{\kappa}$ and $\hat{\kappa}$ are the maximum likelihood estimates of κ under H_0 and H_1 respectively and are given by

$$A_p(\tilde{\kappa}) = \bar{R}, \qquad A_p(\hat{\kappa}) = \frac{1}{n}\sum_{i=1}^{q} R_i. \tag{10.6.7}$$

Under H_0,

$$w \dot{\simeq} \chi^2_{(q-1)(p-1)}, \tag{10.6.8}$$

for large n. From (10.4.2), (10.3.8) and (10.3.9) we have that, for small values of κ,

$$a_p(\kappa) \simeq \frac{\kappa^2}{2p}, \qquad A_p(\kappa) \simeq \frac{\kappa}{p}. \tag{10.6.9}$$

Using (10.6.9) in (10.6.6) gives

$$w \simeq \frac{p}{n}\left\{\left(\sum_{i=1}^{q} R_i\right)^2 - R^2\right\} = U, \tag{10.6.10}$$

say, for small κ. Then, under H_0,

$$U \dot{\simeq} \chi^2_{(q-1)(p-1)}, \tag{10.6.11}$$

for small κ and large n. As in the circular case, this approximation can be improved using a multiplicative correction. It follows from (9.2.13) that

$$\mathrm{E}[R_i^2] = n_i + n_i(n_i - 1)A_p(\kappa)^2. \tag{10.6.12}$$

The approximation

$$\bar{R} \simeq t\left\{1 + \frac{1}{2}\frac{\|\bar{\mathbf{x}}_\perp\|^2}{t^2}\right\}$$

(where $t = \bar{\mathbf{x}}^T \boldsymbol{\mu}$ and $\bar{\mathbf{x}}_\perp = (\mathbf{I}_p - \boldsymbol{\mu}\boldsymbol{\mu}^T)\bar{\mathbf{x}}$) and (10.3.16) give

$$\mathrm{E}[\bar{R}] = A_p(\kappa) + \frac{p-1}{2n\kappa} + O(n^{-2}),$$

and so

$$\mathrm{E}[R_i] = n_i A_p(\kappa) + \frac{p-1}{2\kappa} + O(n_i^{-1}). \tag{10.6.13}$$

Then

$$\mathrm{E}[U] \simeq p(p-1)(q-1)\left\{\frac{A_p(\kappa)}{\kappa} + \frac{q(p-1)}{4n\kappa^2}\right\}. \tag{10.6.14}$$

By (10.3.9), if

$$\frac{1}{c} = 1 - \frac{\kappa^2}{p(p+2)} + \frac{p(p-1)q}{4n\kappa^2} \tag{10.6.15}$$

then

$$cU \mathrel{\dot\sim} \chi^2_{(q-1)(p-1)}. \tag{10.6.16}$$

For $p = 3$, this approximation is found to be satisfactory for small values of n, provided that κ is not very near 0 or 1. In practice, we replace κ by its maximum likelihood estimate $\hat{\kappa}$ given in (10.3.4) and reject H_0 when cU is large.

ANOVA Based on the Embedding Approach

As in the circular case, analysis of variance procedures can also be obtained by taking the embedding approach and considering the observations as elements of $\mathbb{R}^p$. Recall that, for a single sample $\mathbf{x}_1, \ldots, \mathbf{x}_n$ on S^{p-1},

$$\sum_{i=1}^{n} \|\mathbf{x}_i - \bar{\mathbf{x}}\| = n(1 - \bar{R}^2).$$

Similarly, for q samples $\mathbf{x}_{11}, \ldots, \mathbf{x}_{1n_1}; \ldots; \mathbf{x}_{q1}, \ldots, \mathbf{x}_{qn_q}$, the basic ANOVA decomposition

$$\sum_{i=1}^{q}\sum_{j=1}^{n_i} \|\mathbf{x}_{ij} - \bar{\mathbf{x}}_{..}\|^2 = \sum_{i=1}^{q}\sum_{j=1}^{n_i} \|\mathbf{x}_{ij} - \bar{\mathbf{x}}_{i.}\|^2 + \sum_{i=1}^{q} n_i \|\bar{\mathbf{x}}_{i.} - \bar{\mathbf{x}}_{..}\|^2 \tag{10.6.17}$$

can be rewritten as

$$n(1 - \bar{R}^2) = \left(\sum_{i=1}^{q} n_i \bar{R}_i^2 - n\bar{R}^2\right) + \left(n - \sum_{i=1}^{q} n_i \bar{R}_i^2\right), \tag{10.6.18}$$

which is reminiscent of (9.6.25). In (10.6.18) the terms represent respectively the total variation, the variation between samples and the variation within samples. This leads to the test which rejects H_0 for large values of

$$\frac{(\sum_{i=1}^{q} n_i \bar{R}_i^2 - \bar{R}^2)/(q-1)(p-1)}{(n - \sum_{i=1}^{q} n_i \bar{R}_i^2)/(n-q)(p-1)}. \tag{10.6.19}$$

Under H_0, the high-concentration asymptotic distribution of (10.6.19) is $F_{(q-1)(p-1),(n-q)(p-1)}$.

Multiple Comparisons

If (10.6.1) is rejected then the problem arises of how to make multiple comparisons of directions. If the concentration κ is high then one possibility is to compare

$$\frac{\left(n_i\|\bar{\mathbf{x}}_{i.}\| + n_j\|\bar{\mathbf{x}}_{j.}\| - \|n_i\bar{\mathbf{x}}_{i.} + n_j\bar{\mathbf{x}}_{j.}\|\right)/(p-1)}{\left(n - \sum_{i=1}^{q} n_i\|\bar{\mathbf{x}}_{i.}\|\right)/(n-1)}$$

with $F_{(q-1)(p-1),(q-1)(n-1)}$, as suggested by Watson (1994).

10.6.2 Tests for the Homogeneity of Concentrations

Because the tests of Section 10.6.1 are based on the assumption of equal concentrations, it is often necessary to test the hypothesis

$$H_0 : \kappa_1 = \ldots = \kappa_q. \tag{10.6.20}$$

Following the circular multi-sample case of Section 7.4.2, the two-sample tests for equality of concentration parameters of Section 10.5.2 can be extended easily to the multi-sample situation. The results are as follows.

For concentrated von Mises–Fisher distributions, we can approximate them by $(p-1)$-dimensional normal distributions using (9.3.15) and then apply Bartlett's test of homogeneity (Stuart & Ord, 1991, pp. 875–876). Thus we reject homogeneity for large values of

$$B = \frac{1}{1+d}\left\{\nu \log \frac{n\sum_{i=1}^{q} R_i}{\nu} - \sum_{i=1}^{q} \nu_i \log \frac{(n_i - R_i)}{\nu_i}\right\}, \tag{10.6.21}$$

where

$$\nu_i = 2(n_i - 1), \qquad \nu = 2(n-q), \qquad d = \frac{1}{3(q-1)}\left\{\sum_{i=1}^{q} \frac{1}{\nu_i} - \frac{1}{\nu}\right\}.$$

Under H_0,

$$B \mathrel{\dot\sim} \chi^2_{q-1}, \tag{10.6.22}$$

for $n_1, \ldots, n_q$ large. When $p = 3$, this approximation is satisfactory for $\bar{R} \geq 0.67$.

When the distributions are not concentrated, homogeneity can be tested using suitable variance-stabilising transformations of the mean resultant length. We now give appropriate transformations for $p = 3$.

Case I. $\bar{R} < 0.44$. Put

$$U_1 = \sum_{i=1}^{q} w_i g_1(3\bar{R}_i)^2 - \frac{\{\sum_{i=1}^{q} w_i g_1(3\bar{R}_i)\}^2}{\sum_{i=1}^{q} w_i}, \tag{10.6.23}$$

where

$$\frac{1}{w_i} = \frac{3}{5(n_i - 5)}, \qquad g_1(r) = \sin^{-1}\left(\frac{r}{\sqrt{5}}\right). \tag{10.6.24}$$

Then, under H_0,

$$U_1 \mathrel{\dot{\sim}} \chi^2_{q-1}, \tag{10.6.25}$$

for $n_1, \ldots, n_q$ large.

Case II. $0.44 \leq \bar{R} \leq 0.67$. Put

$$U_2 = \sum_{i=1}^{q} w_i g_2(\bar{R}_i)^2 - \frac{\sum_{i=1}^{q} w_i g_2(\bar{R}_i)^2}{\sum_{i=1}^{q} w_i}, \tag{10.6.26}$$

where now

$$\frac{1}{w_i} = \frac{0.394}{n_i - 4}, \qquad g_2(r) = \sin^{-1}\left(\frac{r + 0.176}{1.029}\right). \tag{10.6.27}$$

Then, under H_0,

$$U_2 \mathrel{\dot{\sim}} \chi^2_{q-1}, \tag{10.6.28}$$

for $n_1, \ldots, n_q$ large.

Example 10.10

We test the homogeneity of the concentration parameters for the Torridonian Sandstone Series data of Example 10.9.

From Example 10.9, we have $\bar{R} > 0.67$. Thus it is appropriate to use the statistic B of (10.6.21). We have $q = 3$ and

$$\nu_1 = 18, \qquad n_1 - R_1 = 3.010, \qquad \nu_2 = 20, \qquad n_2 - R_2 = 2.788,$$

$$\nu_3 = 28, \qquad n_3 - R_3 = 2.806, \qquad \nu = 66, \qquad n - \sum_{i=1}^{q} R_i = 8.604,$$

so that

$$\nu \log \frac{n \sum_{i=1}^{q} R_i}{\nu} = -134.470, \qquad \sum_{i=1}^{q} \nu_i \frac{\log(n_i - R_i)}{\nu_i} = -136.012.$$

Then $d = 0.021$, and so $B = 1.51$. Since $\chi^2_{2;0.05} = 5.99$, the null hypothesis of equal concentrations is accepted at the 5% level.

Two-way and multi-way ANOVA for spherical data can be carried out by the methods of Stephens (1982) or by straightforward generalisation of the methods suggested by Harrison, Kanji & Gadsden (1986), Harrison & Kanji (1988) and Anderson & Wu (1995) for the circular case.

10.6.3 The Heterogeneous Case

If it is not reasonable to make the homogeneity assumption $\kappa_1 = \ldots = \kappa_q$ then it is not appropriate to use the methods of Section 10.6.1 to test

$$H_0 : \boldsymbol{\mu}_1 = \ldots = \boldsymbol{\mu}_q \tag{10.6.29}$$

against the alternative that at least one of the equalities does not hold. We now consider tests of (10.6.29) in the heterogeneous case.

If $\kappa_1, \ldots, \kappa_q$ are known then the likelihood ratio statistic for (10.6.29) is

$$2\left(\sum_{i=1}^{q} \kappa_i n_i \|\bar{\mathbf{x}}_i\| - \left\|\sum_{i=1}^{q} \kappa_i n_i \bar{\mathbf{x}}_i\right\|\right), \tag{10.6.30}$$

and its large-sample asymptotic distribution under (10.6.29) is $\chi^2_{(q-1)(p-1)}$ (see Watson, 1983a, p. 155; 1983b). When $p = 2$, the statistic (10.6.30) reduces to (7.4.27).

The geometrical interpretation underlying the test based on the statistic (10.6.30) is the same as that underlying the Watson–Williams test: under (10.6.1), the vector sample means $\bar{\mathbf{x}}_1, \ldots \bar{\mathbf{x}}_q$ tend to be almost parallel, so that the length $\|\sum_{i=1}^{q} \kappa_i n_i \bar{\mathbf{x}}_i\|$ of the weighted sum of the sample means is almost equal to the weighted sum $\sum_{i=1}^{q} \kappa_i n_i \|\bar{\mathbf{x}}_i\|$ of their lengths.

If $\kappa_1, \ldots, \kappa_q$ are not known then they can be replaced by their maximum likelihood estimates to give

$$2\left(\sum_{i=1}^{q} \hat{\kappa}_i n_i \|\bar{\mathbf{x}}_i\| - \left\|\sum_{i=1}^{q} \hat{\kappa}_i n_i \bar{\mathbf{x}}_i\right\|\right), \tag{10.6.31}$$

where $\hat{\kappa}_i = A_p^{-1}(\bar{R}_i)$. Under (10.6.29), the large-sample asymptotic distribution of (10.6.31) is $\chi^2_{(q-1)(p-1)}$. A simulation study by Watson & Debiche (1992) in the case $p = 3$ and $q = 2$ showed that this is a good approximation and that when $n_1 \simeq n_2$ the test based on (10.6.31) performs well compared with various competitors.

In the cases $p = 2$ and $p = 3$, nonparametric simultaneous confidence regions for rotations taking $\boldsymbol{\mu}_i$ to $\boldsymbol{\mu}_j$ (for $1 \leq i < j \leq q$) were introduced by Beran & Fisher (1998).

10.6.4 Tests for Mean Directions in a Subspace

Sometimes the null hypothesis specifies that the population mean directions $\boldsymbol{\mu}_1, \ldots, \boldsymbol{\mu}_q$ of q von Mises–Fisher distributions lie in an s-dimensional subspace V of $\mathbb{R}^p$. In some cases V is prescribed; in others it is unknown. An example of the latter occurs in geomagnetism, in which a certain geophysical model suggests that the mean directions of natural remnant magnetism at various associated sites should be coplanar.

Prescribed Subspace

We wish to test

$$H_0 : \boldsymbol{\mu}_i \in V,\ i = 1, \ldots, q, \quad \text{against } H_1 : \boldsymbol{\mu}_i \notin V, \text{ for some } i. \tag{10.6.32}$$

Let $\mathbf{P}_V$ denote orthogonal projection onto V. For $i = 1, \ldots, q$, the vector mean $\bar{\mathbf{x}}_i$ of the ith sample can be decomposed as

$$\bar{\mathbf{x}}_i = \bar{\mathbf{x}}_{iV} + \bar{\mathbf{x}}_{i\perp}, \tag{10.6.33}$$

where

$$\bar{\mathbf{x}}_{iV} = \mathbf{P}_V \bar{\mathbf{x}}_i \qquad \text{and} \qquad \bar{\mathbf{x}}_{i\perp} = (\mathbf{I}_p - \mathbf{P}_V)\bar{\mathbf{x}}_i \tag{10.6.34}$$

are the components of $\bar{\mathbf{x}}_i$ along V and orthogonal to V, respectively.

Large-Sample Tests

When the concentrations $\kappa_1, \ldots, \kappa_q$ are known, the likelihood ratio statistic is

$$2 \sum_{i=1}^{q} \kappa_i n_i (\|\bar{\mathbf{x}}_i\| - \|\bar{\mathbf{x}}_{iV}\|). \tag{10.6.35}$$

Under H_0, the large-sample asymptotic distribution of (10.6.35) is $\chi^2_{q(p-s)}$.

Since both $\|\bar{\mathbf{x}}_i\|$ and $\|\bar{\mathbf{x}}_{iV}\|$ tend in probability to $A_p(\kappa_i)$ as $n \to \infty$,

$$n_i \|\bar{\mathbf{x}}_{i\perp}\|^2 = n_i(\|\bar{\mathbf{x}}_i\| + \|\bar{\mathbf{x}}_{iV}\|)(\|\bar{\mathbf{x}}_i\| - \|\bar{\mathbf{x}}_{iV}\|) \to 2A_p(\kappa_i)(\|\bar{\mathbf{x}}_i\| - \|\bar{\mathbf{x}}_{iV}\|). \tag{10.6.36}$$

Thus (10.6.35) is asymptotically equivalent to

$$\sum_{i=1}^{q} \frac{\kappa_i n_i}{A_p(\kappa_i)} \|\bar{\mathbf{x}}_{i\perp}\|^2. \tag{10.6.37}$$

Further details are given in Watson (1983a, p. 156; 1983b). In particular, it is shown that the non-null asymptotic distribution of (10.6.37) is a non-central χ^2.

If the concentrations $\kappa_1, \ldots, \kappa_q$ are unknown then suitable statistics are obtained by replacing the κ_i in (10.6.35) and (10.6.37) by their maximum likelihood estimates $A_p^{-1}(\bar{R}_i)$.

A High-Concentration F Test

If $\kappa_1 = \ldots = \kappa_q = \kappa$ then a straightforward extension of the argument leading to (10.6.4) yields

$$\frac{[\sum_{i=1}^{q} n_i(R_i - \|\bar{\mathbf{x}}_{iV}\|)]/q(p-s)}{(n - \sum_{i=1}^{q} R_i)/(n-q)(p-1)} \;\dot{\sim}\; F_{q(p-s),(n-q)(p-1)}, \qquad \kappa \to \infty \tag{10.6.38}$$

(Watson, 1983a, pp. 157–165; 1984).

Unspecified Subspace

We wish to test

$$H_0 : \boldsymbol{\mu}_i \in V, \; i = 1, \ldots, q \text{ for some } s\text{-dimensional subspace } V \tag{10.6.39}$$

against

$$H_1 : \boldsymbol{\mu}_1, \ldots \boldsymbol{\mu}_q \text{ do not lie in any } s\text{-dimensional subspace.} \tag{10.6.40}$$

Suitable statistics can be obtained from (10.6.37) and (10.6.38) by minimising over V as follows.

A Large-Sample Test

Since (10.6.37) can be written as

$$\operatorname{tr}\left(\mathbf{P}_V \sum_{i=1}^{q} \frac{\kappa_i n_i}{A_p(\kappa_i)} \bar{\mathbf{x}}_i \bar{\mathbf{x}}_i^T \mathbf{P}_V\right), \tag{10.6.41}$$

the infimum over V of (10.6.37) or (10.6.41) is

$$\sum_{i=s+1}^{p} v_i, \tag{10.6.42}$$

where

$$v_1 \geq \ldots \geq v_p$$

are the eigenvalues of

$$\sum_{i=1}^{q} \frac{\kappa_i n_i}{A_p(\kappa_i)} \bar{\mathbf{x}}_i \bar{\mathbf{x}}_i^T.$$

Under H_0, the large-sample asymptotic distribution of (10.6.42) is $\chi^2_{(p-s)(q-s)}$.

If the concentrations $\kappa_1, \ldots, \kappa_q$ are unknown then a suitable statistic is obtained by replacing the κ_i in (10.6.41) by their maximum likelihood estimates $\hat{\kappa}_i = A_p^{-1}(\bar{R}_i)$. This gives

$$\sum_{i=s+1}^{p} w_i, \tag{10.6.43}$$

where

$$w_1 \geq \ldots \geq w_p$$

are the eigenvalues of

$$\mathbf{W} = \sum_{i=1}^{q} \frac{\hat{\kappa}_i n_i}{\bar{R}_i} \bar{\mathbf{x}}_i \bar{\mathbf{x}}_i^T. \tag{10.6.44}$$

Under H_0, the large-sample asymptotic distribution of (10.6.43) is $\chi^2_{(p-s)(q-s)}$, as shown by Watson (1983b).

A High-Concentration F Test

If $\kappa_1 = \ldots = \kappa_q = \kappa$ then the approximation

$$\bar{R}_i - \|\bar{\mathbf{x}}_{iV}\| \simeq \tfrac{1}{2}\bar{R}_i\|\bar{\mathbf{x}}_{i\perp}\|^2,$$

together with the argument leading to (10.6.42), gives

$$\inf_V \sum_{i=1}^{q} n_i(\bar{R}_i - \|\bar{\mathbf{x}}_{iV}\|) = \sum_{i=1}^{q} n_i\bar{R}_i - \sum_{i=s+1}^{p} w_i.$$

It follows that

$$\frac{[\sum_{i=1}^{q} n_i(\bar{R}_i - \sum_{i=s+1}^{p} w_i)]/(q-s)(p-s)}{(n - \sum_{i=1}^{q} R_i)/(n-q)(p-1)} \,\tilde{\sim}\, F_{(q-s)(p-s),(n-q)(p-1)}, \quad \kappa \to \infty. \tag{10.6.45}$$

Example 10.11

Table 10.6 summarises the directions of samples taken from three populations (Watson, 1960, dealing with some data of K. M. Creer). (A sign in sample 2 of Watson, 1960, has been corrected.) Certain geophysical considerations suggest that the mean directions of the three populations should be coplanar. Do the data support this hypothesis?

Table 10.6 Summaries of directions in three samples (data of K. M. Creer considered by Watson, 1960)

Sample number	size	$\bar{\mathbf{x}}$	R
1	35	$(-0.070, -0.959, -0.275)^T$	33.172
2	9	$(-0.857, 0.258, -0.446)^T$	8.567
3	6	$(0.547, -0.730, -0.410)^T$	5.786

By Appendix 3.2, $\hat{\kappa}_1 = 19.1$, $\hat{\kappa}_2 = 20.8$, $\hat{\kappa}_3 = 28.0$, which are fairly large. On using the test given by (10.6.21), it is found that the samples may be regarded as drawn from populations having the same concentration parameter κ. For this data set, the matrix $\mathbf{W}$ defined in (10.6.44) is

$$\mathbf{W} = \begin{pmatrix} 8.187 & 2.640 & 5.208 \\ 2.640 & 34.154 & 9.494 \\ 5.208 & 9.494 & 5.183 \end{pmatrix}.$$

The eigenvalues and unit eigenvectors of $\mathbf{W}$ are

$$\begin{aligned} w_1 &= 37.548, & \pm\mathbf{v}_1 &= \pm(0.138, 0.944, 0.299)^T, \\ w_2 &= 9.950, & \pm\mathbf{v}_2 &= \pm(0.866, -0.262, 0.426)^T, \\ w_3 &= 0.027, & \pm\mathbf{v}_3 &= \pm(-0.480, -0.200, 0.854)^T. \end{aligned}$$

Here $p = 3$, $q = 3$, and $s = 2$, since the subspace V mentioned in (10.6.39) is a plane. Calculation gives the F statistic of (10.6.45) as 0.509. Since $F_{1,94;0.05} = 3.95$, the hypothesis of coplanarity of the three populations is accepted at the 5% level.

10.7 TESTS ON AXIAL DISTRIBUTIONS

10.7.1 Tests of Uniformity

The Bingham Test

Perhaps the simplest test of uniformity of axial data is the Bingham test introduced by Bingham (1974). The intuitive idea behind the Bingham test is that uniformity is rejected if the sample scatter matrix

$$\bar{\mathbf{T}} = \frac{1}{n}\sum_{i=1}^{n} \mathbf{x}_i \mathbf{x}_i^T$$

is far from its expected value $p^{-1}\mathbf{I}_p$. More precisely, uniformity is rejected for large values of

$$S = \frac{p(p+2)}{2} n \left\{ \operatorname{tr}(\bar{\mathbf{T}}^2) - \frac{1}{p} \right\}. \tag{10.7.1}$$

This test is also the score test of uniformity in the angular central Gaussian model with probability density functions (10.3.45). Under uniformity,

$$S \stackrel{\cdot}{\sim} \chi^2_{(p-1)(p+2)/2}, \tag{10.7.2}$$

with error of order $O(n^{-1})$. This approximation can be improved by using the modified Bingham test statistic

$$S^* = S\left\{1 - \frac{1}{n}[B_0 + B_1 S + B_2 S^2]\right\}, \tag{10.7.3}$$

where

$$\begin{aligned} B_0 &= \frac{p^2 + 6p + 20}{12(p+4)}, \\ B_1 &= \frac{-(p^2 + 3p + 8)}{3(p+4)(p^2 + p + 2)}, \\ B_2 &= \frac{p^2 - 4}{3(p+4)(p^2 + p + 2)(p^2 + p + 6)}. \end{aligned}$$

Then

$$S^* \dot{\sim} \chi^2_{(p-1)(p+2)/2} \tag{10.7.4}$$

with error of order $O(n^{-3/2})$ (Jupp, 1999).

In the case $p = 2$, the Bingham statistic of $\pm\mathbf{x}_1, \ldots, \pm\mathbf{x}_n$ is the Rayleigh statistic (6.3.3) of the points $(\cos 2\theta_1, \sin 2\theta_1)^T, \ldots, (\cos 2\theta_n, \sin 2\theta_n)^T$ obtained by 'doubling the angles' $\theta_1, \ldots, \theta_n$ defined by $\mathbf{x}_i = (\cos\theta_i, \sin\theta_i)^T$.

Example 10.12

For the calcite grains data of Example 10.3, we test the hypothesis of uniformity.

From Example 10.3, $n = 30$ and the eigenvalues of $\bar{\mathbf{T}}$ are

$$\bar{t}_1 = 0.495 \qquad \bar{t}_2 = 0.365 \qquad \bar{t}_3 = 0.139.$$

The Bingham statistic (10.7.1) is $S = 14.61$ and the modified statistic (10.7.3) is $S^* = 14.87$. Since the 5% value of χ^2_5 is 11.07, the hypothesis of uniformity is rejected at the 5% significance level.

The Bingham test is based on the 'squared distance' (10.7.1) between $\bar{\mathbf{T}}$ and $p^{-1}\mathbf{I}_p$. Anderson & Stephens (1972) used an alternative approach, in which the distance between $\bar{\mathbf{T}}$ and $p^{-1}\mathbf{I}_p$ is measured by $\bar{t}_1 - 1/p$ or $1/p - \bar{t}_p$, where $\bar{t}_1$ and $\bar{t}_p$ are the largest and smallest eigenvalues of $\bar{\mathbf{T}}$. For testing against the alternative of a bipolar distribution, their test rejects uniformity for large values of $\bar{t}_1$; for testing against the alternative of a girdle distribution, their test rejects uniformity for small values of $\bar{t}_p$. In the case $p = 3$, some quantiles of these statistics are given in Anderson & Stephens (1972).

Giné's G_n Test

A disadvantage of the Bingham test is that it is not consistent against alternatives with $\mathrm{E}[\mathbf{x}\mathbf{x}^T] = p^{-1}\mathbf{I}_p$. A test of uniformity on $\mathbb{R}P^{p-1}$ which is consistent against all alternatives is Giné's G_n test. This rejects uniformity for large values of

$$G_n = \frac{n}{2} - \frac{p-1}{2n}\left(\frac{\Gamma((p-1)/2)}{\Gamma(p/2)}\right)^2 \sum_{i<j} \sin\Psi_{ij}, \tag{10.7.5}$$

where Ψ_{ij} is the smaller of the two angles between $\mathbf{x}_i$ and $\mathbf{x}_j$ and is given by (10.4.9). This test was introduced for $p = 3$ by Giné (1975) and extended to the general case by Prentice (1978). The large-sample asymptotic distribution of G_n is that of an infinite weighted sum of independent χ^2 variables (see Prentice, 1978, pp. 172–173). For $p = 3$, selected quantiles of G_n are given by Keilson *et al.* (1983). In particular, the limiting large-sample 10%, 5% and 1%

quantiles of G_n are 1.816, 2.207, and 3.090, respectively. For $p = 3$, Prentice (1978) derived the approximation

$$\Pr\left(G_n > \frac{x}{16}\right) \simeq 4.638\left[\Pr(\chi_5^2 > x) - 1.593\Pr(\chi_3^2 > x) + 1.323\Pr(\chi_1^2 > x)\right], \tag{10.7.6}$$

for large n and x.

For testing uniformity of axial data against Watson alternatives, Bingham's (1974) test of uniformity and Giné's (1975) G_n have comparable power and are preferable to a third test (based on the number of pairs of axes less than a certain distance apart) given in Diggle, Fisher & Lee (1985).

10.7.2 Testing Rotational Symmetry

The Watson distributions are precisely those Bingham distributions which have rotational symmetry about some axis. Thus it is often of interest to test that a Bingham distribution is indeed a Watson distribution, i.e. to test

$$H_0 : \operatorname{rk}\mathbf{A} = 1, \tag{10.7.7}$$

where $\mathbf{A}$ is the parameter matrix of the Bingham distribution (without the restriction $\operatorname{tr}\mathbf{A} = 0$). It follows from standard results on regular exponential models that the likelihood ratio statistic w for (10.7.7) satisifies the large-sample approximation

$$w \simeq n\operatorname{tr}\left\{\left(\hat{\mathbf{A}} - \tilde{\mathbf{A}}\right)\left(\bar{\mathbf{T}} - \mathrm{E}_{\tilde{\mathbf{A}}}[\mathbf{x}\mathbf{x}^T]\right)\right\}, \tag{10.7.8}$$

where $\hat{\mathbf{A}}$ and $\tilde{\mathbf{A}}$ are the maximum likelihood estimates of $\mathbf{A}$ in the Bingham and Watson models, and $\mathrm{E}_{\tilde{\mathbf{A}}}[\mathbf{x}\mathbf{x}^T]$ refers to the expectation in the fitted Watson distribution. Under (10.7.7),

$$w \;\dot{\sim}\; \chi^2_{(p+1)(p-2)/2}$$

for large n. The case $p = 3$ was considered by Bingham (1974). For more explicit versions of (10.7.8) we consider separately the cases $\tilde{\kappa} > 0$ (the bipolar case) and and $\tilde{\kappa} < 0$ (the girdle case), where $\tilde{\kappa}$ denotes the concentration of the fitted Watson distribution.

If $\tilde{\kappa} > 0$ then (10.7.8) can be expressed as

$$w \simeq w_b, \tag{10.7.9}$$

where

$$w_b = n\sum_{j=2}^{p}(\hat{\kappa}_j - \bar{\kappa})(\bar{t}_j - \bar{t}) \tag{10.7.10}$$

and

$$\bar{\kappa} = \frac{1}{p-1}\sum_{j=2}^{p}\hat{\kappa}_j, \qquad \bar{t} = \frac{1}{p-1}\sum_{j=2}^{p}\bar{t}_j, \tag{10.7.11}$$

$\bar{t}_1, \ldots, \bar{t}_p$ being the eigenvalues of $\bar{\mathbf{T}}$ and $\hat{\kappa}_1, \ldots, \hat{\kappa}_p$ being given by (10.3.37) with $\bar{t}_1 \geq \ldots \geq \bar{t}_p$. and $\hat{\kappa}_1 \geq \ldots \geq \hat{\kappa}_p$.

Similarly, if $\tilde{\kappa} < 0$ then (10.7.8) can be expressed as

$$w \simeq w_g, \tag{10.7.12}$$

where

$$w_g = n \sum_{j=1}^{p-1} (\hat{\kappa}_j - \bar{\kappa})(\bar{t}_j - \bar{t}) \tag{10.7.13}$$

and now

$$\bar{\kappa} = \frac{1}{p-1} \sum_{j=1}^{p-1} \hat{\kappa}_j, \qquad \bar{t} = \frac{1}{p-1} \sum_{j=1}^{p-1} \bar{t}_j. \tag{10.7.14}$$

Evaluation of $\hat{\kappa}_1, \ldots, \hat{\kappa}_p$ in w_b and w_g can be avoided by using the large-sample approximations

$$w_b \simeq \frac{n(p^2-1)\left\{\operatorname{tr}(\bar{\mathbf{T}}^2) - \bar{t}_1^2 - (1-\bar{t}_1^2)/(p-1)\right\}}{2(1-2\bar{t}_1 + \hat{c}_{11})} \tag{10.7.15}$$

$$w_g \simeq \frac{n(p^2-1)\left\{\operatorname{tr}(\bar{\mathbf{T}}^2) - \bar{t}_p^2 - (1-\bar{t}_p^2)/(p-1)\right\}}{2(1-2\bar{t}_p + \hat{c}_{pp})}, \tag{10.7.16}$$

where

$$\hat{c}_{11} = \frac{1}{n} \sum_{i=1}^{p} \left(\mathbf{x}_i^T \bar{\mathbf{t}}_1\right)^4, \qquad \hat{c}_{pp} = \frac{1}{n} \sum_{i=1}^{p} \left(\mathbf{x}_i^T \bar{\mathbf{t}}_p\right)^4,$$

$\mathbf{t}_1$ and $\mathbf{t}_p$ being unit eigenvectors corresponding to the eigenvalues $\bar{t}_1$ and $\bar{t}_p$ of $\bar{\mathbf{T}}$. Indeed, as pointed out by Prentice (1984), no distributional assumptions are required for the large-sample asymptotic results

$$\frac{n(p^2-1)\left\{\operatorname{tr}(\bar{\mathbf{T}}^2) - \bar{t}_1^2 - (1-\bar{t}_1^2)/(p-1)\right\}}{2(1-2\bar{t}_1 + \hat{c}_{11})} \;\dot{\sim}\; \chi^2_{(p+1)(p-2)/2}, \tag{10.7.17}$$

$$\frac{n(p^2-1)\left\{\operatorname{tr}(\bar{\mathbf{T}}^2) - \bar{t}_p^2 - (1-\bar{t}_p^2)/(p-1)\right\}}{2(1-2\bar{t}_p + \hat{c}_{pp})} \;\dot{\sim}\; \chi^2_{(p+1)(p-2)/2}, \tag{10.7.18}$$

which provide tests of rotational symmetry in the bipolar and girdle cases, respectively.

Example 10.13

For the calcite grains data of Example 10.3, we test the hypothesis of rotational symmetry.

From Examples 10.3 and 10.4,

$$\bar{t}_1 = 0.495, \quad \bar{t}_2 = 0.365, \quad \bar{t}_3 = 0.139 \tag{10.7.19}$$

and

$$\hat{\kappa}_1 = 0, \quad \hat{\kappa}_2 = -0.68, \quad \hat{\kappa}_3 = -3.62. \tag{10.7.20}$$

Substituting these values into (10.7.10) and (10.7.13) gives

$$w_g = 1.33, \qquad w_b = 9.97.$$

Because $\bar{t}_1 \simeq \bar{t}_2 > \bar{t}_3$, it is appropriate to use w_g. Since $\chi^2_{2;0.05} = 5.99$, the hypothesis of rotational symmetry is accepted at the 5% level.

Use of the approximate maximum likelihood estimates given in (10.3.44) yields

$$w_g \simeq 1.31,$$

giving the same conclusion as above.

Another test of rotational symmetry is given in Jupp & Spurr (1983).

10.7.3 One-Sample Tests on Watson Distributions

Consider the problem of testing

$$H_0 : \pm\boldsymbol{\mu} = \pm\boldsymbol{\mu}_0, \tag{10.7.21}$$

for the axis of the Watson distribution $W(\boldsymbol{\mu}, \kappa)$, where $\pm\boldsymbol{\mu}_0$ is a prescribed axis. We shall give tests which are suitable when the distribution is concentrated, i.e. when κ is either very large (the bipolar case) or very negative (the girdle case). The following high-concentration approximations will be required.

Suppose that $\mathbf{x} \sim W(\boldsymbol{\mu}, \kappa)$. If $t = \mathbf{x}^T \boldsymbol{\mu}$ then t has density function proportional to

$$(1 - t^2)^{(p-3)/2} e^{\kappa t^2} \tag{10.7.22}$$

on $[-1, 1]$. Hence

$$2|\kappa| t^2 \;\dot{\sim}\; \chi^2_1, \qquad \kappa \to -\infty. \tag{10.7.23}$$

Similarly, (10.7.22) is proportional to

$$(1 - t^2)^{(p-3)/2} e^{-\kappa(1-t^2)}, \tag{10.7.24}$$

so that

$$2\kappa(1 - t^2) \;\dot{\sim}\; \chi^2_{p-1}, \qquad \kappa \to \infty. \tag{10.7.25}$$

The Girdle Case

Assume that $\kappa < 0$, so that the Watson distribution is of girdle type. Let $\pm\mathbf{x}_1, \ldots, \pm\mathbf{x}_n$ be a random sample from $W(\boldsymbol{\mu}, \kappa)$. It follows from (10.7.23) that, under H_0,

$$2n|\kappa|\boldsymbol{\mu}_0^T\bar{\mathbf{T}}\boldsymbol{\mu}_0 = 2|\kappa|\sum_{i=1}^{n}(\boldsymbol{\mu}_0^T\mathbf{x}_i)^2 \ \dot{\sim}\ \chi_n^2, \qquad \kappa \to -\infty. \tag{10.7.26}$$

The likelihood ratio statistic for testing $\pm\boldsymbol{\mu} = \pm\boldsymbol{\mu}_0$ is

$$2n\kappa\left\{\hat{\boldsymbol{\mu}}^T\bar{\mathbf{T}}\hat{\boldsymbol{\mu}} - \boldsymbol{\mu}_0^T\bar{\mathbf{T}}\boldsymbol{\mu}_0\right\} = 2n|\kappa|\left\{\boldsymbol{\mu}_0^T\bar{\mathbf{T}}\boldsymbol{\mu}_0 - \bar{t}_p\right\}, \tag{10.7.27}$$

where $\bar{t}_p$ is the smallest eigenvalue of $\bar{\mathbf{T}}$. General high-concentration results on exponential dispersion models (see Jørgensen, 1987, Section 4) show that

$$2n|\kappa|(\boldsymbol{\mu}_0^T\bar{\mathbf{T}}\boldsymbol{\mu}_0 - \bar{t}_p) \ \dot{\sim}\ \chi_{p-1}^2, \qquad \kappa \to -\infty, \tag{10.7.28}$$

There is an 'analysis of variance' decomposition

$$2n|\kappa|\boldsymbol{\mu}_0^T\bar{\mathbf{T}}\boldsymbol{\mu}_0 = 2n|\kappa|\bar{t}_p + 2n|\kappa|\left(\boldsymbol{\mu}_0^T\bar{\mathbf{T}}\boldsymbol{\mu}_0 - \bar{t}_p\right), \tag{10.7.29}$$

which is analogous to (9.6.16). The terms in (10.7.29) represent the dispersion of the data about the hyperplane normal to $\boldsymbol{\mu}_0$, the dispersion of the data about the hyperplane normal to $\hat{\boldsymbol{\mu}}$, and the deviation of $\hat{\boldsymbol{\mu}}$ from $\boldsymbol{\mu}_0$, respectively. We deduce from (10.7.26) and (10.7.28) that

$$2n|\kappa|\bar{t}_p \ \dot{\sim}\ \chi_{n-p+1}^2, \qquad \kappa \to -\infty. \tag{10.7.30}$$

and that $2n|\kappa|\bar{t}_p$ and $2n|\kappa|(\boldsymbol{\mu}_0^T\bar{\mathbf{T}}\boldsymbol{\mu}_0 - \bar{t}_p)$ are approximately independent. This gives the approximation

$$\frac{(\boldsymbol{\mu}_0^T\bar{\mathbf{T}}\boldsymbol{\mu}_0 - \bar{t}_p)/(p-1)}{\bar{t}_p/(n-p+1)} \ \dot{\sim}\ F_{p-1,n-p+1}, \qquad \kappa \to -\infty. \tag{10.7.31}$$

The appropriate test rejects H_0 for large values of the statistic in (10.7.31). An approximate $100(1-\alpha)\%$ confidence region for $\boldsymbol{\mu}$ based on (10.7.31) is

$$\left\{\boldsymbol{\mu} : \boldsymbol{\mu}^T\bar{\mathbf{T}}\boldsymbol{\mu} < \bar{t}_p\left(1 + \frac{p-1}{n-p+1}F_{p-1,n-p+1;\alpha}\right)\right\}.$$

This region is the intersection of S^{p-1} and the (double) cone with axis $\pm\mathbf{t}_p$. This cone is nearly rotationally symmetric, since $\bar{t}_1 \simeq \ldots \simeq \bar{t}_{p-1}$ in the girdle case.

The Bipolar Case

Assume that $\kappa > 0$, so that the Watson distribution is of bipolar type. Let $\pm\mathbf{x}_1, \ldots, \pm\mathbf{x}_n$ be a random sample from $W(\boldsymbol{\mu}, \kappa)$. It follows from (10.7.25) that, under H_0,

$$2n\kappa\left(1 - \boldsymbol{\mu}_0^T\bar{\mathbf{T}}\boldsymbol{\mu}_0\right) = 2\kappa\sum_{i=1}^{n}\left(1 - (\boldsymbol{\mu}_0^T\mathbf{x}_i)^2\right) \ \dot{\sim}\ \chi_{n(p-1)}^2, \qquad \kappa \to \infty. \tag{10.7.32}$$

The likelihood ratio statistic for testing $\pm\boldsymbol{\mu}_1 = \ldots = \pm\boldsymbol{\mu}_n$ in the saturated model with $\mathbf{x}_i \sim W(\boldsymbol{\mu}_i, \kappa)$ (with κ known) is

$$2\kappa\left\{n - n\hat{\boldsymbol{\mu}}^T\bar{\mathbf{T}}\hat{\boldsymbol{\mu}}\right\} = 2n\kappa\left(1 - \hat{\boldsymbol{\mu}}^T\bar{\mathbf{T}}\hat{\boldsymbol{\mu}}\right) = 2n\kappa\left(1 - \bar{t}_1\right),$$

where $\bar{t}_1$ is the largest eigenvalue of $\bar{\mathbf{T}}$. General high-concentration results on exponential dispersion models (see Jørgensen, 1987, Section 4) show that

$$2n\kappa\left(1 - \bar{t}_1\right) \ \dot{\sim}\ \chi^2_{(n-1)(p-1)}, \qquad \kappa \to \infty. \tag{10.7.33}$$

In the 'analysis of variance' decomposition

$$2n\kappa\left(1 - \boldsymbol{\mu}_0^T\bar{\mathbf{T}}\boldsymbol{\mu}_0\right) = 2n\kappa\left(1 - \bar{t}_1\right) + 2n\kappa\left(\bar{t}_1 - \boldsymbol{\mu}_0^T\bar{\mathbf{T}}\boldsymbol{\mu}_0\right), \tag{10.7.34}$$

the terms represent the dispersion of the data about $\boldsymbol{\mu}_0$, the dispersion about $\hat{\boldsymbol{\mu}}$, and the deviation of $\hat{\boldsymbol{\mu}}$ from $\boldsymbol{\mu}_0$, respectively. We deduce from (10.7.32) and (10.7.33) that

$$2n\kappa\left(\bar{t}_1 - \boldsymbol{\mu}_0^T\bar{\mathbf{T}}\boldsymbol{\mu}_0\right) \ \dot{\sim}\ \chi^2_{p-1}, \qquad \kappa \to \infty \tag{10.7.35}$$

and that $2n\kappa(1 - \bar{t}_1)$ and $2n\kappa(\bar{t}_1 - \boldsymbol{\mu}_0^T\bar{\mathbf{T}}\boldsymbol{\mu}_0)$ are approximately independent. This gives the approximation

$$\frac{(\bar{t}_1 - \boldsymbol{\mu}_0^T\bar{\mathbf{T}}\boldsymbol{\mu}_0)/(p-1)}{(1 - \bar{t}_1)/(n-1)(p-1)} \ \dot{\sim}\ F_{p-1,(n-1)(p-1)}, \qquad \kappa \to \infty. \tag{10.7.36}$$

The appropriate test rejects H_0 for large values of the statistic in (10.7.36). An approximate $100(1-\alpha)\%$ confidence region for $\boldsymbol{\mu}$ based on (10.7.36) is

$$\left\{\boldsymbol{\mu} : \boldsymbol{\mu}^T\bar{\mathbf{T}}\boldsymbol{\mu} > \bar{t}_1 - \frac{(1 - \bar{t}_1)F_{p-1,(n-1)(p-1);\alpha}}{n-1}\right\}.$$

This region is the intersection of S^{p-1} and the (double) cone with axis $\pm\mathbf{t}_1$. This cone is nearly rotationally symmetric, since $\bar{t}_2 \simeq \ldots \simeq \bar{t}_p$ in the bipolar case.

10.7.4 Multi-sample Tests on Watson Distributions

Suppose that $\pm\mathbf{x}_{i1}, \ldots, \pm\mathbf{x}_{in_i}$ $(i = 1, \ldots, q)$ are q independent random samples of sizes $n_1, \ldots, n_q$ from $W(\boldsymbol{\mu}_i, \kappa_i)$, $i = 1, \ldots, q$. Put $n = n_1 + \ldots + n_q$. The ith sample and the combined sample are summarised by their respective scatter matrices $\bar{\mathbf{T}}_i$ and $\bar{\mathbf{T}}$. Let $\bar{t}_{i1} \geq \ldots \geq \bar{t}_{ip}$ and $\bar{t}_1 \geq \ldots \geq \bar{t}_p$ denote the eigenvalues of $\bar{\mathbf{T}}_i$ and $\bar{\mathbf{T}}$, respectively.

We wish to test

$$H_0 : \pm\boldsymbol{\mu}_1 = \ldots = \pm\boldsymbol{\mu}_q \tag{10.7.37}$$

against the alternative that at least one of the equalities is not satisfied. For simplicity, we shall assume that $\kappa_1 = \ldots = \kappa_q$, where the common concentration κ is unknown.

Intuition suggests that it is reasonable to reject H_0 when some appropriate weighted sum of the eigenvalues of the $\bar{\mathbf{T}}_i$ is very different from a similar function of the eigenvalues of $\bar{\mathbf{T}}$. We present two tests of this form: one suitable for large samples, the other for concentrated (girdle or bipolar) distributions.

A Large-Sample Test

The following test is a special case of the multi-sample test introduced by Watson (1983c) for testing equality of q distributions on S^{p-1} with density functions (9.3.36).

The Girdle Case

If $\kappa < 0$ then application of the (matrix) central limit theorem to the $\bar{\mathbf{T}}_i$ and to $\bar{\mathbf{T}}$, together with the rotational symmetry of Watson distributions, leads eventually to the result

$$\frac{(1 - p\bar{t}_p)(n\bar{t}_p - \sum_{i=1}^q n_i \bar{t}_{ip})}{2(\bar{t}_p - \hat{\mathrm{E}}[t^4])} \dot{\approx} \chi^2_{(q-1)(p-1)}, \qquad n \to \infty, \tag{10.7.38}$$

where

$$\hat{\mathrm{E}}[t^4] = \frac{1}{n} \sum_{i=1}^{q} \sum_{j=1}^{n_i} \left(\mathbf{x}_{ij}^T \mathbf{t}_p\right)^2,$$

$\mathbf{t}_p$ being a unit eigenvector corresponding to the smallest eigenvalue $\bar{t}_p$ of $\bar{\mathbf{T}}$. The null hypothesis is rejected for large values of the statistic in (10.7.38).

The Bipolar Case

If $\kappa > 0$ then an argument similar to that used in the girdle case yields

$$\frac{(p\bar{t}_1 - 1)(\sum_{i=1}^q n_i \bar{t}_{i1} - n\bar{t}_1)}{2(\bar{t}_1 - \hat{\mathrm{E}}[t^4])} \dot{\approx} \chi^2_{(q-1)(p-1)}, \qquad n \to \infty, \tag{10.7.39}$$

where

$$\hat{\mathrm{E}}[t^4] = \frac{1}{n} \sum_{i=1}^{q} \sum_{j=1}^{n_i} \left(\mathbf{x}_{ij}^T \mathbf{t}_1\right)^2, \tag{10.7.40}$$

$\mathbf{t}_1$ being a unit eigenvector corresponding to the largest eigenvalue $\bar{t}_1$ of $\bar{\mathbf{T}}$. The null hypothesis is rejected for large values of the statistic in (10.7.40).

A High-Concentration F Test

Among the multi-sample tests for q von Mises–Fisher distributions which were considered in Section 10.6.1 was a high-concentration F test based on (10.6.4).

We now give an analogue for Watson distributions.

The Girdle Case

If $\kappa < 0$ then it follows from (10.7.30) and the 'analysis of variance' decomposition

$$2n|\kappa|\bar{t}_p = 2|\kappa| \sum_{i=1}^{q} n_i \bar{t}_{ip} + 2|\kappa| \left(n\bar{t}_p - \sum_{i=1}^{q} n_i \bar{t}_{ip} \right) \tag{10.7.41}$$

that

$$\frac{(n\bar{t}_p - \sum_{i=1}^{q} n_i \bar{t}_{ip})/(q-1)(p-1)}{(\sum_{i=1}^{q} n_i \bar{t}_{ip})/(n - q(p-1))} \dot{\sim} F_{(q-1)(p-1), n-q(p-1)}, \qquad \kappa \to -\infty. \tag{10.7.42}$$

The appropriate test rejects H_0 for large values of the statistic in (10.7.42).

The Bipolar Case

If $\kappa > 0$ then it follows from (10.7.33) and the 'analysis of variance' decomposition

$$2n\kappa (1 - \bar{t}_1) = 2\kappa \sum_{i=1}^{q} n_i (1 - \bar{t}_{i1}) + 2\kappa \left(\sum_{i=1}^{q} n_i \bar{t}_{i1} - n\bar{t}_1 \right) \tag{10.7.43}$$

that

$$\frac{(\sum_{i=1}^{q} n_i \bar{t}_{i1} - n\bar{t}_1)/(q-1)(p-1)}{(\sum_{i=1}^{q} n_i (1 - \bar{t}_{i1}))/(n-q)(p-1)} \dot{\sim} F_{(q-1)(p-1),(n-q)(p-1)}, \qquad \kappa \to -\infty. \tag{10.7.44}$$

The appropriate test rejects H_0 for large values of the statistic in (10.7.44).

In the cases $p = 2$ and $p = 3$, nonparametric simultaneous confidence regions for rotations taking $\pm\boldsymbol{\mu}_i$ to $\pm\boldsymbol{\mu}_j$ (for $1 \leq i < j \leq q$) were introduced by Beran & Fisher (1998).

10.8 A GENERAL FRAMEWORK FOR TESTING UNIFORMITY

Many tests of uniformity on S^{p-1} fit into the general framework established by Giné (1975) on more general sample spaces (such as the Stiefel manifolds and Grassmann manifolds considered in Sections 13.2 and 13.3). This framework extends the embedding approach described in Section 9.1 by transforming directional problems into infinite-dimensional multivariate problems as follows.

Naturally associated with the sphere S^{p-1} is the Hilbert space $L^2(S^{p-1})$ of square-integrable functions on S^{p-1}. Any suitable mapping $\mathbf{t}$ of S^{p-1} into

$L^2(S^{p-1})$ transforms distributions on S^{p-1} into distributions on $L^2(S^{p-1})$. Since the Hilbert space $L^2(S^{p-1})$ behaves just like the familiar Euclidean spaces, infinite-dimensional versions of ordinary (non-normal) multivariate methods can then be applied. Desirable properties of $\mathbf{t}$ are continuity and equivariance. *Equivariance* means that for every rotation $\mathbf{U}$ of S^{p-1}

$$\mathbf{t}(\mathbf{Ux}) = \mathbf{Ut}(\mathbf{x}), \qquad \mathbf{x} \in S^{p-1},$$

where on the right-hand side $\mathbf{U}$ denotes the corresponding rotation of $L^2(S^{p-1})$ given by $f \mapsto f_{\mathbf{U}}$, where $f_{\mathbf{U}}(x) = f(\mathbf{U}x)$. Note that if $\mathbf{t}$ is equivariant then $||\mathbf{t}(\mathbf{x})||$ is constant as $\mathbf{x}$ runs through S^{p-1}, so that $\mathbf{t}(\mathbf{x})$ is a point in an infinite-dimensional sphere.

One way of constructing such mappings $\mathbf{t}$ is the following method due to Giné (1975) and based on the eigenfunctions of the Laplacian on S^{p-1}. (See also Jupp & Spurr, 1985, for the Hilbert space approach given here.) Let E_k denote the space of eigenfunctions corresponding to the kth eigenvalue, for $k \geq 1$. Then there is a well-defined mapping $\mathbf{t}_k$ of S^{p-1} into E_k given by

$$\mathbf{t}_k = \sum_{i=1}^{n_k} f_i(\mathbf{x}) f_i, \tag{10.8.1}$$

where $\{f_i : 1 \leq i \leq n_k\}$ is any orthonormal base of E_k. If $\{a_k\}$ is a sequence which converges sufficiently rapidly to 0 then

$$\mathbf{x} \mapsto \mathbf{t}(\mathbf{x}) = \sum_{k=1}^{\infty} a_k \mathbf{t}_k(\mathbf{x}) \tag{10.8.2}$$

defines an equivariant mapping $\mathbf{t}$ of S^{p-1} into $L^2(S^{p-1})$. For example, on the circle, provided that

$$\sum_{k=1}^{\infty} a_k^2 < \infty,$$

the angle θ is transformed into the function $\mathbf{t}(\theta) = \mathbf{t}_\theta$ defined by

$$\mathbf{t}_\theta(x) = \sum_{k=1}^{\infty} a_k \cos k(x - \theta).$$

Under mild conditions on the sequence $\{a_k\}$, the mapping $\mathbf{t}$ defined by (10.8.2) is continuous. Then $\mathbf{t}$ gives rise to a mapping $\boldsymbol{\tau}$ of the set $P(S^{p-1})$ of all probability distributions on S^{p-1} into $L^2(S^{p-1})$ defined by

$$\boldsymbol{\tau}(\nu) = \int_{S^{p-1}} \mathbf{t}(\mathbf{x}) d\nu(\mathbf{x}).$$

If all the coefficients a_k are non-zero then $\boldsymbol{\tau}$ is a one-to-one function (indeed, a topological embedding) and so gives rise to tests which are consistent against all alternatives. Another way of looking at the transformation $\mathbf{t}$ given by (10.8.2) is to regard it as a 'weighted characteristic function', sending a distribution on S^{p-1} into the sequence $\{a_k \mathrm{E}[\mathbf{t}_k(\mathbf{x})]\}_{k=1,2,...}$ of its 'Fourier coefficients'.

Any suitable mapping $\mathbf{t}$ as above gives a test for uniformity which rejects uniformity for large values of the 'resultant length' $\|\mathbf{t}(\mathbf{x}_1)+\ldots+\mathbf{t}(\mathbf{x}_n)\|$ of the transformed observations, where $\|\cdot\|$ denotes the L_2 norm on $L^2(S^{p-1})$. Thus such a test is an infinite-dimensional version of Rayleigh's test. Note that the equivariance of $\mathbf{t}$ ensures that the test is invariant under rotations. Note also that the test rejects uniformity for large values of

$$T_n = n^{-1}\|\mathbf{t}(\mathbf{x}_1) + \ldots + \mathbf{t}(\mathbf{x}_n)\|^2 = n^{-1}\sum_{i=1}^{n}\sum_{j=1}^{n}\langle \mathbf{t}(\mathbf{x}_i), \mathbf{t}(\mathbf{x}_j)\rangle, \tag{10.8.3}$$

where $\langle \cdot, \cdot \rangle$ denotes the inner product on $L^2(S^{p-1})$ given by

$$\langle f, g\rangle = \int_{S^{p-1}} f(\mathbf{x})g(\mathbf{x})d\mathbf{x},$$

the integration being with respect to the uniform distribution on S^{p-1}. Under uniformity, the distribution of T_n tends as $n \to \infty$ to that of

$$\sum_{k=1}^{\infty} a_k^2 U_k,$$

where the U_k have independent $\chi^2_{d(k)}$ distributions with $d(k) = \dim E_k$.

In the case $p = 2$, the statistic T_n defined by (10.8.3) is related to the corresponding statistic B_n in Beran's class by

$$T_n = \frac{B_n}{n}, \tag{10.8.4}$$

where

$$B_n = 2n^2 \sum_{k=1}^{\infty} a_k^2 \bar{R}_k^2, \tag{10.8.5}$$

as in (6.3.52), and $\bar{R}_k$ is the mean resultant length of $k\theta_1, \ldots, k\theta_n$ with $\theta_1, \ldots, \theta_n$ being the observations on S^1.

As we have seen in Section 6.3.7 for the circular case, appropriate choices of the coefficients a_k in (10.8.2) yield tests with suitable properties of consistency, ease of computation, etc.

The construction of tests of uniformity using mappings $\mathbf{t}$ into $L^2(S^{p-1})$ as above generalises readily to sample spaces which are compact Riemannian manifolds. Giné's (1975) presentation of these tests in this general case is equivalent to that above but used Sobolev spaces of generalised functions with square-integrable derivatives of appropriate order (see Giné, 1975, pp. 1246–1247). Hence these tests are sometimes called 'Sobolev tests'.

In the case of sample spaces (such as spheres) which are compact homogeneous spaces (being acted on transitively by a group G), Beran (1968) obtained these tests using statistics of the form

$$T_n = \frac{1}{n} \int_G \left[\sum_{i=1}^{n} f(g\mathbf{x}_i) - n \right]^2 dg \tag{10.8.6}$$

as locally most powerful invariant tests of uniformity against the alternative of a density $f(g\mathbf{x})$, where g is an unknown element of G,

$$f(\mathbf{x}) = 1 + \sum_{k=1}^{\infty} a_k \sum_{i=1}^{n_k} f_i(\mathbf{x})$$

and $\{f_i : 1 \leq i \leq n_k\}$ are as in (10.8.1). If the sample space is the circle then (10.8.6) reduces to (6.3.70).

The framework for testing uniformity which has been described in this section also forms the basis of permutation tests of (i) equality of two distributions (Wellner, 1979), (ii) symmetry (Jupp & Spurr, 1983), and (iii) independence (Jupp & Spurr, 1985).

11

Correlation and Regression

11.1 INTRODUCTION

This chapter is concerned with the relationships between directional random variables. Many correlation coefficients have been proposed for measuring the strengths of such relationships. The most important correlation coefficients are presented in Section 11.2. Various regression models for describing these relationships are given in Section 11.3. Section 11.4 provides some bivariate models. Directional time series are discussed briefly in Section 11.5.

11.2 MEASURES OF CORRELATION

11.2.1 Linear–Circular Correlation

A common problem is that of measuring the association between a linear random variable X and a circular random variable Θ on the basis of a random sample $(x_1, \theta_1), \ldots, (x_n, \theta_n)$ of observations on (X, Θ). In a typical meteorological example, X is ozone concentration and Θ is wind direction.

A Correlation Coefficient Based on the Embedding Approach

The embedding approach described in Section 9.1 suggests that Θ should be regarded as the random vector $\mathbf{u} = (\cos\Theta, \sin\Theta)^T$ in the plane. An appropriate measure of dependence between X and $\mathbf{u}$ based on $(x_1, \theta_1), \ldots, (x_n, \theta_n)$ is the sample multiple correlation coefficient $R_{x\theta}$ of X and $\mathbf{u}$, i.e. the maximum sample correlation between X and linear functions $\mathbf{a}^T\mathbf{u}$ of $\mathbf{u}$ (see Mardia, Kent & Bibby, 1979, Section 6.5.2.) A straightforward calculation shows that

$$R_{x\theta}^2 = \frac{r_{xc}^2 + r_{xs}^2 - 2r_{xc}r_{xs}r_{cs}}{1 - r_{cs}^2}, \tag{11.2.1}$$

where

$$r_{xc} = \text{corr}(x, \cos\theta), \quad r_{xs} = \text{corr}(x, \sin\theta), \quad r_{cs} = \text{corr}(\cos\theta, \sin\theta)$$

are the sample correlation coefficients. The linear–circular correlation coefficient $R_{x\theta}$ was introduced by Mardia (1976) and Johnson & Wehrly (1977). If X and Θ are independent and X is normally distributed then

$$\frac{(n-3)R_{x\theta}^2}{1-R_{x\theta}^2} \sim F_{2,n-3}. \tag{11.2.2}$$

The exact distribution of this statistic under a specific alternative to uniformity has been derived by Liddell & Ord (1978).

Example 11.1

In studies of atmospheric pollution, it is often of interest to know whether or not the concentration of a pollutant depends on the wind direction. Readings of ozone concentration and wind direction at 4-day intervals at a weather station in Milwaukee are given in Table 11.1. Are ozone concentration and wind direction independent?

Table 11.1 Measurements of ozone concentration x and wind direction (in degrees) θ (Johnson & Wehrly, 1977)

x:	28.0	85.2	80.5	4.7	45.9	12.7	72.5	56.6	31.5	112.0
θ:	327	91	88	305	344	270	67	21	281	8
x:	20.0	72.5	16.0	45.9	32.6	56.6	52.6	91.8	55.2	
θ:	204	86	333	18	57	6	11	27	84	

Calculation gives $R_{x\theta}^2 = 0.522$, and so $(n-3)R_{x\theta}^2/(1-R_{x\theta}^2) = 17.47$. The observed significance level of the Anderson–Darling test of normality applied to the observations of ozone concentration is 0.88. Thus normality can be assumed and it is appropriate to use (11.2.2). Since $\Pr(F_{2,16} \geq 17.47) < 0.001$, the hypothesis of independence is rejected at the 1% level.

Distribution-Free Correlation Coefficients

A Linear–Circular Rank Correlation Coefficient

The construction of Spearman's rho as the product moment correlation coefficient applied to ranks suggests that $R_{x\theta}^2$ should be calculated for the data obtained by replacing the observations $(x_1, \theta_1), \ldots, (x_n, \theta_n)$ by their 'ranks'. More precisely, we reorder the observations so that $x_1 \leq \ldots \leq x_n$. Let $r_1, \ldots, r_n$ be the corresponding circular ranks of $\theta_1, \ldots, \theta_n$. The uniform scores $\beta_1, \ldots, \beta_n$ are defined by

$$\beta_i = \frac{2\pi r_i}{n},$$

as in (8.3.2). Then the value of $nR^2_{x\theta}$ for $(1, \beta_1), \ldots, (n, \beta_n)$ is U_n, where

$$U_n = \frac{24(T_c^2 + T_s^2)}{n^2(n+1)} \tag{11.2.3}$$

with

$$T_c = \sum_{i=1}^{n} i \cos \beta_i, \qquad T_s = \sum_{i=1}^{n} i \sin \beta_i.$$

The rank correlation coefficient U_n was introduced by Mardia (1976). The coefficient U_n has the following geometrical interpretation. Since

$$\frac{T_c^2 + T_s^2}{n^2} = \left| \sum_{j=1}^{n} \frac{j}{n} e^{i\beta_j} \right|^2,$$

U_n is proportional to the squared modulus of a weighted sum of the unit complex numbers represented by the uniform scores.

Because U_n is based on ranks and uniform scores, it is invariant under homeomorphisms (continuous transformations with continuous inverses) of the line and of the circle. An important consequence is that, under independence of X and Θ, the distribution of U_n does not depend on the marginal distributions of X and Θ. The test of independence based on U_n rejects independence for large values of U_n.

Standard results on rank statistics (e.g. Hájek & Šidák, 1967, pp. 57–58, 163–164) show that, under independence,

$$U_n = \frac{24(T_c^2 + T_s^2)}{n^2(n+1)} \mathrel{\dot\sim} \chi^2_2, \qquad n \to \infty,$$

provided that X and Θ have continuous distributions. Some quantiles of U_n are tabulated in Appendix 2.17.

Sometimes it is helpful to scale U_n, in order to have a correlation coefficient which lies in the range $[0, 1]$. This scaled correlation coefficient D_n is

$$D_n = a_n(T_c^2 + T_s^2),$$

where

$$a_n = \begin{cases} 1/\{1 + 5\cot^2(\pi/n) + 4\cot^4(\pi/n)\}, & n \text{ even}, \\ 2\sin^4(\pi n)/\{1 + \cos(\pi/n)\}^3, & n \text{ odd}. \end{cases} \tag{11.2.4}$$

Example 11.2
For the data set on wind direction and ozone concentration which was considered in Example 11.1, $n = 19$ and $U_n = 0.398$. Since (from Appendix 2.17) the upper 10% quantiles of U_n for $n = 15$ and $n = 20$ are 4.59 and 4.6, respectively, the hypothesis of independence is accepted at the 10% level.

A Correlation Coefficient Based on C-association

An alternative approach, due to Fisher & Lee (1981), is based on considering alternatives to independence of X and Θ for which the real-valued function $\mathrm{E}[X|\Theta = \theta]$ of θ is unimodal on the circle. Such association between X and Θ is called *C-association* or *cylindrical association*. One sample measure of C-association is the proportion of 4-tuples of the points $(x_1, \theta_1), \ldots, (x_n, \theta_n)$ on the cylinder $\mathbb{R} \times S^1$ which have a 'zig-zag' configuration. More precisely, put

$$\psi_{i,j,k,l} = \begin{cases} 1 & \text{if } x_i - x_j, x_j - x_k, x_k - x_l, x_l - x_i \text{ alternate in sign} \\ & \text{when } 0 \leq \theta_i \leq \theta_j \leq \theta_k \leq \theta_l \leq 2\pi, \\ 0 & \text{otherwise,} \end{cases} \tag{11.2.5}$$

and

$$\hat{\lambda}_n = \binom{n}{4}^{-1} \sum_{\{i,j,k,l\}} \psi_{i,j,k,l}, \tag{11.2.6}$$

where the sum is over all subsets $\{i, j, k, l\}$ of size 4 of $\{1, \ldots, n\}$. Under independence, the population analogue of $\hat{\lambda}_n$ is 2/3. Thus an appropriate test rejects independence if $\hat{\lambda}_n$ is far from 2/3. Some quantiles of $\hat{\lambda}_n$ for $n \leq 8$ are given in Fisher (1993, Appendix A11). For moderate sample sizes, a randomisation test based on $\hat{\lambda}_n$ is appropriate. For the treatment of ties and the use of random subsets to calculate $\hat{\lambda}_n$, see Fisher & Lee (1981) and Fisher (1993, Section 6.2.2).

11.2.2 Circular–Circular Correlation

Often we are interested in measuring the association between two circular random variables Θ and Φ on the basis of a random sample $(\theta_1, \phi_1), \ldots, (\theta_n, \phi_n)$. A wide variety of correlation coefficients have been proposed for this purpose.

Correlation Coefficients Based on the Embedding Approach

The embedding approach described in Section 9.1 suggests that Θ and Φ should be regarded as the random vectors $\mathbf{u} = (\cos\Theta, \sin\Theta)^T$ and $\mathbf{v} = (\cos\Phi, \sin\Phi)^T$ in the plane.

A Correlation Coefficient Based on Canonical Correlations

An appropriate measure of dependence between **u** and **v** based on $(\theta_1, \phi_1), \ldots, (\theta_n, \phi_n)$ is the the sum

$$r^2 = \text{tr}(S_{11}^{-1} S_{12} S_{22}^{-1} S_{21}) \tag{11.2.7}$$

of the squared canonical correlation coefficients (see Mardia, Kent & Bibby, 1979, Section 10.2.1). A straightforward calculation shows that

$$\begin{aligned} r^2 \quad = \quad & [(r_{cc}^2 + r_{cs}^2 + r_{sc}^2 + r_{ss}^2) + 2(r_{cc} r_{ss} - r_{cs} r_{sc}) r_1 r_2 \\ & -2(r_{cc} r_{cs} + r_{sc} r_{ss}) r_2 - 2(r_{cc} r_{sc} + r_{cs} r_{ss}) r_1]/ \\ & [(1 - r_1^2)(1 - r_2^2)], \end{aligned} \tag{11.2.8}$$

where $r_{cc} = \text{corr}(\cos\theta, \cos\phi)$ etc., $r_1 = \text{corr}(\cos\theta, \sin\theta)$, $r_2 = \text{corr}(\cos\phi, \sin\phi)$ are the ordinary sample correlation coefficients. The correlation coefficient r^2 was introduced by Jupp & Mardia (1980) in a more general context. Independence of Θ and Φ is rejected for large values of r^2. Under independence,

$$nr^2 \ \dot{\sim}\ \chi_4^2, \qquad n \to \infty, \tag{11.2.9}$$

provided that the variance matrix of $(\cos\Theta, \sin\Theta, \cos\Phi, \sin\Phi)$ is non-singular.

Example 11.3

In a medical experiment, various measurements were taken on 10 medical students several times daily for a period of several weeks (Downs, 1974). The estimated peak times (converted into angles θ and ϕ) for two successive measurements of diastolic blood pressure are given in Table 11.2. Is there dependence between the two peak times (as is plausible on medical grounds)?

Table 11.2 Angles (θ, ϕ) in degrees representing estimated peak times for two successive measurements of diastolic blood pressure (Downs, 1974, reproduced by permission of John Wiley & Sons, Inc.)

θ: 30	15	11	4	348	347	341	333	332	285
ϕ: 25	5	349	358	340	347	345	331	329	287

Calculation gives $r_{cc}^2 = 0.974$, $r_{cs}^2 = 0.213$, $r_{sc}^2 = 0.152$, $r_{ss}^2 = 0.933$, $r_1 = 0.661$, $r_2 = 0.714$, so that $r^2 = 6.64$. Although the sample size of 10 is rather small, the large-sample result (11.2.9) should give a reasonable indication of the significance of the observed value of r^2. Since $nr^2 = 66.4$ and $\chi_{4;0.01}^2 = 13.3$, there is strong evidence for dependence.

A Measure of Rotational Dependence

The strongest form of dependence between the circular random variables Θ and Φ occurs when $\Phi = \Theta + \alpha$ or $\Phi = -\Theta + \alpha$, for some constant angle α. In this case, one of the mean resultant lengths ρ_+ and ρ_- of the circular random variables $\Phi - \Theta$ and $\Phi + \Theta$ is 1. Consideration of $2(\rho_+^2 - \rho_-^2)$ led Fisher & Lee (1982) to introduce the sample version

$$\hat{\rho}_T = \frac{\sum_{i \leq j} \sin(\theta_i - \theta_j) \sin(\phi_i - \phi_j)}{\left(\sum_{i \leq j} \sin^2(\theta_i - \theta_j) \sum_{i \leq j} \sin^2(\phi_i - \phi_j)\right)^{1/2}}. \tag{11.2.10}$$

An alternative formula for computing $\hat{\rho}_T$ is

$$\hat{\rho}_T = \frac{|\mathbf{S}_{12}^*|}{(|\mathbf{S}_{11}^*||\mathbf{S}_{22}^*|)^{1/2}}, \tag{11.2.11}$$

where

$$\begin{aligned}
\mathbf{S}_{12}^* &= \frac{1}{n} \sum_{i=1}^{n} \begin{pmatrix} \cos\theta_i \cos\phi_i & \cos\theta_i \sin\phi_i \\ \sin\theta_i \cos\phi_i & \sin\theta_i \sin\phi_i \end{pmatrix}, \\
\mathbf{S}_{11}^* &= \frac{1}{n} \sum_{i=1}^{n} \begin{pmatrix} \cos\theta_i \cos\theta_i & \cos\theta_i \sin\theta_i \\ \sin\theta_i \cos\theta_i & \sin\theta_i \sin\theta_i \end{pmatrix}, \\
\mathbf{S}_{22}^* &= \frac{1}{n} \sum_{i=1}^{n} \begin{pmatrix} \cos\phi_i \cos\phi_i & \cos\phi_i \sin\phi_i \\ \sin\phi_i \cos\phi_i & \sin\phi_i \sin\phi_i \end{pmatrix}.
\end{aligned} \tag{11.2.12}$$

The corresponding test rejects independence for large values of $|\hat{\rho}_T|$. For small values of n a randomisation test is appropriate. Large-sample normal approximations are given in Fisher & Lee (1983) and Fisher (1993, Section 6.3.3).

Distribution-Free Correlation Coefficients

A Circular–Circular Rank Correlation Coefficient

A very strong form of association between the circular random variables Θ and Φ occurs when $\Phi = g(\Theta)$, for some homeomorphism (continuous transformation with continuous inverse) g of the circle into itself. In this case, the circular random variable $\Phi - g(\Theta)$ has mean resultant length 1, and for all random samples $(\theta_1, \phi_1), \ldots, (\theta_n, \phi_n)$, the circular ranks of $\phi_1, \ldots, \phi_n$ are either the same as those of $\theta_1 \ldots, \theta_n$ (if g is orientation-preserving) or the same as those of $\theta_n \ldots, \theta_1$ (if g is orientation-reversing). Then two intuitively reasonable measures of correlation based on $(\theta_1, \phi_1), \ldots, (\theta_n, \phi_n)$ are the mean resultant lengths $n^2 \bar{R}_+^2$ and $n^2 \bar{R}_-^2$ of $\beta_1 - \gamma_1, \ldots, \beta_n - \gamma_n$ and $\beta_1 + \gamma_1, \ldots, \beta_n + \gamma_n$, where $\beta_1, \ldots, \beta_n$ and $\gamma_1, \ldots, \gamma_n$ are the uniform scores of $\theta_1, \ldots, \theta_n$ and $\phi_1, \ldots, \phi_n$ respectively, defined by

$$\beta_i = \frac{2\pi r_i}{n}, \qquad \gamma_i = \frac{2\pi s_i}{n}, \qquad i = 1, \ldots, n, \tag{11.2.13}$$

with $r_1, \ldots, r_n$ and $s_1, \ldots, s_n$ the circular ranks of $\theta_1 \ldots, \theta_n$ and $\phi_1, \ldots, \phi_n$. Thus

$$n^2 \bar{R}_{\pm}^2 = \left(\sum_{i=1}^{n} \cos(\beta_i \mp \gamma_i) \right)^2 + \left(\sum_{i=1}^{n} \sin(\beta_i \mp \gamma_i) \right)^2 . \tag{11.2.14}$$

Note that $\bar{R}_{\pm}^2 \leq 1$ and that $\bar{R}_{+}^2 = 1$ when $\phi_1, \ldots, \phi_n$ can be obtained from $\theta_1 \ldots, \theta_n$ by an orientation-preserving transformation of the circle, while $\bar{R}_{-}^2 = 1$ when $\phi_1, \ldots, \phi_n$ can be obtained from $\theta_1 \ldots, \theta_n$ by an orientation-reversing transformation. The quantities $\bar{R}_{\pm}^2$ were introduced by Mardia (1975a).

Because $\bar{R}_{\pm}^2$ is based on uniform scores, it is invariant under separate continuous invertible transformations of the two copies of the circle. An important consequence is that, under independence of Θ and Φ, the distribution of $\bar{R}_{\pm}^2$ does not depend on the marginal distributions of Θ and Φ.

Manipulation of (11.2.14) shows that

$$\bar{R}_{\pm}^2 = (T_{cc} \pm T_{ss})^2 + (T_{sc} \mp T_{cs})^2,$$

where

$$T_{cc} = \frac{1}{n} \sum_{i=1}^{n} \cos \beta_i \cos \gamma_i, \qquad T_{cs} = \frac{1}{n} \sum_{i=1}^{n} \cos \beta_i \sin \gamma_i,$$

etc. Since T_{cc}, T_{cs}, T_{sc} and T_{ss} are linear rank statistics, it follows from general results (e.g., Hájek & Šidák, 1967, pp. 57–58, 163–164) that

$$2(n-1)^{1/2}(T_{cc}, T_{ss}, T_{cs}, T_{sc})^T \stackrel{\cdot}{\sim} N_4(0, \mathbf{I}_4), \qquad n \to \infty. \tag{11.2.15}$$

Hence, for large n, $2(n-1)\bar{R}_{+}^2$ and $2(n-1)\bar{R}_{-}^2$ are approximately independently distributed as χ_2^2.

The circular–circular rank correlation coefficient

$$r_0 = \max(\bar{R}_{+}^2, \bar{R}_{-}^2) \tag{11.2.16}$$

was introduced by Mardia (1975a). Since, under independence,

$$\bar{R}_{+}^2 \simeq \bar{R}_{-}^2 \simeq 0, \qquad n \to \infty,$$

independence is rejected for large values of r_0. Some quantiles of r_0 for $5 \leq n \leq 10$ are given in Appendix 2.18. For $n > 10$, the approximation

$$\Pr(2(n-1)r_0 > u) \simeq 1 - (1 - e^{-u/2})^2 \qquad n \to \infty, \tag{11.2.17}$$

(which follows from (11.2.15)) is adequate.

Example 11.4
For the data of Example 11.3, calculation gives $\bar{R}_+^2 = 0.731$, $\bar{R}_-^2 = 0.004$, and so $r_0 = 0.731$. Since the 5% value of r_0 is 0.41, the null hypothesis of independence is rejected at the 5% level. The high value of $\bar{R}_+^2$ indicates positive dependence. Inspection of the individual differences $i - r_i$ reinforces this indication.

A Signed Rank Correlation Coefficient
Taking the difference of $\bar{R}_+^2$ and $\bar{R}_-^2$ (instead of their maximum, as in r_0) gives the signed correlation coefficient

$$\hat{\Pi}_n = \bar{R}_+^2 - \bar{R}_-^2, \tag{11.2.18}$$

which was introduced by Fisher & Lee (1982). The statistic $\hat{\Pi}_n$ takes values in $[-1, 1]$ and can be regarded as an analogue of Spearman's rho. Values of $\hat{\Pi}_n$ near 1 indicate positive dependence, while values of $\hat{\Pi}_n$ near -1 indicate negative dependence. Under independence, $\hat{\Pi}_n \simeq 0$. However, small values of $\hat{\Pi}_n$ need not indicate independence, since $\hat{\Pi}_n \simeq 0$ whenever $\bar{R}_+^2 \simeq \bar{R}_-^2$. For example, for $n = 6$ and $(r_1, r_2, r_3, r_4, r_5, r_6) = (4, 1, 6, 5, 2, 3)$, we find that $\bar{R}_+^2 = \bar{R}_-^2 = 0.25$, so the data exhibit both positive and negative dependence but $\hat{\Pi}_n = 0$.

Fisher & Lee (1983) showed that $\hat{\Pi}_n$ can be written as

$$\hat{\Pi}_n = \frac{4}{n^2} \sum_{i \leq j} \sin(\beta_i - \beta_j) \sin(\gamma_i - \gamma_j), \tag{11.2.19}$$

where $\beta_1, \ldots, \beta_n$ and $\gamma_1, \ldots, \gamma_n$ are the uniform scores of $\theta_1, \ldots, \theta_n$ and $\phi_1, \ldots, \phi_n$ respectively, defined by (11.2.13). Thus $\hat{\Pi}_n$ is $\hat{\rho}_T$ calculated for the uniform scores $(\beta_1, \gamma_1), \ldots, (\beta_n, \gamma_n)$. A useful formula for computing $\hat{\Pi}_n$ is

$$\hat{\Pi}_n = \frac{4}{n^2}(AB - CD), \tag{11.2.20}$$

where

$$A = \sum_{i=1}^{n} \cos\beta_i \cos\gamma_i, \qquad B = \sum_{i=1}^{n} \sin\beta_i \sin\gamma_i,$$

$$C = \sum_{i=1}^{n} \cos\beta_i \sin\gamma_i, \qquad D = \sum_{i=1}^{n} \sin\beta_i \cos\gamma_i.$$

The test based on $\hat{\Pi}_n$ rejects independence for large values of $|\hat{\Pi}_n|$. Further details are given in Fisher & Lee (1982; 1983) and Fisher (1993, Section 6.3.2). Some quantiles of $\hat{\Pi}_n$ are tabulated in Fisher (1993, Appendix A13).

A Test of Independence Consistent against all Alternatives

A natural way of measuring the dependence between circular random variables Θ and Φ on the basis of a sample $(\theta_1, \phi_1), \ldots, (\theta_n, \phi_n)$ is by some suitable distance between the empirical distribution function and the product of its marginals. One such distance is C_n, where

$$C_n = \frac{1}{n^4} \sum_{i=1}^{n} \sum_{j=1}^{n} \left(T_{i,i} - T_{i,j} - T_{j,i} + T_{j,j}\right)^2, \qquad (11.2.21)$$

with

$$T_{i,j} = n \min(r_i, s_j) - r_i s_j, \qquad i, j = 1, \ldots, n, \qquad (11.2.22)$$

$r_1, \ldots, r_n$ and $s_1, \ldots, s_n$ being the circular ranks of $\theta_1 \ldots, \theta_n$ and $\phi_1, \ldots, \phi_n$. Rothman (1971) introduced the test which rejects independence for large values of C_n. Rothman's test is consistent against all alternatives to independence. For large n and x, the approximation (Jupp & Spurr, 1985)

$$\Pr\left(16\pi^4 C_n > x\right) \simeq (1.466x - 0.322)e^{-x/2} \qquad (11.2.23)$$

can be used.

A General Class of Correlation Coefficients Based on Uniform Scores

We have seen in Sections 6.3.7 and 10.8 that each square-summable sequence $\{a_k\}_{k=1,2,\ldots}$ determines a function $\mathbf{t} : S^1 \to L^2(S^1)$ and a corresponding test of uniformity. Such functions $\mathbf{t}$ also give rise to correlation coefficients, as follows. The definition (11.2.14) generalises (Jupp & Spurr, 1985) to

$$\bar{R}_{\pm}^2 = \left\| \frac{1}{n} \sum_{i=1}^{n} \mathbf{t}(\beta_i \pm \gamma_i) \right\|^2 \qquad (11.2.24)$$

and appropriate correlation coefficients are $\bar{R}_{+}^2 + \bar{R}_{-}^2$ and $\bar{R}_{+}^2 - \bar{R}_{-}^2$. Particular cases of this construction are:

(i) if $a_1 = 1$ and $a_k = 0$ for $k > 1$ then $\mathbf{t}$ can be regarded as the usual embedding of the circle in the plane, (11.2.24) reduces to (11.2.14), and $\bar{R}_{+}^2 + \bar{R}_{-}^2 = 2r^2$, where r^2 is calculated for the uniform scores;

(ii) if $a_k = 1/k$ (so that the corresponding test of uniformity is Watson's U^2 test based on (6.3.31)) then $\bar{R}_{+}^2 + \bar{R}_{-}^2 = 16\pi^4 C_n$, where C_n is Rothman's (1971) statistic defined by (11.2.21).

An Analogue of Kendall's Tau

For pairs $(x_1, y_1), \ldots, (x_n, y_n)$ of points on the line, Kendall's tau can be constructed from the proportion of pairs (x_i, y_i), (x_j, y_j) for which the sign of $x_j - x_i$ is the same as that of $y_j - y_i$. This suggests that for points

$(\theta_1, \phi_1), \ldots, (\theta_n, \phi_n)$ on the torus $S^1 \times S^1$, it would be useful to consider the proportion of triples $(\theta_i, \phi_i), (\theta_j, \phi_j), (\theta_k, \phi_k)$ for which the circular ordering of $(\theta_i, \theta_j, \theta_k)$ is the same as that of (ϕ_i, ϕ_j, ϕ_k). The following signed correlation coefficient, introduced by Fisher & Lee (1982), is based on this idea. Put

$$\begin{aligned} \delta_{i,j,k} &= \operatorname{sign}(\theta_i - \theta_j) \operatorname{sign}(\theta_j - \theta_k) \operatorname{sign}(\theta_k - \theta_i) \\ &\quad \times \operatorname{sign}(\phi_i - \phi_j) \operatorname{sign}(\phi_j - \phi_k) \operatorname{sign}(\phi_k - \phi_i) \end{aligned} \tag{11.2.25}$$

and

$$\hat{\Delta}_n = \binom{n}{3}^{-1} \sum_{\{i,j,k\}} \delta_{i,j,k}, \tag{11.2.26}$$

where the sum is over all subsets $\{i, j, k\}$ of size 3 of $\{1, \ldots, n\}$. (Note the similarity between $\hat{\Delta}_n$ and $\hat{\lambda}_n$, which was defined in (11.2.6).) Under independence, the population analogue of $\hat{\Delta}_n$ is zero. Thus an appropriate test rejects independence for large values of $|\hat{\Delta}_n|$. For further details, including the treatment of ties and the use of random subsets to calculate $\hat{\Delta}_n$, see Fisher & Lee (1982) and Fisher (1993, Section 6.3.2). Some quantiles of $\hat{\Delta}_n$ for $n < 8$ are given in Fisher (1993, Appendix A12).

11.2.3 Spherical–Spherical Correlation

Correlation coefficients for spherical random variables can be obtained by simple generalisations of those circular–circular correlation coefficients of Section 11.2.2 which are based on the embedding approach. Let $(\mathbf{x}_1, \mathbf{y}_1), \ldots, (\mathbf{x}_n, \mathbf{y}_n)$ be independent observations on random vectors $\mathbf{X}$ and $\mathbf{Y}$ on S^{p-1} and S^{q-1}, respectively. Let

$$\mathbf{S} = \begin{pmatrix} \mathbf{S}_{11} & \mathbf{S}_{12} \\ \mathbf{S}_{21} & \mathbf{S}_{22} \end{pmatrix}$$

and

$$\mathbf{S}^* = \frac{1}{n} \sum_{i-1}^{n} \begin{pmatrix} \mathbf{x}\mathbf{x}^T & \mathbf{x}\mathbf{y}^T \\ \mathbf{y}\mathbf{x}^T & \mathbf{y}\mathbf{y}^T \end{pmatrix} = \begin{pmatrix} \mathbf{S}_{11}^* & \mathbf{S}_{12}^* \\ \mathbf{S}_{21}^* & \mathbf{S}_{22}^* \end{pmatrix},$$

denote the sample variance matrix and the sample mean of products matrices of $(\mathbf{X}, \mathbf{Y})$, both partitioned in the usual way.

The circular–circular correlation coefficient r^2 defined in (11.2.7) extends to

$$r^2 = \operatorname{tr}(\mathbf{S}_{11}^{-1} \mathbf{S}_{12} \mathbf{S}_{22}^{-1} \mathbf{S}_{21}) \tag{11.2.27}$$

(provided that $\mathbf{S}_{11}$ and $\mathbf{S}_{22}$ are non-singular). Under independence,

$$nr^2 \ \dot{\sim}\ \chi^2_{pq}, \qquad n \to \infty \tag{11.2.28}$$

(provided that the variance matrix of $(\mathbf{x}, \mathbf{y})$ is non-singular). For large n, the statistic nr^2 is asymptotically equivalent to the score statistic (and so the likelihood ratio statistic) for testing independence of Θ and Φ in the bivariate von Mises–Fisher model (11.4.1). For large n, nr^2 is also asymptotically equivalent to a case of Kent's (1983b) general correlation coefficient, which measures the information gain on modelling $\mathbf{x}$ and $\mathbf{y}$ as dependent rather than independent.

When $p = q$, (11.2.11) generalises to

$$\hat{\rho}_T = \frac{|\mathbf{S}_{12}^*|}{(|\mathbf{S}_{11}^*||\mathbf{S}_{22}^*|)^{1/2}} \tag{11.2.29}$$

(see Fisher & Lee, 1986). The corresponding test rejects independence for large values of $|\hat{\rho}_T|$. Fisher & Lee (1986) also generalised the form (11.2.10) of (11.2.11) to a U-statistic based on determinants.

The circular–circular correlation coefficient $\hat{\Delta}_n$ defined in (11.2.26) generalises readily to the spherical case when $p = q$. For pairs $(\mathbf{x}_1, \mathbf{y}_1), \ldots, (\mathbf{x}_n, \mathbf{y}_n)$ of points on S^{p-1}, define

$$\hat{\Delta}_n = \binom{n}{p+1}^{-1} \sum_{\{i_1,\ldots,i_{p+1}\}} \delta_{i_1,\ldots,i_{p+1}}, \tag{11.2.30}$$

where the sum is over all subsets $\{i_1, \ldots, i_{p+1}\}$ of size $p+1$ of $\{1, \ldots, n\}$ and

$$\delta_{i_1,\ldots,i_{p+1}} = \text{sign}|\mathbf{x}_{i_1} \ldots \mathbf{x}_{i_{p+1}}| \times \text{sign}|\mathbf{y}_{i_1} \ldots \mathbf{y}_{i_{p+1}}|. \tag{11.2.31}$$

The geometrical interpretation of (11.2.31) is that $\delta_{i_1,\ldots,i_{p+1}}$ is 1 if the (oriented) simplex with (ordered) vertices $\mathbf{x}_{i_1}, \ldots, \mathbf{x}_{i_{p+1}}$ has the same orientation as the (oriented) simplex with (ordered) vertices $\mathbf{y}_{i_1}, \ldots, \mathbf{y}_{i_{p+1}}$, and is -1 otherwise. The statistic $\hat{\Delta}_n$ is resistant (in that large changes to a small part of the data have little effect) and robust (in that small changes to the data have little effect), as shown in Jupp (1987). Independence is rejected for large values of $|\hat{\Delta}_n|$. Since $\hat{\Delta}_n$ is not distribution-free under uniformity (for $p \geq 3$), permutation tests are appropriate.

Various correlation coefficients based on $\mathbf{S}$ and $\mathbf{S}^*$ are listed in Table 11.3. All these correlation coefficients are invariant under separate rotations of $\mathbf{x}$ and $\mathbf{y}$.

The idea behind the correlation coefficient of Stephens (1979) is to match $\mathbf{x}_1, \ldots, \mathbf{x}_n$ to $\mathbf{y}_1, \ldots, \mathbf{y}_n$ by rotation, i.e. to measure correlation by

$$\max_{\mathbf{U} \in SO(p)} \frac{1}{n} \sum_{i=1}^{n} \mathbf{y}_i^T \mathbf{U} \mathbf{x}_i \tag{11.2.32}$$

(see also Mackenzie, 1957). A straightforward calculation shows that a signed version of (11.2.32) is

$$\frac{1}{2} \left\{ \min_{\mathbf{U} \in SO(p)} \frac{1}{n} \sum_{i=1}^{n} \mathbf{y}_i^T \mathbf{U} \mathbf{x}_i - \min_{\mathbf{U} \in O(p) \setminus SO(p)} \frac{1}{n} \sum_{i=1}^{n} \mathbf{y}_i^T \mathbf{U} \mathbf{x}_i \right\} = (\text{sign}|\mathbf{S}_{12}^*|)\ 2 l_p, \tag{11.2.33}$$

Table 11.3 Some correlation coefficients for directional data.

(a) Coefficients based on the sample variance matrix $\mathbf{S}$	
$\mathrm{tr}(\mathbf{S}_{11}^{-1}\mathbf{S}_{12}\mathbf{S}_{22}^{-1}\mathbf{S}_{21})$	Mardia (1976); Jupp & Mardia (1980)
$\mathrm{tr}[(\mathbf{S}_{12}\mathbf{S}_{21})^{1/2}]/\{\mathrm{tr}\,\mathbf{S}_{11}\,\mathrm{tr}\,\mathbf{S}_{22}\}^{1/2}$	Downs (1974); Mardia (1975a)
largest eigenvalue of $\mathbf{S}_{11}^{-1}\mathbf{S}_{12}\mathbf{S}_{22}^{-1}\mathbf{S}_{21}$	Johnson & Wehrly (1977)
$\mathrm{tr}(\hat{\mathbf{\Sigma}}_{11}^{-1}\mathbf{S}_{12}\hat{\mathbf{\Sigma}}_{22}^{-1}\mathbf{S}_{21})$, where $\hat{\mathbf{\Sigma}}_{ii} = A_p'(\hat{\kappa}_i)\bar{\mathbf{x}}_{0i}\bar{\mathbf{x}}_{0i}^T + A_p(\hat{\kappa}_i)(\mathbf{I}_p - \bar{\mathbf{x}}_{0i}\bar{\mathbf{x}}_{0i}^T)$	Mardia & Puri (1978)
$\mathrm{tr}(\check{\mathbf{\Sigma}}_{11}^{-1}\mathbf{S}_{12}\check{\mathbf{\Sigma}}_{22}^{-1}\mathbf{S}_{21})$ with $\check{\mathbf{\Sigma}}_{ii}$ obtained from $\mathbf{S}_{ii}$ (special to $p = q = 2, 3$)	Mardia & Puri (1978)
$\mathrm{sign}\lvert\mathbf{S}_{12}\rvert\,\{(\mathrm{tr}(\mathbf{S}\mathbf{S}^T) + 2\lvert\mathbf{S}_{12}\rvert)/(\mathrm{tr}\,\mathbf{S}_{11}\,\mathrm{tr}\,\mathbf{S}_{22})\}^{1/2}$	Hanson *et al.* (1992)
(b) Coefficients based on the means of products matrix $\mathbf{S}^*$ (for $p = q$)	
$\mathrm{tr}\,\mathbf{S}_{12}^*$	Watson & Beran (1967)
$\mathrm{tr}[(\mathbf{S}_{12}^*\mathbf{S}_{21}^*)^{1/2}]$	Mackenzie (1957); Stephens (1979)
smallest singular value of $\mathbf{S}_{12}^*$, suitably normalised	Rivest (1982)
$\lvert\mathbf{S}_{11}^{*\,-1/2}\mathbf{S}_{12}^*\mathbf{S}_{22}^{*\,-1/2}\rvert$	Fisher & Lee (1983; 1986)

where l_p is the smallest singular value of $\mathbf{S}_{12}^*$ (see Rivest, 1982; Fisher & Lee, 1986).

Further correlation coefficients can be obtained by using the machinery described in Section 10.8. Given continuous equivariant functions $\mathbf{t} : S^{p-1} \to L^2(S^{p-1})$ and $\mathbf{u} : S^{q-1} \to L^2(S^{q-1})$, corresponding correlation coefficients are defined as in Table 11.3, where now $\mathbf{S}$ and $\mathbf{S}^*$ are the sample variance matrix and the sample mean of products matrix of $(\mathbf{t}(\mathbf{X}), \mathbf{u}(\mathbf{Y}))$. For further details, see Jupp & Spurr (1985).

11.3 REGRESSION MODELS

11.3.1 Linear Response

A natural model for regression of a linear variable X on an angular variable θ is

$$X|\theta \sim N(\alpha + \beta_1 \cos\theta + \beta_2 \sin\theta, \sigma^2), \tag{11.3.1}$$

as introduced by Mardia (1976). Since (11.3.1) is multiple linear regression of X on $(\cos\theta, \sin\theta)$, maximum likelihood estimates can be obtained using almost any standard statistical package. Note that the cylindrical model (3.7.2) of Mardia & Sutton (1978) has regression functions of the form

$$\hat{x} = \alpha + \beta_1 \cos\theta + \beta_2 \sin\theta,$$

as in (11.3.1).

A slight extension of (11.3.1) and an application to air pollution studies are given in Johnson & Wehrly (1978).

11.3.2 Circular Response

Consider a circular response variable Θ which is measured at values $\mathbf{x}_1, \ldots, \mathbf{x}_n$ of a k-dimensional explanatory variable $\mathbf{X}$.

Helical Regression Functions

In the case where $k = 1$, the idea of wrapping the line onto the circle leads to Gould's (1969) regression model in which

$$\Theta_i \sim M(\mu + \beta x_i, \kappa), \qquad i = 1, \ldots, n, \tag{11.3.2}$$

with $\Theta_1, \ldots, \Theta_n$ independent and with μ, β and κ unknown parameters (see also Laycock, 1975). Because the regression curves given by (11.3.2) are helical, this model is sometimes known as the 'barber's pole' model. A major disadvantage of (11.3.2) is that the likelihood function has infinitely many maxima of comparable size. Further, if $x_1, \ldots, x_n$ are equally spaced then β is not identifiable.

Regression Using Link Functions

The problems with (11.3.2) are due to the fact that the helical regression function $x \mapsto \beta x \pmod{2\pi}$ wraps the real line infinitely many times round the circle. One way (suggested by Fisher & Lee, 1992) of avoiding these problems is to replace this function by a suitable one-to-one function. It is useful to define a *link function* as a one-to-one function g which maps the line onto $(-\pi, \pi)$ and satisfies $g(0) = 0$. Two useful link functions are the *inverse tan link*

$$g(x) = 2\tan^{-1} x \tag{11.3.3}$$

and the *scaled probit link*

$$g(x) = 2\pi(\Phi(x) - 0.5). \tag{11.3.4}$$

Links of the form

$$g(x) = 2\pi F(x), \tag{11.3.5}$$

where F is a cumulative distribution function, were used by Johnson & Wehrly (1978).

Given a link function g, a corresponding regression model with constant concentration is

$$\Theta_i \sim M(\mu_i, \kappa), \tag{11.3.6}$$

where

$$\mu_i = g(\boldsymbol{\beta}^T \mathbf{x}_i), \tag{11.3.7}$$

with $\boldsymbol{\beta}$ a k-dimensional parameter vector. A variant of (11.3.6) in which the mean direction is constant is

$$\Theta_i \sim M(\mu, \kappa_i), \tag{11.3.8}$$

where

$$\kappa_i = h(\boldsymbol{\gamma}^T \mathbf{x}_i), \tag{11.3.9}$$

with $\boldsymbol{\gamma}$ a k-dimensional parameter vector. Models (11.3.6) and (11.3.8) can be combined into the mixed model

$$\Theta_i \sim M(\mu_i, \kappa_i), \tag{11.3.10}$$

where μ_i and κ_i are given by (11.3.7) and (11.3.9), respectively. Details of maximum likelihood estimation in models (11.3.6), (11.3.8) and (11.3.10) can be found in Fisher & Lee (1992) and Fisher (1993, Section 6.4).

11.3.3 Spherical Response

Regression Using Rotation

Regression of a spherical variate $\mathbf{y}$ in S^{p-1} on a spherical predictor $\mathbf{x}$ in S^{p-1} occurs in various fields. The following five examples give some idea of the variety of these contexts. An important crystallographic problem (Mackenzie, 1957) is to relate an axis $\mathbf{y}$ of a crystal to an axis $\mathbf{x}$ of a standard coordinate system. Determination of the orientation of a satellite (Wahba, 1966) involves comparing directions $\mathbf{y}$ of stars in the satellite's coordinate system with corresponding directions $\mathbf{x}$ in a terrestrial coordinate system. Assessing 'geometric integrity' in industrial quality control (Chapman, Chen & Kim, 1995) is achieved by comparing directions $\mathbf{y}$ normal to a machined part with the corresponding directions $\mathbf{x}$ in a computer-aided design specification. Many applications of machine vision (Kanatani, 1993, Chapter 5) involve comparison of directions $\mathbf{y}$ of objects detected by one sensor with the corresponding directions $\mathbf{x}$ detected by another sensor. An important problem in geophysics is to estimate the rotation of one tectonic plate relative to another. This is done by comparing the positions $\mathbf{y}$ on the earth's surface of points (intersections of fracture zones with magnetic anomalies) on one plate with the positions $\mathbf{x}$ of the corresponding positions of points on the other plate (see Chang, 1993).

The simplest regression functions for regressing a spherical variate $\mathbf{y}$ in S^{p-1} on a spherical predictor $\mathbf{x}$ in S^{p-1} are those of the form

$$\hat{\mathbf{y}} = \mathbf{A}\mathbf{x} \tag{11.3.11}$$

for some rotation $\mathbf{A}$ in $SO(p)$. A simple class of regression models with regression functions of the form (11.3.11) are Chang's (1986) models in which the conditional distribution of $\mathbf{y}$ given $\mathbf{x}$ is circularly symmetric with mean direction $\mathbf{Ax}$, so that the conditional density has the form

$$f(\mathbf{y}; \mathbf{A}|\mathbf{x}) = g(\mathbf{y}^T\mathbf{A}\mathbf{x}). \tag{11.3.12}$$

Given paired observations $(\mathbf{x}_1, \mathbf{y}_1), \ldots, (\mathbf{x}_n, \mathbf{y}_n)$, put

$$\mathbf{S}_{12}^* = \frac{1}{n}\sum_{i=1}^{n} \mathbf{x}_i\mathbf{y}_i^T \tag{11.3.13}$$

and let $\mathbf{U\Lambda V}^T$ be the modified singular value decomposition of $\mathbf{S}_{12}^*$, i.e. $\mathbf{S}_{12}^* = \mathbf{U\Lambda V}^T$, with $\mathbf{U}$ and $\mathbf{V}$ in $SO(p)$, and $\mathbf{\Lambda} = \text{diag}(\lambda_1, \ldots, \lambda_{p-1}, \lambda_p)$ with $\lambda_1 \geq \ldots \geq |\lambda_p|$. If g in (11.3.12) is an increasing function on $[-1, 1]$ then the least-squares estimate of $\mathbf{A}$, i.e. the value of $\mathbf{A}$ which minimises

$$\sum_{i=1}^{n} \|\mathbf{y}_i - \mathbf{A}\mathbf{x}_i\|^2,$$

is

$$\hat{\mathbf{A}} = \mathbf{V}\mathbf{U}^T. \tag{11.3.14}$$

This is one type of Procrustes matching (see Mardia, Kent & Bibby, 1979, Section 14.7). In particular, if the conditional distribution is von Mises–Fisher with constant concentration,

$$\mathbf{y}|\mathbf{x} \sim M_p(\mathbf{Ax}, \kappa), \tag{11.3.15}$$

then the maximum likelihood estimate $\hat{\mathbf{A}}$ of $\mathbf{A}$ is given by (11.3.14). Large-sample asymptotic results on the distribution of $\hat{\mathbf{A}}$ under (11.3.12) were given by Chang (1986). In particular, he showed that, under (11.3.15),

$$2n\kappa\,(r - \mathrm{tr}(\mathbf{AS}_{12}^*)) \;\dot{\sim}\; \chi^2_{p(p-1)/2}, \qquad n \to \infty, \tag{11.3.16}$$

where

$$r = \mathrm{tr}\left(\hat{\mathbf{A}}\mathbf{S}_{12}^*\right)$$

is the Mackenzie–Stephens correlation coefficient (11.2.32). High-concentration asymptotic results on the distribution of $\hat{\mathbf{A}}$ under (11.3.15) were given by Rivest (1989). In particular, he showed that

$$2n\kappa(r - \mathrm{tr}(\mathbf{AS}_{12}^*)) \;\dot{\sim}\; \chi^2_{p(p-1)/2}, \qquad \kappa \to \infty, \tag{11.3.17}$$

$$2n\kappa(1 - r) \;\dot{\sim}\; \chi^2_{n(p-1)-p(p-1)/2}, \qquad \kappa \to \infty, \tag{11.3.18}$$

and that the statistics on the left-hand sides of (11.3.17) and (11.3.18) are asymptotically independent for large κ. These results are analogous to (9.6.15) and (9.6.17) and can be obtained using the decomposition

$$2n\kappa(1 - \mathrm{tr}(\mathbf{AS}_{12}^*)) = 2n\kappa(1 - r) + 2n\kappa(r - \mathrm{tr}(\mathbf{AS}_{12}^*)),$$

which is analogous to (9.6.16). When $p = 2$ and $\mathbf{u}_1 = \ldots = \mathbf{u}_n = \boldsymbol{\mu}$, (11.3.17) and (11.3.18) reduce to (4.8.28) and (4.8.31), respectively. The approximation (11.3.17) was improved by Bingham, Chang & Richards (1992, pp. 319, 330), using the fact that the distribution of $\hat{\mathbf{A}}$ conditional on $(\mathbf{U}, \Lambda)$ is matrix Fisher (13.2.15) with parameter matrix $n\kappa\mathbf{AU}\Lambda\mathbf{U}^T$, where $\mathbf{U}\Lambda\mathbf{V}^T$ is the modified singular value decomposition of $\mathbf{S}_{12}^*$.

The spherical regression model (11.3.12) extends readily to a model for regressing a $p \times r$ random matrix $\mathbf{Y}$ in the Stiefel manifold $V_r(\mathbb{R}^p)$ (defined in (13.2.1)) on a predictor $\mathbf{X}$ in $V_r(\mathbb{R}^p)$. The conditional distribution of $\mathbf{Y}$ given $\mathbf{X}$ has the form

$$f(\mathbf{Y}; \mathbf{A}_1, \mathbf{A}_2|\mathbf{X}) = g(\mathbf{Y}^T\mathbf{A}_1^T\mathbf{X}\mathbf{A}_2), \tag{11.3.19}$$

with $\mathbf{A}_1$ in $SO(p)$ and $\mathbf{A}_2$ in $SO(r)$. Many of the distributional results for $\hat{\mathbf{A}}$ in (11.3.12) extend to results for $(\hat{\mathbf{A}}_1, \hat{\mathbf{A}}_2)$ in (11.3.19). For further details, see Prentice (1989).

In various contexts (in particular for the problem in plate tectonics) a more appropriate model (Chang, 1987; 1989) is the associated errors-in-variables model in which $(\mathbf{x}_i, \mathbf{y}_i)$ $(i = 1, \ldots, n)$ are independent with densities

$$f(\mathbf{x}_i, \mathbf{y}_i; \xi_i, \mathbf{A}) = g(\mathbf{x}_i^T\xi_i)g(\mathbf{y}_i^T\mathbf{A}\xi_i).$$

Here g is a known function, while the rotation $\mathbf{A}$ and the unit vectors $\xi_1, \ldots, \xi_n$ are unknown. Chang provided asymptotic tests and confidence regions for $\mathbf{A}$ in the cases in which either the sample size is large or $\mathbf{x}_i$ and $\mathbf{y}_i$ have concentrated Fisher distributions. Detailed consideration of a related problem in plate tectonics led Chang (1988) to the model in which the spherical random variables $\mathbf{x}_{ij}$ and $\mathbf{y}_{ik}$ $(1 \leq j \leq m_i, 1 \leq k \leq n_i, 1 \leq i \leq s)$ have certain independent concentrated Fisher–Bingham or wrapped multivariate normal distributions with respective means $\boldsymbol{\alpha}_{ij}$ and $\boldsymbol{\beta}_{ik}$, where $\boldsymbol{\alpha}_{ij}^T \boldsymbol{\eta}_i = \mathbf{0}$ and $\boldsymbol{\beta}_{ik}^T \mathbf{A} \boldsymbol{\eta}_i = \mathbf{0}$ for some unkown unit vectors $\boldsymbol{\eta}_1, \ldots, \boldsymbol{\eta}_s$. He obtained high-concentration asymptotic confidence regions for the rotation $\mathbf{A}$. See also Chang (1993).

Residuals for spherical data can be constructed using the wrapping approach described in Section 9.1. When points $\hat{\mathbf{y}}_1, \ldots, \hat{\mathbf{y}}_n$ are fitted to observations $\mathbf{y}_1, \ldots, \mathbf{y}_n$ on S^{p-1}, the corresponding crude residuals are the tangent vectors $\mathbf{y}_i - (\mathbf{y}_i^T \hat{\mathbf{y}}_i)\hat{\mathbf{y}}_i$ to S^{p-1} at $\hat{\mathbf{y}}_i$. These and more refined residuals were considered by Jupp (1988) and Rivest (1989).

Regression Models for Finding a Centre

In some contexts, such as finding the source $\mathbf{a}$ of a signal or the centre of an explosion, directions $\mathbf{y}_i$ in S^{p-1} are observed at positions $\mathbf{x}_i$ in $\mathbb{R}^p$. A convenient model for this is

$$\mathbf{y}|\mathbf{x} \sim M_p(\|\mathbf{x} - \mathbf{a}\|^{-1}(\mathbf{x} - \mathbf{a}), \kappa_0 + \kappa\|\mathbf{x} - \mathbf{a}\|^c), \tag{11.3.20}$$

so that $\mathbf{y}|\mathbf{x}$ has a von Mises–Fisher distribution with mean direction pointing towards the unknown 'centre' $\mathbf{a}$. Various particular cases of (11.3.20) with $\kappa_0 = 0$ have been applied by Lenth (1981b) (who also considers robust estimation in this context), Jupp *et al.* (1987) and Jupp & Spurr (1989). A useful extension of (11.3.20) is the model in which the conditional density of $\mathbf{y}$ given $\mathbf{x}$ has the form

$$f(\mathbf{y}|\mathbf{x}) = g\left(\mathbf{y}; \frac{1}{\|\mathbf{x} - \mathbf{a}\|}\mathbf{A}(\mathbf{x} - \mathbf{a})\right),$$

where $g(\cdot; \boldsymbol{\mu})$ is a density function with mean direction $\boldsymbol{\mu}$ and $\mathbf{A}$ is in $SO(p)$. In the case $p = 2$, Rivest (1997) considered the large-sample behaviour of the estimates of $\mathbf{A}$ which maximise $\operatorname{tr}(\mathbf{A}\mathbf{S}_{12}^*)$, where $\mathbf{S}_{12}^*$ is defined by (11.3.13).

Other Regression Models

A wide variety of regression models are obtained as generalised linear models with error distributions in an appropriate exponential model, so that the conditional density of $\mathbf{y}$ given $\mathbf{x}$ has the form

$$f(\mathbf{y}; \mathbf{A}|\mathbf{x}) \propto \exp\{[\boldsymbol{\phi} + \mathbf{A}^T \mathbf{t}(\mathbf{x})]^T \mathbf{u}(\mathbf{y})\}, \tag{11.3.21}$$

for suitable functions $\mathbf{t}$ and $\mathbf{u}$. In particular, if the joint distribution of $(\mathbf{x}, \mathbf{y})$ has density (11.4.2) then $\mathbf{y}|\mathbf{x}$ has density of the form (11.3.21). Some examples of (11.3.21) were given by Johnson & Wehrly (1978).

For regressing a spherical variate $\mathbf{y}$ in S^2 on a circular predictor θ, a class of simple regression functions consists of those of the form (Jupp & Mardia, 1980)

$$\hat{\mathbf{y}} = \mathbf{A}\mathbf{x} + \mathbf{b}, \tag{11.3.22}$$

where $\mathbf{x} = (\cos\theta, \sin\theta)^T$, $\mathbf{A}^T\mathbf{A} = \mathbf{I}_2$ and $\mathbf{A}\mathbf{b} = \mathbf{0}$.

11.4 BIVARIATE MODELS

A natural model for bivariate spherical data is the generalisation to $S^{p-1} \times S^{p-1}$ of the bivariate von Mises model (3.7.1). This generalisation is the exponential model with densities

$$f(\mathbf{x}, \mathbf{y}; \kappa_1, \kappa_2, \boldsymbol{\mu}_1, \boldsymbol{\mu}_2, \mathbf{A}) = a(\kappa_1, \kappa_2, \boldsymbol{\mu}_1, \boldsymbol{\mu}_2, \mathbf{A}) \exp\{\kappa_1 \boldsymbol{\mu}_1^T \mathbf{x} + \kappa_2 \boldsymbol{\mu}_2^T \mathbf{y} + \mathbf{x}^T \mathbf{A} \mathbf{y}\}, \tag{11.4.1}$$

introduced by Mardia (1975a; 1975c). Although the conditional distributions of $\mathbf{x}$ given $\mathbf{y}$ and $\mathbf{y}$ given $\mathbf{x}$ are von Mises–Fisher, the marginal distributions are more complicated. For the special case in which the dependency matrix $\mathbf{A}$ is a multiple of a rotation matrix, Jupp & Mardia (1980) found an expansion in terms of products of Bessel functions for the norming constant $a(\kappa_1, \kappa_2, \boldsymbol{\mu}_1, \boldsymbol{\mu}_2)$.

The model (11.4.1) has $p(p+2)$ parameters, so it can be useful to consider submodels with fewer parameters. Sometimes it is appropriate to consider submodels of (11.4.1) for which the densities have some symmetry. Rivest (1988) introduced the $((p^2+p+6)/2)$-parameter submodel of (11.4.1) for which the densities have rotational symmetry, in that they are invariant under $(\mathbf{x}, \mathbf{y}) \mapsto (\mathbf{U}\mathbf{x}, \mathbf{R}\mathbf{U}\mathbf{R}^{-1}\mathbf{y})$, for rotations $\mathbf{U}$ about $\boldsymbol{\mu}_1$, where $\mathbf{R}$ is a rotation such that $\mathbf{R}\boldsymbol{\mu}_1 = \boldsymbol{\mu}_2$. Thus the matrix $\mathbf{A}$ has the form

$$\mathbf{A} = \mathbf{H}_1^T \operatorname{diag}(\alpha, \beta\mathbf{Q})\mathbf{H}_2,$$

where $\mathbf{Q} \in O(p-1)$, α and β are real numbers, and $\mathbf{H}_i$ rotates $\boldsymbol{\mu}_i$ to $(1, 0, \ldots, 0)^T$ for $i = 1, 2$.

The model (11.4.1) can be generalised to models with densities of the form

$$f(\mathbf{x}, \mathbf{y}; \kappa_1, \kappa_2, \boldsymbol{\mu}_1, \boldsymbol{\mu}_2, \mathbf{A}) = \\ a(\kappa_1, \kappa_2, \boldsymbol{\mu}_1, \boldsymbol{\mu}_2, \mathbf{A}) \exp\{\kappa_1 \boldsymbol{\mu}_1^T \mathbf{t}(\mathbf{x}) + \kappa_2 \boldsymbol{\mu}_2^T \mathbf{u}(\mathbf{y}) + \mathbf{t}(\mathbf{x})^T \mathbf{A} \mathbf{u}(\mathbf{y})\}, \tag{11.4.2}$$

where $\mathbf{t}$ and $\mathbf{u}$ are suitable vector-valued functions. For example, in the case where $\mathbf{x}$ is scalar and $\mathbf{y}$ is circular (so that the sample space is a cylinder), a submodel of (11.4.2) with $\mathbf{t}(x) = (x, x^2)$, $\mathbf{u}(\mathbf{y}) = \mathbf{y}$ and convenient restrictions on $\mathbf{A}$ gives the model (3.7.2) of Mardia & Sutton (1978). Other models of the form (11.4.2) were obtained by Johnson & Wehrly (1978).

In the circular case there is a special method of constructing bivariate models with given marginals. Let f_1 and f_2 be probability density functions on the circle with cumulative distribution functions F_1 and F_2. Johnson & Wehrly (1978) showed that any probability density function g on the circle gives rise to a distribution on the torus with probability density function

$$p(\theta_1, \theta_2) = 2\pi g\,(2\pi\,[F_1(\theta_1) - F_2(\theta_2)])\, f_1(\theta_1) f_2(\theta_2). \tag{11.4.3}$$

The marginal probability density functions of θ_1 and θ_2 are f_1 and f_2.

11.5 DIRECTIONAL TIME SERIES

Data $(\mathbf{x}_1, t_1), \ldots, (\mathbf{x}_n, t_n)$ which consist of directions $\mathbf{x}_1, \ldots, \mathbf{x}_n$ at times $t_1, \ldots, t_n$ occur frequently in various fields. For example, time series of wind directions are important in meteorology. A major application of directional time series to wind directions was given by Breckling (1989). For some data sets of the form $(\mathbf{x}_1, t_1), \ldots, (\mathbf{x}_n, t_n)$, the observations $\mathbf{x}_1, \ldots, \mathbf{x}_n$ are directions but $t_1, \ldots, t_n$ are positions on a line, rather than times. In a typical palaeomagnetic example, the $\mathbf{x}_i$ are directions of palaeomagnetism and the t_i are depths.

11.5.1 Assessing Serial Dependence

A sensible first step when analysing directional time series is to test the null hypothesis of independence against the alternative of serial dependence. This can be done using any of the correlation coefficients described in Sections 11.2.2–11.2.3. If the observed directions are $\mathbf{x}_1, \ldots, \mathbf{x}_n$, the lag–1 correlation coefficient is the correlation coefficient calculated for the pairs $(\mathbf{x}_1, \mathbf{x}_2), \ldots, (\mathbf{x}_{n-1}, \mathbf{x}_n)$. Independence is rejected for large absolute values of the lag–1 correlation coefficient. The observed significance level can be obtained from (sampling from) the permutation distribution of the correlation coefficient. A correlation coefficient which is calculated easily is that of Watson & Beran (1967) (see Table 11.3). Use of their coefficient is equivalent to that of the statistic

$$U = \sum_{i=1}^{n-1} \mathbf{x}_i^T \mathbf{x}_{i+1}. \tag{11.5.1}$$

A large-sample normal approximation to the distribution of U was given by Epp, Tukey & Watson (1971).

11.5.2 Time Series Models

Models for Time Series on the Circle

There are several ways of constructing time series models on the circle from time series models on the line. The following brief description is based on those in Fisher (1993, Section 7.2) and in Fisher & Lee (1994). Details of model selection, model identification and fitting can be found in these references.

Projected Normal Processes

The embedding approach described in Section 9.1 suggests the following construction. Let $\{(X_t, Y_t)\}_{t=1,2,\ldots}$ be a process on the plane. Then radial projection to the unit circle produces a corresponding process $\{\Theta_t\}_{t=1,2,\ldots}$ on the circle, defined by

$$X_t = R_t \cos\Theta_t, \qquad Y_t = R_t \sin\Theta_t. \tag{11.5.2}$$

If $\{(X_t, Y_t)\}_{t=1,2,\ldots}$ is a stationary bivariate Gaussian process then Θ_t has a projected normal distribution. In particular, if $\{X_t\}_{t=1,2,\ldots}$ and $\{Y_t\}_{t=1,2,\ldots}$ are independent realisations of a stationary zero-mean Gaussian process then Θ_t has the uniform distribution. Since the radial part $\{R_t\}_{t=1,2,\ldots}$ of a projected normal process is not observed, fitting such processes can be regarded as a missing data problem, and can be carried out using the EM algorithm.

Wrapped Processes

The wrapping approach described in Section 9.1 leads to the following construction. Let $\{X_t\}_{t=1,2,\ldots}$ be a process on the line. Then wrapping the line onto the unit circle by (3.5.54) produces a corresponding process $\{\Theta_t\}_{t=1,2,\ldots}$ on the circle, defined by

$$\Theta_t = X_t \pmod{2\pi}. \tag{11.5.3}$$

Useful processes constructed this way include the *wrapped autoregressive processes* $WAR(p)$ obtained when $\{X_t\}_{t=1,2,\ldots}$ is a $AR(p)$ process. As pointed out by Breckling (1989), the linear process $\{X_t\}_{t=1,2,\ldots}$ which gives rise to a wrapped process $\{\Theta_t\}_{t=1,2,\ldots}$ can be decomposed as

$$X_t = \Theta_t + 2\pi k_t, \tag{11.5.4}$$

where k_t is an unobserved integer. Thus fitting such processes can be regarded as a missing data problem, and can be carried out using the EM algorithm. Coles (1998) showed how Markov chain Monte Carlo methods can be used to fit $WAR(p)$ models.

Direct Linked Processes

Processes can be transferred from the line to the circle by using a link function (a one-to-one function g which maps the line onto $(-\pi, \pi)$ and satisfies $g(0) = 0$) instead of the wrapping map (3.5.54). Useful link functions include (11.3.3) and (11.3.4). If $\{X_t\}_{t=1,2,\ldots}$ is a process on the line, g is a link, and μ is a point on the circle then the corresponding *direct linked process* $\{\Theta_t\}_{t=1,2,\ldots}$ on the circle is defined by

$$\Theta_t = g(X_t) + \mu. \tag{11.5.5}$$

Useful processes constructed this way include the *direct linked autoregressive moving-average processes* $LARMA(p,q)$ obtained by taking $\{X_t\}_{t=1,2,\ldots}$ to be an $ARMA(p,q)$ process. These processes (with the scaled probit link (11.3.4)) are useful for analysing concentrated series.

Circular Autoregressive Processes

Circular analogues of autoregressive processes on the line can be constructed by using conditional distributions. If g is a link function, μ is a point on the circle, $\kappa \geq 0$, and $\alpha_1, \ldots, \alpha_p$ are real-valued coefficients then the corresponding *circular autoregressive process* (or *inverse linked process*) $CAR(p)$ is defined by

$$\Theta_t|(\theta_{t-1}, \ldots, \theta_{t-p}) \sim M(\mu_t, \kappa), \tag{11.5.6}$$

where

$$\mu_t = \mu + g\left(\alpha_1 g^{-1}(\theta_{t-1} - \mu) + \ldots + \alpha_p g^{-1}(\theta_{t-p} - \mu)\right). \tag{11.5.7}$$

These processes are useful for analysing dispersed series ($\kappa < 2$). When $\alpha_1 = 1$, the $CAR(1)$ model reduces to

$$\Theta_t|\theta_{t-1} \sim M(\theta_{t-1}, \kappa), \tag{11.5.8}$$

as considered by Accardi, Cabrera & Watson (1987).

Other Processes

Taking $f_1 = f_2$ in (11.4.3) leads to the stationary Markov processes of Wehrly & Johnson (1980) with transition densities

$$p(\theta_2|\theta_1) = 2\pi g\left(2\pi\left[F_1(\theta_1) - F_2(\theta_2)\right]\right). \tag{11.5.9}$$

The $CAR(1)$ model (11.5.8) was generalised by Breckling (1989) to the *von Mises process* in which

$$\Theta_t|(\theta_{t-1}, \ldots, \theta_{t-p}) \sim M(\mu_t, \lambda_t), \tag{11.5.10}$$

where

$$\lambda_t e^{i\mu_t} = \kappa_0 e^{i\mu_0} + \sum_{j=1}^{p} \kappa_j e^{i\theta_{t-j}} \tag{11.5.11}$$

for some point μ_0 on the circle and some $\kappa_0, \ldots, \kappa_n \geq 0$.

Models for Time Series on Spheres

Some discrete-time stationary Markov processes on spheres were proposed by Accardi, Cabrera & Watson (1987). Their models are based on the generalised linear model construction (11.3.21), so that $\mathbf{x}_t$ conditional on $\mathbf{x}_{t-1}$ has a von Mises–Fisher or Watson distribution with canonical parameter which is an affine function of $\mathbf{x}_{t-1}$. In particular, they considered processes with conditional distributions

$$\mathbf{x}_t|\mathbf{x}_{t-1} \sim M_p(\mathbf{x}_{t-1}, \kappa) \qquad (11.5.12)$$

and

$$\pm\,\mathbf{x}_t|\pm\,\mathbf{x}_{t-1} \sim W(\mathbf{x}_{t-1}, \kappa). \qquad (11.5.13)$$

They explored the behaviour of some of these processes using simulation.

12

Modern Methodology

12.1 INTRODUCTION

This chapter is concerned with various topics outside the mainstream of likelihood-based inference for von Mises–Fisher, Watson and Bingham distributions. Sections 12.2, 12.3 and 12.4 consider respectively outliers, goodness-of-fit tests and robust methods for these distributions. Bootstrap methods for directional data are useful when (large-sample or high-concentration) asymptotic results are not appropriate. These methods are described very briefly in Section 12.5. In Section 12.6 we present some methods of density estimation. Bayesian inference is discussed very briefly in Section 12.7. Curve fitting and smoothing on spheres are important in applications to the earth sciences and are not quite as straightforward as on the plane. Some methods for fitting and smoothing curves on spheres are considered in Section 12.8.

12.2 OUTLIERS

Since observations which are far from the main body of a sample may have an undue influence on inferences made, it is sensible to try to detect such outliers and to investigate them. Note that in the spherical case, the effect of outliers on estimates of location may not be as great as the effect on estimates of variability. This contrasts with the effects of outliers on data on the line.

For data on the sphere S^2, an informal method of detecting outliers is to use the plots described in Section 10.2. Formal tests of discordancy of observations from a von Mises–Fisher distribution fall into three main groups.

12.2.1 Tests Based on Mean Resultant Length

Let $\mathbf{x}_1, \ldots, \mathbf{x}_n$ be observations on S^{p-1} with mean resultant length $\bar{R}$. An intuitively appealing approach to the detection of outliers is to consider as a potential outlier the observation most influential on the mean resultant length.

Thus two appropriate statistics are

$$\begin{aligned} E &= \min_{1 \le i \le n} \frac{1 - \bar{R}_{(-i)}}{1 - \bar{R}}, \\ C &= \max_{1 \le i \le n} \frac{\bar{R}_{(-i)}}{\bar{R}}, \end{aligned}$$

where $\bar{R}_{(-i)}$ denotes the mean resultant length of the $n - 1$ observations obtained by omitting $\mathbf{x}_i$. The statistics E and C on the circle were introduced by Mardia (1975a, p. 390) and Collett (1980), respectively. The versions for S^2 are due to Fisher, Lewis & Willcox (1981). For the circular case, some upper quantiles of C are given by Collett (1980). For the case of S^2, the approximation

$$\Pr(E < t) \simeq \frac{[(n-1)t]^{n-2}}{n^{n-2}} - I\left(t > \frac{n}{2(n-1)}\right)\left\{\frac{2(n-1)t}{n}\right\}^{n-2} \tag{12.2.1}$$

(where I is the appropriate indicator function) was given by Fisher, Lewis & Willcox (1981). Simulation comparisons by Collett (1980) and Fisher, Lewis & Willcox (1981) of various tests for a single outlier found consistently good performance from E and C.

For the case of several outliers, tests of this type based on (9.6.18) were proposed by Fisher, Lewis & Willcox (1981).

12.2.2 Likelihood Ratio Tests

Another way of detecting outliers is to use the likelihood ratio test for slippage in a given parametric model. For the circle, Collett (1980) considered the likelihood ratio test of location slippage in the von Mises model, so that $n - 1$ observations come from $M(\mu, \kappa)$ and one observation comes from $M(\mu^*, \kappa)$. For the sphere S^2, Fisher, Lewis & Willcox (1981) considered the likelihood ratio test of concentration slippage in the Fisher model, so that $n-1$ observations come from $F(\boldsymbol{\mu}, \kappa)$ and one observation comes from $F(\boldsymbol{\mu}, \kappa^*)$.

For the case of several outliers, a test of this type was given by Kimber (1985).

12.2.3 Tests Based on Exponential Distributions

For the sphere S^2, the result (9.3.19) relating the colatitude of a Fisher random variable to an exponential random variable can be exploited to convert the problem of testing for discordancy of an observation from a Fisher distribution into one of testing for discordancy of an observation from an exponential distribution. Several tests of this type were introduced by Fisher, Lewis & Willcox (1981). Analogous tests for discordancy of an observation from a Watson distribution on $\mathbb{R}P^2$ were given by Best & Fisher (1986).

A detailed survey of the detection and accommodation of outliers in circular and spherical data is given by Barnett & Lewis (1994, Chapter 11). Bagchi & Guttmann (1990) gave a Bayesian method of handling data consisting of $n-k$ observations from $M_p(\boldsymbol{\mu}, \kappa)$ and k observations from $M_p(\boldsymbol{\mu}^*, \kappa)$, where k is unknown. A test for discordancy of a single observation in the spherical regression model (11.3.15) was given by Rivest (1989).

12.3 GOODNESS-OF-FIT

Goodness-of-fit tests for circular data were considered in Section 6.4. These were based on the cumulative distribution function, and so have no analogue in higher dimensions. Goodness-of-fit tests for spherical data are of three types.

12.3.1 Tests Based on the Tangent-Normal Decomposition

For data on S^2, tests of goodness-of-fit to a Fisher distribution can be based on the tangent-normal decomposition (9.1.2). These tests can be regarded as formal versions of the graphical tests obtained from the plots of Section 10.2. Let $\mathbf{x}_1, \ldots, \mathbf{x}_n$ be observations on S^2. As in Section 10.2, let (θ_i', ϕ_i') denote the spherical polar coordinates of $\mathbf{x}_i$ in some coordinate system in which the sample mean direction is the pole $\theta' = 0$. Similarly, let (θ_i'', ϕ_i'') denote the spherical polar coordinates of $\mathbf{x}_i$ in the coordinate system in which the sample mean direction $\bar{\mathbf{x}}_0$ has spherical polar coordinates $(\theta'', \phi'') = (\pi/2, 0)$, as in (10.2.2).

(i) The *colatitude test* is based on (9.3.19) and examines the fit of $X_1, \ldots, X_n$ to an exponential distribution, where $X_i = 1 - \cos\theta_i'$. It rejects the hypothesis of a Fisher distribution for large values of

$$M_E(D_n) = \left(D_n - \frac{0.2}{n}\right)\left(\sqrt{n} + 0.26 + \frac{0.2}{\sqrt{n}}\right), \tag{12.3.1}$$

where D_n is the Kolmogorov–Smirnov statistic

$$D_n = \max\left\{\max_{1\le i\le n}\left[\frac{i}{n} - F(X_{(i)})\right], \max_{1\le i\le n}\left[F(X_{(i)}) - \frac{i-1}{n}\right]\right\} \tag{12.3.2}$$

with

$$F(x) = 1 - \exp(\hat{\kappa}x) \tag{12.3.3}$$

and $\hat{\kappa}$ as the approximate maximum likelihood estimate

$$\hat{\kappa} = \frac{n-1}{\sum_{i=1}^n \cos\theta_i'}$$

of κ. Some quantiles of $M_E(D_n)$ are given in Table 12.1.

(ii) The *longitude test* is a test of rotational symmetry and examines the fit of the longitudes $\phi'_1, \ldots, \phi'_n$ to the uniform distribution. It rejects the hypothesis of a Fisher distribution for large values of the modification $M_U(V_n)$ of Kuiper's V_n statistic (6.3.21), where

$$M_U(V_n) = V_n \left(\sqrt{n} - 0.567 + \frac{1.623}{\sqrt{n}} \right). \tag{12.3.4}$$

Some quantiles of $M_U(V_n)$ are given in Table 12.1.

(iii) The *two-variable test* is a test of normality and examines the fit of $z_1, \ldots, z_n$ to a normal distribution, where $z_i = \phi''_i \sqrt{\theta''_i}$. It rejects the hypothesis of a Fisher distribution for large values of

$$M_N(D_n) = D_n \left(\sqrt{n} - 0.01 + \frac{0.85}{\sqrt{n}} \right), \tag{12.3.5}$$

where D_n is the Kolmogorov–Smirnov statistic

$$D_n = \max \left\{ \max_{1 \le i \le n} \left[\frac{i}{n} - F(z_{(i)}) \right], \max_{1 \le i \le n} \left[F(z_{(i)}) - \frac{i-1}{n} \right] \right\}$$

with

$$F(x) = \Phi \left(\frac{x}{s} \right) \quad \text{and} \quad s^2 = \frac{1}{n} \sum_{i=1}^{n} z_i^2,$$

Φ being the cumulative distribution of the standard normal distribution. Some quantiles of $M_E(D_n)$ are given in Table 12.1.

Further details are given in Fisher & Best (1984) (together with a power study) and by Fisher, Lewis & Embleton (1987, pp. 122–125).

Table 12.1 Upper quantiles of statistics for testing the goodness-of-fit of a Fisher distribution. After Stephens (1974) and Fisher & Best (1984). Reproduced by permission the publishers of *J. Amer. Statist. Assoc.* and *Austral. J. Statist.*

	α		
Statistic	0.10	0.05	0.01
$M_E(D_n)$	0.990	1.094	1.308
$M_U(V_n)$	1.138	1.207	1.347
$M_N(D_n)$	0.819	0.895	1.035

Example 12.1

For the remnant-magnetisation data set of Example 9.1, calculation gives the following.

(i) *Colatitude test*

$$M_E(D_n) = 0.530 < 0.990$$

and so the hypothesis of a Fisher distribution is accepted at the 10% level.

(ii) *Longitude test*

$$M_U(V_n) = 0.953 < 1.138$$

and so rotational symmetry is accepted at the 10% level.

(iii) *Two-variable test*

$$M_N(D_n) = 0.250 < 0.819$$

and so the hypothesis of a Fisher distribution is accepted at the 10% level.

All three tests confirm the conclusions reached from the graphical analysis in Example 10.1.

Likewise, tests of goodness-of-fit of data on $\mathbb{R}P^2$ to a Watson distribution can be obtained as formal versions of the graphical tests obtained from the plots of Section 10.2. Let $\pm\mathbf{x}_1, \ldots, \pm\mathbf{x}_n$ be observations on $\mathbb{R}P^2$.

Bipolar Case

As in Section 10.2, let (θ_i', ϕ_i') be spherical polar coordinates of $\mathbf{x}_i$ about the dominant eigenvector $\mathbf{t}_1$ of $\bar{\mathbf{T}}$.

(i) The *colatitude test* examines the fit of $X_1, \ldots, X_n$ to an exponential distribution, where $X_i = 1 - \cos^2\theta_i'$. It rejects the hypothesis of a Watson distribution for large values of

$$M_B(D_n) = \left(D_n - \frac{0.2}{n}\right)\left(\sqrt{n} + 0.26 + \frac{0.2}{\sqrt{n}}\right), \tag{12.3.6}$$

where D_n is the Kolmogorov–Smirnov statistic (12.3.2) with fitted cumulative distribution function given by (12.3.3) and

$$\hat{\kappa} = D_3^{-1}(\bar{t}_1),$$

$\bar{t}_1$ being the largest eigenvalue of $\bar{\mathbf{T}}$. (The function D_3^{-1} on $(0.34, 1)$ is tabulated in Appendix 3.5.) Quantiles of $M_B(D_n)$ are the same as those of $M_E(D_n)$, some of which are given in Table 12.1.

(ii) The *longitude test* is a test of rotational symmetry and examines the fit of the longitudes $2\phi_1', \ldots, 2\phi_n'$ to the uniform distribution. It rejects the hypothesis of a Fisher distribution for large values of the modification (12.3.4) of Kuiper's V_n statistic (6.3.21). Some quantiles of $M_U(V_n)$ are given in Table 12.1.

Girdle Case

As in Section 10.2, let (θ_i', ϕ_i') be spherical polar coordinates of $\mathbf{x}_i$ about the eigenvector $\mathbf{t}_3$ of $\bar{\mathbf{T}}$ corresponding to the smallest eigenvalue.

(i) The *colatitude test* examines the fit of $X_1, \ldots, X_n$ to an exponential distribution, where $X_i = \cos^2 \theta_i'$. It rejects the hypothesis of a Watson distribution for large values of

$$M_G(D_n) = \left(D_n - \frac{0.2}{n}\right)\left(\sqrt{n} + 0.26 + \frac{0.2}{\sqrt{n}}\right), \tag{12.3.7}$$

where D_n is the Kolmogorov–Smirnov statistic (12.3.2) with fitted cumulative distribution function given by (12.3.3) and

$$\hat{\kappa} = D_3^{-1}(\bar{t}_3),$$

$\bar{t}_3$ being the smallest eigenvalue of $\bar{\mathbf{T}}$. (The function D_3^{-1} on $(0, 0.333)$ is tabulated in Appendix 3.4.) The 10%, 5% and 1% quantiles of $M_G(D_n)$ are 1.04, 1.15 and 1.36, respectively.

(ii) The *longitude test* is a test of rotational symmetry and examines the fit of the longitudes $2\phi_1', \ldots, 2\phi_n'$ to the uniform distribution. It rejects the hypothesis of a Fisher distribution for large values of the modification (12.3.4) of Kuiper's V_n statistic (6.3.21). Some quantiles of $M_U(V_n)$ are given in Table 12.1.

Details can be found in Best & Fisher (1986) and Fisher, Lewis & Embleton (1987, pp. 168–170).

12.3.2 Score Tests against Specified Larger Models

Von Mises–Fisher versus Fisher–Bingham

A convenient way of testing Fisherness is by the score test of $\mathbf{A} = \mathbf{0}$ in the Fisher–Bingham densities (9.3.22). An explicit form of this test was given by Mardia, Holmes & Kent (1984). This test can be regarded as an omnibus test of Fisherness which is designed to detect axial asymmetry. In the case $p = 2$, the test reduces to Cox's (1975) test of von Misesness which was given in Section 7.5.

Von Mises–Fisher versus Kent

For $p = 3$, a test with a simpler form than the above is obtained by restricting the alternative to be a Kent distribution. Let $\mathbf{x}_1, \ldots, \mathbf{x}_n$ be observations on the sphere S^2. Let $\mathbf{H}(\bar{\mathbf{x}}_0, \mathbf{n})$ be the rotation given by (10.2.1) which takes the sample mean direction $\bar{\mathbf{x}}_0$ to the north pole $\mathbf{n} = (0, 0, 1)^T$. Denote by $\mathbf{y}_i$ the

2-vector consisting of the last two components of $\mathbf{H}(\bar{\mathbf{x}}_0, \mathbf{n})\mathbf{x}_i$. Let $\hat{l}_1$ and $\hat{l}_2$ be the eigenvalues of

$$\frac{1}{n}\sum_{i=1}^{n} \mathbf{y}_i \mathbf{y}_i^T.$$

Then the score statistic for testing $\mathbf{A} = \mathbf{0}$ in the Kent densities (9.3.23) takes the form

$$\begin{aligned} \hat{T} &= n\left(\frac{\hat{\kappa}}{2}\right)^2 \frac{I_{1/2}(\hat{\kappa})}{I_{5/2}(\hat{\kappa})}(\hat{l}_1 - \hat{l}_2)^2 \\ &= n\frac{\hat{\kappa}^3}{4(\hat{\kappa} - 3\bar{R})}(\hat{l}_1 - \hat{l}_2)^2, \end{aligned} \tag{12.3.8}$$

where $\hat{\kappa}$ is the maximum likelihood estimate of κ in the Fisher model. Fisherness is rejected for large values of $\hat{T}$. See Kent (1982) and Rivest (1986). Under the null hypothesis of a Fisher distribution, the large-sample asymptotic distribution of $\hat{T}$ is

$$\hat{T} \,\dot{\sim}\, \chi_2^2. \tag{12.3.9}$$

The modification

$$\hat{T}^* = \begin{cases} -(n-2)\log\left(1 - \hat{T}/n\right), & \hat{T} < n, \\ \hat{T}, & \hat{T} \geq n, \end{cases} \tag{12.3.10}$$

of $\hat{T}$ satisfies

$$\hat{T}^* \,\dot{\sim}\, \chi_2^2 \tag{12.3.11}$$

for large samples or large concentrations (Rivest, 1986). Let $\hat{T}_\lambda$ and $\hat{T}_\lambda^*$ be the statistics obtained from $\hat{T}$ and $\hat{T}^*$ by replacing $\hat{l}_1$ and $\hat{l}_2$ by the two smallest eigenvalues $\bar{t}_2$ and $\bar{t}_3$ of $\bar{\mathbf{T}}$. Rivest pointed out that

$$\hat{T}_\lambda = \hat{T} + O_p(\kappa^{-1}) \qquad \hat{T}_\lambda^* = \hat{T}^* + O_p(\kappa^{-1}), \qquad \kappa \to \infty.$$

Bingham versus Fisher–Bingham

Sometimes it is appropriate to test goodness-of-fit of a Bingham distribution to data on S^{p-1} which are not necessarily axial. A convenient test is the score test of $\boldsymbol{\mu} = \mathbf{0}$ in the Fisher–Bingham densities (9.3.22). Binghamness is rejected for large values of

$$n\bar{\mathbf{x}}\mathbf{S}^{-1}\bar{\mathbf{x}}, \tag{12.3.12}$$

where $\bar{\mathbf{x}}$ and $\mathbf{S}$ are the sample (vector) mean and variance matrix, respectively (Kent, 1982). Under the null hypothesis of a Bingham distribution,

$$n\bar{\mathbf{x}}\mathbf{S}^{-1}\bar{\mathbf{x}} \,\dot{\sim}\, \chi_p^2, \qquad n \to \infty.$$

If the distribution is uniform then (12.3.12) is approximately the Rayleigh test statistic (10.4.4).

General Alternatives

The above tests can be put in the following general framework. Suppose that a parametric model is given which consists of rotationally symmetric distributions on S^{p-1}. Then appropriate goodness-of-fit tests are the score tests of this model against a larger model with log-densities in a suitable $O(p)$-invariant subspace of an associated Hilbert space. See Boulerice & Ducharme (1997).

12.3.3 Tests Based on Density Estimates

Tests which compare a fitted density in the specified class to a suitable non-parametric density estimate were introduced by Bowman (1992).

12.4 ROBUST METHODS

Because outliers can have a considerable effect on estimates and on observed significance levels of tests, robust methods are useful. We now give an outline of robust methods for the principal directional distributions. A survey of robust methods on spheres was given by He (1992). For the cases of the circle and the sphere S^2, see also Barnett & Lewis (1994, Chapter 11).

12.4.1 Estimation of Mean Direction

One quantitative measure of robustness of an estimator is its influence function (which can be regarded as the bias in the presence of infinitesimal contamination). The influence functions of the mean and median directions of circular distributions were found by Wehrly & Shine (1981). They concluded that the mean direction is quite robust. This is not surprising, since the circle is compact. For rotationally symmetric distributions on S^{p-1}, i.e. those with densities of the form (9.3.31), the influence functions of the mean direction and the dominant eigenvector $\mathbf{t}_1$ of the scatter matrix $\bar{\mathbf{T}}$ were calculated by Watson (1986). He pointed out that, for sufficiently concentrated distributions, the latter estimator can be more robust than the mean direction.

Influence functions place little emphasis on small infinitesimal bias. Since a small bias can be important if it is considerably larger than the standard deviation of an estimate, it is appropriate to consider *standardised influence functions* of estimators, in which the influence functions are standardised with respect to some other estimator (Ko & Guttorp, 1988) or with respect to Fisher information (He & Simpson, 1992; Ko & Chang, 1993). Similarly, it is appropriate to consider *standardised gross error sensitivity* (to measure the relative asymptotic effect of a small contamination). An estimator is *SB-robust* (*standardised bias robust*) if its standardised gross error sensitivity is bounded. Ko & Guttorp (1988) showed that, for a very wide class of families of distributions on S^{p-1}, the mean direction has infinite standardised gross error sensitivity, i.e. the asymptotic effect of a small contamination can be very large compared with the dispersion.

Several robust estimators of mean direction are forms of median. The intrinsic approach described in Section 9.1 leads to the *spherical median* $\mathbf{m}$ of a set of observations $\mathbf{x}_1, \ldots, \mathbf{x}_n$ on S^{p-1} as the point minimising the mean arc length to the observations, i.e. the point $\mathbf{a}$ in S^{p-1} minimising

$$\frac{1}{n}\sum_{i=1}^{n} \cos^{-1}(\mathbf{x}_i^T \mathbf{a}).$$

In the circular case, $\mathbf{m}$ is the circular median, as defined in Section 2.2.2. Fisher (1985) introduced the spherical median and proposed associated test procedures. Some large-sample asymptotic properties of the circular median were investigated by Ducharme & Milasevic (1987b). The embedding approach described in Section 9.1 leads to two forms of median. One is the *normalised spatial median* $\tilde{\boldsymbol{\mu}}$, defined by

$$\tilde{\boldsymbol{\mu}} = \|\tilde{\mathbf{m}}\|^{-1}\tilde{\mathbf{m}},$$

where $\tilde{\mathbf{m}}$ is the spatial median in $\mathbb{R}^p$, i.e. the point $\mathbf{a}$ in $\mathbb{R}^p$ minimising

$$\frac{1}{n}\sum_{i=1}^{n} \cos^{-1} \|\mathbf{x}_i - \mathbf{a}\|. \tag{12.4.1}$$

The normalised spatial median was introduced by Ducharme & Milasevic (1987a). The other form of median is the L_1-estimator $\check{\boldsymbol{\mu}}$ of He & Simpson (1992), which is the point $\mathbf{a}$ in S^{p-1} minimising (12.4.1). Chan & He (1993) calculated the asymptotic distributions, efficiencies, influence functions and sensitivities of $\mathbf{m}$, $\tilde{\boldsymbol{\mu}}$ and $\check{\boldsymbol{\mu}}$. Their overall conclusion was that the normalised spatial median $\tilde{\boldsymbol{\mu}}$ is to be preferred.

A natural generalisation of these medians is given by M-estimators of the mean direction $\boldsymbol{\mu}$. These estimate $\boldsymbol{\mu}$ on the basis of observations $\mathbf{x}_1, \ldots, \mathbf{x}_n$ on S^{p-1} by minimising

$$\sum_{i=1}^{n} \rho(t(\mathbf{x}_i^T \boldsymbol{\mu}; \kappa)),$$

where ρ and t are suitable functions (and if κ is unknown it is replaced by a suitable robust estimate). On the circle, such M-estimators were considered by Lenth (1981a), using $t(u;\kappa) = \operatorname{sign}(u)\, 2\kappa(1-u)^{1/2}$ and taking ρ to be Huber's (1964) function defined by

$$\rho(t) = \begin{cases} t^2/2, & |t| \leq c, \\ c|t| - c^2/2, & |t| \geq c, \end{cases} \tag{12.4.2}$$

or Andrews's (1974) function

$$\rho(t) = \begin{cases} -c^2 \cos(t/c), & |t| \leq c\pi, \\ 0, & |t| \geq c\pi, \end{cases} \tag{12.4.3}$$

for a suitable value of c. Influence functions, large-sample asymptotic distributions, and conditions for SB-robustness of M-estimators of the mean direction $\boldsymbol{\mu}$ and the concentration parameter κ in von Mises-Fisher distributions on spheres were considered by Ko & Chang (1993). Influence functions and large-sample asymptotic distributions of M-estimators of the rotation matrix $\mathbf{A}$ in the rotation regresssion model (11.3.15) were given by Chang & Ko (1995). The use of median axes in robust estimation of the axis of symmetry of a Watson distribution was considered by Fisher, Lunn & Davies (1993).

12.4.2 Estimation of Concentration

The maximum likelihood estimator of the concentration κ in a von Mises–Fisher distribution is far from robust, as it has infinite standardised gross error sensitivity. Thus robust estimation of concentration is important.

In the case of S^2, (9.3.19) can be exploited to transfer familiar robust methods for the mean of an exponential distribution into robust methods for the concentration parameter κ of a Fisher distribution. In particular, Fisher (1982) suggested the use of the L-estimator (i.e. linear combination of order statistics)

$$\hat{\kappa}_R^{-1} = \sum_{i=1}^{n} l_i c_{(i)}$$

of κ^{-1}, where the $c_{(i)}$ are the order statistics of $1 - \mathbf{x}_1^T \hat{\boldsymbol{\mu}}, \ldots, 1 - \mathbf{x}_n^T \hat{\boldsymbol{\mu}}$ and $l_1, \ldots, l_n$ are suitable weights. He also considered a Winsorised version which was modified by Kimber (1985) to reduce its bias. Other robust estimators of κ include those of Ducharme & Milasevic (1990) and of Ko (1992). Ronchetti (1992) considered optimal robust estimation of κ.

A robust method of testing the equality of the concentration parameters of several von Mises–Fisher or Watson distributions was suggested by Fisher (1986a; 1986b). His test is based on the high-concentration asymptotic $(p-1)$-variate normality of the tangential part $\mathbf{x} - (\mathbf{x}^T \boldsymbol{\mu})\boldsymbol{\mu}$ of $\mathbf{x}$ (cf. (9.3.15)) and is carried out as a one-way analysis of variance on the absolute values of the components of the sample analogues $\mathbf{x}_i - (\mathbf{x}_i^T \hat{\boldsymbol{\mu}})\hat{\boldsymbol{\mu}}$ of $\mathbf{x} - (\mathbf{x}^T \boldsymbol{\mu})\boldsymbol{\mu}$.

12.5 BOOTSTRAP METHODS

For inferential problems in which (i) there are no grounds for adopting a simple parametric model, or (ii) the sample size and concentration are too small for asymptotic methods to applied, or (iii) the assumption of independence is unwarranted, it is appropriate to use bootstrap methods for variance estimation. In a typical use of such methods, the variability of a statistic is assessed by resampling B times *with replacement* from the data. The statistic is evaluated for each of these B ***bootstrap samples***, and the variability of these B values is taken as the variability of the statistic over the population. A general reference for bootstrap methods is Efron & Tibshirani (1993).

Since the distributions of the statistics commonly used for inference on directional distributions are more complex than those arising in standard normal theory, bootstrap methods are particularly useful in the directional context. A detailed discussion of bootstrap methods for circular data is given by Fisher (1993, Chapter 8). Watson (1983e) and Fisher, Lewis & Embleton (1987) called attention to the use of bootstrap methods for small samples of spherical data with low concentration. A simulation study comparing bootstrap confidence cones for the mean direction $\boldsymbol{\mu}$ of a spherical distribution with various parametric competitors was given by Ducharme *et al.* (1985). A short review of bootstrap methods for directional data was given by Fisher & Hall (1992).

The coverage properties of bootstrap confidence regions are improved when they are based on asymptotically pivotal statistics. By using asymptotically pivotal statistics, Fisher & Hall (1989) obtained a variety of bootstrap confidence regions for the mean direction $\boldsymbol{\mu}$ with good coverage properties. The alternative pivotal approaches of Fisher *et al.* (1996) give improved bootstrap confidence regions for the mean direction $\boldsymbol{\mu}$ of a distribution on S^{p-1}, as well as bootstrap confidence regions for the principal or polar axis of an axial distribution.

12.6 DENSITY ESTIMATION

One way of obtaining a grasp of the message given by a data set is to estimate the underlying density and to produce a corresponding contour plot.

12.6.1 Kernel Density Estimation

A simple method of density estimation on the sphere S^{p-1} is kernel density estimation based on the von Mises–Fisher kernel, i.e. replacing each data point $\mathbf{x}_i$ by the von Mises–Fisher distribution $M_p(\mathbf{x}_i, \kappa)$. Given observations $\mathbf{x}_1, \ldots, \mathbf{x}_n$, the corresponding density estimate is given by

$$\hat{f}_F(\mathbf{x}; \kappa) = n^{-1} a_p(\kappa) \sum_{i=1}^{n} \exp\left(\kappa \mathbf{x}^T \mathbf{x}_i\right), \tag{12.6.1}$$

where the constant κ determines the degree of smoothing, and $a_p(\kappa)$ is the normalising constant (defined in (10.4.2)) for the von Mises–Fisher density (9.3.4) (see Watson, 1983a, pp. 9, 37; 1983e). In the case $p = 3$, a computer program to produce these density estimates was given by Diggle & Fisher (1985).

Two useful generalisations of the kernel density estimators given by (12.6.1) are those of the forms

$$\hat{f}_K(\mathbf{x};\kappa) = n^{-1}c_0(\kappa)\sum_{i=1}^{n} K(\kappa\mathbf{x}^T\mathbf{x}_i) \tag{12.6.2}$$

and

$$\hat{f}_L(\mathbf{x};\kappa) = n^{-1}d_0(\kappa)\sum_{i=1}^{n} L(\kappa(1-\mathbf{x}^T\mathbf{x}_i)), \tag{12.6.3}$$

where K and L are known kernel functions and κ determines the degree of smoothing. Hall, Watson & Cabrera (1987) investigated the bias and variance of these estimators, as well as their expected squared-error and Kullback–Leibler losses. They showed that all estimators of the form (12.6.2) are asymptotically equivalent for large samples and they gave an asymptotically optimal kernel of the type (12.6.3). Conditions on L and f for consistency of $\hat{f}_L$ as an estimator of the underlying density f were given by Bai, Rao & Zhao (1988). Generalisations of the estimators (12.6.3) to density estimators on Stiefel manifolds (as considered in Section 13.2) were given by Chikuse (1998).

Perhaps the simplest density estimators are the 'naive' estimators $\hat{f}_{\rho_n}$, defined by

$$\hat{f}_{\rho_n}(\mathbf{x}) = \frac{\text{number of } \mathbf{x}_i \text{ in } C_{\rho_n}(\mathbf{x})}{n\text{vol } C_{\rho_n}(\mathbf{x})}, \tag{12.6.4}$$

where $C_r(\mathbf{x})$ denotes the cap on S^{p-1} with centre $\mathbf{x}$ and radius r (see Ruymgaart, 1989). The estimators (12.6.4) are kernel density estimators of the form (12.6.2) with

$$K(x) = \begin{cases} 1, & x \geq \cos\rho_n, \\ 0, & x < \cos\rho_n. \end{cases}$$

Ruymgaart (1989) showed that if the ρ_n tend to zero at a suitable rate then the $\hat{f}_{\rho_n}$ are strongly consistent. Generalisations of the estimators (12.6.4) to strongly consistent density estimators on compact Euclidean manifolds (in particular, Stiefel and Grassmann manifolds, as considered in Sections 13.2–13.3) were given by Hendriks, Janssen & Ruymgaart (1993). Generalisations to estimators of real-valued functions on Stiefel manifolds were considered by Lee & Ruymgaart (1996).

12.6.2 Density Estimators Based on the Embedding Approach

A different construction of density estimators was given by Hendriks (1990) for quite general Riemannian manifolds. It is based on the infinite-dimensional embedding approach of Section 10.8. For $T = 1, 2, \ldots$, let $\mathbf{t}_T$ denote the function from S^{p-1} into $L^2(S^{p-1})$ given by (10.8.2) with $a_k = 1$ for $k < T$ and $a_k = 0$ for $k \geq T$. Then the corresponding density estimator $\hat{f}_T$ is defined by

$$\hat{f}_T(\mathbf{x}) = n^{-1} \sum_{i=1}^{n} \mathbf{t}_T(\mathbf{x}_i)$$

or equivalently by

$$\hat{f}_T(\mathbf{x}) = n^{-1} \sum_{i=1}^{n} \langle \mathbf{t}_T(\mathbf{x}), \mathbf{t}_T(\mathbf{x}_i) \rangle,$$

where $\langle \cdot, \cdot \rangle$ denotes the inner-product on $L^2(S^{p-1})$. Hendriks gave bounds in terms of T for the expected squared-error and supremum-error losses of these estimators.

Deconvolution density estimation on $SO(p)$ has been developed by Kim (1998), using Fourier analysis on $L^2(SO(p))$. Deconvolution density estimation on S^{p-1} was considered by Healy, Hendriks & Kim (1998).

12.7 BAYESIAN METHODS

Bayesian inference on von Mises–Fisher distributions has been considered by Mardia & El-Atoum (1976) and Bagchi & Guttmann (1988). Bayes procedures for inference on unimodal rotationally symmetric distributions on S^{p-1} were introduced by Brunner & Lo (1994). A Bayesian analysis of the regression model (11.3.20) with $\kappa_0 = c = 0$ was given by Guttorp & Lockhart (1988).

Laplace approximations to posterior moments were considered by Bagchi & Kadane (1991).

Conditions for consistency of Bayes procedures for estimating (i) the mean direction of a rotationally symmetric distribution on S^{p-1}, and (ii) the axis of symmetry of a rotationally symmetric distribution on $\mathbb{R}P^{p-1}$, were given by Lo & Cabrera (1987)

Empirical Bayes estimation of the mean direction of von Mises–Fisher distributions on S^{p-1} was considered by Bagchi (1994). Healy & Kim (1996) used Fourier analysis on $L^2(S^2)$ to derive a consistent nonparametric empirical Bayes estimator of the prior density of the mean direction of rotationally symmetric distributions on S^2.

Bayes estimators (relative to squared-error loss) of the rotation matrix $\mathbf{A}$ in the spherical regression model (11.3.12) were characterised by Kim (1991). In particular, he showed that the Bayes estimator with respect to the uniform prior is the least-squares estimator $\hat{\mathbf{A}}$ given in (11.3.14).

12.8 CURVE FITTING AND SMOOTHING

12.8.1 Scalar Predictor and Spherical Response

Because of its geophysical importance, regression of a spherical variable (such as the position of the palaeomagnetic north pole) on a scalar predictor (such as time) has been the subject of much attention. Perhaps the simplest parametric family of regression functions for regressing a spherical variable on a scalar (or an unrestricted vector) is

$$f(t; \mathbf{A}, c) = \mathbf{A}(\cos ct, \sin ct, 0)^T, \tag{12.8.1}$$

where the parameter $\mathbf{A}$ is a rotation matrix and c is the constant speed of movement along the corresponding great circle path. A major problem with (12.8.1) (as with the 'barber's pole' regression model (11.3.2)) is that one can obtain an arbitrarily close fit to the data by increasing the speed c. Consequently, almost all the work on curve fitting and smoothing for spherical data has been non-parametric.

Apart from a 'kernel-type' smoothing algorithm given by Watson (1985), most of the methods proposed for non-parametric spherical regression have been concerned with constructing suitable sphere-valued analogues of cubic splines in the plane.

The intrinsic geometrical approach described in Section 9.1 leads to the following construction of smoothing splines on the sphere. For fitting data $\{(t_i, \mathbf{v}_i) : 1 \leq i \leq n\}$ with $\|\mathbf{v}_i\| = 1$, Jupp & Kent (1987) proposed the curve $\mathbf{f}$ which minimises

$$\sum_{i=1}^{n} w_i \{\cos^{-1}(\mathbf{f}(t_i)^T \mathbf{v}_i)\}^2 + \lambda \int \|[\mathbf{I} - \mathbf{f}(t)\mathbf{f}(t)^T]\mathbf{f}''(t)\|^2 dt \tag{12.8.2}$$

(by analogy with minimising

$$\sum_{i=1}^{n} w_i \|\mathbf{f}(t_i) - \mathbf{v}_i\|^2 + \lambda \int \|\mathbf{f}''(t)\|^2 dt \tag{12.8.3}$$

for cubic splines in $\mathbb{R}^3$). Here $w_1, \ldots, w_n$ are weights and λ is a smoothing parameter. The first term in (12.8.2) and (12.8.3) penalises lack of fit; the second term penalises curvature of the fitted path. An alternative which is easier to implement uses an extension of the wrapping approach described in Section 9.1 to 'unwrap' the sphere onto the plane (using a given path rather than a given point), thus converting a spherical problem into a standard planar one. Details are given in Jupp & Kent (1987). A specific application of this work occurs in studying apparent polar wander paths (see, e.g. Jupp & Kent, 1987). The data are usually obtained through a pre-processing step which extracts the remanent magnetism in given rocks through finding linear segments given a set of ordered points; see Kent, Briden & Mardia (1983) for further details.

Two other forms of spherical spline were introduced by Fisher & Lewis (1985). One of these uses piecewise smooth functions $\mathbf{f}$ with $||\mathbf{f}(t)|| = 1$ for which the third derivative $\mathbf{f}^{(3)}$ satisfies $||\mathbf{f}^{(3)}(t)|| =$ constant (by analogy with $\mathbf{f}^{(3)}(t) =$ constant for cubic splines in $\mathbb{R}^3$). The other uses portions of loxodromes (curves of constant 'compass bearing' with respect to some pole). For interpolating curves (i.e. curves with $\mathbf{f}(t_i) = \mathbf{v}_i$), another form of spherical spline was given by Watson (1985). He tackled the problem of unbounded speed which arises in fitting (12.8.1) by using instead curves which minimise

$$\int ||[\mathbf{I} - \mathbf{f}(t)\mathbf{f}(t)^T]\mathbf{f}''(t)||^2 dt + \nu \int ||\mathbf{f}'(t)||^2 dt$$

for an appropriate value of the parameter ν, thus penalising speed $\mathbf{f}'$ as well as curvature.

12.8.2 Scalar Predictor and Rotational Response

For non-parametric regression of a rotation matrix on a scalar, the wrapping approach leads to the use of the exponential map (13.2.6) to transform splines in $\mathbb{R}^p$ to 'splines' in $SO(p)$. For further details when $p = 3$, see Prentice (1987).

12.8.3 Planar Predictor and Circular Response

Non-parametric regression of an angular variate on a planar variate can be interpreted as smoothing a unit vector field (a field of directions at points on the plane). Mendoza (1986) has shown how thin plate splines can be used for this purpose.

12.8.4 Circular Predictor and Scalar Response

Following Wahba (1990), Mardia *et al.* (1996) developed forms of spline which map the circle to the line, as part of a generalisation of splines to a wider context. These *periodic splines* are defined in terms of minimum squared prediction error subject to linear constraints (e.g. on derivatives of order $1, \ldots, m$) at points $\theta_1, \ldots, \theta_n$ on the circle. The periodic splines have explicit representations of the form

$$f(\theta) = \sum_{i=1}^{n} b_i \rho_\alpha(\theta - \theta_i) + \sum_{k=1}^{m} a_k g_k(\theta), \tag{12.8.4}$$

where

$$\rho_\alpha(h) = \sum_{j=1}^{\infty} \frac{1}{(2\pi j)^{2\alpha+1}} \cos jh, \tag{12.8.5}$$

with α a positive parameter and $g_1, \ldots, g_m$ certain trend functions on the circle. When $\alpha = r + 1/2$ with r an integer, (12.8.5) can be written more explicitly as

$$\rho_{r+1/2}(h) = \frac{(-1)^r}{(2r+2)!} B_{2r}(|h|), \tag{12.8.6}$$

where B_{2r} denotes the Bernoulli polynomial of order $2r$.

A general discussion of splines and local regression methods for directional regression was given by Hansen & Mount (1990) in a geological context.

12.9 OTHER METHODS

Most techniques for linear data have analogues for directional data. In particular, discriminant analysis of spherical data has been considered by Morris & Laycock (1974), while sequential methods for circular data have been developed by Gadsden & Kanji (1980; 1982). Optimal designs for several regression models involving circular variates were described by Laycock (1975).

13

General Sample Spaces

13.1 INTRODUCTION

In the previous chapters we have considered mainly observations which are unit vectors or axes. However, other types of observation occur in directional statistics, the most important of these from the practical point of view being rotations, orthonormal frames, and subspaces. To see how these might arise, consider the following example.

Example 13.1
The orbit of a periodic comet is an ellipse with the sun at one focus (see Fig. 13.1).

There are various directional aspects of such orbits in which we might be interested.

(i) We might be interested in just the perihelion direction (the direction from the sun to the point of nearest approach of the comet) of each orbit. Thus each orbit provides a unit vector $\mathbf{x}_1$, so that we have spherical data. A question of astronomical interest is whether or not the perihelion directions are uniformly distributed on the sphere.

(ii) We might know the plane of each orbit but not necessarily the sense of rotation of the comet. Thus each orbit would provide a plane through the origin (the sun) of three-dimensional space. A question of interest is whether or not these planes are uniformly distributed.

(iii) For most comets we know the perihelion direction, the plane and the sense of rotation. Thus each orbit provides three orthonormal unit vectors, $\mathbf{x}_1, \mathbf{x}_2$ and $\mathbf{x}_3$, where $\mathbf{x}_1$ is the perihelion direction, $\mathbf{x}_2$ is the directed unit normal to the orbit, and $\mathbf{x}_3$ is the vector product $\mathbf{x}_1 \wedge \mathbf{x}_2$. Then the 3×3 matrix $(\mathbf{x}_1, \mathbf{x}_2, \mathbf{x}_3)$ is a rotation matrix. A question of astronomical interest is whether or not these matrices are uniformly distributed on the group $SO(3)$ of all rotations of $\mathbb{R}^3$ (Mardia, 1975a; Jupp & Mardia, 1979).

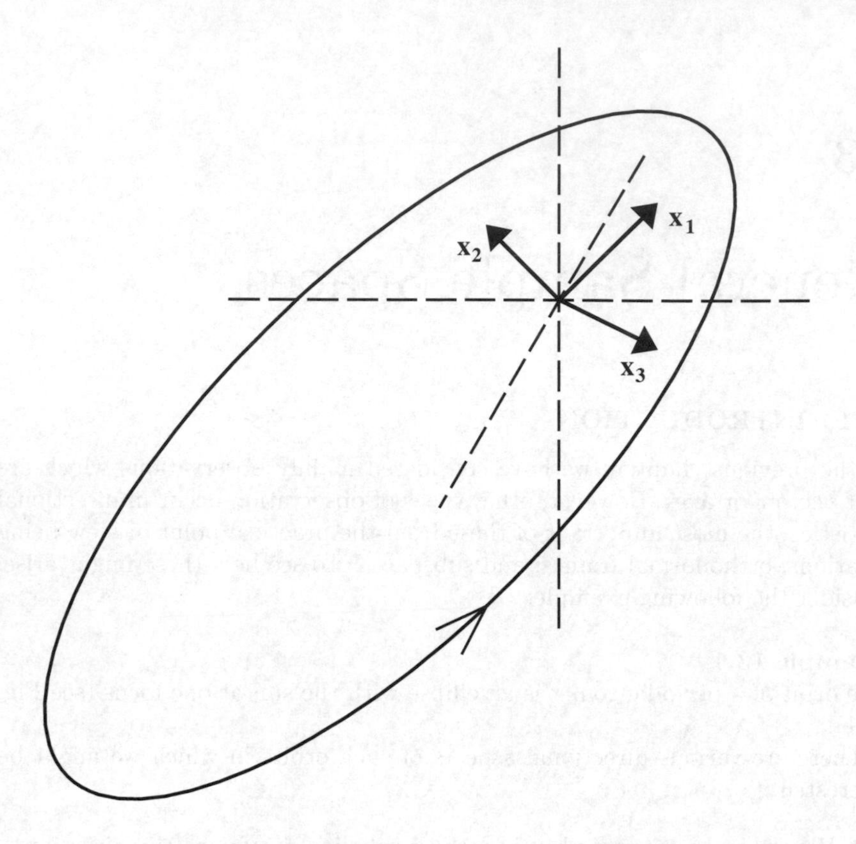

Figure 13.1 The rotation matrix $(\mathbf{x}_1, \mathbf{x}_2, \mathbf{x}_3)$ of a cometary orbit. The sun is at the origin, which is the focus of the ellipse.

Observations which are rotations arise also in vectorcardiography (Downs, 1972), since a vectorcardiogram can be regarded as a roughly cardioid-shaped planar 'orbit' in 3-space. Distributions on the group $SO(3)$ of rotations arise also as distributions of estimates of rotations. For example, as pointed out by Bingham, Chang & Richards (1992), Chang's (1986) spherical regression model (11.3.15) is parameterised by a rotation matrix. The distribution of its maximum likelihood estimate is a distribution on $SO(3)$.

In order to analyse data of the forms arising in Example 13.1(ii) and (iii), it is necessary to develop models and techniques generalising those used in the previous chapters. Such generalisation is comparatively straightforward and will be described in Sections 13.2 and 13.3. In Section 13.4 we present analogous models and techniques for distributions on hyperboloids and we discuss briefly the study of statistics on general manifolds.

13.2 FRAMES AND ROTATIONS

13.2.1 Stiefel Manifolds

Example 13.1(iii) indicates that it is necessary to consider orthonormal frames. An orthonormal r-frame in $\mathbb{R}^p$ is a set $(\mathbf{x}_1, \ldots, \mathbf{x}_r)$ of orthonormal vectors in $\mathbb{R}^p$, i.e.

$$\mathbf{x}_i^T \mathbf{x}_j = \begin{cases} 1 & \text{if } i = j, \\ 0 & \text{if } i \neq j. \end{cases}$$

The space of orthonormal r-frames in $\mathbb{R}^p$ is called the *Stiefel manifold* $V_r(\mathbb{R}^p)$. In terms of $p \times r$ matrices $\mathbf{X}$,

$$V_r(\mathbb{R}^p) = \{\mathbf{X} : \mathbf{X}^T\mathbf{X} = \mathbf{I}_r\}. \tag{13.2.1}$$

Three important special cases are of interest:

(i) a 1-frame is just a unit vector, so $V_1(\mathbb{R}^p) = S^{p-1}$;
(ii) an orthonormal p-frame is equivalent to an orthogonal matrix, so $V_p(\mathbb{R}^p) = O(p)$, the orthogonal group consisting of all orthogonal $p \times p$ matrices;
(iii) an orthonormal $(p-1)$-frame $(\mathbf{x}_1, \ldots, \mathbf{x}_{p-1})$ can be extended uniquely to an orthonormal p-frame $(\mathbf{x}_1, \ldots, \mathbf{x}_p)$ with matrix of determinant 1, so $V_{p-1}(\mathbb{R}^p) = SO(p)$, the special orthogonal group consisting of all $p \times p$ rotation matrices.

Two of the rotation groups can be identified with more familiar sample spaces. Firstly, the mapping

$$\theta \mapsto \begin{pmatrix} \cos\theta & \sin\theta \\ -\sin\theta & \cos\theta \end{pmatrix}$$

identifies the circle S^1 with $SO(2)$. Secondly, there is a mapping $\mathbf{M}$ from the sphere S^3 to the rotation group $SO(3)$ which sends $\mathbf{u} = (u_1, \ldots, u_4)^T$ in S^3 to

$$\mathbf{M}(\mathbf{u}) = \begin{pmatrix} u_1^2 + u_2^2 - u_3^2 - u_4^2 & -2(u_1u_4 - u_2u_3) & 2(u_1u_3 + u_2u_4) \\ 2(u_1u_4 + u_2u_3) & u_1^2 + u_3^2 - u_2^2 - u_4^2 & -2(u_1u_2 - u_3u_4) \\ -2(u_1u_3 - u_2u_4) & 2(u_1u_2 + u_3u_4) & u_1^2 + u_4^2 - u_2^2 - u_3^2 \end{pmatrix}. \tag{13.2.2}$$

An alternative expression for $\mathbf{M}(\mathbf{u})$ is

$$\mathbf{M}(\mathbf{u}) = \mathbf{I}_3 + 2u_1\mathbf{A}(\mathbf{v}) + 2\mathbf{A}(\mathbf{v})^2,$$

where $\mathbf{v} = (u_2, u_3, u_4)^T$ and $\mathbf{A}(\mathbf{v})$ is defined by (13.2.7) below. Note that $\mathbf{M}(\mathbf{u})$ represents the rotation through angle $2\cos^{-1}(u_1)$ about $\mathbf{v}$. Since $\mathbf{M}(-\mathbf{u}) = \mathbf{M}(\mathbf{u})$, $\mathbf{M}$ defines a map of $\mathbb{R}P^3$ into $SO(3)$. It is straightforward to check that this map identifies $\mathbb{R}P^3$ with $SO(3)$. The transformation $\mathbf{M}$ can be described neatly in terms of quaternions (Prentice, 1986). This transformation has been exploited by Moran (1975) and Prentice (1987).

For handling distributions on the sphere a basic role is played by polar decomposition of vectors, e.g. (i) the decomposition (9.2.2) $\bar{\mathbf{x}} = \bar{R}\bar{\mathbf{x}}_0$ of the sample mean, and (ii) the polar decomposition $\kappa\boldsymbol{\mu}$ of the canonical parameter of a Fisher distribution. (Note that if κ and $\boldsymbol{\mu}$ are regarded as 1×1 and $p \times 1$ matrices, respectively, then it is appropriate to regard the vector $\kappa\boldsymbol{\mu}$ as the $p \times 1$ matrix $\kappa\boldsymbol{\mu}$.) The analogous tool for distributions on $V_r(\mathbb{R}^p)$ is polar decomposition of matrices. Any $p \times r$ matrix $\mathbf{X}$ can be decomposed as

$$\mathbf{X} = \mathbf{MK}, \tag{13.2.3}$$

where $\mathbf{M}$ is in $V_r(\mathbb{R}^p)$ and $\mathbf{K}$ is $r \times r$ symmetric positive semi-definite. The matrix $\mathbf{M}$ is called the *polar part* (or *orientation*) and $\mathbf{K}$ is the *elliptical part* (or *concentration*) of $\mathbf{X}$. If $\mathbf{X}$ has rank r then $\mathbf{K}$ is positive definite and this decomposition is unique. (Take $\mathbf{K} = (\mathbf{X}^T\mathbf{X})^{1/2}$, the positive definite symmetric square root, and $\mathbf{M} = \mathbf{XK}^{-1}$.) Further, as pointed out by Chikuse (1990a), there is an analogue for Stiefel manifolds of the tangent-normal decomposition (9.1.2) on spheres. In matrix terms, it says that for $r \leq q \leq p - r$, any $\mathbf{X}$ in $V_r(\mathbb{R}^p)$ can be decomposed as

$$\mathbf{X} = \begin{pmatrix} \mathbf{T} \\ \mathbf{U}(\mathbf{I}_r - \mathbf{T}^T\mathbf{T})^{1/2} \end{pmatrix}, \tag{13.2.4}$$

where $\mathbf{T}$ is a $q \times r$ matrix with $\mathbf{I}_r - \mathbf{T}^T\mathbf{T}$ positive semi-definite and $\mathbf{U}$ is an element of $V_r(\mathbb{R}^{p-q})$ (see also Chikuse, 1993c, Section 2).

The concept of wrappping distributions from a tangent space onto a sphere extends to Stiefel manifolds. We describe it just for rotation groups. The tangent space to $SO(p)$ at $\mathbf{X}$ is

$$T_{\mathbf{X}}SO(p) = \{\mathbf{XA} : \mathbf{A} \in so(p)\},$$

where $so(p)$ denotes the set of skew-symmetric $p \times p$ matrices. The analogue on $SO(p)$ of the wrapping map (9.1.3) is the *exponential map* from $T_{\mathbf{X}}SO(p)$ given by

$$\mathbf{XA} \mapsto \mathbf{X}\exp(\mathbf{A}), \tag{13.2.5}$$

where

$$\exp(\mathbf{A}) = \sum_{n=0}^{\infty} \frac{1}{n!}\mathbf{A}^n. \tag{13.2.6}$$

There is a useful identification of $so(3)$ with $\mathbb{R}^3$, in which $\mathbf{v} = (v_1, v_2, v_3)^T$ in $\mathbb{R}^3$ corresponds to the matrix

$$\mathbf{A}(\mathbf{v}) = \begin{pmatrix} 0 & -v_3 & v_2 \\ v_3 & 0 & -v_1 \\ -v_2 & v_1 & 0 \end{pmatrix} \tag{13.2.7}$$

in $so(3)$. Then (13.2.6) can be expressed in the simple form

$$\exp(\mathbf{A}(\mathbf{v})) = \mathbf{I}_3 + \frac{\sin\|\mathbf{v}\|}{\|\mathbf{v}\|}\mathbf{A}(\mathbf{v}) + \frac{1-\cos\|\mathbf{v}\|}{\|\mathbf{v}\|^2}\mathbf{A}(\mathbf{v})^2 \quad (13.2.8)$$

$$= (\cos\|\mathbf{v}\|)\mathbf{I}_3 + \frac{\sin\|\mathbf{v}\|}{\|\mathbf{v}\|}\mathbf{A}(\mathbf{v}) + \frac{1-\cos\|\mathbf{v}\|}{\|\mathbf{v}\|^2}\mathbf{v}\mathbf{v}^T. \quad (13.2.9)$$

The matrix $\exp(\mathbf{A}(\mathbf{v}))$ represents the rotation through angle $\|\mathbf{v}\|$ about $\mathbf{v}$.

13.2.2 Uniformity

Like the sphere, the Stiefel manifold $V_r(\mathbb{R}^p)$ has a *uniform distribution.* This is the unique distribution which is invariant both under rotations and reflections of $\mathbb{R}^p$ and under rotations and reflections of the r-dimensional subspace spanned by $\mathbf{X}$, i.e. $\mathbf{UXV}$ has the same distribution as $\mathbf{X}$ for all $\mathbf{U}$ in $O(p)$ and $\mathbf{V}$ in $O(r)$. Thus it has the greatest symmetry of all distributions on $V_r(\mathbb{R}^p)$. Because of this symmetry, a natural null hypothesis about a distribution on $V_r(\mathbb{R}^p)$ is that of uniformity.

As in the spherical case, the simplest test of uniformity is the Rayleigh test. Let $\mathbf{X}_1, \ldots, \mathbf{X}_n$ be observations on $V_r(\mathbb{R}^p)$ with sample mean

$$\bar{\mathbf{X}} = \frac{1}{n}\sum_{i=1}^{n}\mathbf{X}_i.$$

The intuitive idea of the Rayleigh test is that uniformity is rejected if the sample mean $\bar{\mathbf{X}}$ is far from the population mean $\mathbf{0}$. More precisely, uniformity is rejected for large values of

$$S = pn\,\mathrm{tr}(\bar{\mathbf{X}}^T\bar{\mathbf{X}}). \quad (13.2.10)$$

If $r = 1$ this reduces to the Rayleigh statistic (10.4.4) for testing uniformity on the sphere S^{p-1}. Under uniformity, the asymptotic large-sample distribution of S is

$$S \,\tilde{\sim}\, \chi^2_{rp}. \quad (13.2.11)$$

The error in the approximation (13.2.11) is of order $O(n^{-1/2})$. The modified Rayleigh statistic S^* on $V_r(\mathbb{R}^p)$ given by

$$S^* = S\left\{1 - \frac{1}{2n}\left(1 - \frac{S}{pr+2}\right)\right\} \quad (13.2.12)$$

(generalising (6.3.5) and (10.4.6)), has a χ^2_{rp} distribution with error of order $O(n^{-1})$ (Jupp, 1999). Even for moderate sample sizes, the upper tail of the null distribution of S^* is close to that of the χ^2_{rp} distribution. This modification is based on the fact that the Rayleigh test is the score test of the simple hypothesis $\mathbf{F} = \mathbf{0}$ in the matrix Fisher model (13.2.15).

The general machinery of Sobolev tests described in Section 10.8 can be used to provide a very large class of tests of uniformity on $V_r(\mathbb{R}^p)$ which are invariant under $\mathbf{X} \mapsto \mathbf{UXV}$ for $\mathbf{U}$ in $O(p)$ and $\mathbf{V}$ in $O(r)$. The Rayleigh test based on (13.2.10) is the simplest of these. Another simple Sobolev test on $V_r(\mathbb{R}^p)$ is the Bingham test considered in Section 13.3.2. The map

$$(\mathbf{x}_1, \ldots, \mathbf{x}_r) \mapsto \mathbf{x}_1$$

which takes an orthonormal frame to its first vector maps $V_r(\mathbb{R}^p)$ to S^{p-1}. The methods of Jupp & Spurr (1983) can be used to decompose each Sobolev test of uniformity on $V_r(\mathbb{R}^p)$ into the sum of a Sobolev test of uniformity on S^{p-1} and a test of $O(r-1)$-symmetry of $(\mathbf{x}_2, \ldots, \mathbf{x}_r)$ about $\mathbf{x}_1$.

Example 13.2
As explained in Example 13.1(iii), the orbit of a periodic comet provides a rotation matrix. A question of astronomical interest is whether or not these rotation matrices $\mathbf{X}$ are uniformly distributed on $SO(3)$. The data set of orbits of 240 comets considered by Jupp & Mardia (1979) can be summarised by the mean rotation matrix

$$\bar{\mathbf{X}} = \begin{pmatrix} 0.014 & -0.012 & 0.001 \\ 0.523 & -0.015 & 0.113 \\ 0.148 & -0.025 & -0.013 \end{pmatrix}. \tag{13.2.13}$$

Since the Rayleigh statistic (13.2.10) is $3n\mathrm{tr}(\bar{\mathbf{X}}^T\bar{\mathbf{X}}) = 223.1 > 21.666 = \chi^2_{9;0.01}$, the hypothesis of uniformity is rejected strongly. Note that the modified Rayleigh statistic (13.2.12) is $S^* = 232.0$, so the modification has negligible effect.

The Rayleigh statistic S on $SO(3)$ can be decomposed as

$$3n\mathrm{tr}(\bar{\mathbf{X}}^T\bar{\mathbf{X}}) = 3n\bar{\mathbf{x}}_1^T\bar{\mathbf{x}}_1 + 3n\mathrm{tr}(\bar{\mathbf{X}}_{-1}^T\bar{\mathbf{X}}_{-1}), \tag{13.2.14}$$

where $\bar{\mathbf{x}}_1$ and $\bar{\mathbf{X}}_{-1}$ denote respectively the first column and the last two columns of $\bar{\mathbf{X}}$. The first term on the right-hand side of (13.2.14) is the Rayleigh statistic (10.4.4) for testing uniformity of the perihelion directions $\mathbf{x}_1$. Since $3n\bar{\mathbf{x}}_1^T\bar{\mathbf{x}}_1 = 213.0 > 11.345 = \chi^2_{3;0.01}$, the hypothesis of uniformity of perihelion directions is rejected strongly. The second term on the right-hand side of (13.2.14) tests rotational symmetry of orbital planes about perihelion directions (more precisely that, for a given perihelion direction $\mathbf{x}_1$, the normals to orbits with perihelion direction $\mathbf{x}_1$ are distributed uniformly on the circle normal to $\mathbf{x}_1$). Under uniformity of $\mathbf{x}_1$, the asymptotic large-sample distribution of this term is χ^2_6. A more useful test of symmetry is based on c_s, defined by

$$c_s = 2n\mathrm{tr}\left(\bar{\mathbf{X}}_{-1}^T(\mathbf{I}_3 - \bar{\mathbf{T}}_1)^{-1}\bar{\mathbf{X}}_{-1}\right).$$

where $\bar{\mathbf{T}}_1$ denotes the scatter matrix of $\mathbf{x}_1$. Note that if $\mathbf{x}_1$ is uniformly distributed then $c_s \simeq 3n\mathrm{tr}(\bar{\mathbf{X}}_{-1}^T\bar{\mathbf{X}}_{-1})$ for large n. Under rotational symmetry of the orbital planes, the asymptotic large-sample distribution of c_s is χ_6^2. For this data set, $c_s = 3.866$. Since $\Pr(\chi_6^2 \geq 3.866) = 0.694$, the hypothesis of symmetry of orbital planes can certainly be accepted. Thus the non-uniformity of the orbits comes largely from the non-uniformity of the perihelion directions.

In the special case of $SO(3)$, testing for uniformity can be reduced to testing for uniformity on the projective space $\mathbb{R}P^3$. Since the mapping (13.2.2) takes the uniform distribution on S^3 to the uniform distribution on $SO(3)$, the Sobolev tests of uniformity on $\mathbb{R}P^3$ considered in Section 10.8 give rise to tests of uniformity on $SO(3)$. In particular, the Bingham test with statistic (10.7.1) on $\mathbb{R}P^3$ yields the Rayleigh test with statistic (13.2.10) on $SO(3)$. Similarly, Giné's statistic (10.7.5) on $\mathbb{R}P^3$ yields the statistic

$$G_n = \frac{n}{2} - \frac{3\pi}{16n} \sum_{i=1}^{n} \sum_{j=i+1}^{n} [\mathrm{tr}(\mathbf{I}_3 - \mathbf{X}_i^T \mathbf{X}_j)]^{1/2}.$$

This gives a test of uniformity on $SO(3)$ which is easy to compute and is consistent against all alternatives.

13.2.3 Matrix Fisher Distributions

Perhaps the simplest non-uniform distributions on $V_r(\mathbb{R}^p)$ are the matrix Fisher (or matrix Langevin) distributions introduced by Downs (1972). These form an exponential model with canonical statistic the matrix $\mathbf{X}$ itself. Thus the density functions with respect to the uniform distribution are

$$f(\mathbf{X}; \mathbf{F}) = \left\{ {}_0F_1 \left(\frac{p}{2}; \frac{1}{4}\mathbf{F}^T\mathbf{F} \right) \right\}^{-1} \exp\{\mathrm{tr}(\mathbf{F}^T\mathbf{X})\}. \qquad (13.2.15)$$

Here, $\mathbf{F}$ is an $p \times r$ parameter matrix and ${}_0F_1$ is the hypergeometric function of matrix argument defined by (A.28) of Appendix 1. We shall denote the distribution with density (13.2.15) by $L(\mathbf{F})$. The density (13.2.15) has a mode at $\mathbf{X} = \mathbf{M}$, where $\mathbf{M}$ is the polar part of $\mathbf{F}$. In the case $r = 1$, (13.2.15) is just the density (10.3.1) of a Fisher distribution on the sphere S^{p-1}. The matrix Fisher distribution $L(\mathbf{F})$ can be obtained by suitable conditioning of a multivariate normal distribution (cf. Section 9.3.2).

The mapping from $\mathbb{R}P^3$ to $SO(3)$ obtained from (13.2.2) can be regarded as taking axes $\pm\mathbf{x}$ to rotation matrices $\mathbf{M}(\mathbf{x})$. The random matrix $\mathbf{M}(\mathbf{x})$ has a Fisher matrix distribution on $SO(3)$ if and only if $\pm\mathbf{x}$ has a Bingham distribution on $\mathbb{R}P^3$, as shown by Prentice (1986). If $\pm\mathbf{x}$ has a Bingham distribution on $\mathbb{R}P^3$ then the density of $\mathbf{M}(\mathbf{x})$ depends only the distance from the mode if and only if $\pm\mathbf{x}$ has a Watson distribution on $\mathbb{R}P^3$, as proved by Schaeben (1993).

Now consider maximum likelihood estimation in the model with densities (13.2.15), based on observations $\mathbf{X}_1, \ldots, \mathbf{X}_n$. Since this model is an exponential model, the maximum likelihood estimator of the canonical parameter $\mathbf{F}$ is given implicitly by

$$\mathrm{E}_{\hat{\mathbf{F}}}[\mathbf{X}] = \bar{\mathbf{X}}.$$

To make use of this, let $\bar{\mathbf{R}}$ be the elliptical part of $\bar{\mathbf{X}}$, as in (13.2.3). Note that $\bar{\mathbf{R}}$ is a generalisation to $V_r(\mathbb{R}^p)$ of the mean resultant length $\bar{R}$ of observations on a sphere. Then

$$\bar{\mathbf{R}} = (\bar{\mathbf{X}}^T\bar{\mathbf{X}})^{1/2}$$

and equivariance arguments show that, in terms of the polar decomposition $\mathbf{F} = \mathbf{MK}$, we have

$$\hat{\mathbf{M}} = \bar{\mathbf{X}}\bar{\mathbf{R}}^{-1}, \tag{13.2.16}$$

i.e. the maximum likelihood estimate of the polar part of $\mathbf{F}$ is the polar part of $\bar{\mathbf{X}}$. We can then obtain $\hat{\mathbf{K}}$ from $\bar{\mathbf{R}}$ as follows (Khatri & Mardia, 1977). If

$$\bar{\mathbf{R}} = \mathbf{U}^T \mathrm{diag}(g_1, \ldots, g_r)\mathbf{U}, \quad \text{with } \mathbf{U} \in O(r) \text{ and } g_1 \geq \ldots \geq g_r,$$

then

$$\hat{\mathbf{K}} = \mathbf{U}^T \mathrm{diag}(\hat{\phi}_1, \ldots, \hat{\phi}_r)\mathbf{U}, \quad \text{with } \hat{\phi}_1 \geq \ldots \geq \hat{\phi}_r,$$

and

$$g_i = \frac{\partial}{\partial \phi_i} {}_0F_1\left(\frac{p}{2}; \frac{1}{4}\mathrm{diag}(\phi_1^2, \ldots, \phi_r^2)\right)\Bigg|_{(\phi_1,\ldots,\phi_r)=(\hat{\phi}_1,\ldots,\hat{\phi}_r)}. \tag{13.2.17}$$

Further, approximate solutions of (13.2.17) are given by

$$\hat{\phi}_i \simeq p g_i, \qquad i = 1, \ldots, r,$$

for $(\phi_1, \ldots, \phi_r)$ small, and (provided that $\mathbf{F}$ has rank r)

$$\left(p - r - \tfrac{1}{2}\right)\hat{\phi}_i^{-1} + \sum_{j=1}^{r}\left(\hat{\phi}_i + \hat{\phi}_j\right)^{-1} \simeq 2(1 - g_i), \qquad i = 1, \ldots, r,$$

for $(\phi_1, \ldots, \phi_r)$ large.

General theory for exponential models shows that the conditional distribution of $\hat{\mathbf{M}}$ given $\hat{\mathbf{K}}$ is

$$\hat{\mathbf{M}}|\hat{\mathbf{K}} \sim L(n\mathbf{MK}\bar{\mathbf{R}}),$$

generalising to $V_r(\mathbb{R}^p)$ the conditional distribution (9.5.10) of $\bar{\mathbf{x}}_0$ given $\bar{R}$ for a von Mises–Fisher distribution on a sphere.

The von Mises–Fisher distributions on $V_r(\mathbb{R}^p)$ are characterised among distributions with densities of the form

$$f(\mathbf{X}; \mathbf{A}) = g(\mathrm{tr}(\mathbf{X}^T \mathbf{A})), \qquad \mathbf{A} \in V_r(\mathbb{R}^p)$$

(with g a given lower semi-continuous function), by the property that the polar part of $\mathbf{X}$ is a maximum likelihood estimate of $\mathbf{A}$. This result is due to Purkayastha & Mukerjee (1992). For $r = 1$, it is the maximum likelihood characterisation of von Mises–Fisher distributions on S^{p-1} which was given in Section 9.3.2.

There are various high-concentration asymptotic results. The most basic of these is that for concentrated matrix Fisher distributions (i.e. with the eigenvalues of $\mathbf{K}$ tending to infinity) the approximation

$$\mathbf{K}^{1/2}(\mathbf{X} - \mathbf{M})^T \;\dot\sim\; N(\mathbf{0}, \mathbf{I}_r \otimes \mathbf{I}_\nu) \tag{13.2.18}$$

holds, where $\mathbf{I}_r \otimes \mathbf{I}_\nu$ denotes the identity matrix in the tangent space to $V_r(\mathbb{R}^p)$ at $\mathbf{M}$ and

$$\nu = p - (r+1)/2. \tag{13.2.19}$$

This yields the Wishart approximation

$$\mathbf{K}^{1/2}(\mathbf{X} - \mathbf{M})^T(\mathbf{X} - \mathbf{M})\mathbf{K}^{1/2} \;\dot\sim\; W(\nu, \mathbf{I}_r)\,. \tag{13.2.20}$$

See Downs (1972) and Khatri & Mardia (1977).

Let $\mathbf{X}_1, \ldots, \mathbf{X}_n$ be a sample from (13.2.15). It follows from (13.2.20) that, for $\mathbf{K} \to \infty$,

$$\begin{aligned}\sum_{i=1}^{n} \mathbf{K}^{1/2}(2\mathbf{I}_r - \mathbf{X}_i^T\mathbf{M} - \mathbf{M}^T\mathbf{X}_i)\mathbf{K}^{1/2} &= \sum_{i=1}^{n} \mathbf{K}^{1/2}(\mathbf{X}_i - \mathbf{M})^T(\mathbf{X}_i - \mathbf{M})\mathbf{K}^{1/2} \\ &\;\dot\sim\; W(n\nu, \mathbf{I}_r),\end{aligned} \tag{13.2.21}$$

generalising (4.8.23) and (9.6.13). Using the decomposition

$$\begin{aligned}&\sum_{i=1}^{n} \mathbf{K}^{1/2}(\mathbf{X}_i - \mathbf{M})^T(\mathbf{X}_i - \mathbf{M})\mathbf{K}^{1/2} \\ &= 2n\mathbf{K}^{1/2}(\mathbf{I}_r - \bar{\mathbf{R}})\mathbf{K}^{1/2} + n\mathbf{K}^{1/2}(2\bar{\mathbf{R}} - \bar{\mathbf{X}}^T\mathbf{M} - \mathbf{M}^T\,\bar{\mathbf{X}})\mathbf{K}^{1/2},\end{aligned} \tag{13.2.22}$$

which generalises (4.8.30) and (9.6.16), we see that

$$2n\mathbf{K}^{1/2}(\mathbf{I}_r - \bar{\mathbf{R}})\mathbf{K}^{1/2} \;\dot\sim\; W((n-1)\nu, \mathbf{I}_r), \tag{13.2.23}$$

$$n\mathbf{K}^{1/2}(2\bar{\mathbf{R}} - \bar{\mathbf{X}}^T\mathbf{M} - \mathbf{M}^T\,\bar{\mathbf{X}})\mathbf{K}^{1/2} \;\dot\sim\; W(\nu, \mathbf{I}_r), \tag{13.2.24}$$

and that these are asymptotically independent. Since the expectation of $W((n-1)\nu, \mathbf{I}_r)$ is $(n-1)\nu\mathbf{I}_r$, (13.2.23) leads to the approximation

$$\hat{\mathbf{K}} \simeq \frac{1}{2}\,\frac{n-1}{n}\left(p - \frac{r+1}{2}\right)(\mathbf{I}_r - \bar{\mathbf{R}})^{-1}$$

to the maximum likelihood estimator of $\mathbf{K}$, cf. (5.3.9) and (10.3.7). The approximation

$$\hat{\mathbf{F}} \simeq \frac{1}{2}\frac{n-1}{n}\left(p - \frac{r+1}{2}\right)(\mathbf{I}_r - \bar{\mathbf{X}})^{-1}$$

follows.

Taking the trace of (13.2.20) yields,

$$-2\text{tr}\,[\mathbf{F}^T(\mathbf{X}-\mathbf{M})] \stackrel{\cdot}{\sim} \chi^2_{r\nu}.$$

It follows from general results of Jørgensen (1987) on exponential dispersion models and was shown explicitly by Bingham, Chang & Richards (1992) that this statistic can be Bartlett-corrected, i.e. multiplying it by a constant to make its mean equal to νr ensures that all the cumulants differ from those of the limiting distribution by terms of order $O(\mathbf{K}^{-2})$. Various one- and two-sample tests) are given in Downs (1972) and Khatri & Mardia (1977).

Let $\mathbf{Z} = (np)^{1/2}\bar{\mathbf{X}}$. Then, under uniformity, $\mathbf{Z} \stackrel{\cdot}{\sim} N(\mathbf{0}, \mathbf{I}_p \otimes \mathbf{I}_r)$ for large n. Expansions to order $O(n^{-2})$ of the probability density functions of $\mathbf{Z}$, $\mathbf{Z}^T\mathbf{Z}$ and $\text{tr}(\mathbf{Z}^T\mathbf{Z})$ under uniformity and suitable local alternatives are given in Chikuse (1991b). (See also Mardia & Khatri, 1977.)

Various high-dimensional asymptotic properties are known. Let $\mathbf{X}$ in $V_r(\mathbb{R}^p)$ have a matrix Fisher distribution with parameter matrix $\mathbf{F}$ and let the singular value decomposition of $\mathbf{F}$ be $\mathbf{F} = \mathbf{\Gamma\Lambda\Delta}$, where $\mathbf{\Gamma}$ is in $V_r(\mathbb{R}^p)$, $\mathbf{\Lambda}$ is diagonal and $\mathbf{\Delta}$ is in $O(r)$. Put $\mathbf{Y} = p^{1/2}\mathbf{\Gamma}^T\mathbf{X}$. Then as $p \to \infty$ the distribution of $\mathbf{Y}$ tends to normality. See Watson (1983d) and Chikuse (1991a, 1993c). Asymptotic expansions to order $O(p^{-2})$ of the probability density functions of $\mathbf{Y}$, $\mathbf{YY}^T$ and $\text{tr}(\mathbf{YY}^T)$ in the cases $\mathbf{X} \sim L(\mathbf{F})$ and $\mathbf{X} \sim L(p^{1/2}\mathbf{F})$ are given by Chikuse (1993a).

The following generalisation is given by Chikuse (1993c). Let $\mathbf{P}$ be a $p \times p$ matrix representing a projection from $\mathbb{R}^p$ onto an r-dimensional subspace. Let $\mathbf{X}$ be distributed on $V_r(\mathbb{R}^p)$ with a probability density function which depends only on $\mathbf{PX}$. Then, as $p \to \infty$, the distribution of $p^{1/2}\mathbf{PX}$ tends to $N(\mathbf{0}, \mathbf{P} \otimes \mathbf{I}_r)$.

13.2.4 Other Distributions on Stiefel Manifolds

A natural extension of the matrix Fisher and Bingham distributions is the exponential model with canonical statistic $\mathbf{X} \otimes \mathbf{X}$ introduced by Prentice (1982). He proposed subfamilies with appropriate symmetry as models for X-shapes and T-shapes in p-space. Khatri & Mardia (1977) introduced the submodel with density function

$$f(\mathbf{X}; \mathbf{K}, \mathbf{V}, \boldsymbol{\mu}) = c(\mathbf{K}, \mathbf{V}, \boldsymbol{\mu}) \exp\{\text{tr}[\mathbf{K}(\mathbf{X}-\boldsymbol{\mu})\mathbf{V}(\mathbf{X}-\boldsymbol{\mu})^T]\},$$

where $\mathbf{K}$ and $\mathbf{V}$ are positive definite matrices. A series expansion in terms of matrix polynomials for the normalising constant $c(\mathbf{K}, \mathbf{V}, \boldsymbol{\mu})$ was given by de Waal (1979).

The complex analogue of the orthogonal group $O(p)$ is the unitary group $U(p)$ defined in terms of $p \times p$ complex matrices as

$$U(p) = \{\mathbf{X} : \mathbf{X}^*\mathbf{X} = \mathbf{I}_p\},$$

where $\mathbf{X}^*$ denotes the complex conjugate transpose of $\mathbf{X}$. The complex matrix Fisher distributions on $U(p)$ introduced by Bingham, Chang & Richards (1992) have probability density functions

$$f(\mathbf{X};\mathbf{F}) = \left\{ {}_0\tilde{F}_1\left(\frac{p}{2};\frac{1}{4}\mathbf{F}^*\mathbf{F}\right)\right\}^{-1} \exp\{\operatorname{Re}\operatorname{tr}(\mathbf{F}^T\mathbf{X})\}, \tag{13.2.25}$$

where the normalising constant is given by

$${}_0\tilde{F}_1\left(\frac{p}{2};\frac{1}{4}\mathbf{F}^*\mathbf{F}\right) = \frac{\det\mathbf{G}}{\prod_{1\le i<j\le p}(\lambda_i^2 - \lambda_j^2)},$$

with $\mathbf{G}$ being the matrix with (i,j)th entry $\lambda_i^{2p-2j}\,{}_0F_1(p+1-j;\lambda_j^2/4)$ and $\lambda_1, \lambda_2, \ldots, \lambda_k$ (with $\lambda_1 > \lambda_2 > \ldots > \lambda_k$) denoting the eigenvalues of $\mathbf{F}^*\mathbf{F}$.

Various random walks on the rotation group $SO(3)$ were considered by Roberts & Winch (1984). In particular, they considered the cases where (i) each rotation is through a fixed angle about an arbitrary axis, and (ii) the angle of rotation is arbitrary but the axis of rotation makes a given angle θ with a fixed axis. If $\theta = \pi/2$ in (ii) then the set of such rotations can be identified (via the image of the north pole) with S^2 and the distributions obtained on $SO(3)$ correspond to those obtained from random walks on the sphere.

13.3 SUBSPACES

13.3.1 *Grassmann Manifolds*

Example 13.1(ii) indicates that it is necessary to consider subspaces of $\mathbb{R}^p$. The set of r-dimensional subspaces of $\mathbb{R}^p$ is called the *Grassmann manifold* $G_r(\mathbb{R}^p)$. Since a subspace is specified by the orthogonal projection onto it, $G_r(\mathbb{R}^p)$ can be described in in terms of $p \times p$ matrices $\mathbf{Y}$ as

$$G_r(\mathbb{R}^p) = \{\mathbf{Y} : \mathbf{Y} = \mathbf{Y}^T = \mathbf{Y}^T\mathbf{Y}, \operatorname{rk}\mathbf{Y} = r\}.$$

In particular, because a one-dimensional subspace is an axis, we have $G_1(\mathbb{R}^p) = \mathbb{R}P^{p-1}$, $(p-1)$-dimensional real projective space.

Assigning to each subspace its orthogonal complement identifies $G_r(\mathbb{R}^p)$ with $G_{p-r}(\mathbb{R}^p)$. This identification is given in matrix terms by

$$\mathbf{Y} \mapsto \mathbf{I}_p - \mathbf{Y}.$$

By assigning to each frame the subspace which it spans, we obtain a map from $V_r(\mathbb{R}^p)$ to $G_r(\mathbb{R}^p)$, given in matrix terms by

$$\mathbf{X} \mapsto \mathbf{X}\mathbf{X}^T. \tag{13.3.1}$$

13.3.2 Uniformity

The Grassmann manifold $G_r(\mathbb{R}^p)$ has a distinguished distribution, the *uniform distribution*, which is the unique distribution invariant under rotation, i.e. $\mathbf{UYU}^T$ has the same distribution as $\mathbf{Y}$ for all $\mathbf{U}$ in $SO(p)$. The map (13.3.1) takes the uniform distribution on $V_r(\mathbb{R}^p)$ to the uniform distribution on $G_r(\mathbb{R}^p)$.

Again, the central role of the uniform distribution means that the basic problem on $G_r(\mathbb{R}^p)$ is that of testing for uniformity. The simplest test is the Bingham test introduced by Mardia & Khatri (1977). Let $\mathbf{Y}_1, \ldots, \mathbf{Y}_n$ be observations on $G_r(\mathbb{R}^p)$ with sample mean

$$\bar{\mathbf{Y}} = \frac{1}{n}\sum_{i=1}^{n} \mathbf{Y}_i.$$

The intuitive idea behind the Bingham test is that uniformity is rejected if the sample mean $\bar{\mathbf{Y}}$ is far from its expected value $(r/p)\mathbf{I}_p$. More precisely, uniformity is rejected for large values of

$$S = \frac{(p-1)p(p+2)}{2r(p-r)} n \left\{ \operatorname{tr}(\bar{\mathbf{Y}}^2) - \frac{r^2}{p} \right\}. \tag{13.3.2}$$

If $r = 1$ then this reduces to Bingham's statistic (10.7.1) for testing uniformity on the projective space $\mathbb{R}P^{p-1}$. Under uniformity, the asymptotic large-sample distribution of S is

$$S \mathrel{\dot\sim} \chi^2_{(p-1)(p+2)/2}. \tag{13.3.3}$$

with error of order $O(n^{-1/2})$.

Because the Bingham test is also the score test of the simple hypothesis $\mathbf{A} = \mathbf{0}$ in the matrix Bingham model (13.3.5), there is a cubic modification of S generalising (10.7.3). The modified Bingham statistic on $G_r(\mathbb{R}^p)$ is

$$S^* = S\left\{1 - \frac{1}{n}\left(B_0 + B_1 S + B_2 S^2\right)\right\}, \tag{13.3.4}$$

where

$$\begin{aligned}
B_0 &= \frac{p^2(p^2+p-2) + 2r(p-r)(p^2+4p-20)}{12r(p-r)(p-2)(p-1)(p+4)(p+2)}, \\
B_1 &= -\left[\frac{p^2(p^2+p-2) - r(p-r)(p^2-2p+16)}{3r(p-r)(p^2+p+2)(p-2)(p+4)}\right], \\
B_2 &= \frac{(p-2r)^2(p-1)(p+2)}{3r(p-r)(p-2)(p+4)(p^2+p+2)(p^2+p+6)}.
\end{aligned}$$

The modified Bingham statistic S^* has a $\chi^2_{(p-1)(p+2)/2}$ distribution with error of order $O(n^{-1})$ (Jupp, 1999).

13.3.3 Matrix Bingham Distributions

The matrix Bingham distributions on $G_r(\mathbb{R}^p)$ are generalisations of the Bingham distributions on $\mathbb{R}P^{p-1}$. They have density functions

$$f(\mathbf{Y};\mathbf{A}) = \left\{ {}_1F_1\left(\frac{r}{2};\frac{p}{2};\mathbf{A}\right)\right\}^{-1} \exp\{\mathrm{tr}(\mathbf{AY})\}, \quad \mathbf{Y} \in G_r(\mathbb{R}^p) \qquad (13.3.5)$$

with respect to the uniform distribution, where $\mathbf{A}$ is a symmetric $p \times p$ parameter matrix and ${}_1F_1$ is the hypergeometric function of matrix argument defined by (A.26) of Appendix 1. The matrices $\mathbf{A}$ and $\mathbf{A} + \alpha\mathbf{I}_p$ yield the same distribution for any real α (as for the Bingham distributions (9.4.3) on the projective space $\mathbb{R}P^{p-1}$). Thus we may assume without loss of generality that

$$\mathrm{tr}\,\mathbf{A} = 0. \qquad (13.3.6)$$

Let $\mathbf{A}$ have singular value decomposition $\mathbf{A} = \mathbf{U\Delta U}^T$ with $\mathbf{U}$ in $V_r(\mathbb{R}^p)$ and $\mathbf{\Delta}$ diagonal. Then (13.3.5) has a mode at $\mathbf{UU}^T$.

Use of the map (13.3.1) from $V_r(\mathbb{R}^p)$ to $G_r(\mathbb{R}^p)$ enables the Bingham test to be used on $V_r(\mathbb{R}^p)$.

High-concentration asymptotics for matrix Bingham distributions were considered by Chikuse (1993b). She showed that if $\mathbf{Y}$ has a matrix Bingham distribution with parameter $\mathbf{A} = \mathbf{U\Delta U}^T$ with $\mathbf{U}$ in $V_r(\mathbb{R}^p)$ and $\mathbf{\Delta}$ diagonal, then for large $\mathbf{A}$ (i.e. the absolute values of all non-zero eigenvalues of $\mathbf{A}$ tending to infinity),

$$V = 2\mathbf{\Delta}^{1/2}(\mathbf{I}_r - \mathbf{U}^T\mathbf{YU})\mathbf{\Delta}^{1/2} \;\dot{\sim}\; W(p-r, \mathbf{I}_r).$$

She gave an asymptotic expansion up to order $O(\mathbf{\Delta}^{-1})$ for the distribution of V.

Complex Bingham distributions on complex projective spaces $\mathbb{C}P^{p-1}$ are considered in Section 14.6.

13.3.4 Other Distributions on Grassmann Manifolds

The angular central Gaussian distributions on projective spaces have analogues on Grassmann manifolds. These are the matrix angular central Gaussian distributions on $G_r(\mathbb{R}^p)$ which were introduced by Chikuse (1990b). They are given by the probability density functions

$$f(\mathbf{X};\mathbf{\Sigma}) = |\mathbf{\Sigma}|^{-r/2}|\mathbf{X}^T\mathbf{\Sigma}^{-1}\mathbf{X}|^{-p/2}, \qquad \mathbf{X} \in V_r(\mathbb{R}^p), \qquad (13.3.7)$$

on $V_r(\mathbb{R}^p)$, where $\mathbf{\Sigma}$ is a $p \times p$ positive definite matrix. The matrix $\mathbf{\Sigma}$ is identifiable up to multiplication by a positive scalar. Note that, because (13.3.7) satisfies

$$f(\mathbf{X};\mathbf{\Sigma}) = f(\mathbf{XU};\mathbf{\Sigma}), \qquad \mathbf{U} \in O(r),$$

it may be considered as a probability density function on $G_r(\mathbb{R}^p)$. For $r = 1$, (13.3.7) gives the density (9.4.7) of an angular central Gaussian distribution on $\mathbb{R}P^{p-1}$. The action of the unimodular group $SL_p(\mathbb{R})$ on $\mathbb{R}^p$ given by (9.4.8) generalises to an action of $SL_p(\mathbb{R})$ on $V_r(\mathbb{R}^p)$: for $\mathbf{A}$ in $SL_p(\mathbb{R})$, the corresponding transformation $\varphi_{\mathbf{A}}$ of $V_r(\mathbb{R}^p)$ is given by

$$\varphi_{\mathbf{A}}(\mathbf{X}) = \mathbf{A}\mathbf{X}\,\mathbf{K}(\mathbf{A}\mathbf{X})^{-1}, \tag{13.3.8}$$

where $\mathbf{K}(\mathbf{AX})$ denotes the elliptical part of $\mathbf{AX}$, i.e. $\varphi_{\mathbf{A}}(\mathbf{X})$ is the polar part of $\mathbf{AX}$. A simple calculation shows that the matrix angular central Gaussian distributions on $V_r(\mathbb{R}^p)$ form a transformation model under this action of $SL_p(\mathbb{R})$.

The matrix angular central Gaussian distributions have the following characterisation: an (unrestricted) $p \times r$ random matrix $\mathbf{X}$ has a distribution which is invariant under $\mathbf{X} \mapsto \mathbf{XU}$ for $\mathbf{U}$ in $O(r)$ and with density of the form

$$f(\mathbf{X}; \mathbf{\Sigma}) = |\mathbf{\Sigma}|^{-r/2} g(\mathbf{X}^T \mathbf{\Sigma}^{-1} \mathbf{X})$$

for some function g if and only if the polar part of $\mathbf{X}$ has a matrix angular central Gaussian distribution (Chikuse, 1990b).

The maximum likelihood estimate $\hat{\mathbf{\Sigma}}$ of $\mathbf{\Sigma}$ in (13.3.7) based on observations $\mathbf{X}_1, \ldots, \mathbf{X}_n$ satisfies

$$\hat{\mathbf{\Sigma}} = \frac{p}{nr} \sum_{i=1}^{n} \mathbf{X}_i (\mathbf{X}_i^T \hat{\mathbf{\Sigma}}^{-1} \mathbf{X}_i)^{-1} \mathbf{X}_i^T. \tag{13.3.9}$$

Putting $r = 1$ in (13.3.9) leads to equation (10.3.46) for the maximum likelihood estimate of the parameter matrix of an angular central Gaussian distribution on S^{p-1}.

A general class of distributions on $G_r(\mathbb{R}^p)$ has been introduced by Chikuse & Watson (1995). These distributions have probability density functions of the form

$$f(\mathbf{Y}; \mathbf{F}) = c(\mathbf{F}) g(\mathbf{Y}^{1/2} \mathbf{F} \mathbf{Y}^{1/2}),$$

where g is a function on the space of symmetric $p \times p$ matrices satisfying

$$g(\mathbf{UFU}^T) = g(\mathbf{F}), \qquad \mathbf{U} \in O(p).$$

Chikuse & Watson (1995) proved the large-sample asymptotic normality of

$$n^{1/2} \left(\bar{\mathbf{Y}} - \frac{r}{p} \mathbf{I}_p \right)$$

and gave asymptotic expansions to order $O(n^{-1})$ for its density under local alternatives to uniformity of the form $\mathbf{F}_n = n^{-1} \mathbf{F}$.

13.4 OTHER SAMPLE SPACES

13.4.1 Hyperboloids

The definition

$$S^{p-1} = \{\mathbf{x} \in \mathbb{R}^p : \mathbf{x}.\mathbf{x} = 1\}$$

of S^{p-1} uses the Euclidean inner product given by

$$\mathbf{x}.\mathbf{y} = x_1 y_1 + x_2 y_2 + \ldots + x_p y_p.$$

If the Euclidean inner product on $\mathbb{R}^p$ is replaced by the indefinite symmetric bilinear form $*$ given by

$$\mathbf{x} * \mathbf{y} = x_1 y_1 - x_2 y_2 - \ldots - x_p y_p$$

for $\mathbf{x} = (x_1, \ldots, x_p)$ and $\mathbf{y} = (y_1, \ldots, y_p)$ then the analogue of S^{p-1} is the unit hyperboloid H^{p-1} in $\mathbb{R}^p$ defined by

$$H^{p-1} = \{\mathbf{x} \in \mathbb{R}^p : \mathbf{x} * \mathbf{x} = 1,\ x_1 > 0\}.$$

Since the function $(x_1, \ldots, x_p) \mapsto (x_1 x_2, \ldots, x_1 x_p)$ identifies H^{p-1} with $\mathbb{R}^{p-1}$ and takes the distinguished point $(1, 0, \ldots, 0)$ to the origin, distributions on H^{p-1} can be useful for modelling random $(p-1)$-dimensional vectors, especially where zero has a special role, e.g. (for $p = 3$) the wind speeds and directions considered by Jensen (1981).

Just as the orthogonal group $O(p)$ is defined by

$$O(p) = \{\mathbf{A} : \mathbf{A}\mathbf{x}.\mathbf{A}\mathbf{x} = \mathbf{x}.\mathbf{x}\},$$

so the ***pseudo-orthogonal group*** $O(1, p-1)$ is defined by

$$O(1, p-1) = \{\mathbf{A} : \mathbf{A}\mathbf{x} * \mathbf{A}\mathbf{x} = \mathbf{x} * \mathbf{x}\}.$$

Analogous to $SO(p)$ is the subgroup $SO^{\uparrow}(1, p-1)$ of $O(1, p-1)$ defined by

$$SO^{\uparrow}(1, p-1) = \{\mathbf{A} \in O(1, p-1) : |\mathbf{A}| = 1, A_{11} = 1\}$$

The action $\mathbf{x} \mapsto \mathbf{A}\mathbf{x}$ of $SO^{\uparrow}(1, p-1)$ on H^{p-1} is analogous to the action of $SO(p)$ on S^{p-1}. There is an $SO^{\uparrow}(1, p-1)$-invariant measure on H^{p-1} (unique up to a scalar multiple). As such a measure has infinite mass, there is no uniform probability distribution on H^{p-1} (in contrast to the situation on S^{p-1}).

There is a family of distributions on H^{p-1} analogous to the von Mises–Fisher family on S^{p-1}. The ***hyperboloid distributions*** $H_p(\boldsymbol{\mu}, \kappa)$ have probability density functions (with respect to an $SO^{\uparrow}(1, p-1)$-invariant measure) of the form

$$f(\mathbf{x}; \boldsymbol{\mu}, \kappa) = a(\kappa)^{-1} \exp\{-\kappa \boldsymbol{\mu} * \mathbf{x}\}, \tag{13.4.1}$$

where the parameters κ and $\boldsymbol{\mu}$ satisfy $\kappa > 0$ and $\boldsymbol{\mu} \in H^{p-1}$. This family is an exponential transformation model under the action of $SO^{\uparrow}(1, p-1)$. These distributions were introduced by Barndorff-Nielsen (1978b) and their properties (which are analogous to those of von Mises–Fisher distributions) were investigated by Jensen (1981). Three examples of these properties are given in the next paragraph.

Let $\mathbf{x}_1, \ldots, \mathbf{x}_n$ be observations on H^{p-1}. Their vector mean $\bar{\mathbf{x}}$ can be written in the form

$$\bar{\mathbf{x}} = \bar{R}\bar{\mathbf{x}}_H,$$

where $\bar{\mathbf{x}}_H \in H^{p-1}$ and $\bar{R} = \sqrt{\bar{\mathbf{x}} * \bar{\mathbf{x}}}$. This is the analogue of (9.2.2) on S^{p-1}. Note that the *hyperbolic mean resultant length* $\bar{R}$ satisfies $\bar{R} \geq 1$. The *hyperbolic resultant length* is $n\bar{R}$. The analogue for hyperboloid distributions on H^{p-1} of the high-concentration approximation (9.6.17) is the high-concentration approximation

$$2n\kappa(\bar{R} - 1) \overset{\cdot}{\sim} \chi^2_{(n-1)(p-1)}. \qquad (13.4.2)$$

Let R_1, R_2 and R be the hyperbolic resultant lengths of two independent random samples from the same hyperboloid distribution on H^{p-1}, and of the combined sample of size n, respectively. Then the distribution of $R_1 + R_2 | R$ does not depend on κ. This is analogous to the corresponding property of von Mises–Fisher distributions on S^{p-1} which forms the basis of the two-sample Watson–Williams test given in Section 10.5.1.

In the case $p = 3$, we may write

$$\begin{aligned} \mathbf{x} &= (\cosh u, \sinh u \cos w, \sinh u \sin w)^T, \\ \boldsymbol{\mu} &= (\cosh \chi, \sinh \chi \cos \theta, \sinh \chi \sin \theta)^T \end{aligned}$$

and the density (13.4.1) reduces to

$$f(u, w; \chi, \theta, \kappa) = \frac{\kappa \exp(\kappa)}{2\pi} \sinh u \exp\{-\kappa[\cosh \chi \cosh u - \sinh \chi \sinh u \cos(w - \theta)]\}. \qquad (13.4.3)$$

Several properties of the hyperboloid distributions on H^2 can be expressed more neatly than the analogous properties of the Fisher distributions on S^2. Here are three such properties. Firstly, the analogue for hyperboloid distributions on H^2 of the high-concentration approximation (10.3.7) for the maximum likelihood estimate $\hat{\kappa}$ of κ is the exact result

$$\hat{\kappa} = 1/(\bar{R} - 1). \qquad (13.4.4)$$

Secondly, the analogue for hyperboloid distributions on H^2 of the high-concentration approximation (9.6.17) is the exact result

$$n\kappa(\bar{R} - 1) \sim \Gamma(n - 1, 1). \qquad (13.4.5)$$

Thirdly, if R_1, R_2 and R are the hyperbolic resultant lengths of two independent random samples from the same hyperboloid distribution on H^2 then the observed significance level is

$$\left(\frac{R_1 + R_2 - n}{R - n}\right)^{n-2}. \tag{13.4.6}$$

In contrast, for the spherical case, no explicit expression for the observed significance level is known.

Example 13.3

Jensen (1981) used hyperboloid distributions to analyse three sets of data on wind speed and direction. Two of these data sets can be summarised by

$$n_1 = 8, \qquad \bar{R}_1 = 1.70, \qquad \sinh \hat{\chi}_1 = 0.36, \qquad \hat{\theta}_1 = 198°,$$
$$n_2 = 8, \qquad \bar{R}_2 = 1.72, \qquad \sinh \hat{\chi}_2 = 0.36, \qquad \hat{\theta}_2 = 196°.$$

The hyperbolic resultant length of the combined sample is $R = 22.71$. Jensen found that hyperboloid distributions $H_3(\boldsymbol{\mu}_1, \kappa_1)$ and $H_3(\boldsymbol{\mu}_2, \kappa_2)$ fitted these data sets reasonably well. From (13.4.4), $\hat{\kappa}_1 = 1.44$ and $\hat{\kappa}_2 = 1.39$. The hypothesis $\kappa_1 = \kappa_2$ can be tested using (13.4.5). The observed significance level is 0.83, so equality of κ_1 and κ_2 is accepted. Under this assumption of equal concentrations, the hypothesis $\boldsymbol{\mu}_1 = \boldsymbol{\mu}_2$ can be tested using the conditional distribution of R_1, R_2 given R. From (13.4.6), the observed significance level is 0.85, and so equality of $\boldsymbol{\mu}_1$ and $\boldsymbol{\mu}_2$ is accepted.

13.4.2 General Manifolds

Almost all the sample spaces used in directional statistics are *manifolds*, i.e. spaces which look locally like some Euclidean space. Thus an appropriate general setting for directional statistics is the study of statistics on general manifolds. Many manifolds of interest have some extra structure, e.g. a Riemannian metric or a group action. On a compact Riemannian manifold there is a uniform distribution and it is possible to define Sobolev tests of uniformity, generalising those considered in Section 10.8 (see Giné, 1975). As in the spherical case discussed in Section 10.8, the same machinery can be used to construct permutation tests of (i) equality of two distributions (Wellner, 1979), (ii) symmetry (Jupp & Spurr, 1983), (iii) independence (Jupp & Spurr, 1985). On a manifold with a group action one can consider composite transformation models. An account of their rich structure is given in Barndorff-Nielsen, Blæsild & Eriksen (1989).

Riemannian manifolds have exponential maps which extend the 'wrapping' map (9.1.3) and the exponential map (13.2.5) used on spheres and rotation groups. For each point m of a Riemannian manifold M, the exponential map maps $T_m M$ (the tangent space to M at m) to M. The corresponding wrapping approach leads to a definition (Oller, 1993) of bias of an estimator taking values in M. The same approach leads to general versions of the Cramér–Rao, Rao–Blackwell and Lehmann–Scheffé theorems (Oller & Corcuera, 1995). A more intrinsic version of the Cramér–Rao theorem for general manifolds M and general smooth loss functions (which generalises (5.2.5)) has been given by Hendriks (1991). The ingredients are:

(i) a family of probability density functions on M parameterised by Θ,
(ii) a smooth map $\phi : \Theta \to \Omega$,
(iii) an estimator $\mathbf{t} : M \to \Omega$ of $\phi(\theta)$,
(iv) a loss function $r : \Omega \times \Omega \to \mathbb{R}$.

The estimator $\mathbf{t}$ is said to be *unbiased* if

$$\mathrm{E}_\theta[r(\omega, \mathbf{t}(x))] \text{ has a minimum at } \omega = \phi(\theta). \tag{13.4.7}$$

Given local coordinates $(\theta^1, \ldots, \theta^d)$ on Θ and $(\omega^1, \ldots, \omega^r)$ on Ω, the matrices $\mathbf{C}$, $\mathbf{H}$ and $\mathbf{\Phi}$ are defined by

$$\begin{aligned} \mathbf{C}_{ab} &= \mathrm{E}\left[\frac{\partial r(\omega, \mathbf{t}(x))}{\partial \omega^a}\frac{\partial r(\omega, \mathbf{t}(x))}{\partial \omega^b}\right], \quad a, b = 1, \ldots, r, \\ \mathbf{H}_{ab} &= \mathrm{E}\left[\frac{\partial^2 r(\omega, \mathbf{t}(x))}{\partial \omega^a \partial \omega^b}\right], \quad a, b = 1, \ldots, r \\ \mathbf{\Phi}_{aj} &= \frac{\partial \phi^a}{\partial \theta^j}, \quad a = 1, \ldots, r, \quad j = 1, \ldots, d. \end{aligned}$$

Hendriks's result is that if $\mathbf{t}$ is unbiased then

$$\mathbf{C} \geq \mathbf{H\Phi I}^{-1}\mathbf{\Phi}^T\mathbf{H} \tag{13.4.8}$$

(meaning that $\mathbf{C} - \mathbf{H\Phi I}^{-1}\mathbf{\Phi}^T\mathbf{H}$ is positive semi-definite), where $\mathbf{I}$ denotes the Fisher information matrix. For the case in which $\Omega = S^{p-1}$ and $r(\omega, \omega') = \|\omega - \omega'\|^2$, the unbiasedness condition (13.4.7) becomes

$$\phi(\theta) = \frac{\mathrm{E}_\theta[\mathbf{t}(x)]}{\|\mathrm{E}_\theta[\mathbf{t}(x)]\|} \tag{13.4.9}$$

and (if $\mathbf{H}$ is invertible) the generalised Cramér–Rao inequality (13.4.8) can be expressed as

$$\|\mathrm{E}_\theta[\mathbf{t}(x)]\|^{-2} \left(\mathbf{I}_p - \phi(\theta)\phi(\theta)^T\right) \mathrm{var}(\mathbf{t}(x)) \left(\mathbf{I}_p - \phi(\theta)\phi(\theta)^T\right) \geq \mathbf{\Phi I}^{-1}\mathbf{\Phi}^T, \tag{13.4.10}$$

where $\mathbf{I}_p$ denotes the $p \times p$ identity matrix. Note that the left-hand side of (13.4.10) is proportional to the variance matrix of the tangential part of $\mathbf{t}(x)$. When $p = 2$, (13.4.9) and (13.4.10) reduce to (5.2.2) and (5.2.5), respectively.

For probability distributions on general manifolds embedded in a Euclidean space (or a space of matrices), there is a generalisation (Hendriks & Landsman,

1996a) of the concept of mean direction. The *mean location* of a random point X on a submanifold M of $\mathbb{R}^k$ is the point $\boldsymbol{\mu}$ of M which minimises

$$\int_M ||\boldsymbol{\mu} - \mathbf{x}||^2 dP(\mathbf{x}),$$

where P denotes the corresponding probability distribution on M. Under weak conditions the sample mean location is defined almost surely. A large-sample test of given mean location (generalising the score test based on (10.4.18)) was given in Hendriks & Landsman (1996b). A corresponding two-sample test was given in Hendriks & Landsman (1998).

Random walks on general Riemannian manifolds were studied by Roberts & Ursell (1960).

Not only do certain manifolds provide interesting sample spaces but so do certain quotients of these manifolds by appropriate group actions. Examples of particular practical importance are the shape spaces, which will be considered in Chapter 14.

14

Shape Analysis

14.1 INTRODUCTION

This chapter is concerned with the shapes of sets of points in Euclidean space. There have been various developments in shape analysis in the last decade. For a full treatment of the subject, see Dryden & Mardia (1998). We describe here some relationships of shape analysis with directional statistics. In particular, certain distributions of directional statistics have emerged in shape analysis. These include the complex Bingham distributions considered in Section 14.6. This chapter first gives some background to shape analysis in the plane and then goes on to directional distributions and their applications to shape analysis. Note that the idea of projecting concentrated data onto a tangent space is used in both contexts. Generalisations to higher dimensions are considered in Section 14.10.

By shape we mean 'the description which remains after *location*, *scale* and *rotational* effects are filtered out' (Kendall, 1977). For example, consider the six triangles A–F given in Fig. 14.1. The triangles A, B and C are of the same shape. Triangle D does not have the same shape as A, B, C, although it does after relabelling of the vertices. Triangle E is not of the same shape as C unless we also allow reflection as shape preserving. Triangle F is not of the same shape as any of A, B, C. We will also deal with invariance under labelling and reflection, and in that case the five triangles A–E will be regarded as having the same shape. We shall concentrate mainly on the shapes of sets of points in two-dimensional Euclidean space. Such *configurations* can be described by sets of k points (known in various applied contexts as *landmarks*, *vertices*, *anchor points*, *control points* etc.). For convenience, *complex notation* will be used for these points $z_1^0, \ldots, z_k^0$ in $\mathbb{C}$ and we call them *raw landmarks*. We put $\mathbf{z}^0 = (z_1^0, \ldots, z_k^0)^T$ in $\mathbb{C}^k$; in the case of triangles, $k = 3$.

For any complex number c, the configuration $(z_1^0 + c, \ldots, z_k^0 + c)$ has the same shape as $(z_1^0, \ldots, z_k^0)$. Hence to describe shape it is sufficient to look at any set of $k - 1$ independent contrasts of $(z_1^0, \ldots, z_k^0)$; (a *contrast* is a function of the form $\sum_{i=1}^k a_i z_i^0$ with $\sum_{i=1}^k a_i = 0$). These contrasts can be constructed in various ways. For example, we could use the k dependent

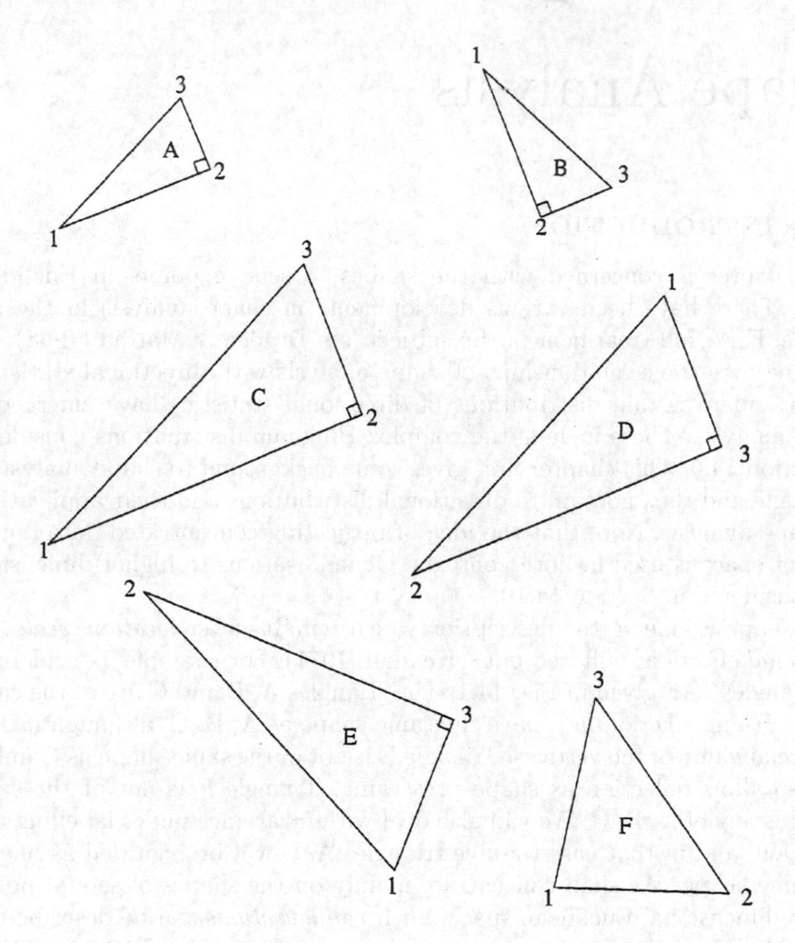

Figure 14.1 Six labelled triangles: A, B and C have the same shape; D has the same shape after relabelling; E can be reflected to C; F has a different shape from the rest (Dryden & Mardia, 1998. © John Wiley & Sons, Ltd, reproduced with permission).

contrasts $u_i = z_i^0 - \bar{z}^0, i = 1, \ldots, k$, where $\bar{z}^0 = \sum_{i=1}^k z_i^0/k$. We call these *centred landmarks.* Alternatively, a convenient set of $k-1$ orthogonal contrasts can be obtained as follows. Recall that the standard $k \times k$ Helmert matrix has its first row with elements equal to $k^{-\frac{1}{2}}$ and the remaining rows are orthogonal to this first row. Let $\mathbf{H}$ denote the $(k-1) \times k$ matrix which consists of the last $k-1$ rows of the Helmert matrix. We call this matrix $\mathbf{H}$ the *Helmert sub-matrix.* Explicitly, the jth row of $\mathbf{H}$ is given by

$$(\underbrace{h_j, \ldots, h_j}_{j}, -jh_j, 0, \ldots, 0), \; h_j = \{j(j+1)\}^{-\frac{1}{2}}, \;\; j = 1, \ldots, k-1, \quad (14.1.1)$$

where h_j is repeated j times. Thus a convenient set of $k-1$ orthogonal contrasts consists of the rows of $\mathbf{H}$. Let $\mathbf{z}_H = \mathbf{H}\mathbf{z}^0$ be these new landmarks obtained from $\mathbf{z}^0$. We call $\mathbf{z}_H$ the vector of *Helmertised landmarks.*

Further, $\mathbf{z}_H$ and $\alpha\mathbf{z}_H$ are equivalent in shape for any real number $\alpha > 0$. To filter out this scaling effect we can simply scale $\mathbf{z}_H$, i.e. we take $\mathbf{z}_H / \parallel \mathbf{z}_H \parallel = \mathbf{z}$, say, where $\mathbf{z} = (z_1, \ldots, z_{k-1})^T$. We have

$$|z_1|^2 + \ldots + |z_{k-1}|^2 = 1, \quad (14.1.2)$$

so that $\mathbf{z}$ lies on the unit sphere $S^{2k-3} = \mathbb{C}S^{k-2}$ in $\mathbb{C}^{k-1}$. This formulation indicates that we require statistics of observations on a complex sphere. Such a $\mathbf{z}$ is called a *preshape* and S^{2k-3} is called the *preshape space* of configurations of k points in the plane. Specifically, for the triangle case ($k = 3$), define

$$\begin{aligned} z_0 &= (z_1^0 + z_2^0 + z_3^0)/\sqrt{3}, \\ z_1 &= (z_2^0 - z_1^0)/(\sqrt{2}||\mathbf{H}\mathbf{z}^0||), \\ z_2 &= (2z_3^0 - z_1^0 - z_2^0)/(\sqrt{6}||\mathbf{H}\mathbf{z}^0||), \end{aligned} \quad (14.1.3)$$

where

$$||\mathbf{H}\mathbf{z}^0||^2 = \tfrac{1}{2}|z_2^0 - z_1^0|^2 + \tfrac{1}{6}|2z_3^0 - z_1^0 - z_2^0|^2.$$

Then (z_1, z_2) is the preshape determined by the raw landmarks z_1^0, z_2^0 and z_3^0.

Note that the preshape $\mathbf{z}$ lies on a complex sphere in $(k-1)$ complex dimensions, although we started with a configuration which can be regarded as a point in k complex dimensions. We will see below that sometimes it is convenient to preserve the k dimensions. This can be done by using

$$\mathbf{z}_C = \mathbf{u}/||\mathbf{u}|| = \mathbf{H}^T\mathbf{z}, \quad (14.1.4)$$

where $\mathbf{u} = (u_1, \ldots, u_k)$ with $u_i = z_i^0 - \bar{z}^0$, $i = 1, \ldots, k$. Then $\mathbf{z}_C$ lies on the unit sphere in $S^{2k-1} = \mathbb{C}S^{k-1}$ in $\mathbb{C}^k$. We call $\mathbf{z}_C$ the *centred preshape.* Of course, there is some redundancy in this representation. Mostly we will be using the preshape $\mathbf{z}$, which will be called the *Helmertised* preshape if there is any confusion with the centred preshape.

We have yet to remove the effect of rotation on $\mathbf{z}$. The shape of $\mathbf{z}$ is the same as that of $e^{i\psi}\mathbf{z}$ for any real ψ. The set of equivalence classes $[\mathbf{z}]$ of points $\mathbf{z}$ on the complex sphere where $\mathbf{z}$ is regarded as equivalent to $e^{i\psi}\mathbf{z}$ for all ψ is known as *complex projective space* $\mathbb{C}P^{k-2}$; $[\mathbf{z}]$ is the *shape* of $\mathbf{z}$ and $\mathbb{C}P^{k-2}$ is known as the *shape space*. We also denote the shape of $\mathbf{z}^0$ by $[\mathbf{z}^0]$. Note that two configurations $\mathbf{z}_1^0$ and $\mathbf{z}_2^0$ have the same shape if

$$\mathbf{z}_1^0 = \alpha \mathbf{1}_k + \delta e^{i\psi}\mathbf{z}_2^0 \tag{14.1.5}$$

for some α in $\mathbb{C}, \psi$ in $S^1, \delta > 0$ and $\mathbf{1}_k = (1, 1, \ldots, 1)^T$. Here α represents a change of location, δ a change of scale, and ψ a rotation matching $\mathbf{z}_2^0$ to $\mathbf{z}_1^0$. We could rewrite this expression as

$$\mathbf{z}_1^0 = \alpha \mathbf{1}_k + \beta \mathbf{z}_2^0, \quad \alpha \in \mathbb{C}, \beta \in \mathbb{C}\backslash\{0\}. \tag{14.1.6}$$

One possible coordinate system on $\mathbb{C}P^{k-2}$ is given by

$$w_j^K = z_j / z_1, \quad j = 1, \ldots, k-1, \tag{14.1.7}$$

when $z_1 \neq 0$, and by adding a point at infinity when $z_1 = 0$. These are called *Kendall's shape coordinates.* Alternatively, we have Bookstein's (1986; 1991) complex shape coordinates

$$w_j^B = (z_j^0 - z_1^0)/(z_2^0 - z_1^0), \qquad j = 1, 2, 3, \ldots, k, \tag{14.1.8}$$

provided that $z_1^0 \neq z_2^0$. Since $w_1^B = 0$, $w_2^B = 1$, this is effectively a $(k-2)$-dimensional complex representation. Landmarks 1 and 2 are called the *base* with respect to which the objects are *registered* i.e. the objects are standardised with respect to the common base line. Note that this is not a symmetrical method of *registration.*

We now describe some practical examples.

Example 14.1
For 21 microscopic fossils (*Globorotalia truncatulinoides*, see Fig. 14.2) three landmarks are given in Table 14.1 below. Do the triangles given by these landmarks tend to be almost equilateral?

Example 14.2
Given a set of a mouse vertebrae (2nd thoracic bone: T2) with six mathematical landmarks (see Fig. 14.3), what is the sample mean shape? How can its variation be measured? If we take additional landmarks (*pseudo-landmarks*), (say) seven equally spaced points between each pair of adjacent landmarks, what is the effect on the analysis?

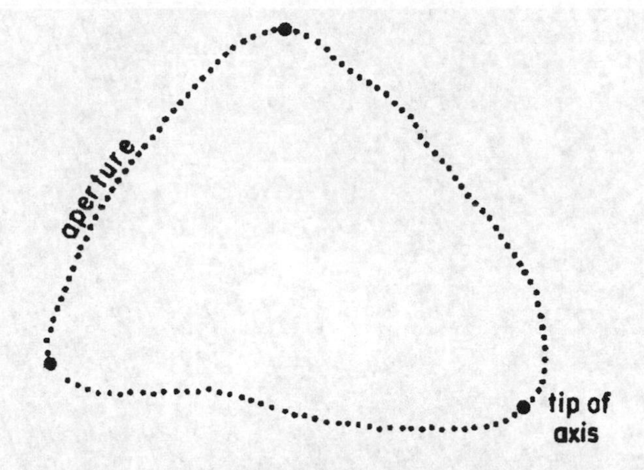

Figure 14.2 The outline of a microfossil with three landmarks (from Bookstein, 1986; reproduced by permission of the Institute of Mathematical Studies).

Example 14.3
Given unlabelled triangles arising in central place theory (see Example 14.5), how should one assess the evidence for the tendency of the triangles to be equilateral?

Other practical examples include: object recognition in images through shape representation, testing whether or not some prehistoric sites tend to lie on ley-lines, testing whether or not a given set of quasars lies on a great circle on the celestial sphere.

We end this section with a few comments on why some of the most straightforward coordinate systems are not as useful. For the triangle case two internal angles are the most obvious choice of coordinates that are invariant under the similarity transformations. However, it soon becomes apparent that using angles to describe shape can be problematic. For cases such as that of very flat triangles (three points in a straight line) there are many different arrangements of three points. For larger numbers of points ($k > 3$) one could sub-divide the configuration into triangles and so $2k - 4$ angles would be needed. Also, probability distributions of the angles themselves are not easy to work with (see Mardia, Edwards & Puri, 1977). If the angles of the triangles are x_1, x_2 and x_3, then the use of $\log(x_1/x_3)$ and $\log(x_2/x_3)$ (where $x_1 + x_2 + x_3 = 180°$) has some potential for analysing triangle shape (Aitchison, 1986; Pukkila & Rao, 1988); we can follow the standard procedure for compositional data analysis and this approach can be adapted to higher dimensions.

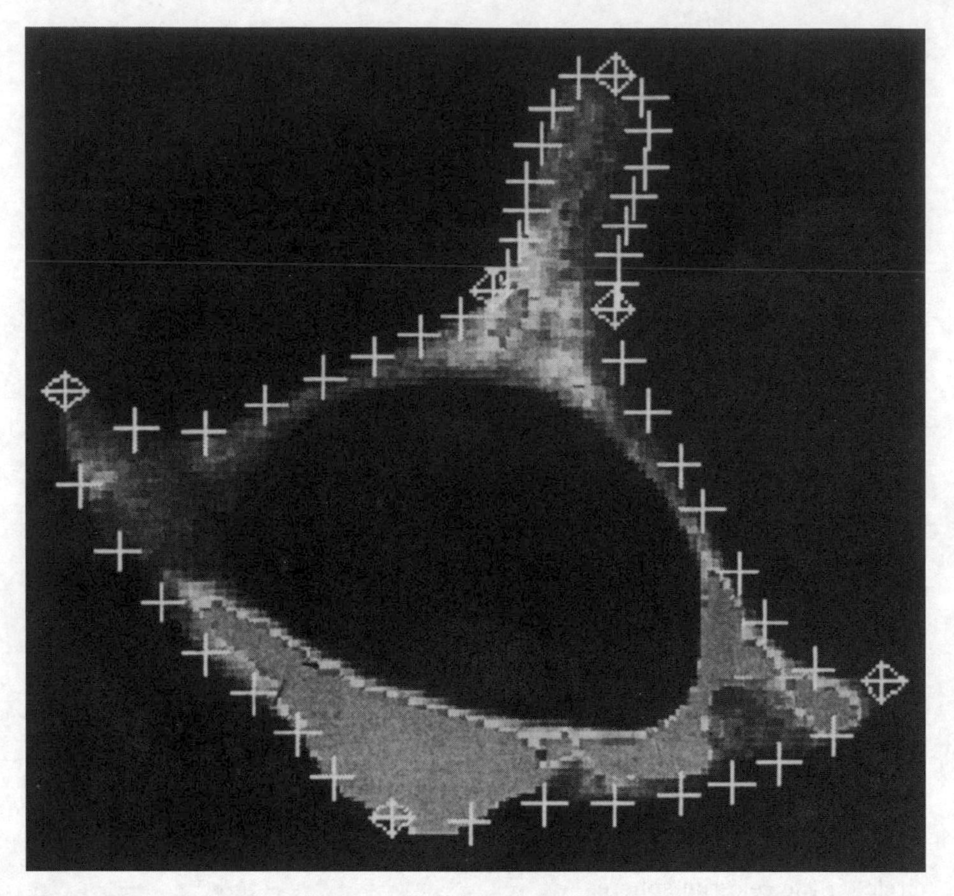

Figure 14.3 Grey-level image of a T2 mouse vertebra with six mathematical landmarks (diamond round a +) and 42 pseudo-landmarks (+), equally-spaced between adjacent pairs of landmarks (Dryden & Mardia, 1998. © John Wiley & Sons, Ltd. Reproduced with permission).

14.2 MEAN SHAPE AND VARIATION

Suppose that we are given n centred configurations $\mathbf{u}_j$ in $\mathbb{C}^k$, $j = 1, \ldots, n$. We give in this section an intuitively reasonable definition of mean shape.

First consider only two configurations $\mathbf{u}_1$ and $\mathbf{u}_2$. In order to compare these two configurations in shape, we need to estimate a measure of distance between their shapes. A suitable procedure is to match $\mathbf{u}_2$ to $\mathbf{u}_1$ by using the similarity transformations, and the differences between the fitted $\mathbf{u}_2$ and observed $\mathbf{u}_1$ indicate the magnitude of the difference in shape between $\mathbf{u}_1$ and $\mathbf{u}_2$. From (14.1.6), we can express the matching by the complex regression equation

$$\mathbf{u}_1 = \alpha \mathbf{1}_k + \beta \mathbf{u}_2 + \varepsilon,$$

where $\alpha,\ \beta \in \mathbb{C}$ with $\beta \neq 0$ and ε is a complex error vector.

We can obtain the least-squares estimates $\hat{\alpha}$ and $\hat{\beta}$ of α and β by minimising

$$\| \mathbf{u}_1 - \alpha \mathbf{1}_k - \beta \mathbf{u}_2 \|^2 = (\mathbf{u}_1 - \alpha \mathbf{1}_k - \beta \mathbf{u}_2)^* (\mathbf{u}_1 - \alpha \mathbf{1}_k - \beta \mathbf{u}_2), \tag{14.2.1}$$

where $\mathbf{u}^*$ denotes the transpose of the complex conjugate of $\mathbf{u}$. The optimal representation of $\mathbf{u}_1$ in terms of $\mathbf{u}_2$ is given by $\mathbf{u}_1 = \hat{\alpha}\mathbf{1} + \hat{\beta}\mathbf{u}_2$ and we say that $\mathbf{u}_2$ is *registered* with respect to $\mathbf{u}_1$. Indeed, if $\mathbf{u}_1, \mathbf{u}_2$ are centred landmarks, it can be shown by standard differentiation that

$$\hat{\alpha} = 0, \ \hat{\beta} = \mathbf{u}_2^* \mathbf{u}_1 / \mathbf{u}_2^* \mathbf{u}_2. \tag{14.2.2}$$

Furthermore, the residual sum of squares is given by

$$d(\mathbf{u}_1, \mathbf{u}_2)^2 = \| \mathbf{u}_1 \|^2 - \| \mathbf{u}_2 \|^{-2} |\mathbf{u}_1^* \mathbf{u}_2|^2. \tag{14.2.3}$$

Since $d(\mathbf{u}_1, \mathbf{u}_2)$ is not symmetric in $\mathbf{u}_1$ and $\mathbf{u}_2$ we replace the problem of minimising (14.2.1) by the problem of minimising

$$\left\| \frac{\mathbf{u}_1}{\| \mathbf{u}_1 \|} - \alpha \mathbf{1}_k - \beta \frac{\mathbf{u}_2}{\| \mathbf{u}_2 \|} \right\|^2 \tag{14.2.4}$$

with respect to α and β. This leads to

$$d_F(\mathbf{u}_1, \mathbf{u}_2)^2 = 1 - \{|\mathbf{u}_1^* \mathbf{u}_2|\}^2 / \{\| \mathbf{u}_1 \| \| \mathbf{u}_2 \|\}^2 = 1 - |\mathbf{z}_{1C}^* \mathbf{z}_{2C}|^2, \tag{14.2.5}$$

where $\mathbf{z}_{1C}$ and $\mathbf{z}_{2C}$ are the centred preshapes corresponding to $\mathbf{u}_1$ and $\mathbf{u}_2$, respectively. The quantity $d_F(\mathbf{u}_1, \mathbf{u}_2)$ is termed the *full Procrustes distance* between the shapes $[\mathbf{u}_1]$ and $[\mathbf{u}_2]$. For a proof that d_F represents a distance see, for example, Kent (1992). The distance is qualified by the word 'full' since we have minimised with respect to both scale ($|\beta|$) and rotation ($\arg \beta$). Alternatively, we could have set up the problem of minimising

$$\| \mathbf{H}^T \mathbf{z}_1 - \alpha \mathbf{1}_k - \beta \mathbf{H}^T \mathbf{z}_2 \|^2 = \| \mathbf{z}_1 - \beta \mathbf{z}_2 \|^2,$$

where $\mathbf{z}_1$ and $\mathbf{z}_2$ are preshapes corresponding to $\mathbf{u}_1$ and $\mathbf{u}_2$ respectively, so that

$$d_F(\mathbf{u}_1, \mathbf{u}_2)^2 = 1 - |\mathbf{z}_1^* \mathbf{z}_2|^2. \tag{14.2.6}$$

We will denote this distance also by $d_F(\mathbf{z}_1, \mathbf{z}_2)$. If $\rho(\mathbf{z}_1, \mathbf{z}_2)$ denotes the *shortest* great circle distance between $\mathbf{z}_1$ and $\mathbf{z}_2$ on the preshape sphere in $\mathbb{C}^{k-2}$, then (see, for example, Dryden & Mardia, 1998, pp. 72–73)

$$d_F(\mathbf{z}_1, \mathbf{z}_2) = \sin \rho(\mathbf{z}_1, \mathbf{z}_2). \tag{14.2.7}$$

Kendall (1984) has shown that $\rho(\mathbf{z}_1, \mathbf{z}_2)$ is the Riemanian distance on $\mathbb{C}P^{k-2}$ between $\mathbf{z}_1$ and $\mathbf{z}_2$ (given by half the Fubini–Study metric).

Bearing this type of registration in mind, we define the mean shape $[\hat{\boldsymbol{\mu}}]$ of the centred configurations $[\mathbf{u}_1], \ldots, [\mathbf{u}_n]$ as the shape of the element $\boldsymbol{\mu}$ of $\mathbb{C}S^{k-2}$ which minimises the objective function

$$\min_{\alpha_j, \beta_j} \sum_{j=1}^{n} \| \boldsymbol{\mu} - \alpha_j \mathbf{1}_k - \beta_j \mathbf{u}_j \|^2, \tag{14.2.8}$$

where $||\boldsymbol{\mu}|| = 1$ and α_j and β_j are included because the shape of $\mathbf{u}_j$ is defined only up to location, scale and rotation. Without any loss of generality, we centre $\boldsymbol{\mu}$ so that $\boldsymbol{\mu}^* \mathbf{1} = 0$. Thus we aim to minimise

$$\sum_{j=1}^{n} (\boldsymbol{\mu} - \alpha_j \mathbf{1}_k - \beta_j \mathbf{u}_j)^* (\boldsymbol{\mu} - \alpha_j \mathbf{1}_k - \beta_j \mathbf{u}_j) \tag{14.2.9}$$

with respect to $\boldsymbol{\mu}, \alpha_j$ and β_j. Assuming $\boldsymbol{\mu}$ to be known and using $\mathbf{1}^T \mathbf{u}_j = 0$, the least-squares estimates for α_j and β_j are

$$\hat{\alpha}_j = 0, \quad \hat{\beta}_j^* = \boldsymbol{\mu}^* \mathbf{u}_j / (\mathbf{u}_j^* \mathbf{u}_j), \quad j = 1, \ldots, n. \tag{14.2.10}$$

Note that from (14.2.10)

$$\boldsymbol{\mu}^* - \hat{\alpha}_j^* \mathbf{1}_k - \hat{\beta}_j^* \mathbf{u}_j^* = \boldsymbol{\mu}^* \mathbf{H}_j, \tag{14.2.11}$$

where $\mathbf{H}_j = \mathbf{I} - (\mathbf{u}_j^* \mathbf{u}_j)^{-1} \mathbf{u}_j \mathbf{u}_j^*$. Further, from (14.2.11) we have $\boldsymbol{\mu}^* \mathbf{H}_j \mathbf{H}_j \boldsymbol{\mu} = \boldsymbol{\mu}^* \mathbf{H}_j \boldsymbol{\mu}$, so that the minimum of (14.2.9) with respect to $\alpha_1, \ldots, \alpha_n$ and $\beta_1, \ldots, \beta_n$ becomes

$$n \boldsymbol{\mu}^* \boldsymbol{\mu} - \boldsymbol{\mu}^* \mathbf{S}_u \boldsymbol{\mu}, \tag{14.2.12}$$

where

$$\mathbf{S}_u = \sum_{j=1}^{n} (\mathbf{u}_j \mathbf{u}_j^*) / (\mathbf{u}_j^* \mathbf{u}_j). \tag{14.2.13}$$

Thus, under the constraint $\boldsymbol{\mu}^* \mathbf{1} = 0$, $||\boldsymbol{\mu}^*|| = 1, ||\mathbf{u}_j|| = 1$, $j = 1, \ldots, n$ the value of $\hat{\boldsymbol{\mu}}$ which minimises (14.2.8) is the dominant eigenvector of $\mathbf{S}_u$ up to rotation. This $[\hat{\boldsymbol{\mu}}]$ is called the *Procrustes mean shape.* We have followed a general Procrustes strategy for matching configurations (see, for example, Mardia, Kent & Bibby, 1979, pp. 417–419). This result is due to Kent (1992) and further details are given in Kent (1994). A full treatment of shape analysis based on Procrustes methods has been developed by Goodall (1991).

Using (14.1.4), we can rewrite (14.2.13) in terms of the preshapes $\mathbf{z}_j$ as

$$\mathbf{S}_u = \mathbf{H}^T \mathbf{S} \mathbf{H}, \tag{14.2.14}$$

with $\mathbf{S} = \sum_{j=1}^{n} \mathbf{z}_j \mathbf{z}_j^*$, i.e. $\mathbf{S}$ is the matrix of the complex sum of squares and products. Let $\hat{\boldsymbol{\nu}}$ be the dominant eigenvector of $\mathbf{S}$. Then $\hat{\boldsymbol{\mu}} = \mathbf{H}^T \hat{\boldsymbol{\nu}}$ up to a rotation. Similar arguments apply to the other eigenvectors of $\mathbf{S}_u$ and $\mathbf{S}$. For shape analysis, we choose to work on $\mathbf{S}$.

Let $l_1, \ldots, l_{k-1}$ be the eigenvalues of $\mathbf{S}$. Note these are all real and non-negative, and

$$l_1 + \ldots + l_{k-1} = n. \tag{14.2.15}$$

Let $\mathbf{g}_1, \ldots, \mathbf{g}_{k-1}$ be the eigenvectors of $\mathbf{S}$ corresponding to $l_1, \ldots, l_{k-1}$, respectively. If $l_1 \simeq \ldots \simeq l_{k-1}$ then the shapes are 'diffuse', i.e. $[\mathbf{z}_1], \ldots, [\mathbf{z}_n]$ are highly dispersed on $\mathbb{C}P^{k-2}$. Let $0 < l_1 < \ldots < l_{k-1}$; then $\mathbf{g}_{k-1} = \hat{\boldsymbol{\nu}}$, where $[\boldsymbol{\nu}]$ is the mean shape. Since $||\mathbf{g}_{k-1}|| = 1, \mathbf{g}_{k-1}$ is defined only up to multiplication by a unit complex number, i.e. a rotation of the plane. For visualisation purposes we can select a suitable rotation. If l_{k-1} is very large compared with $l_1, \ldots, l_{k-2}$, i.e. l_{k-1} is nearly n, then the data set is highly concentrated.

Example 14.4

Let us consider the data consisting of six landmarks on the T2 vertebrae from 23 mice; the full data set is given in Dryden & Mardia (1998, pp. 313–314, small-group case). The eigenvalues of $\mathbf{S}$ are

$$l_1 = 0.004, \quad l_2 = 0.005, \quad l_3 = 0.012, \quad l_4 = 0.072, \quad l_5 = 22.905.$$

Thus there is an extremely high concentration around the mean shape, as 22.905 is very close to 23. The mean shape is given by the eigenvector corresponding to l_5, which is

$$\begin{aligned} \mathbf{g}_5 &= (0.041 + 0.246i, 0.404 + 0.072i, \\ &\quad -0.495 - 0.716i, -0.056 + 0.037i, 0.089)^T. \end{aligned}$$

Note that $\mathbf{1}^T(\mathbf{H}^T\mathbf{g}_5) = 0$, so that the configuration is centred (in addition to being of unit size, as $\mathbf{g}_5^*\mathbf{H}\mathbf{H}^T\mathbf{g}_5 = \mathbf{g}_5^*\mathbf{g}_5 = 1$). The eigenvector is defined only up to a rotation, so we can plot $e^{i\alpha}\mathbf{H}^T\mathbf{g}_5$ for any suitable α. Here α is selected by rotating the mean shape so that the line joining the two landmarks furthest apart is horizontal. This mean shape $e^{i\alpha}\mathbf{H}^T\mathbf{g}_5$ is given by

$$\begin{aligned} &(-0.507 - 0.143i, 0.506 - 0.143i, 0.085 + 0.154i, \\ &0.012 + 0.424i, -0.070 + 0.160i, -0.026 - 0.451i)^T \end{aligned}$$

and its six components are plotted in Fig. 14.4, together with four sample shapes which are farthest away from this mean in terms of the Procrustes distance, the largest distance being 0.13. We call this mean shape the *sub-mean shape*. If we use the mean shape using the 42 pseudo-landmarks as well (shown in Fig. 14.3), the new mean shape is, of course, closer to the continuous bone outline than the sub-mean shape, as shown in Fig.14.5. Figure 14.5 also shows the mean shape of the six landmarks used before; the Procrustes distance of these points to the previous mean is 0.006, which is very small.

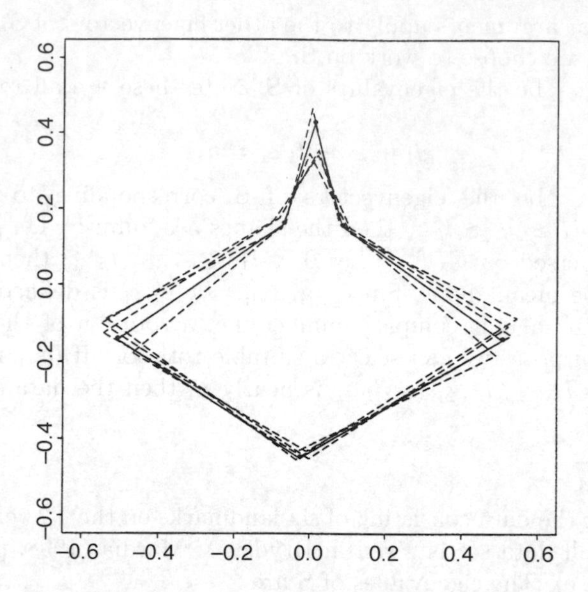

Figure 14.4 Procrustes mean shape (solid line) for the mouse T2 vertebrae (six landmarks) with the four furthest sample shapes superimposed (dotted line).

When data are concentrated, we can use an appropriate tangent space, as in directional statistics. Recall from Chapter 9 that if $\boldsymbol{\mu}$ ($\boldsymbol{\mu} \in S^{p-1}$) is the mean direction of $\mathbf{x}$, $||\mathbf{x}|| = 1, \boldsymbol{\mu}^T\mathbf{x} = \cos\theta$ then as in the tangent-normal decomposition (9.1.2), we have

$$\mathbf{x} = (\cos\theta)\boldsymbol{\mu} + \boldsymbol{\xi}, \quad \boldsymbol{\xi} = (\mathbf{I}_k - \boldsymbol{\mu}\boldsymbol{\mu}^T)\mathbf{x}, \tag{14.2.16}$$

so that $\boldsymbol{\xi}^T\boldsymbol{\mu} = 0$ and the vector $\boldsymbol{\xi}$ is the projected value of $\mathbf{x}$. Note that in the case ($p = 2$) of the circle with $\boldsymbol{\mu}^T = (\cos\mu, \sin\mu)$, $\mathbf{x}^T = (\cos\theta, \sin\theta)^T$, we have $\boldsymbol{\xi}^T = (-\sin\mu, \cos\mu)^T \sin(\theta - \mu)$. That is, for $\mu = 0$, we can work on θ, since $\sin\theta \simeq \theta$ for $\theta \simeq 0$. So, for concentrated data we can use standard statistical tools on $\theta_1 - \mu, \ldots, \theta_n - \mu$.

In shape analysis on the plane, let $\mathbf{z} = (z_1, \ldots, z_{k-1})^T$ be preshape landmarks and let γ be a mean shape. Then we can select a tangent projection such that the configuration is rotated to be as close as possible to γ before projection (Kent, 1994), i.e. we use the tangent shape coordinates $\mathbf{v}$ defined by

$$\mathbf{v} = (\mathbf{I}_{k-1} - \gamma\gamma^*)\mathbf{z}e^{i\psi}, \tag{14.2.17}$$

where ψ minimises $(\boldsymbol{\gamma} - e^{i\psi}\mathbf{z})^*(\boldsymbol{\gamma} - e^{i\psi}\mathbf{z})$, i.e. $\psi = -\arg(\boldsymbol{\gamma}^*\mathbf{z})$.

It is common in multivariate analysis to use principal components to obtain a parsimonious summary of the data, and we could apply the technique to

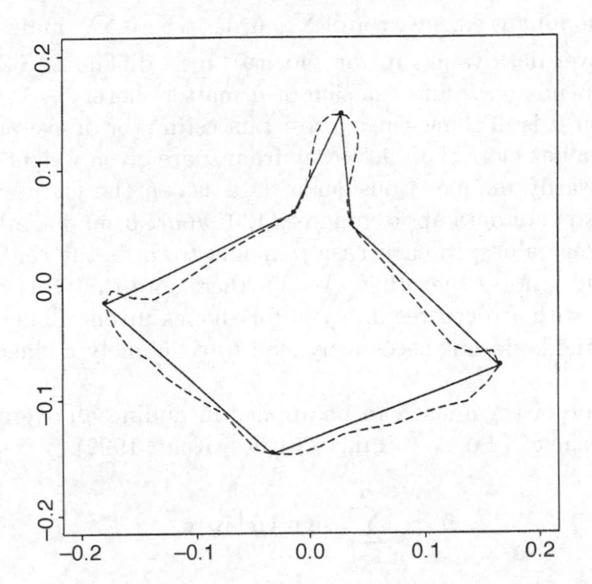

Figure 14.5 Procrustes mean for the mouse T2 vertebrae (i) 48 landmarks (dotted line), and (ii) 6 landmarks (solid line).

landmark data using the principal components of tangent shape coordinates (see also Example 14.8). It can be shown that an isotropic distribution for landmarks about a mean gives rise to an isotropic distribution in the tangent space to the mean (Mardia, 1995), which is a desirable property since the covariance structure in the tangent space with respect to the orthonormal basis is preserved. However this property breaks down if we use Bookstein's coordinates (see Bookstein, 1991; Kent & Mardia, 1994).

14.2.1 Directional Averages

The Procrustes procedure used above can be applied to directions (Kent, 1992). That is, consider an equivalence class of vectors in $\mathbb{R}^p$ in which a non-zero vector $\mathbf{x}$ is identified with the ray $\{r\mathbf{x} : r > 0\}$. Given data $\mathbf{u}_1, \ldots, \mathbf{u}_n$ on S^{p-1}, we can define the average $\hat{\boldsymbol{\mu}}$ in S^{p-1} to minimise

$$\sum_{i=1}^{n} \| r_i \mathbf{u}_i - \boldsymbol{\mu} \|^2 = \sum_{i=1}^{n} (r_i^2 + 1 - 2r_i \boldsymbol{\mu}^T \mathbf{u}_i)$$

over r_i and $\boldsymbol{\mu}$ in S^{p-1}. Minimising first over $r_i = (\boldsymbol{\mu}^T \mathbf{u}_i)_+$ (where $\alpha_+ = \alpha$ if $\alpha > 0$ and $\alpha_+ = 0$ if $\alpha < 0$), the objective function becomes

$$\sum_{i=1}^{n} [1 - (\boldsymbol{\mu}^T \mathbf{u}_i)_+^2] \,. \tag{14.2.18}$$

Here $\hat{\boldsymbol{\mu}}$ is the dominant eigenvector of $\sum_+ \mathbf{u}_i \mathbf{u}_i^T$, where $\sum_+$ indicates that the sum is taken over data values $\mathbf{u}_i$ for which $\hat{\boldsymbol{\mu}}^T \mathbf{u}_i > 0$. The selection of terms in this sum depends on $\hat{\boldsymbol{\mu}}$ and the sign of $\hat{\boldsymbol{\mu}}$ matters here.

The direction $\hat{\boldsymbol{\mu}}$ is in some sense a resistant estimator of average direction, because data values more than 90° away from $\hat{\boldsymbol{\mu}}$ are given weight 0. However, $\hat{\boldsymbol{\mu}}$ is not necessarily unique. Consider a data set on the circle consisting of three tight clusters of data approximately 120° apart from one another. Then $\hat{\boldsymbol{\mu}}$ can take three values; in each case $\hat{\boldsymbol{\mu}}$ points towards the centre of one of the clusters and ignores the other two. Further, note that this construction of $\hat{\boldsymbol{\mu}}$ coincides with Procrustes analysis for shapes in one dimension. Thus, Procrustes methods do not necessarily lead to a uniquely defined average in this case.

A similar type of argument can be applied in finding an average for axial data. The average of $(\pm\mathbf{u}_1, \ldots, \pm\mathbf{u}_n)$ satisfies (Kent, 1992)

$$\hat{\boldsymbol{\mu}} \propto \sum_{i=1}^{n} \text{sign } (\hat{\boldsymbol{\mu}}^T \mathbf{u}_i)\mathbf{u}_i \tag{14.2.19}$$

and is obtained by minimising

$$\sum_{i=1}^{n} \| r_i \sigma_i \mathbf{u}_i - \boldsymbol{\mu} \|^2$$

over $r_i > 0$, $\sigma_i = \pm 1$ and $\boldsymbol{\mu}$ in S^{p-1}. Thus $\hat{\boldsymbol{\mu}}$ is proportional to the resultant vector of the $\mathbf{u}_i$ after the signs of the $\mathbf{u}_i$ have been switched so that they all lie in the same hemisphere. Again, $\hat{\boldsymbol{\mu}}$ is not unique.

14.2.2 Form Average

Note that we can define the Procrustes size-and-shape (form) average (e.g. Kent, 1992). Two configurations $\mathbf{u}_1$ and $\mathbf{u}_2$ have the same *form* if $\mathbf{u}_1 = \alpha \mathbf{1}_k + e^{i\psi}\mathbf{u}_2$ for some α in $\mathbb{C}$ and ψ in S^1. Following the method described earlier for the Procrustes shape average, we find that the Procrustes form average is

$$\hat{\boldsymbol{\nu}} = \arg\min_{\boldsymbol{\nu}} \sum_{j=1}^{n} \{\boldsymbol{\nu}^* \boldsymbol{\nu} - 2|\boldsymbol{\nu}^* \mathbf{v}_j| + \mathbf{v}_j^* \mathbf{v}_j\}, \tag{14.2.20}$$

where $\mathbf{v}_j = \mathbf{H}\mathbf{u}_j$, $\boldsymbol{\nu} = \mathbf{H}\boldsymbol{\mu}$, and $\hat{\boldsymbol{\nu}}$ has to be found by a numerical procedure.

14.3 SHAPE COORDINATE SYSTEMS AND UNIFORM DISTRIBUTIONS

We now discuss uniform distributions on shape spaces. Let $|z_1|^2 + \ldots + |z_{k-1}|^2 = 1$. Since $\mathbb{C}S^{k-2} = S^{2k-3}$, we can use spherical polar coordinates for $(\text{Re}(z_1),\ \text{Im}(z_1),\ \ldots,\ \text{Re}(z_{k-1}),\ \text{Im}(z_{k-1}))$ on S^{2k-3}. However, various alternatives are available.

14.3.1 The General Case

Kent's Polar Coordinates

Kent (1994) has proposed some non-standard polar coordinates on the preshape sphere. Given a point $(z_1, \ldots, z_{k-1})^T$ on $\mathbb{C}S^{k-2}$, we transform to $(s_1, \ldots, s_{k-2}, \theta_1, \ldots, \theta_{k-1})$ by

$$\text{Re}(z_j) = s_j^{\frac{1}{2}} \cos\theta_j, \quad \text{Im}(z_j) = s_j^{\frac{1}{2}} \sin\theta_j,$$
$$s_j \geq 0,\ 0 \leq \theta_j < 2\pi,\ j = 1, \ldots, k-1, \tag{14.3.1}$$

where $s_{k-1} = 1 - \sum_{j=1}^{k-2} s_j$. The coordinates $\mathbf{s} = (s_1, \ldots, s_{k-2})^T$ are on the $(k-2)$-dimensional unit simplex,

$$\Delta_{k-2} = \{\mathbf{s} \in \mathbb{R}^{k-2} : s_1 + \ldots + s_{k-2} \leq 1, s_j \geq 0,\ j = 1, \ldots, k-2\}.$$

Noting that the volume of the unit simplex Δ_{k-2} in $\mathbb{R}^{k-2}$ is $1/(k-2)!$, it can be shown that the uniform distribution on $\mathbb{C}S^{k-2} = S^{2k-3}$ has probability density function

$$f(\mathbf{s}, \theta_1, \ldots, \theta_{k-1}) = \frac{(k-2)!}{(2\pi)^{k-1}},$$
$$\mathbf{s} \in \Delta_{k-2}, \theta_j \in S^1, j = 1, \ldots, k-1. \tag{14.3.2}$$

Thus $\theta_1, \ldots, \theta_{k-1}$ are independently and identically distributed uniform variates on S^1, independent of $\mathbf{s}$. To obtain the uniform distribution on $\mathbb{C}P^{k-2}$, we proceed as follows. Define

$$\phi_j = \theta_j - \theta_{k-1},\quad j = 1, \ldots, k-2. \tag{14.3.3}$$

We can represent any element of $\mathbb{C}P^{k-2}$ (i.e. a shape) by $(s_1, \phi_1), \ldots, (s_{k-2}, \phi_{k-2})$. The uniform distribution on $\mathbb{C}P^{k-2}$ has density

$$f_1(s_1, \phi_1, \ldots, s_{k-2}, \phi_{k-2}) = \int_0^{2\pi} f(\mathbf{s}, \theta_1, \ldots, \theta_{k-1}) d\theta_{k-1}, \tag{14.3.4}$$

where f is given by (14.3.2). Note that $\phi_1, \ldots, \phi_{k-2}$ and θ_{k-1} are identically and independently distributed as uniform variables on S^1. Hence from (14.3.4) the uniform distribution on $\mathbb{C}P^{k-2}$ is given by

$$d\gamma = \frac{(k-2)!}{(2\pi)^{k-2}} ds_1 \ldots ds_{k-2} d\phi_1 \ldots d\phi_{k-2}, \tag{14.3.5}$$

where $\mathbf{s}$ is in Δ_{k-2} and $\phi_1, \ldots, \phi_{k-1}$ are in S^1. Note that the probability density function (14.3.5) is constant with respect to $ds_1 \ldots ds_{k-2}$ $d\phi_1 \ldots d\phi_{k-2}$. The measure dj in (14.3.5) is the *uniform shape measure.* In the triangle case ($k = 3$), the uniform shape measure is simply

$$(2\pi)^{-1} ds_1 d\phi_1, \quad 0 \leq s_1 \leq 1, \; 0 \leq \phi_1 < 2\pi. \tag{14.3.6}$$

Kendall's Coordinates

Recall the Kendall coordinate system defined in (14.1.7):

$$z_j / z_{k-1} = r_j e^{i\phi_j}, \tag{14.3.7}$$

where $r_j > 0$ and $0 \leq \phi_j < 2\pi$, $j = 1, \ldots, k-2$. On substituting $z_j = s_j^{\frac{1}{2}} e^{i\theta_j}$ in (14.3.7) and using (14.1.2), we have

$$r_j^2 = s_j \Big/ \left(1 - \sum_{i=1}^{k-1} s_i\right), \quad j = 1, \ldots, k-2. \tag{14.3.8}$$

Summing (14.3.8) over all values of j, we find that

$$s_j = r_j^2 / (1 + A), \quad A = \sum_{j=1}^{k-2} r_j^2, \tag{14.3.9}$$

and

$$\frac{\partial s_j}{\partial r_i} = \begin{cases} 2r_j\{(1+A) - r_j^2\}(1+A)^{-2} & \text{if } \; i = j, \\ -2r_i r_j^2 (1+A)^{-2} & \text{if } \; i \neq j. \end{cases} \tag{14.3.10}$$

Using (14.3.10) and the result that for a $(k-2) \times (k-2)$ matrix $\mathbf{B}$ with $(\mathbf{B})_{ij} = 1 + b_i, i = j, (\mathbf{B})_{ij} = 1$ for $i \neq j$, we have

$$|\mathbf{B}| = \left(\prod_{i=1}^{k-2} b_i\right) \left(1 + \sum_{i=1}^{k-2} b_i^{-1}\right),$$

it is found that the Jacobian $J = |\partial s_i / dr_j|$ of the transformation is given by

$$J = 2^{k-2} \left(\prod_{i=1}^{k-2} r_i\right) \Big/ \; (1+A)^{k-1}.$$

Consequently, from (14.3.5), the uniform distribution on $\mathbb{C}P^{k-2}$, in terms of the shape variables defined by (14.3.7), is given by

$$\frac{(k-2)!}{\pi^{k-2}} \left(1 + \sum_{j=1}^{k-2} r_j^2\right)^{1-k} \prod_{j=1}^{k-2} (r_j dr_j d\phi_j). \tag{14.3.11}$$

Thus, in the Kendall coordinate system $r_j e^{i\phi_j} = x_j + iy_j,\ \ j = 1, \ldots, k-2$ (see also Mardia & Dryden, 1989b), the uniform measure on the shape space $\mathbb{C}P^{k-2}$ given by (14.3.11) reduces to

$$\frac{(k-2)!}{\pi^{k-2}} \left\{ 1 + \sum_{j=1}^{k-2} (x_j^2 + y_j^2) \right\}^{1-k} \prod_{j=1}^{k-2} (dx_j dy_j), \quad -\infty < x_j, y_j < \infty. \tag{14.3.12}$$

14.3.2 The Triangle Case

Since $\mathbb{C}P^1$ is equivalent to S^2, the shape of a triangle can be studied through directional statistics for the sphere. We give here a specific description (Mardia, 1999). For $k = 3$, the transformation (14.3.1) can be written as

$$z_1 = s^{\frac{1}{2}} e^{i(\phi+\psi)}, \quad z_2 = (1-s)^{\frac{1}{2}} e^{i\psi}, \quad 0 \le s \le 1, 0 \le \phi < 2\pi, \tag{14.3.13}$$

on letting $\theta_2 = \psi, \phi_1 = \phi, s_1 = s$. The preshape probability element of (s, ϕ, ψ) from (14.3.2) is given by

$$(2\pi)^{-2} ds\ d\phi\ d\psi, \tag{14.3.14}$$

so that s, ϕ and ψ are all independently distributed, where s is uniform on $[0,1]$ and ϕ and ψ are each uniform on S^1. Let $s^{\frac{1}{2}} = \cos\theta, 0 \le \theta < \pi/2$, so that from (14.3.13) we have

$$z_1 = \cos\theta e^{i(\phi+\psi)}, \quad z_2 = \sin\theta e^{i\psi} \tag{14.3.15}$$

and the preshape probability element from (14.3.14) of (θ, ϕ, ψ) is given by

$$(2\pi)^{-2} \sin 2\theta\ d\theta\ d\phi\ d\psi, \quad 0 \le \theta < \pi/2, \quad 0 \le \phi, \quad \psi < 2\pi. \tag{14.3.16}$$

Also, note that (θ, ϕ) are the Kendall shape coordinates in $\mathbb{C}P^1$ and the shape probability element of (θ, ϕ) is simply

$$(2\pi)^{-1} \sin 2\theta\ d\theta\ d\phi, \quad 0 \le \theta < \pi/2, \quad 0 \le \phi < 2\pi. \tag{14.3.17}$$

On 'doubling' the angle θ, we see that the point with spherical polar coordinates $(2\theta, \phi)$ is uniformly distributed on S^2 and the Cartesian coordinates for $(2\theta, \phi)$ are given by

$$x = \sin 2\theta \cos\phi, \quad y = \sin 2\theta \sin\phi, \quad z = \cos 2\theta, \quad x^2 + y^2 + z^2 = 1\ . \tag{14.3.18}$$

From (14.3.15) it can be shown that we can rewrite (14.3.18) as (Kendall, 1984; Kent 1994)

$$x = 2\mathrm{Re}(\bar{z}_1 z_2), \quad y = 2\ \mathrm{Im}(\bar{z}_1 z_2), \quad z = |z_1|^2 - |z_2|^2; \tag{14.3.19}$$

we will call this transformation the *spherical isometric transformation.*

Set $\Phi(\mathbf{z}) = (x, y, z)^T$. Then it can be seen that $\Phi(\mathbf{z}) = \Phi(e^{i\psi}\mathbf{z})$, so that $\Phi(\mathbf{z})$ depends only on the equivalence class of $\mathbf{z}$ in $\mathbb{C}P^1$. Also note that the Kendall coordinate z_1/z_2 in this representation from (14.3.15) is $e^{i\phi} \cot\theta$ which, as expected, depends only on (θ, ϕ).

14.3.3 The Shape Sphere

We now give some insight into the spherical shape space. Consider the Bookstein coordinates (u, v) with base $z_1^0 = -\frac{1}{2}, z_2^0 = \frac{1}{2}$ and $z_3^0 = u + iv$. The components z_1 and z_2 given by (14.1.3) of the corresponding preshape are found to be

$$z_1 = (1 + r^2)^{-\frac{1}{2}}, \quad z_2 = 2(\sqrt{3})^{-1}(1 + r^2)^{-\frac{1}{2}}(u + iv),$$

where $r^2 = 4(u^2 + v^2)/3$. Substituting these values of z_1 and z_2 into (14.3.19), and using (14.3.18) with 2θ replaced by θ, we find that

$$\begin{aligned} \sin\theta\cos\phi &= 4(\sqrt{3})^{-1}(1 + r^2)^{-1}u, \\ \sin\theta\sin\phi &= 4(\sqrt{3})^{-1}(1 + r^2)^{-1}v, \\ \cos\theta &= (1 - r^2)/(1 + r^2), \end{aligned} \tag{14.3.20}$$

where $0 < \theta \leq \pi, 0 < \phi \leq 2\pi$. We can now use (14.3.20) to map the shapes of triangles from (u, v) in $\mathbb{R}^2$ to the spherical shape coordinates (θ, ϕ). For example, for the equilateral triangle (with labelling anticlockwise) we have $(u, v) = (0, \sqrt{3}/2)$ which leads to the 'north pole' $(\theta = 0)$ on S^2 from (14.3.20) and for the reflected equilateral triangle (with labelling clockwise), we have $(u, v) = (0, -\sqrt{3}/2)$ which leads to the 'south pole' $(\theta = \pi)$ on S^2. Continuing in this way, Fig. 14.6 shows various triangle shapes located on spherical shape space. Thus the equilateral triangle and its reflection are at the north pole $(\theta = 0)$ and the south pole $(\theta = \pi)$, respectively. The flat triangles (three collinear points) lie around the equator $(\theta = \pi/2)$. The isosceles triangles lie on the meridians $\phi = 0, \pi/3, 2\pi/3, \pi, 4\pi/3, 5\pi/3$. The right-angled triangles lie on three small circles given by

$$\sin\theta\cos\left(\phi - \frac{2k\pi}{3}\right) = \frac{1}{2}, \quad k = 0, 1, 2,$$

and we see the arc of unlabelled right-angled triangles on the front *half-lune* in Fig. 14.6, where we note that the sphere can be partitioned into 6 lunes and 12 half-lunes; one example of a full lune is the region defined by $0 \leq \phi \leq \pi/3, 0 \leq \theta \leq \pi$ and one example of a half-lune is the region defined by $0 \leq \phi \leq \pi/3, 0 \leq \theta \leq \pi/2$. Reflections of triangles in the upper hemisphere at (θ, ϕ) are located in the lower hemisphere at $(\pi - \theta, \phi)$. In addition, permuting the triangle labels gives rise to points in each of the six equal half-lunes in each hemisphere. Thus, if invariance under labelling and reflection were required, then we would be restricted to one of these half-lunes, e.g. the region defined by $0 \leq \phi \leq \pi/3, 0 < \theta < \pi/2$.

Consider a triangle with labels A, B and C, and edge lengths AB, BC and AC. If the labelling and reflection of the points was unimportant, then we could relabel each triangle so that, for example, $AB \geq AC \geq BC$ and point C is above the baseline AB.

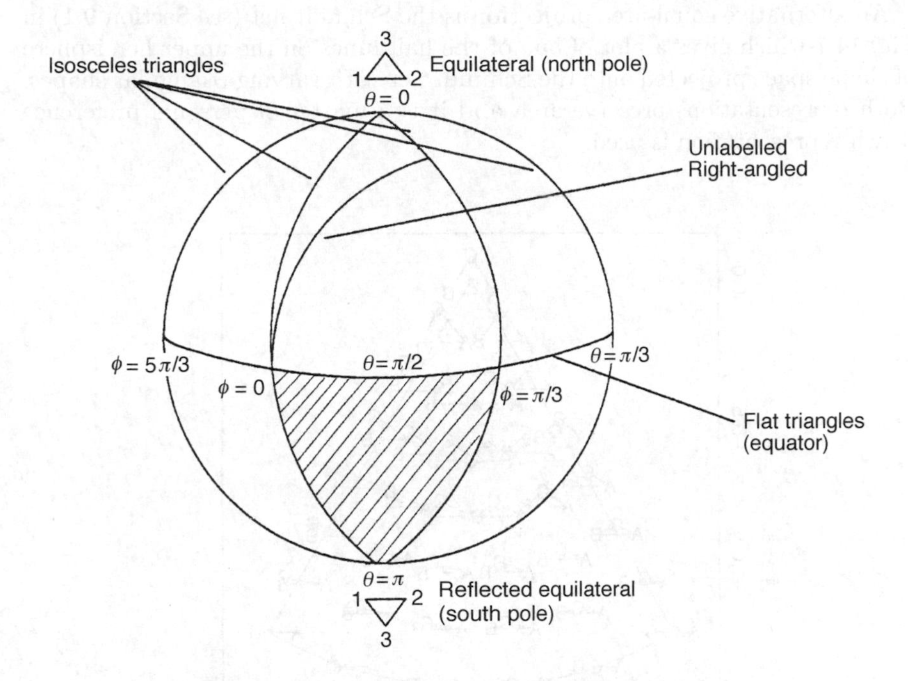

Figure 14.6 Kendall's spherical shape space for triangles in two dimensions. The shape coordinates are the latitude θ (with zero at the north pole) and the longitude ϕ; the shaded area is a half-lune (Dryden & Mardia, 1998. © John Wiley & Sons, Ltd. Reproduced with permission).

For practical analysis and the presentation of data it is often desirable to use a suitable projection of the sphere for triangle shapes. Kendall (1983) defined an equal-area projection of one of the half-lunes of the shape sphere for displaying unlabelled triangle shapes. The projected lune is bell-shaped and this graphical tool is also known as 'Kendall's bell' or the spherical blackboard. Let (x, y, z) be a point on a half-lune with vertices at

$$L = (1, 0, 0), \quad M = (1/2, \sqrt{3}/2, 0), \quad N = (0, 0, 1).$$

Consider the point P such that $LP = MP = NP$. Points on the half-lune are mapped to the cylinder which touches the sphere at P, and this cylinder is unwrapped, with P at the origin to give the blackboard. Explicitly, a given point (x, y, z) on S^2 is mapped to (X, Y) on the bell (see, for example, Mardia & Walder, 1988),

$$X = \tan^{-1}\left\{\frac{\sqrt{7}}{2}\,\frac{-x + \sqrt{3}y}{2\sqrt{3}x + y + \sqrt{3}z}\right\}, \quad Y = \frac{1}{2\sqrt{7}}\{-3x - \sqrt{3}y + 4z\}.$$

For an example, see Fig. 14.7 below.

An alternative equal-area projection is the Schmidt net (see Section 9.1) in Fig. 14.7 which gives a plot of one of the half-lunes on the upper hemisphere of shape space projected onto the Schmidt net with varying triangular shapes. Both representations preserve area and it is a matter of personal preference which representation is used.

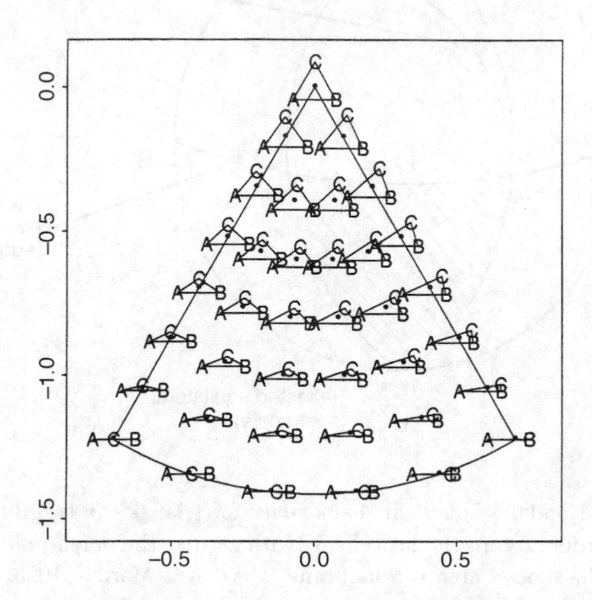

Figure 14.7 Part of the shape space of triangles (from a half-lune of Fig. 14.6) projected onto the equal-area projection Schmidt net. If labelling and reflection were not important, then all triangles could be projected into this sector (Dryden & Mardia, 1998. © John Wiley & Sons, Ltd. Reproduced with permission).

Example 14.5

We now illustrate the graphical method for data where the triangles are unlabelled and reflection invariant. Central place theory is concerned with the pattern of human settlement (see Okabe, Boots & Sugihara, 1992), and is the situation where towns are distributed on a regular hexagonal lattice over a homogeneous area with towns at centres of hexagons. Mardia, Edwards & Puri (1977) consider this hypothesis for a map of 44 places in six counties in Iowa, namely Union, Ringgold, Clarke, Decatur, Lucas and Wayne Counties.

In order to examine whether or not central place theory holds, one could examine the shapes of the triangles formed by a town and its neighbours to see if they are more equilateral than expected under a hypothesis of randomness. A convenient triangulation of the towns is a Delaunay triangulation (Mardia,

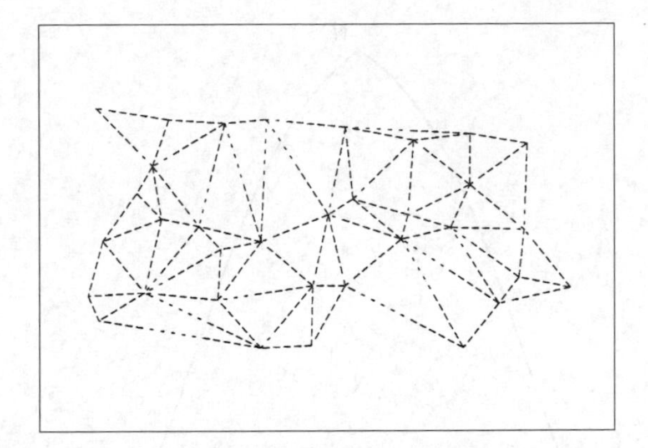

Figure 14.8 Delaunay triangles for the Iowa towns. Triangles with one or two points on the boundary have been removed.

Edwards & Puri, 1977) which is given in Fig. 14.8 for this data set. For ideal central places, Delaunay triangles would be equilateral triangles. An important question to ask is whether the Delaunay triangles are more equilateral than expected by chance.

To answer this question by exploratory analysis, points have been plotted in Fig. 14.9 on the bell according to the shapes of the Delaunay triangles. This figure also gives an average shape of the data and the mean under the hypothesis of the uniform distribution. Figure 14.9 indicates that there is a tendency for shapes to concentrate towards the top of the bell, i.e. to be more equilateral. For various analyses of the data, see Dryden, Faghihi & Taylor (1995), Kendall (1989), Mardia, Edwards & Puri (1977) and Mardia (1989).

14.4 A TEST OF UNIFORMITY

We have noted in Section 14.2 that if all the eigenvalues of $\mathbf{S}$ are nearly equal then the shapes will be very dispersed. We now provide a test for assessing such a situation. However, it should be mentioned at the outset that one does not usually come across such shape data in practice. To formalise the problem, let $\mathbf{z}_1, \dots, \mathbf{z}_n$ be a random sample on $\mathbf{z}$ uniformly distributed on $\mathbb{C}S^{k-2}$. It follows from symmetry or (9.6.1) that under this assumption all the eigenvalues of $\mathrm{E}[\mathbf{z}\mathbf{z}^*]$ are equal.

Let us write $\mathrm{Re}(\mathbf{z}_i) = \mathbf{x}_i$, $\mathrm{Im}(\mathbf{z}_i) = \mathbf{y}_i, i = 1, \dots, n$ and

$$\mathbf{S}/n = \mathbf{T} + i\mathbf{U} \ ,$$

where

$$\mathbf{T} = \frac{1}{n}\sum_{i=1}^{n}(\mathbf{x}_i\mathbf{x}_i^T + \mathbf{y}_i\mathbf{y}_i^T), \quad \mathbf{U} = \frac{1}{n}\sum_{i=1}^{n}(\mathbf{y}_i\mathbf{x}_i^T - \mathbf{x}_i\mathbf{y}_i^T). \tag{14.4.1}$$

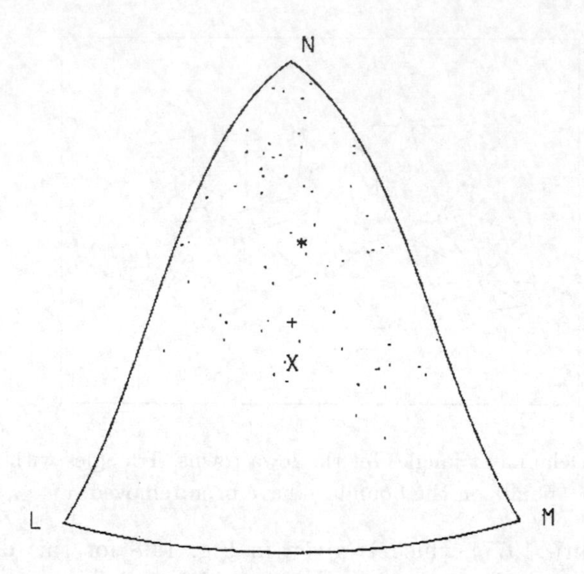

Figure 14.9 The Iowa central place data on Kendall's spherical blackboard (from Mardia, 1989). Some special points are marked, an average shape ($*$), the mean of the uniform distribution ($\times$) and the centre of the bell ($+$). (Reproduced by permission of the Royal Statistical Society.)

Let $l_1, \ldots, l_{k-1}$ be the eigenvalues of $\mathbf{S}$. For large n, we have that

$$Q = \frac{k(k-1)}{n} \sum_{j=1}^{k-1} \left(l_j - \frac{n}{k-1} \right)^2 \dot{\sim} \chi^2_{k^2-2k}. \tag{14.4.2}$$

To prove this, consider the stacked vector

$$\begin{aligned} \mathbf{v}^T &= (t_{11}, t_{22}, \ldots, t_{k-1,k-1}, t_{12}, t_{13}, \ldots, t_{1k-2}, \ldots, t_{k-2,k-1}; \\ &\quad u_{12}, u_{13}, \ldots, u_{k-2,k-1}). \end{aligned} \tag{14.4.3}$$

Let $\mathbf{z} = \mathbf{x} + i\mathbf{y}$, so that $(x_1, \ldots, x_{k-1},\ y_1, \cdots, y_{k-1})$ is uniform on S^{2k-3}. For the uniform distribution, we have $\mathrm{E}[x_i^2] = \{2[k-1]\}^{-1}, \mathrm{E}[x_i x_j] = 0$, so that

$$\mathrm{E}[t_{ii}] = \mathrm{E}[x_i^2 + y_i^2] = 1/(k-1), \mathrm{E}[t_{ij}] = \mathrm{E}[x_i x_j + y_i y_j] = 0, \quad i \neq j.$$

Similarly, $\mathrm{E}[u_{ij}] = 0$, $i \neq j$. Hence

$$\mathrm{E}[\mathbf{v}^T] = ((k-1)^{-1}\mathbf{1}_{k-1}^T, \mathbf{0}_p^T, \mathbf{0}_p^T), \tag{14.4.4}$$

where $\mathbf{1}_{k-1}$ is a $(k-1) \times 1$ vector of ones, and $\mathbf{0}_p$ is a vector of zeros with

$p = (k-1)(k-2)/2$. Now

$$\mathrm{var}(t_{ii}) = \frac{1}{n}\mathrm{var}(x_i^2 + y_i^2) = \frac{1}{n}\left\{\mathrm{E}[(x_i^2+y_i^2)^2] - \mathrm{E}[(x_i^2+y_i^2)]^2\right\}.$$

From the result (9.6.2) on the moments of the uniform distribution on S^{2k-3}, we have

$$\mathrm{E}[x_i^4] = 3\{4k(k-1)\}^{-1}, \quad \mathrm{E}[x_i^2 y_i^2] = \{4k(k-1)\}^{-1}.$$

This leads to

$$\mathrm{var}(t_{ii}) = a(k-2)/(k-1), \quad \mathrm{cov}(t_{ii}, t_{jj}) = -a/(k-1),$$

where $a = \{nk(k-1)\}^{-1}$. Similarly, we find that $\mathrm{var}(t_{ij}) = \mathrm{var}(u_{ij}) = a/2$, $\mathrm{cov}(t_{ij}, t_{i'j'}) = \mathrm{cov}(u_{ij}, u_{i'j'}) = 0, i \neq i', j \neq j'$, $\mathrm{cov}(t_{ij}, u_{i'j'}) = 0$ for all i, i', j, j'.

Using these moments, we find that

$$\mathrm{cov}(\mathbf{v}) = a\ \mathrm{diag}(\mathbf{I}_{k-1} - \mathbf{1}_{k-1}\mathbf{1}'_{k-1}(k-1)^{-1}, \tfrac{1}{2}\mathbf{I}_p, \tfrac{1}{2}\mathbf{I}_{k-1}) = \mathbf{A}, \text{ say.} \tag{14.4.5}$$

It can be shown that

$$\mathbf{A}^- = a^{-1}(\mathbf{I}_{k-1}, 2\mathbf{I}_p, 2\mathbf{I}_p) \tag{14.4.6}$$

and the rank of $\mathbf{A}$ is $(k-1)^2 - 1 = k^2 - 2k$. Hence for large n, by the central limit theorem,

$$\{\mathbf{v} - \mathrm{E}[\mathbf{v}]\}^T \mathbf{A}^- \{\mathbf{v} - \mathrm{E}[\mathbf{v}]\} \mathrel{\dot\sim} \chi^2_{k^2-2k}. \tag{14.4.7}$$

Using (14.4.3)–(14.4.6), we can write the left-hand side of (14.4.7) as

$$a^{-1}\left\{\sum_{i=1}^{k-1}\left(t_{ii} - \frac{1}{k-1}\right)^2 + 2\sum_{i<j} t_{ij}^2 + 2\sum_{i<j} u_{ij}^2\right\}$$
$$= a^{-1}\ \mathrm{tr}\left(\frac{\mathbf{S}}{n} - \frac{1}{k-1}\mathbf{I}_{k-1}\right)\left(\frac{\mathbf{S}}{n} - \frac{1}{k-1}\mathbf{I}_{k-1}\right). \tag{14.4.8}$$

Hence the result follows.

Whereas the approximation (14.4.2) has error of order $O(n^{-1/2})$, the modification Q^* of Q given by (Jupp, 1999)

$$Q^* = Q\left\{1 - \frac{2k^2-3k+3}{6(k+1)n} + \frac{4k^2-9k+3}{6(k^2-2k+2)(k+1)n}Q\right.$$
$$\left. - \frac{k(k-3)}{3(k+1)(k^2-2k+2)(k^2-2k+4)n}Q^2\right\},$$

has a $\chi^2_{k^2-2k}$ distribution with error of order $O(n^{-1})$. Also, note that under uniformity $\mathrm{E}[Q] = k^2 - 2k$. Further, Q is the score statistic for testing the uniformity of a complex Bingham distribution (see Section 14.6.1).

For the triangle case ($k = 3$), we note from (14.4.2) that

$$Q = 6n\{(\bar{l}_1 - \tfrac{1}{2})^2 + (\bar{l}_2 - \tfrac{1}{2})^2\} = 3n(\bar{l}_1 - \bar{l}_2)^2 \ \dot{\sim}\ \chi^2_3 \ , \qquad (14.4.9)$$

since $\bar{l}_1 + \bar{l}_2 = 1$, where $\bar{l}_i = l_i/n$ for $i = 1, 2$. Let $\bar{R}$ be the mean resultant length of the data on S^2 obtained by using the isometric transformation (14.3.19). Calculating the eigenvalues of the 2×2 matrix $\mathbf{S}$ explicitly, it can be shown that $\bar{R} = |\bar{l}_1 - \bar{l}_2|$ and $Q = 3n\bar{R}^2$. Hence the test given by (14.4.9) of uniformity on $\mathbb{C}P^1$ is the same as the Rayleigh test of uniformity on S^2 given by (10.4.5). The reason for this result is that the complex Bingham distributions are equivalent to the Fisher distributions for $k = 3$ (see Section 14.6.5).

Example 14.6

For the full T2 vertebra data set ($k = 6, n = 23$), using the eigenvalues of $\mathbf{S}$ given in Example 14.4, namely,

$$l_1 = 0.004, \quad l_2 = 0.005, \quad l_3 = 0.012, \quad l_4 = 0.072, \quad l_5 = 22.905,$$

we find that $Q = 546.3$. Now $\Pr(\chi^2_{24} > 100) = 3 \times 10^{-11}$, hence the full data provide very strong evidence of non-uniformity. The use of Q^* is inappropriate here, since Q is very large. This is a typical practical example in shape analysis which indicates that the data are extremely far from uniform.

14.5 SHAPE DISTRIBUTIONS

Our main emphasis will be on distributions on $\mathbb{C}P^{k-2}$. We have seen that there are two main approaches in directional statistics which have produced directional distributions from multivariate normal distributions:

(i) the marginal approach, where we integrate out the non-directional variables (as in the derivation of the offset normal distributions; see Section 3.5.6);

(ii) the conditional approach, where the non-directional variables are held constant (as for a von Mises density, we fix the length in a suitable bivariate normal distribution; see Section 3.5.4).

Recently both approaches have produced useful shape distributions, starting with the distributions of Mardia and Dryden (Mardia & Dryden, 1989a; 1989b) following the marginal approach. Kent (1994) adopted the conditional approach and introduced the complex Bingham distributions (Section 14.6). However, we can construct shape distributions directly from directional distributions themselves.

1. For the triangle case, the identification of $\mathbb{C}P^1$ with S^2 using the isometric transformation (14.3.19) sends any shape distribution to a spherical distribution (see Section 14.5.2 below). Since the

mapping (14.3.19) is an isometry (Kendall, 1984), we call the resulting distributions on S^2 'distributions obtained by isometry'.

2. For $k > 3$, we can again use a directional distribution of $\mathbf{z}$ on a preshape space $\mathbb{C}S^{k-2}$ and integrate out, say, ψ in $z_{k-1} = re^{i\psi}, r > 0, 0 < \psi \leq 2\pi$ in Kent's coordinates, to obtain a shape density, as in (14.3.4). However, a simpler approach is to consider a density on the preshape sphere $\mathbb{C}S^{k-2}$ which satisfies the complex symmetry condition

$$f(\mathbf{z}) = f(e^{i\psi}\mathbf{z}) \quad \text{for all } \psi \text{ in } S^1, \quad \mathbf{z} \in \mathbb{C}S^{k-2}, \tag{14.5.1}$$

so that integrating out over ψ is not necessary. In particular, complex symmetric distributions with densities of the form $f(\mathbf{z}^*\mathbf{A}\mathbf{z})$ are automatically shape distributions; we shall mostly discuss such distributions (see Sections 14.6–14.9).

14.5.1 Offset Shape Distributions

Let $\mathbf{X}_1, \ldots, \mathbf{X}_k$ be random points in $\mathbb{R}^2$. Then the shape of $(\mathbf{X}_1, \ldots, \mathbf{X}_k)$ is a random point in $\mathbb{C}P^{k-2}$. A case of particular interest is that in which $\mathbf{X}_1, \ldots, \mathbf{X}_k$ are independent and $\mathbf{X}_i \sim N(\boldsymbol{\mu}_i, \boldsymbol{\Sigma})$ for $i = 1, \ldots, k$. For this case Mardia & Dryden (1989a; 1989b), Mardia (1989) and Dryden & Mardia (1991) have given expressions for the probability density function of the corresponding shape distribution on $\mathbb{C}P^{k-2}$. Dryden & Mardia (1991) also allow correlations between landmarks. In general, these distributions are known in the literature as Mardia–Dryden distributions.

In the case $k = 3$, the shape space is $\mathbb{C}P^1$, which can be identified with the sphere S^2. Mardia (1989) showed that the distribution on S^2 of the shape of $(\mathbf{X}_1, \mathbf{X}_2, \mathbf{X}_3)$ with $\mathbf{X}_i \sim N(\boldsymbol{\mu}_i, \sigma^2\mathbf{I}), i = 1, 2, 3$, has probability density function

$$f(\mathbf{x}; \boldsymbol{\lambda}, \kappa) = \{1 + \kappa(\boldsymbol{\lambda}'\mathbf{x} + 1)\} \exp \kappa(\boldsymbol{\lambda}'\mathbf{x} - \mathbf{1}), \quad \mathbf{x} \in S^2, |\boldsymbol{\lambda}| = 1, \tag{14.5.2}$$

and $\boldsymbol{\lambda}$ is the 'mean shape'. For all κ this density is quite close to a Fisher density with the same mean and hence the use of Fisher distributions for shapes was recommended in Mardia (1989). Furthermore, Goodall & Mardia (1993) have obtained shape distributions in higher dimensions following the marginal approach. For further details, see Dryden & Mardia (1998).

14.5.2 Distributions of Triangle Shapes obtained by Isometry

A simple procedure for obtaining shape distributions for triangles is to use the mapping of the preshape space $\mathbb{C}P^1$ to S^2 given by (14.3.19). For given raw landmarks $\mathbf{z}^0$, the preshape coordinates z_1 and z_2 are obtained explicitly from (14.1.3).

We first consider the case where a Fisher distribution $F(\boldsymbol{\mu}, \kappa)$ is appropriate. Mardia (1989) has shown that the offset shape distribution (14.5.2) can be approximated by a Fisher distribution. Hence when the landmarks are perturbed by a normal distribution as described in Section 14.5.1, a Fisher distribution will be appropriate. However, under the assumption of a general multivariate normal distribution for the configuration, a Kent distribution given in Section 9.3.3 will be more appropriate. Various hypotheses related to triangle shape are of interest under the assumption of a $F(\boldsymbol{\mu}, \kappa)$ distribution.

(i) The hypothesis that the shapes are uniformly distributed corresponds to $\kappa = 0$.
(ii) The hypothesis that the mean shape is equilateral is equivalent to $\boldsymbol{\mu} = (0,0,1)^T$. We can take, without any loss of generality, $\boldsymbol{\mu} = (0,0,1)^T$ instead of $\boldsymbol{\mu} = (0,0,-1)^T$.
(iii) Similarly, if $\boldsymbol{\mu}$ corresponds to an isosceles triangle, we can relabel the triangles so that $\phi = 0$ and then $\boldsymbol{\mu} = (\sin\delta, 0, \cos\delta)^T$ for some δ.
(iv) The hypothesis that the triangles are flat can be expressed as $\mu^2 + \nu^2 = 1$ or equivalently $\eta = 0$, where $\boldsymbol{\mu}^T = (\mu, \nu, \eta)$.

Example 14.7
Bookstein (1991, p. 406) has given landmarks for 21 triangles for the microscopic fossil data of Lohmann. (Each observation is a mean value.) For these landmarks, Table 14.1 gives the Bookstein shape variables $(u_i, v_i), i = 1, \ldots, 21$ with base line $(-\frac{1}{2}, 0), (\frac{1}{2}, 0)$. The corresponding isometric spherical coordinates $(x, y, z) = (\sin\theta\cos\phi, \sin\theta\cos\phi, \cos\theta)$ are obtained using (14.3.20) and are given in Table 14.1. We find that the mean vector is $\bar{\mathbf{x}} = (0.29, -0.14, 0.94)^T$ and the mean resultant length is $\bar{R} = 0.990$. Thus, $\bar{R}$ is large. Assuming that $(x_1, y_1, z_1), \ldots, (x_n, y_n, z_n)$ is a random sample from the Fisher distribution $F(\boldsymbol{\mu}, \kappa)$, we find that $\hat{\kappa} = 101$, so that the data set is highly concentrated. Further, we can test the hypothesis that the mean shape is equilateral, i.e.

$$H_0 : \boldsymbol{\mu} = (0,0,1)^T.$$

An appropriate test is the Watson–Williams test, which is based on (10.4.23), i.e.

$$(n-1)(\bar{R} - \bar{C})/(1 - \bar{R}) \dot{\sim} F_{2,2n-2},$$

where $\bar{C} = \bar{\mathbf{x}}^T(0,0,1)^T$. Here $\bar{C} = 0.936$ and $F_{2,40} = 108.0$ and the 0.1% point of $F_{2,40}$ is 8.3. Hence this hypothesis is clearly rejected.

14.6 COMPLEX BINGHAM DISTRIBUTIONS

14.6.1 The Distributions

The preshape $\mathbf{z} = (z_1, z_2, \ldots, z_{k-1})^T$ lies on the complex sphere $\mathbb{C}S^{k-2}$ and one way of constructing an appropriate distribution is by conditioning the

Table 14.1 The fossil data in Bookstein coordinates (u, v) base $(-\frac{1}{2}, 0)$ and $(\frac{1}{2}, 0)$ and the isometric spherical coordinates (x, y, z).

u	v	x	y	z
−0.07	0.44	0.58	−0.12	0.80
−0.10	0.48	0.51	−0.16	0.84
−0.07	0.52	0.46	−0.11	0.88
−0.14	0.66	0.24	−0.19	0.95
−0.08	0.55	0.42	−0.12	0.90
−0.08	0.59	0.36	−0.12	0.93
−0.06	0.56	0.41	−0.09	0.91
−0.09	0.63	0.30	−0.12	0.94
−0.11	0.62	0.31	−0.16	0.94
−0.10	0.66	0.25	−0.13	0.96
−0.10	0.64	0.28	−0.14	0.95
−0.08	0.66	0.26	−0.11	0.96
−0.09	0.62	0.31	−0.13	0.94
−0.09	0.71	0.19	−0.11	0.97
−0.08	0.73	0.16	−0.10	0.98
−0.09	0.72	0.18	−0.11	0.98
−0.13	0.72	0.17	−0.17	0.97
−0.17	0.74	0.13	−0.21	0.97
−0.18	0.66	0.23	−0.25	0.94
−0.14	0.68	0.22	−0.19	0.96
−0.09	0.76	0.12	−0.11	0.99

complex multivariate normal distribution with probability density function proportional to $\exp(-\frac{1}{2}\mathbf{z}^*\mathbf{\Sigma}^{-1}\mathbf{z})$, where $\mathbf{\Sigma}$ is Hermitian (i.e. $\mathbf{\Sigma} = \mathbf{\Sigma}^*$). Conditioning it on $\mathbf{z}^*\mathbf{z} = 1$ gives rise to the following complex Bingham distributions. (For a similar construction of the real Bingham distributions see Section 9.4.3.)

The *complex Bingham distribution* with canonical parameter matrix $\mathbf{A}$ has probability density function

$$f(\mathbf{z}) = C(\mathbf{A})^{-1}\exp(\mathbf{z}^*\mathbf{A}\mathbf{z}), \quad \mathbf{z} \in \mathbb{C}S^{k-2}, \tag{14.6.1}$$

where the $(k-1)\times(k-1)$ matrix $\mathbf{A}$ is Hermitian and $C(\mathbf{A})$ is the normalising constant. We write

$$\mathbf{z} \sim \mathbb{C}B_{k-2}(\mathbf{A}).$$

The density (14.6.1) satisfies the invariance property (14.5.1), and so it defines a distribution on $\mathbb{C}P^{k-2}$. Thus the distribution is suitable for the analysis of two-dimensional shapes and was proposed by Kent (1994) with this aim in mind.

Since $\mathbf{z}^*\mathbf{z} = 1$ for $\mathbf{z}$ in $\mathbb{C}S^{k-2}$, the parameter matrices $\mathbf{A}$ and $\mathbf{A} + \alpha\mathbf{I}$ define the same complex Bingham distribution and $C(\mathbf{A} + \alpha\mathbf{I}) = C(\mathbf{A})\exp\alpha$ for any complex number α. It is convenient to remove this non-identifiability by setting $\lambda_{\max}(\mathbf{A}) = 0$, where $\lambda_{\max}(\mathbf{A})$ denotes the largest eigenvalue of $\mathbf{A}$. Let $\gamma_1, \cdots, \gamma_{k-1}$ denote the standardised eigenvectors of $\mathbf{A}$, so that

$$\gamma_j^* \gamma_j = 1, \quad \gamma_i^* \gamma_j = 0, \quad i \neq j, \qquad \mathbf{A}\gamma_j = \lambda_j \gamma_j, \quad j = 1, \ldots, k-1.$$

Each γ_j is defined only up to rotation by a unit complex number. If $\lambda_1 = \ldots = \lambda_{k-1} = 0$, then the distribution reduces to the uniform distribution on $\mathbb{C}P^{k-2}$. Provided that $\lambda_{k-2} < 0$, it can be seen that $\mathbf{z} = \gamma_{k-1}$ maximises the density and γ_{k-1} is unique up to a scalar rotation $\gamma_{k-1} \exp(i\psi)$. Further, if $\lambda_1, \ldots, \lambda_{k-2}$ are far below 0, then the distribution becomes highly concentrated about this modal axis.

A complex Bingham distribution on $\mathbb{C}P^{k-2}$ can be regarded as a special case of a real Bingham distribution on $\mathbb{R}P^{2k-3}$. If $z_j = x_j + iy_j$, define a $2k$-dimensional vector $\mathbf{v} = (x_1, y_1, \ldots, x_k, y_k)^T = V(\mathbf{z})$, say, by splitting each complex number into its real and imaginary parts. Also, if $\mathbf{A} = (a_{hj})$ has entries $a_{hj} = \delta_{hj} \exp(i\psi_{hj})$ with $\psi_{jh} = -\psi_{hj}$, define a $(2k-2) \times (2k-2)$ matrix $\mathbf{B}$ made up of $(k-1)^2$ blocks of size 2×2 given by

$$\mathbf{B}_{hj} = \gamma_{hj} \begin{pmatrix} \cos\psi_{hj} & -\sin\psi_{hj} \\ \sin\psi_{hj} & \cos\psi_{hj} \end{pmatrix}, \quad -\pi \leq \psi_{hj} \leq \pi.$$

Then $\mathbf{z}^* \mathbf{A} \mathbf{z} = \mathbf{v}^T \mathbf{B} \mathbf{v}$, so that $\mathbf{z} \sim \mathbb{C}B_{k-1}(\mathbf{A}) \Leftrightarrow \mathbf{v} \sim B(\mathbf{B})$ where $B(\mathbf{B})$ denotes the real Bingham distribution with canonical parameter matrix $\mathbf{B}$.

It is tempting to define a complex von Mises–Fisher distribution on the complex sphere $\mathbb{C}S^{k-1}$ as having probability density function proportional to

$$\exp\{\kappa \mathrm{Re}(\boldsymbol{\mu}^* \mathbf{z})\}, \quad \mathbf{z} \in \mathbb{C}S^{k-1},$$

where $\boldsymbol{\mu} \in \mathbb{C}^k$. Put $\mathbf{z} = \mathbf{x} + i\mathbf{y}$ and $\boldsymbol{\mu} = \boldsymbol{\nu} + i\boldsymbol{\pi}$. Then $\mathrm{Re}(\boldsymbol{\mu}^* \mathbf{z}) = \boldsymbol{\nu}^T \mathbf{x} + i\boldsymbol{\pi}^T \mathbf{y}$, and so the complex von Mises–Fisher distribution on $\mathbb{C}S^{k-1}$ with canonical parameter vector $\boldsymbol{\mu}$ is just the von Mises–Fisher distribution on S^{2k-1} with canonical parameter vector $(\boldsymbol{\nu}^T, \boldsymbol{\pi}^T)^T$. For $k \geq 3$, it is not rotationally invariant and therefore will not be useful for shape analysis. In contrast to this, the complex matrix von Mises–Fisher distributions on $U(p)$ given by (13.2.25) are different from their counterparts on $O(p)$ given by (13.2.15).

14.6.2 The Normalising Constant

Let $\lambda_1, \cdots, \lambda_{k-1}$ be the eigenvalues of $\mathbf{A}$ with $\lambda_1 < \lambda_2 < \cdots < \lambda_{k-1} = 0$. Then the normalising constant for the complex Bingham distribution $\mathbb{C}B_{k-2}(\mathbf{A})$ is

$$C(\mathbf{A}) = 2\pi^{k-1} \sum_{j=1}^{k-1} a_j \exp(\lambda_j), \quad a_j^{-1} = \prod_{i \neq j} (\lambda_j - \lambda_i). \tag{14.6.2}$$

Further, $C(\mathbf{A}) = C(\boldsymbol{\Lambda})$, where $\boldsymbol{\Lambda} = \mathrm{diag}(\lambda_1, \ldots, \lambda_{k-1})$.

To prove this we proceed as follows. Since $\mathbf{A}$ is Hermitian, we can write $\mathbf{A} = \mathbf{U}\boldsymbol{\Lambda}\mathbf{U}^*$, where $\mathbf{U}$ is unitary ($\mathbf{U}^*\mathbf{U} = \mathbf{I}$), and $\boldsymbol{\Lambda} = \text{diag}(\lambda_1, \ldots, \lambda_{k-1})$, where $\lambda_1, \ldots, \lambda_{k-1}$ are real. If $\mathbf{z} \sim \mathbb{C}B_{k-2}(\mathbf{A})$, then we have $\mathbf{U}^*\mathbf{z} \sim \mathbb{C}B_{k-2}(\boldsymbol{\Lambda})$ so these two distributions have the same normalisation constant, i.e. $C(\mathbf{A}) = C(\boldsymbol{\Lambda})$.

In terms of Kent's preshape polar coordinates given in (14.3.1), we have the complex Bingham density (with respect to $ds_1 \ldots ds_{k-2} d\theta_1 \ldots d\theta_{k-1}$)

$$c(\boldsymbol{\Lambda})^{-1} \exp\left(\sum_{j=1}^{k-1} \lambda_j s_j\right) 2^{2-k}, \tag{14.6.3}$$

since $\mathbf{z}^*\mathbf{A}\mathbf{z} = \sum_{j=1}^{k-1} \lambda_j \mathbf{z}_j^* \mathbf{z}_j = \sum_{j=1}^{k-1} \lambda_j s_j$. Note that $(s_1, \ldots, s_{k-2})$ and $(\theta_1, \ldots, \theta_{k-1})$ are independent. Furthermore, $\theta_1, \ldots, \theta_{k-1}$ are independently uniformly distributed and $(s_1, \ldots, s_{k-2})$ have a joint multivariate exponential distribution truncated to a simplex. It follows that

$$c(\boldsymbol{\Lambda}) = 2^{-k+2} \int_{\triangle_{k-1}} \int_{[0,2\pi]^{k-1}} \exp\left\{\sum_{j=1}^{k-1} \lambda_j s_j\right\} d\theta_1 \ldots d\theta_{k-1} ds_1 \ldots ds_{k-2}$$

$$= 2\pi^{k-1} \int_{\triangle_{k-1}} \exp\left\{\sum_{j=1}^{k-1} \lambda_j s_j\right\} ds_1 \ldots ds_{k-2} = 2\pi^{k-1} \sum_{j=1}^{k-1} a_j \exp(\lambda_j).$$

The last step follows from the following result with $\alpha = 1$:

$$I(\lambda_1, \ldots, \lambda_{k-1}; \alpha) = \int_{\alpha\triangle_{k-2}} \exp\left(\sum_{j=1}^{k-1} \lambda_j s_j\right) ds_1 \ldots ds_{k-2} = \sum_{j=1}^{k-1} a_j \exp(\alpha\lambda_j), \tag{14.6.4}$$

where $\alpha\triangle_{k-2} = \{(s_1, \ldots, s_{k-2})^T : s_j \geq 0 \text{ and } \sum_{j=1}^{k-2} s_j \leq \alpha\}$ denotes the 'scaled' unit simplex. We now prove (14.6.4) by induction. For $k = 3$, the result holds by showing that

$$\int_0^\alpha \exp\{\lambda_1 s_1 + \lambda_2(\alpha - s_1)\} ds_1 = \sum_{j=1}^{2} a_j e^{\alpha\lambda_j},$$

where $a_1^{-1} = \lambda_1 - \lambda_2$, $a_2^{-1} = \lambda_2 - \lambda_1$. For $k > 3$, we proceed as follows. Let $b_j^{-1} = \prod_{i \neq j}^{k-1} (\lambda_j - \lambda_i)$, $j = 2, \ldots, k-2$. Then using $\sum_{j=1}^{k-1} \lambda_j s_j = \lambda_1 s_1 + \sum_{j=2}^{k-1} \lambda_j s_j$, and $\alpha\triangle_{k-2} = \{(s_1, \ldots, s_{k-2})^T \colon 0 \leq s_1 \leq \alpha,\ (s_2, \ldots, s_{k-2}) \in (\alpha - s_1)\triangle_{k-3}\}$, we have

$$\begin{aligned} I(\lambda_1, \ldots, \lambda_{k-1}; \alpha) &= \int_0^\alpha \exp(\lambda_1 s_1) I(\lambda_1, \ldots, \lambda_{k-2}; \alpha - s_1) ds_1 \\ &= \int_0^\alpha \exp(\lambda_1 s_1) \left[\sum\nolimits_{j=2}^{k-1} b_j \exp\{\lambda_j(\alpha - s_1)\}\right] ds_1. \end{aligned}$$

On integrating with respect to s_1, we obtain

$$I(\lambda_1, \ldots, \lambda_{k-1}; \alpha) = \sum_{j=2}^{k-1} b_j(\lambda_1 - \lambda_j)^{-1}\{\exp(\alpha\lambda_1) - \exp(\alpha\lambda_j)\}. \qquad (14.6.5)$$

Using the definition of a_j, the sum of terms in (14.6.5) involving $\exp(\alpha\lambda_j)$ can be rewritten as

$$-\sum_{j=2}^{k-1} b_j(\lambda_1 - \lambda_j)^{-1} \exp(\alpha\lambda_j) = \sum_{j=2}^{k-1} a_j \exp(\alpha\lambda_j),$$

whereas for the first term a partial fraction expansion of a_1 as a function of λ_1 shows that

$$\sum_{j=2}^{k-1} b_j(\lambda_1 - \lambda_j)^{-1} = a_1.$$

Substituting these in (14.6.5) completes the proof.

If equalities exist between some λ_is then $C(\mathbf{A})$ can be obtained using repeated applications of L'Hôpital's rule. For example, if $\lambda_{k-2} = \lambda_{k-1}$ but all other λs are distinct, we obtain (Bingham, Chang & Richards, 1992)

$$C(\mathbf{A}) = (k-2)! \left[\sum_{i=1}^{k-3} \frac{e^{-\lambda_i}}{\Pi_{j \neq i}(\lambda_j - \lambda_i)} + \frac{e^{-\lambda_{k-1}}\{1 - \sum_{i=1}^{k-3}(\lambda_i - \lambda_{k-1})^{-1}\}}{\Pi_{j \leq k-3}(\lambda_j - \lambda_{k-1})} \right]. \qquad (14.6.6)$$

A particularly useful and simple case is that of the complex Watson distributions, which occurs when there are just two distinct eigenvalues, as described in Section 14.7.

Note that various distributional results follow from (14.6.3). The random variables $\mathbf{s} = (s_1, \ldots, s_{k-2})^T$ and $(\theta_1, \ldots, \theta_{k-1})$ are independent, and $\theta_1, \ldots, \theta_{k-1}$ are independently uniformly distributed on the unit circle. Furthermore, $\mathbf{s}$ has a joint multivariate exponential distribution truncated to a simplex and its probability density function is

$$f(\mathbf{s}) = 2\pi^{k-1} C(\mathbf{\Lambda})^{-1} \exp\left(\sum_{j=1}^{k-2} \lambda_j s_j\right), \quad \mathbf{s} \in \triangle_{k-2}, \qquad (14.6.7)$$

so that these distributions of $\mathbf{s}$ form a $(k-2, k-2)$ exponential model with canonical statistic $\mathbf{s}$ (see Section 3.5.1). Thus the moment generating function of $\mathbf{s}$, $\mathrm{E}[\exp(\Sigma_{j=1}^{k-2} \delta_j s_j)] = \phi(\boldsymbol{\delta}; \mathbf{\Lambda})$, say, is given by

$$\phi(\boldsymbol{\delta}; \mathbf{\Delta}) = C(\mathbf{\Lambda} + \mathbf{\Delta})/C(\mathbf{\Lambda}), \qquad (14.6.8)$$

where $\boldsymbol{\delta} = (\delta_1, \ldots, \delta_{k-2})^T \in \mathbb{C}^{k-2}$ and $\mathbf{\Delta} =$ diag $(\boldsymbol{\delta}^T, \mathbf{0})$. Thus, we have the moments

$$\mathrm{E}(s_j) = \{C(\mathbf{\Lambda})\}^{-1} \frac{\partial C(\mathbf{\Lambda})}{\partial \lambda_j}, \quad j = 1, \ldots, k-2. \qquad (14.6.9)$$

14.6.3 High Concentrations

To model high concentration, replace $\mathbf{A}$ by $\kappa\mathbf{A}$, so that from (14.6.7)

$$f(\mathbf{s}) = 2\pi^{k-1}C(\kappa\mathbf{\Lambda})^{-1}\exp\left(\kappa\sum_{j=1}^{\kappa-2}\lambda_j s_j\right), \quad \mathbf{s}\in\Delta_{k-2}, \tag{14.6.10}$$

and we assume that there is a unique $\lambda_{k-1} = 0 > \lambda_{k-2}$.

As $\kappa\to\infty, u_j = \kappa s_j$ $(j = 1,\ldots,k-2)$ tend to independent exponential random variables with densities

$$-\lambda_i^{-1}\exp(\lambda_i u_i), \quad \lambda_i < 0, u_i > 0.$$

Thus, as $\lambda_{k-2}\to-\infty$,

$$C(\mathbf{\Lambda}) \simeq 2\pi^{k-1}\prod_{j=1}^{k-2}(-\lambda_j)^{-1}. \tag{14.6.11}$$

Another way of looking at the density (14.6.1) when there is high concentration about the axis γ_{k-1} is to rotate $\mathbf{z}$ to γ_{k-1} and then to project $\mathbf{z}$ onto the tangent plane at γ_{k-1}. Given $\mathbf{z}$ in $\mathbb{C}S^{k-2}$, we construct $\mathbf{v}$ in $\mathbb{C}^{k-1}$ by

$$\mathbf{v} = \exp(-i\psi)(\mathbf{I}_{k-1} - \gamma_{k-1}\gamma_{k-1}^*)\mathbf{z}, \quad \psi = \arg(\gamma_{k-1}^*\mathbf{z}). \tag{14.6.12}$$

It can be proved that ψ and $\mathbf{v}$ are independently distributed with ψ uniform on the unit circle. Writing $\mathbf{A} = \sum_{j=1}^{k-1}\lambda_j\gamma_j\gamma_j^* = \sum_{j=1}^{k-2}\lambda_j\gamma_j\gamma_j^*$, we have

$$\kappa\mathbf{z}^*\mathbf{A}\mathbf{z} = \kappa\mathbf{v}^*\mathbf{A}\mathbf{v}$$

and therefore $\mathbf{v}$ has a complex normal distribution in $k-1$ complex dimensions with variance matrix

$$\mathbf{\Sigma} = (-2\mathbf{A})^+ = \tfrac{1}{2}\sum_{j=1}^{k-2}(-\lambda_j)^{-1}\gamma_j\gamma_j^*, \tag{14.6.13}$$

where $(-2\mathbf{A})^+$ denotes the Moore–Penrose generalised inverse of $-2\mathbf{A}$. Here $\mathbf{\Sigma}$ is singular because $\mathbf{v}$ lies in the tangent plane $\gamma_{k-1}^*\mathbf{v} = 0$. Thus $\mathbf{z}$ has the complex normal distribution $\mathbb{C}N(\gamma_k, \mathbf{\Sigma})$, so that $\mathbf{\Sigma}$ is determined by the eigenvectors other than γ_k.

The asymptotic distribution shows that the $\mathbf{v}_i$ are approximately jointly distributed as $N_{k-1}(\mathbf{0}, \sigma_i^2\mathbf{I})$. Hence a complex Bingham distribution imposes an isotropic distribution on the marginal distribution. However, it does allow intercorrelation (that of a complex normal covariance matrix) between landmarks.

14.6.4 Inference

Let $\mathbf{z}_1, \ldots, \mathbf{z}_n$ be a random sample from the complex Bingham distribution $\mathbb{C}B_{k-2}(\mathbf{A})$ where $n \geq k-1$. Set

$$\mathbf{S} = \sum_{j=1}^{n} \mathbf{z}_j \mathbf{z}_j^*$$

to be the $(k-1) \times (k-1)$ complex sum of squares and products matrix. Suppose that the eigenvalues of $\mathbf{S}$ are positive and distinct, $0 < l_1 < \ldots < l_{k-1}$, and let $\mathbf{g}_1, \ldots, \mathbf{g}_{k-1}$ denote the corresponding eigenvectors. Note that $\Sigma_{j=1}^{k-1} l_j = n$.

As in the case of a real Bingham distribution, it can be seen that when the eigenvalues $l_1, \ldots, l_{k-1}$ are distinct then

$$\hat{\gamma}_j = \mathbf{g}_j, \quad j = 1, \ldots, k-2, \qquad (14.6.14)$$

since the log-likelihood is

$$\log L(\mathbf{A}; \mathbf{z}_1, \ldots, \mathbf{z}_n) = \sum_{j=1}^{k-1} \operatorname{tr}(\mathbf{AS}) - n \log C(\mathbf{\Lambda}).$$

The maximum likelihood estimates of the eigenvalues are found by solving

$$\frac{\partial \log C(\mathbf{\Lambda})}{\partial \lambda_j} = \frac{1}{n} l_j, \quad j = 1, \ldots, k-2,$$

and the dominant eigenvector $\hat{\gamma}_k$ can be regarded as the 'average axis' of the data. This estimate is the same as that obtained from Procrustes shape analysis (Kent, 1994). Under high concentration we have, from (14.6.11),

$$\log C(\mathbf{\Lambda}) \simeq \text{ constant } - \sum_{j=1}^{k-2} \log(-\lambda_j),$$

so that

$$\hat{\lambda}_j \simeq -n/l_j, \quad j = 1, \ldots, k-2. \qquad (14.6.15)$$

The dominant eigenvector $\hat{\gamma}_{k-1}$ can be regarded as the average axis of the data – an estimate of modal shape. Hence, the average axis from the complex Bingham maximum likelihood estimate is the same as the full Procrustes estimate of mean shape, which was also given by the dominant eigenvector of $\mathbf{S}$ as shown in Section 14.2.

We note finally that, using the canonical representation of the complex Bingham distributions, the statistic Q given in (14.4.2) can be shown to be the score statistic for testing uniformity of a complex Bingham distribution.

Example 14.8
The Procrustes mean shape obtained in Example 14.4 is the maximum likelihood estimate of the population dominant eigenvector of $\mathrm{E}[\mathbf{z}\mathbf{z}^T]$ under a complex Bingham distribution. Since the largest eigenvalue $l_5 = 22.9$ (given in Example 14.4) is nearly $n = 23$, the data are highly concentrated. For large concentration, from (14.6.14)–(14.6.15),

$$\widehat{\mathrm{cov}}(\hat{\mathbf{v}}) = \sum_{j=1}^{k-2} \frac{1}{2\hat{\lambda}_j} \hat{\gamma}_j \hat{\gamma}_j^* = \frac{1}{2n} \sum_{j=1}^{k-2} l_j \mathbf{g}_j \mathbf{g}_j^* = \frac{1}{2n}(\mathbf{S} - l_{k-1}\mathbf{g}_k \mathbf{g}_k^*).$$

Thus we can obtain $\widehat{\mathrm{cov}}(\mathbf{H}\hat{\mathbf{v}})$, to obtain some idea of variances and correlations. It is found from $\widehat{\mathrm{cov}}(\mathbf{H}\hat{\mathbf{v}})$ that the variances $(\times 2 \times 23)$ are

$$0.020, 0.021, 0.005, 0.036, 0.005, 0.007.$$

Thus landmark 4 has most variance, whereas landmarks 1 and 2 have the next highest. There is almost no variability in landmarks 3, 5 and 6. The estimated correlation matrix of $\mathbf{H}\hat{\mathbf{v}}$ is

$$\begin{pmatrix} 1 & & & & & \\ 0.05 - 0.76i & 1 & & & & \\ -0.34 - 0.25i & 0.10 - 0.29i & 1 & & & \\ -0.60 + 0.68i & -0.60 - 0.62i & -0.09 - 0.38i & 1 & & \\ 0.30 + 0.12i & -0.24 + 0.26i & -0.36 - 0.01i & -0.24 - 0.39i & 1 & \\ -0.35 - 0.08i & -0.34 + 0.16i & 0.50 - 0.08i & 0.08 + 0.04i & -0.08 - 0.21i & 1 \end{pmatrix}.$$

This indicates that correlations between landmarks are high. Note that the matrix is necessarily singular.

The above analysis gives some idea of variability, although the restrictions of a complex covariance structure are imposed. In particular, we necessarily have isotropic perturbations at each landmark. A practical method from the shape analysis point of view would be to examine the shape variability in tangent space (see Dryden & Mardia, 1998, pp. 44–51). The next step will be to examine the hypothesis of complex symmetry.

14.6.5 Relationship with the Fisher Distributions

In Section 14.5.2, we used Fisher distributions for shapes of triangles. Following Kent (1994), we now establish links between complex Bingham distributions on $\mathbb{C}P^1$ and real Fisher distributions on S^2. First, consider the parameter matrix $\mathbf{A}$ of a complex Bingham distribution on $\mathbb{C}P^1$. This matrix is Hermitian, so that there appear to be four independent real variables. Since we can replace $\mathbf{A}$ by $\mathbf{A} + \alpha\mathbf{I}$, where α is real, we may reduce the rank of $\mathbf{A}$ to 1 and there are at most three independent real variables. Hence we can take

$$\mathbf{A} = \lambda \begin{pmatrix} u \\ v \end{pmatrix} (u^*, v), \quad |u|^2 + v^2 = 1.$$

Further, we can choose $\lambda > 0$. Multiplying the eigenvector (u, v) by a unit complex scalar does not alter $\mathbf{A}$, so that we can assume that $v \geq 0$. Hence, provided that $v \neq 0, 1$, the one-to-one mapping between (u, v) and (η, ζ) is given by

$$u = e^{i\eta} \cos\zeta, v = \sin\zeta, \quad 0 < \zeta < \frac{\pi}{2}, 0 \leq \eta < 2\pi. \tag{14.6.16}$$

Further, we transform the random variables (z_1, z_2) to (ψ, θ, ϕ) by (14.3.15) so that the exponent in the complex Bingham density (14.6.1) becomes

$$\mathbf{z}^*\mathbf{A}\mathbf{z} = \lambda \left| (u^*, v) \begin{pmatrix} z_1 \\ z_2 \end{pmatrix} \right|^2$$

$$= \lambda\{\cos^2\zeta \cos^2\theta + \sin^2\zeta \sin^2\theta + 2\sin\zeta\cos\zeta\sin\theta\cos\theta\cos(\phi - \eta)\}. \tag{14.6.17}$$

Let $\mathbf{x}$ and $\boldsymbol{\mu}$ be the vectors corresponding to the spherical polar coordinates $(2\theta, \phi)$ and $(2\zeta, \eta)$, respectively. Then (14.6.17) can be written as $\mathbf{z}^*\mathbf{A}\mathbf{z} = \lambda(1 + \boldsymbol{\mu}^T\mathbf{t})/2$. Using (14.3.16) and the above expression for $\mathbf{z}^*\mathbf{A}\mathbf{z}$, the distribution $\mathbb{C}B_1(\mathbf{A})$ transformed to (θ, ϕ, ψ) has the probability element

$$\tfrac{1}{2}\{c(\lambda)\}^{-1} \sin 2\theta e^{\frac{1}{2}\lambda(1+\boldsymbol{\mu}^T\mathbf{t})} d\psi d\theta d\phi,$$

where $0 \leq 2\theta < \pi, 0 \leq \phi < 2\pi, -\pi/2 \leq \psi < \pi/2$ and $c(\lambda) = 2\pi^2(e^\lambda - 1)/\lambda$.

Hence ψ is independent of (θ, ϕ) and uniformly distributed on $(-\pi/2, \pi/2)$, whereas $(2\theta, \phi)$ is distributed as the Fisher distribution $F(\boldsymbol{\mu}, \lambda/2)$. One consequence of this result is that we can work with the Fisher distribution rather than $\mathbb{C}B_1(\mathbf{A})$. For example, we can obtain the estimators of the parameters (u, v) and λ of the complex Bingham distribution from those of the Fisher distribution. Let the maximum likelihood estimators of $\boldsymbol{\mu} = (\alpha, \beta)$ and κ of $F(\boldsymbol{\mu}, \kappa)$ be $(\hat{\alpha}, \hat{\beta})$ and $\hat{\kappa}$, respectively. Then from (14.6.16) the maximum likelihood estimators of the Procrustes mean (u, v) and λ are respectively

$$(\hat{u}, \hat{v}) = (e^{i\hat{\beta}} \cos(\hat{\alpha}/2), \sin(\hat{\alpha}/2)), \quad \hat{\lambda} = 2\hat{\kappa}.$$

This result also explains the relationship between the test of uniformity given by (14.4.9) and the Rayleigh test of uniformity on S^2.

14.7 COMPLEX WATSON DISTRIBUTIONS

14.7.1 The Density

Let $\mathbf{z} = (z_1, z_2, \ldots, z_{k-1})^T$ again be a point on the complex sphere $\mathbb{C}S^{k-2}$. The complex Watson distribution on $\mathbb{C}S^{k-2} = S^{2k-3}$ with parameters $\boldsymbol{\mu}$ and κ has density (Mardia & Dryden, 1999)

$$f(\mathbf{z}) = c(\kappa)^{-1} \exp(\kappa \cos^2 \rho), \tag{14.7.1}$$

where $\boldsymbol{\mu} \in \mathbb{C}S^{k-2}$ and

$$\cos\rho = |\mathbf{z}^*\boldsymbol{\mu}|, \quad 0 \le \rho \le \pi/2.$$

Note that κ is a concentration parameter and, as in the real Watson distributions considered in Section 9.4.2, it can take values over the whole real line. If $\kappa > 0$ then the distribution has modes at $e^{i\theta}\boldsymbol{\mu}$ $(\theta \in S^1)$ and so is useful in shape analysis for representing population modal shape. If $\kappa < 0$ then the distribution has modes at $\boldsymbol{\mu}^{\perp} = \{\boldsymbol{\nu} : \boldsymbol{\nu}^*\boldsymbol{\mu} = 0\}$. If $\kappa = 0$ then the distribution is uniform. As our main motivation is shape analysis we shall concentrate on the case $\kappa \ge 0$.

The role of ρ is similar to that of the angle $\cos^{-1}(\boldsymbol{\mu}^T\mathbf{x})$ in the real Watson distributions. For the case of interest to us $(\kappa \ge 0)$ the analogy is with the bipolar real Watson distributions. We write $\mathbb{C}W_{k-2}(\boldsymbol{\mu}, \kappa)$ to denote the complex Watson distribution with parameters $\boldsymbol{\mu}$ and κ. We have complex symmetry since $f(e^{i\psi}\mathbf{z}) = f(\mathbf{z})$ for all ψ on the unit circle and so $\mathbb{C}W_{k-2}(\boldsymbol{\mu}, \kappa)$ defines a distribution on $\mathbb{C}P^{k-2}$.

14.7.2 The Normalising Constant

We prove that the normalising constant is

$$c(\kappa) = 2\pi^{k-1}\kappa^{2-k}\left\{e^{\kappa} - \sum_{r=0}^{k-3}\frac{\kappa^r}{r!}\right\}.$$

On using Kent's polar coordinates on $\mathbb{C}S^{k-2}$ given by (14.3.1), we have

$$\begin{aligned} c(\kappa) &= 2^{2-k}\int_{\Delta_{k-2}}\underbrace{\int_{S^1}\cdots\int_{S^1}}_{k-1} e^{\kappa s_1}d\theta_1\ldots d\theta_{k-1}ds_1ds_2\ldots ds_{k-2} \\ &= 2^{2-k}(2\pi)^{k-1}\int_0^1 e^{\kappa s_1}f(s_1)ds_1, \end{aligned} \tag{14.7.2}$$

where

$$f(s_1) = \int_{s_2+s_3+\ldots+s_{k-2}<1-s_1} ds_2\ldots ds_{k-2} = (1-s_1)^{k-3}\int_{\Delta_{k-3}} du_1\ldots du_{k-3}.$$

Now, the volume measure of the unit simplex Δ_{k-3} is $1/(k-3)!$. Hence

$$c(\kappa) = \frac{2\pi^{k-1}}{(k-3)!}\int_0^1(1-s)^{k-3}e^{\kappa s}ds = \frac{2\pi^{k-1}e^{\kappa}}{\kappa^{k-2}(k-3)!}\int_0^{\kappa}u^{k-3}e^{-u}du. \tag{14.7.3}$$

To calculate the integral on the right-hand side of (14.7.3), we note that

$$I_k = \int_0^{\kappa}u^{k-3}e^{-u}du = -\kappa^{k-3}e^{-\kappa} + (k-3)I_{k-1} = (k-3)!e^{-\kappa}\left[e^{\kappa} - \sum_{r=0}^{k-3}\frac{\kappa^r}{r!}\right].$$

Hence the result follows.

14.7.3 Relationship with the Complex Bingham Distributions

The complex Watson distributions are particular cases of complex Bingham distributions. When $\mathbf{A}$ has two distinct eigenvalues (one of them can be taken as zero) we can reparameterise $\mathbf{A}$ as

$$\mathbf{A} = -\kappa(\mathbf{I}_{k-1} - \boldsymbol{\mu}\boldsymbol{\mu}^*),$$

so that $\mathbb{C}B_{k-2}(\mathbf{A}) = \mathbb{C}W_{k-2}(\boldsymbol{\mu}, \kappa)$. If $k = 3$ then the matrix $\mathbf{A}$ has at most two distinct values and so the complex Watson distributions are identical to the complex Bingham distributions in this case, and thus are related to the Fisher distributions, in view of Section 14.6.5.

14.7.4 Large Concentration

We show that if κ is large, then the distribution $\mathbb{C}W_{\kappa-2}(\boldsymbol{\mu}, \kappa)$ tends to a complex normal distribution with mean $\boldsymbol{\mu}$ and covariance matrix $(1/2\kappa)(\mathbf{I}_{k-1} - \boldsymbol{\mu}\boldsymbol{\mu}^*)^-$, which is a generalized inverse of $2\kappa(\mathbf{I}_{k-1} - \boldsymbol{\mu}\boldsymbol{\mu}^*)$. We first prove the following equality:

$$(\mathbf{z} - \boldsymbol{\mu})^*(\mathbf{I}_{k-1} - \boldsymbol{\mu}\boldsymbol{\mu}^*)^-(\mathbf{z} - \boldsymbol{\mu}) = 1 - \mathbf{z}^*\boldsymbol{\mu}\boldsymbol{\mu}^*\mathbf{z}, \tag{14.7.4}$$

where $\mathbf{z}^*\mathbf{z} = \boldsymbol{\mu}^*\boldsymbol{\mu} = 1$. Put $\mathbf{A} = \mathbf{I}_{k-1} - \boldsymbol{\mu}\boldsymbol{\mu}^*$. It is easily verified that $\mathbf{A}$ is idempotent and therefore $\mathbf{A}^3 = \mathbf{A}$. Then taking $\mathbf{A}^- = \mathbf{A}$ gives $\mathbf{A}\mathbf{A}^-\mathbf{A}$, i.e. $\mathbf{A}$ is a generalized inverse of $\mathbf{A}$. To prove (14.7.4), expand the left-hand side as

$$\mathbf{z}^*(\mathbf{I}_{k-1} - \boldsymbol{\mu}\boldsymbol{\mu}^*)\mathbf{z} + \boldsymbol{\mu}^*(\mathbf{I}_{k-1} - \boldsymbol{\mu}\boldsymbol{\mu}^*)\boldsymbol{\mu} - \mathbf{z}^*(\mathbf{I}_{k-1} - \boldsymbol{\mu}\boldsymbol{\mu}^*)\boldsymbol{\mu} - \boldsymbol{\mu}^*(\mathbf{I}_{k-1} - \boldsymbol{\mu}\boldsymbol{\mu}^*)\mathbf{z}.$$

Since $\mathbf{z}^*\mathbf{z} = 1$, we have

$$\mathbf{z}^*(\mathbf{I}_{k-1} - \boldsymbol{\mu}\boldsymbol{\mu}^*)\mathbf{z} = \mathbf{z}^*\mathbf{z} - \mathbf{z}^*\boldsymbol{\mu}\boldsymbol{\mu}^*\mathbf{z} = 1 - \mathbf{z}^*\boldsymbol{\mu}\boldsymbol{\mu}^*\mathbf{z}.$$

Further, $(\mathbf{I}_{k-1} - \boldsymbol{\mu}\boldsymbol{\mu}^*)\boldsymbol{\mu} = \boldsymbol{\mu} - \boldsymbol{\mu}(\boldsymbol{\mu}^*\boldsymbol{\mu}) = \mathbf{0}$. Hence (14.7.4) is proved. Thus, the exponential factor in the complex Watson density reduces to

$$\exp\{-\tfrac{1}{2}(2\kappa)(\mathbf{z} - \boldsymbol{\mu})^*(\mathbf{I}_{k-1} - \boldsymbol{\mu}\boldsymbol{\mu}^*)^-(\mathbf{z} - \boldsymbol{\mu})\},$$

and so the complex normal approximation follows. Hence for large κ, we have that

$$2\kappa\left\{(\mathbf{z} - \boldsymbol{\mu})^*(\mathbf{I}_{k-1} - \boldsymbol{\mu}\boldsymbol{\mu}^*)^-(\mathbf{z} - \boldsymbol{\mu})\right\} = 2\kappa\left\{1 - \mathbf{z}^*\boldsymbol{\mu}\boldsymbol{\mu}^*\mathbf{z}\right\}$$

is approximately distributed as χ^2_{2k-4}, since the complex rank of $(\mathbf{I}_{k-1} - \boldsymbol{\mu}\boldsymbol{\mu}^*)$ is $k-2$. Further,

$$\kappa\mathbf{z}^*\boldsymbol{\mu}\boldsymbol{\mu}^*\mathbf{z} = \kappa\cos^2\rho,$$

where ρ is the Procrustes distance between the shapes corresponding to $\mathbf{z}$ and $\boldsymbol{\mu}$, so that, for large κ,

$$2\kappa(1 - \mathbf{z}^*\boldsymbol{\mu}\boldsymbol{\mu}^*\mathbf{z}) = 2\kappa\sin^2\rho \;\dot{\sim}\; \chi^2_{2k-4}, \tag{14.7.5}$$

which is analogous to (10.7.25) for real Watson distributions.

14.7.5 Maximum Likelihood Estimation

Let $\mathbf{z}_1, \ldots, \mathbf{z}_n$ be a random sample from a population modelled by a complex Watson distribution, and suppose that $n \geq k-1$. Set

$$\mathbf{S} = \sum_{i=1}^{n} \mathbf{z}_i \mathbf{z}_i^*$$

to be the $(k-1) \times (k-1)$ complex sum of squares and products matrix. Suppose that the eigenvalues of $\mathbf{S}$ are positive with eigenvalues $l_{k-1} > l_{k-2} \geq \ldots \geq l_1 > 0$ and let $\mathbf{g}_{k-1}, \ldots, \mathbf{g}_1$ denote the corresponding unit eigenvectors. We assume here $\kappa > 0$, the case of interest in shape analysis.

Under the complex Watson distribution $\mathbb{C}W_{k-2}(\boldsymbol{\mu}, \kappa)$ the maximum likelihood estimators of $\boldsymbol{\mu}$ and κ are given by

$$\hat{\boldsymbol{\mu}} = \mathbf{g}_{k-1},$$

and the solution to

$$\frac{c'(\hat{\kappa})}{c(\hat{\kappa})} = \frac{l_{k-1}}{n},$$

where $c'(\kappa) = \partial c(\kappa)/\partial \kappa$. Under high concentrations

$$\hat{\kappa} \simeq \frac{n(k-2)}{n - l_{k-1}}. \tag{14.7.6}$$

We prove this as follows. The log-likelihood for $\boldsymbol{\mu}$ and κ based on the data reduces to

$$\begin{aligned}\log L(\boldsymbol{\mu}, \kappa; \mathbf{z}_1, \ldots, \mathbf{z}_n) &= \kappa \sum_{i=1}^{n} \mathbf{z}_i^* \boldsymbol{\mu}\boldsymbol{\mu}^* \mathbf{z}_i - n \log c(\kappa) \\ &= \kappa \operatorname{tr}(\mathbf{S}\boldsymbol{\mu}\boldsymbol{\mu}^*) - n \log c(\kappa) \\ &= \kappa \operatorname{tr}\left(\left(\sum_{j=1}^{k-1} l_j \mathbf{g}_j \mathbf{g}_j^*\right) \boldsymbol{\mu}\boldsymbol{\mu}^*\right) - n \log c(\kappa).\end{aligned}$$

Holding κ constant, it can be seen that

$$\hat{\boldsymbol{\mu}} = e^{i\alpha} \mathbf{g}_{k-1}$$

provides the maximum, where α is an arbitrary rotation angle, and so

$$\log L(\hat{\boldsymbol{\mu}}, \kappa; \mathbf{z}_1, \ldots, \mathbf{z}_n) = \kappa l_{k-1} - n \log c(\kappa).$$

The maximum likelihood estimator of κ is found by solving

$$\frac{\partial \log c(\kappa)}{\partial \kappa} = \frac{c'(\kappa)}{c(\kappa)} = \frac{1}{n} l_{k-1}.$$

Under high concentration $c(\kappa) \simeq 2\pi^{k-1}\kappa^{2-k}e^{\kappa}$, and so

$$c'(\kappa) \simeq 2\pi^{k-1}\kappa^{2-k}e^{\kappa}\left(1 - \frac{k-2}{\kappa}\right).$$

Therefore

$$\frac{c'(\kappa)}{c(\kappa)} \simeq 1 - \frac{k-2}{\kappa} \simeq \frac{l_{k-1}}{n},$$

and so the high-concentration approximate maximum likelihood estimator (14.7.6) for κ follows.

14.7.6 Hypothesis Testing

Various one- and two-sample tests for concentrated real Watson distributions are given in Sections 10.7.3 and 10.7.4 Analogous tests can be derived in exactly the same way for complex Watson distributions, where a random sample $\mathbf{z}_1, \ldots, \mathbf{z}_n$ is available. Using the approximate distribution (14.7.5) we have

$$2\kappa \sum_{i=1}^{n} \sin^2 \rho(\mathbf{z}_i, \boldsymbol{\mu}) \overset{\cdot}{\sim} \chi^2_{(2k-4)n}, \quad \kappa \to \infty \tag{14.7.7}$$

and

$$2\kappa \sum_{i=1}^{n} \sin^2 \rho(\mathbf{z}_i, \hat{\boldsymbol{\mu}}) \overset{\cdot}{\sim} \chi^2_{(2k-4)(n-1)}, \quad \kappa \to \infty, \tag{14.7.8}$$

as $2k-4$ parameters are estimated to obtain $\hat{\boldsymbol{\mu}}$. Hence,

$$F_1 = (n-1)\frac{\sum_{i=1}^n \sin^2 \rho(\mathbf{z}_i, \boldsymbol{\mu}) - \sum_{i=1}^n \sin^2 \rho(\mathbf{z}_i, \hat{\boldsymbol{\mu}})}{\sum_{i=1}^n \sin^2 \rho(\mathbf{z}_i, \hat{\boldsymbol{\mu}})} \overset{\cdot}{\sim} F_{2k-4,(n-1)(2k-4)}.$$

This result can be used to obtain single-sample tests on $\boldsymbol{\mu}$ and to obtain confidence regions for $\boldsymbol{\mu}$. In particular, a $100(1-\alpha)\%$ confidence region for $\boldsymbol{\mu}$ is given by values of $\boldsymbol{\mu}$ satisfying $\|\boldsymbol{\mu}\| = 1$ and

$$\operatorname{tr}(\mathbf{S}\boldsymbol{\mu}\boldsymbol{\mu}^*) \geq n - \frac{1}{n-1}\left((1 + F_{2k-4,(n-1)(2k-4);1-\alpha})\sum_{i=1}^{n} \sin^2 \rho(\mathbf{z}_i, \hat{\boldsymbol{\mu}})\right),$$

where $F_{p,q;1-\alpha}$ is the $100(1-\alpha)$th percentile of the $F_{p,q}$ distribution.

Now consider two independent random samples $\mathbf{z}_1, \ldots, \mathbf{z}_n$ from $\mathbb{C}W(\boldsymbol{\mu}, \kappa)$ and $\mathbf{y}_1, \ldots, \mathbf{y}_n$ from $\mathbb{C}W(\boldsymbol{\nu}, \kappa)$. We wish to test between

$$H_0 : [\boldsymbol{\mu}] = [\boldsymbol{\nu}] \text{ and } H_1 : [\boldsymbol{\mu}] \neq [\boldsymbol{\nu}],$$

where $[\boldsymbol{\mu}] = \{e^{i\alpha}\boldsymbol{\mu} : 0 \leq \alpha < 2\pi\}$ (i.e. $[\boldsymbol{\mu}]$ represents the shape corresponding to the modal preshape $\boldsymbol{\mu}$). Using (14.7.7) and (14.7.8), it follows that for large κ

$$2\kappa\left\{\sum_{i=1}^{n}\sin^2\rho(\mathbf{z}_i,\boldsymbol{\mu})+\sum_{j=1}^{m}\sin^2\rho(\mathbf{y}_j,\boldsymbol{\nu})\right\}\ \dot{\sim}\ \chi^2_{(2k-4)(n+m)}$$

and

$$2\kappa\left\{\sum_{i=1}^{n}\sin^2\rho(\mathbf{z}_i,\hat{\boldsymbol{\mu}})+\sum_{j=1}^{m}\sin^2\rho(\mathbf{y}_j,\hat{\boldsymbol{\nu}})\right\}\ \dot{\sim}\ \chi^2_{(2k-4)(n+m-2)}.$$

By analogy with analysis of variance we can write

$$\sum_{i=1}^{n}\sin^2\rho(\mathbf{z}_i,\hat{\hat{\boldsymbol{\mu}}})+\sum_{j=1}^{m}\sin^2\rho(\mathbf{y}_j,\hat{\hat{\boldsymbol{\mu}}})=\sum_{i=1}^{n}\sin^2\rho(\mathbf{z}_i,\hat{\boldsymbol{\mu}})+\sum_{j=1}^{m}\sin^2\rho(\mathbf{y}_j,\hat{\boldsymbol{\nu}})+B,$$

where $\hat{\hat{\boldsymbol{\mu}}}$ is the overall maximum likelihood estimator of $\boldsymbol{\mu}$ if the two groups are pooled and

$$B=\sum_{i=1}^{n}\sin^2\rho(\mathbf{z}_i,\hat{\hat{\boldsymbol{\mu}}})+\sum_{j=1}^{m}\sin^2\rho(\mathbf{y}_j,\hat{\hat{\boldsymbol{\mu}}})-\sum_{i=1}^{n}\sin^2\rho(\mathbf{z}_i,\hat{\boldsymbol{\mu}})-\sum_{j=1}^{m}\sin^2\rho(\mathbf{y}_j,\hat{\boldsymbol{\nu}}).$$

Note that B is analogous to the explained sum of squares. Since

$$2\kappa\left\{\sum_{i=1}^{n}\sin^2\rho(\mathbf{z}_i,\hat{\hat{\boldsymbol{\mu}}})+\sum_{j=1}^{m}\sin^2\rho(\mathbf{y}_j,\hat{\hat{\boldsymbol{\mu}}})\right\}\ \dot{\sim}\ \chi^2_{(2k-4)(n+m-1)},$$

it follows that

$$2\kappa B\ \dot{\sim}\ \chi^2_{2k-4},\qquad \kappa\to\infty. \tag{14.7.9}$$

Therefore, under H_0 we have

$$F_2=\frac{(n+m-2)B}{\sum_{i=1}^{n}\sin^2\rho(\mathbf{z}_i,\hat{\boldsymbol{\mu}})+\sum_{j=1}^{m}\sin^2\rho(\mathbf{y}_j,\hat{\boldsymbol{\nu}})}\ \dot{\sim}\ F_{2k-4,(2k-4)(n+m-2)}, \tag{14.7.10}$$

and so we reject H_0 for large values of F_2.

Using Taylor series expansions for large concentrations

$$B\simeq(n^{-1}+m^{-1})^{-1}\sin^2\rho(\hat{\mu},\hat{\nu}),$$

and so for large κ the test statistic F_2 is equivalent to the two-sample test statistic of Goodall (1991).

14.7.7 Likelihood Ratio Tests

An alternative procedure is that based on the likelihood ratio, which does not require large concentrations. Consider two independent random samples $\mathbf{z}_1, \ldots, \mathbf{z}_n$ from $\mathbb{C}W_{k-2}(\boldsymbol{\mu}, \kappa_1)$ and $\mathbf{y}_1, \ldots, \mathbf{y}_m$ from $\mathbb{C}W_{k-2}(\boldsymbol{\nu}, \kappa_2)$. We wish to test

$$H_0 : (\boldsymbol{\mu}, \boldsymbol{\nu}, \kappa_1, \kappa_2) \in \Omega_0, \text{ against } H_1 : (\boldsymbol{\mu}, \boldsymbol{\nu}, \kappa_1, \kappa_2) \in \Omega_1,$$

where $\Omega_0 \subset \Omega_1$ and $\dim(\Omega_0) = p < \dim(\Omega_1) = q \le 4k - 2$. For example, we could test for equality in modal shapes or concentration parameters. Let Λ be the likelihood ratio for testing H_0 against H_1. Then, using Wilks's theorem, we have

$$w = -2 \log \Lambda \simeq \chi^2_{q-p}$$

for large samples.

Alternatively, we could consider a Bayesian approach to inference. Consider modelling a random sample of data $\mathbf{z}_1, \ldots, \mathbf{z}_n$ with a complex Watson distribution with mode $\boldsymbol{\mu}$ and concentration parameter κ. The concentration parameter is assumed to be known. Let the prior distribution of $\boldsymbol{\mu}$ be complex Bingham with known parameter matrix $\mathbf{A}$. Then the posterior density of $\boldsymbol{\mu}$ is given by

$$\pi(\boldsymbol{\mu}|\mathbf{z}_1, \ldots, \mathbf{z}_n) \propto \exp\left\{\boldsymbol{\mu}^* \mathbf{A} \boldsymbol{\mu} + \kappa \sum_{i=1}^{n} \mathbf{z}_i^* \boldsymbol{\mu}\boldsymbol{\mu}^* \mathbf{z}_i\right\} = \exp\{\boldsymbol{\mu}^*(\kappa \mathbf{S} + \mathbf{A})\boldsymbol{\mu}\},$$

where $\mathbf{S} = \sum_{i=1}^{n} \mathbf{z_i z_i^*}$ is the complex sum of squares and products matrix. Since the posterior is also a complex Bingham distribution (but with parameter matrix $\kappa\mathbf{S} + \mathbf{A}$), the complex Bingham prior is a conjugate prior.

Example 14.9

As a complex Watson distribution with $\kappa > 0$ is a complex Bingham distribution with all eigenvalues except the largest equal, under the complex Watson distribution we should expect all eigenvalues $l_1, \ldots, l_{k-2}$ to be approximately equal. In other words, the shape variability for the complex Watson distribution is isotropic and all principal components of shape variability have equal weightings. This observation provides us with a model-checking procedure.

Consider the schizophrenia data described by Bookstein (1996), where 13 landmarks are taken on near-midsagittal two-dimensional slices from magnetic resonance brain scans of $n = 14$ schizophrenic patients and $m = 14$ normal subjects. It is of interest to study any shape differences in the brain between the two groups, in average (modal) shape. If morphometric differences between the two groups can be established then this should enable researchers to gain an increased understanding of the condition. In Fig. 14.10 we see the $k = 13$

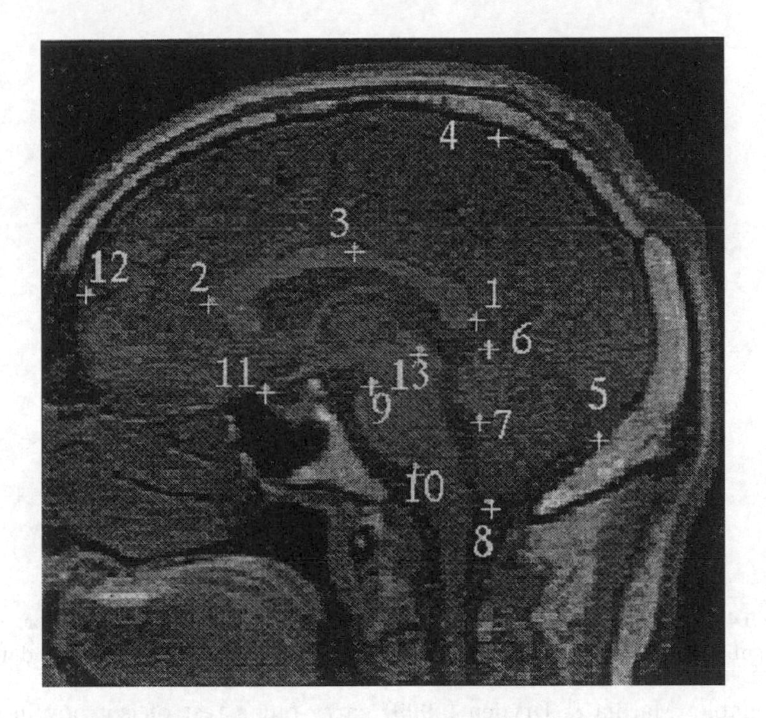

Figure 14.10 The 13 landmarks on a near-midsagittal brain scan of a schizophrenic patient (after Bookstein, 1996). The landmark positions are located approximately at each cross (+). (From Dryden & Mardia, 1998. © John Wiley & Sons, Ltd. Reproduced with permission.)

landmarks on a two-dimensional slice from the scan of a schizophrenic patient.

The Procrustes rotated data for the groups are displayed in Fig. 14.11. We see that there are generally circular scatters of points at each landmark in each group (as required for an isotropic model, such as the complex Watson model). The eigenvalues of the complex sum of squares and products matrix for the controls are

$$(13.935, 0.0222, 0.0162, 0.0090, 0.0063, 0.0044,$$
$$0.0028, 0.0017, 0.0012, 0.0004, 0.0003, 0.0002),$$

and for the schizophrenic patients

$$(13.925, 0.0274, 0.0191, 0.0086, 0.0079, 0.0041,$$
$$0.0030, 0.0020, 0.0013, 0.0007, 0.0005, 0.0001).$$

If a complex Watson distribution is appropriate then all eigenvalues but the largest are approximately equal in each group, which does not seem

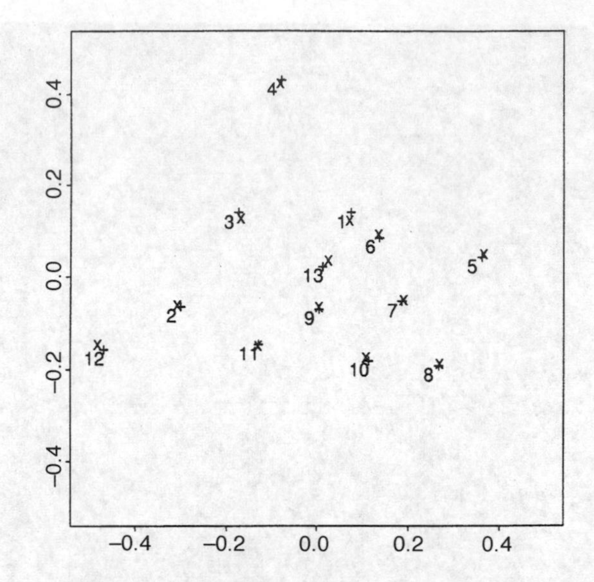

Figure 14.11 The maximum likelihood estimates of the modal shapes for the normal subjects (×) and schizophrenic patients (+) for the brain landmark data.

unreasonable. Mardia & Dryden (1999) carry out a test of isotropy in each group and find that there is no evidence against isotropy. They carry out a Monte Carlo permutation test as well, to verify that the same conclusions hold without such restrictive modelling assumptions.

The maximum likelihood estimates of modal shape for each group are displayed in Fig. 14.11, by plotting $\mathbf{H}^T\hat{\boldsymbol{\mu}}$. The maximum likelihood estimates of the concentration parameter are

$$\hat{\kappa} = 2377.4 \quad \text{(controls)} \quad \text{and} \quad \hat{\kappa} = 2056.7 \quad \text{(schizophrenics)},$$

using the large-concentration approximation of (14.7.6). The concentration parameters of the two groups are fairly similar. The value of B in the numerator of the test statistic F_2 of (14.7.10) is 0.010. The denominator is 0.139, and so $F_2 = 1.89$. Since

$$\Pr(F_{22,572} \geq 1.893) \simeq 0.0085,$$

we conclude that the subjects with schizophrenia have different-shaped modal landmark configurations from the control subjects for this study. Carrying out the likelihood ratio test of equal modal shapes (but not restricting the concentrations to be equal) we have $-2\log\Lambda = 43.294$. Since $\Pr(\chi^2_{22} \geq 43.294) = 0.0043$, there is strong evidence against equal modal shapes. A permutation test carried out by Mardia & Dryden (1999) leads to the same conclusion.

14.8 THE COMPLEX ANGULAR GAUSSIAN MODEL

Another suitable model is an extension (Kent, 1997) of the angular central Gaussian distributions (9.4.7) of Tyler (1987). Consider the distribution on $\mathbb{C}S^{k-2}$ with density

$$f(\mathbf{z}) = |\mathbf{\Sigma}|^{-1}(\mathbf{z}^*\mathbf{\Sigma}^{-1}\mathbf{z})^{-k+1} \tag{14.8.1}$$

with respect to the uniform distribution, where the parameter $\mathbf{\Sigma}$ is a positive definite Hermitian matrix. The density satisfies complex symmetry condition (15.5.1), so (14.8.1) defines a distribution on $\mathbb{C}P^{k-2}$. Note that this distribution also appears as a particular case of the offset normal distribution with equal means and complex covariance matrix (see Dryden & Mardia, 1991).

Also, (14.8.1) is invariant under the rescaling $\mathbf{\Sigma} \mapsto c\mathbf{\Sigma}, c > 0$. We can verify that the normalising constant in (14.8.1) is correct by the same method as used for the complex Bingham distribution. The mode of the density (14.8.1) is the dominant eigenvector of $\mathbf{\Sigma}$. Hence, it is natural to define an average shape from a set of landmark data by the dominant eigenvector of the maximum likelihood estimator of $\mathbf{\Sigma}$. This maximum likelihood estimator $\hat{\mathbf{\Sigma}}$ is given by the solution of the estimating equation,

$$\hat{\mathbf{\Sigma}} = \frac{k-1}{n}\sum_{j=1}^{n} \frac{\mathbf{z}_j\mathbf{z}_j^*}{\mathbf{z}_j^*\hat{\mathbf{\Sigma}}^{-1}\mathbf{z}_j}, \tag{14.8.2}$$

so that $\hat{\mathbf{\Sigma}}$ can be viewed as a weighted complex sum of squares and products matrix. (Note that (14.8.2) is analogous to (10.3.46).) This equation can be solved iteratively by the EM algorithm (Kent, 1997) using

$$\mathbf{\Sigma}^{(\nu+1)} = \frac{1}{n}\sum_{j=1}^{n} v_j^{(\nu)}\mathbf{z}_j\mathbf{z}_j^*, \quad \nu = 1, 2, \ldots, \tag{14.8.3}$$

where

$$v_j^{(\nu)} = (k-1)/\mathbf{z}_j^*\{\mathbf{\Sigma}^{(\nu)}\}^{-1}\mathbf{z}_j. \tag{14.8.4}$$

Under mild regularity conditions on the data the iterations in (14.8.3) converge to the maximum likelihood estimator $\hat{\mathbf{\Sigma}}$, which is unique up to scaling. Note that in (14.8.3) outlying values of $\mathbf{z}$ are downweighted when forming the complex sum of squares and products matrix. In order to link this result to conventional multivariate M-estimators, consider what happens in the case of highly concentrated data. Let the dominant eigenvector of $\mathbf{\Sigma}$ be given by $\boldsymbol{\nu} = (1, 0, \ldots, 0)^T$ for simplicity (any angular central Gaussian distribution can be rotated to this form by a suitable $(k-1)$-dimensional unitary transformation), and suppose that $\mathbf{\Sigma}$ is of the form

$$\mathbf{\Sigma} = \begin{pmatrix} 1 & 0 \\ 0 & \kappa^{-1}\mathbf{A} \end{pmatrix}, \tag{14.8.5}$$

where $\kappa > 0$ is large and $\mathbf{A}$ is a fixed $(k-2)$-dimensional positive definite matrix. In general, the projection map of $\mathbf{z}$ onto the tangent space at $\boldsymbol{\nu}$ of the shape space $\mathbb{C}P^{k-2}$ is given by $\mathbf{z} \mapsto (\mathbf{I} - \boldsymbol{\nu}\boldsymbol{\nu}^*)e^{-i\psi}\mathbf{z} = \mathbf{w}$, say, where $\psi = \arg \boldsymbol{\nu}^*\mathbf{z}$. In this case $\mathbf{w} = e^{-i\psi}(z_2, \ldots, z_{k-1})^T$ and $\psi = \arg z_1$. Under the angular central Gaussian model, ψ is uniformly distributed on S^1 independently of $(z_2, \ldots, z_{k-1})$, and the distribution of $(z_2, \ldots, z_{k-1})$ is invariant under multiplication by $e^{i\psi}$. Further, if κ is large, the distribution of $\mathbf{z}$ becomes concentrated near $\boldsymbol{\nu}$, and the distribution can be studied by projection onto the tangent space to $\mathbb{C}P^{k-2}$ at $\boldsymbol{\nu}$. In particular, if $\mathbf{v} = \sqrt{\kappa}\mathbf{w}$ then, as $\kappa \to \infty$, the limiting distribution of $\mathbf{v}$ has density proportional to

$$\begin{aligned}
\lim_{\kappa\to\infty} (\mathbf{z}^*\boldsymbol{\Sigma}^{-1}\mathbf{z})^{-k+1} &= \lim_{\kappa\to\infty} (|z_1|^2 + \mathbf{v}^*\mathbf{A}^{-1}\mathbf{v})^{-k+1} \\
&= \lim_{\kappa\to\infty} (1 - \kappa^{-1}\mathbf{v}^*\mathbf{v} + \mathbf{v}^*\mathbf{A}^{-1}\mathbf{v})^{-k+1} \\
&= (1 + \mathbf{v}^*\mathbf{A}^{-1}\mathbf{v})^{-k+1}, \quad \mathbf{v} \in \mathbb{C}^{k-2}. \qquad (14.8.6)
\end{aligned}$$

After switching to real coordinates in $\mathbb{R}^{2k-4}$, this density can be viewed as a multivariate t density with 2 degrees of freedom and with scatter matrix given by re-expressing $\frac{1}{2}\mathbf{A}$ in real coordinates (Mardia, Kent & Bibby, 1979, p. 57). This result gives some insight into the extent to which the angular central Gaussian distributions can accommodate outliers in shape analysis.

14.9 A ROTATIONALLY SYMMETRIC SHAPE FAMILY

One very large family of rotationally symmetric shape distributions consists of those in which the densities are functions of the shape distance to an average configuration. (This family is analogous to the family of distributions on S^{p-1} with densities of the form (9.3.31).) This class of densities on the preshape sphere S^{2k-1} is given by

$$c_\phi(\kappa)^{-1} \exp(\kappa\phi(\cos^2\rho)), \qquad (14.9.1)$$

where $\rho = \cos^{-1}|\mathbf{z}^*\boldsymbol{\mu}|$ and ϕ is a suitable increasing function. We have

$$c_\phi(\kappa) = \frac{2\pi^{k-1}}{(k-3)!}\int_0^1 (1-s)^{k-3}e^{\kappa\phi(s)}ds,$$

following calculations similar to those leading to (14.7.3). A particular subclass is given by the densities with

$$\phi(\cos^2\rho) = (1 - \cos^{2h}\rho)/h, \qquad (14.9.2)$$

which give the same maximum likelihood estimator of average shape as a class of shape estimators proposed by Kent (1992). The estimators become more resistant to outliers as h increases, since the densities have heavier tails for larger h. The complex Watson distributions are members of this family with $h = 1$. Other members (with $h = 1/2$) are the distributions with densities proportional to

$$\exp(\kappa \cos \rho). \tag{14.9.3}$$

Here the maximum likelihood estimator of shape is given by the *partial Procrustes mean* (obtained by matching configurations over rotations and translations but not scale); see Dryden (1991). The densities in (14.9.3) have particularly light tails and so the maximum likelihood estimator of modal shape under (14.9.3) will be more affected by outliers than that in the complex Watson shape distributions.

The family (14.9.1) of rotationally symmetric shape distributions is important for the analysis of shape data. The distributions are appropriate when an isotropic covariance structure is plausible. If strong correlations are present, then it would be more appropriate to use more general models, e.g. densities of the form

$$c_\phi(\mathbf{A})^{-1} \exp(\phi(\mathbf{z}^* \mathbf{A} \mathbf{z})),$$

where $\mathbf{A}$ is Hermitian. The complex Bingham distributions discussed in Section 14.6 are in this class. The rotationally symmetric family of distributions (14.9.1) is a sub-class of this general family of distributions. If the complex correlation structure is not appropriate then distributions such as the offset normal distributions of Dryden & Mardia (1991) could be used, although this approach is much more complicated.

14.10 SHAPES IN HIGHER DIMENSIONS

Most of the above discussion generalises to higher dimensions. For practical examples of shapes in $\mathbb{R}^3$, see Goodall & Mardia (1993) and Dryden & Mardia (1998). Let $\mathbf{x}_1, \ldots, \mathbf{x}_k$ and $\mathbf{y}_1, \ldots, \mathbf{y}_k$ be two (not totally coincident) labelled sets of k points in $\mathbb{R}^m$. We regard these two configurations as having the same shape if

$$\mathbf{y}_i = c\mathbf{R}\mathbf{x}_i + \mathbf{b}, \qquad i = 1, \ldots, k$$

for some scale factor $c > 0$, $m \times m$ rotation matrix $\mathbf{R}$ and vector $\mathbf{b}$, i.e. we can transform $\mathbf{x}_1, \ldots, \mathbf{x}_k$ into $\mathbf{y}_1, \ldots, \mathbf{y}_k$ by translation, rotation and scaling. This representation is an extension of (14.1.5). Note that we do not allow $\mathbf{R}$ to be a reflection. Following Kendall (1984), we denote by Σ_m^k the space of shapes of sets of k labelled points in $\mathbb{R}^m$. The Procrustes distance on Σ_2^k given in Section 14.2 can be generalised to a Procrustes distance on Σ_m^k. Although (as explained below) the shape space Σ_m^k is a quotient of the manifold $S^{m(k-1)-1}$ by an action of $SO(m)$, Σ_m^k is not a smooth manifold if $m > 2$ and $k > 3$. Such a shape space contains a singular set consisting of degenerate shapes. The singular set is negligible, in that it has measure zero under the uniform distribution on Σ_m^k. The complement in Σ_m^k of the singular set is a Riemannian manifold with metric given by the Procrustes distance.

For a description of shape in coordinate terms, consider a set of k (not totally coincident) points in $\mathbb{R}^m$, represented by a $k \times m$ matrix $\mathbf{X}$. Location and scale effects are easy to eliminate directly.

Let $\mathbf{H}$ be the Helmert sub-matrix as in Section 14.1. Then the $(k-1) \times m$ matrix $\mathbf{X}_H = \mathbf{HX}$ is invariant under location shifts $\mathbf{X} \mapsto \mathbf{X} + \mathbf{1}_k \mathbf{w}^T$, for $\mathbf{w}$ in $\mathbb{R}^m$. The k rows of $\mathbf{X}$ are again the *raw landmarks* and the $k-1$ rows of $\mathbf{X}_H$ are the *Helmertised landmarks.*

Scale effects can be eliminated by the standardisation $\mathbf{Z} = \mathbf{X}_H / ||\mathbf{X}_H||$, where $||\mathbf{X}_H||^2 = \text{tr}(\mathbf{X}_H^T \mathbf{X}_H)$. Then $\mathbf{Z}$ can be regarded as a point on the unit sphere $S^{m(k-1)-1}$ in $\mathbb{R}^{(k-1)m}$ and is called the *preshape* of $\mathbf{X}$. The *shape* of $\mathbf{X}$ is the equivalence class of its preshape, where a preshape $\mathbf{Z}$ is equivalent to $\mathbf{ZR}$ for any $\mathbf{R}$ in $SO(m)$.

There are two ways of passing from a configuration of k points in $\mathbb{R}^m$ to its shape. In the first way, described above, the effects of location and scale are removed first to give the preshape. Removal of the effect of rotation from the preshape then gives the shape. In the second way, described in the next paragraph, the effects of location and rotation are removed first to give the size-and-shape (or form). Removal of the effect of scale from the size-and-shape then gives the shape.

The method of obtaining the size-and-shape of a configuration of k points in $\mathbb{R}^m$ is based on the QR decomposition, as used widely in multivariate analysis. Again let the $k \times m$ matrix $\mathbf{X}$ contain the raw landmarks and $\mathbf{X}_H = \mathbf{HX}$ denote the Helmertised landmarks. To remove orientation, consider the QR decomposition of $\mathbf{X}_H$, namely,

$$\mathbf{X}_H = \mathbf{T\Gamma}, \quad \mathbf{\Gamma} \in V_n(\mathbb{R}^m), \| \mathbf{T} \| > 0, \tag{14.10.1}$$

where $n = \min(k-1, m)$, and $\mathbf{T}$ is a $(k-1) \times n$ lower triangular matrix with non-negative diagonal elements (i.e. $T_{ii} \geq 0$, $i = 1, \ldots, n$).

The most important case in practice is when $k > m$. In that case, $\mathbf{\Gamma} \in O(m)$ and $|\mathbf{\Gamma}| = \pm 1$, so that the QR decomposition removes orientation and reflection. To remove orientation only, we require $\mathbf{\Gamma} \in SO(m)$, so $|\mathbf{\Gamma}| = +1$ and T_{mm} is unrestricted. Then $\mathbf{T} = \{T_{ij} : 1 \leq i \leq j \leq m\}$ is called the *size-and-shape* (or *form*) if $\mathbf{\Gamma} \in SO(m)$ and the *reflection size-and-shape* if $\mathbf{\Gamma} \in O(m)$.

To remove scale, we divide by the Euclidean norm $\| \mathbf{T} \|$ of $\mathbf{T}$, i.e. we define $\mathbf{W}$ by

$$\mathbf{W} = \mathbf{T} / \| \mathbf{T} \|, \quad \| \mathbf{T} \| > 0. \tag{14.10.2}$$

We call $\mathbf{W}$ the *shape* of our configuration if $\mathbf{\Gamma} \in SO(m)$ and the *reflection shape* if $\mathbf{\Gamma} \in O(m)$. (If $k \leq m$ then these distinctions are irrelevant.) For $m = 2$, the QR decomposition is closely related to the Bookstein coordinates (see Dryden & Mardia, 1998, p. 81).

We can treat the Procrustes mean for higher dimensions as follows. Given $k \times m$ centred data matrices $\mathbf{U}_i$, $i = 1, \ldots, n$, define the average shape $[\hat{\mu}]$ of

$[\mathbf{U}_1], \ldots, [\mathbf{U}_n]$ as the shape of the preshape $\hat{\boldsymbol{\mu}}$ which minimises

$$\sum_{i=1}^{n} \| r_i \mathbf{U}_i \mathbf{R}_i - \boldsymbol{\mu} \|^2 \tag{14.10.3}$$

over $\mathbf{R}_i$ in $SO(m), r_i > 0$ and $\boldsymbol{\mu}$ in $S^{m(k-1)-1}$ (the centred $\hat{\boldsymbol{\mu}}$ is defined up to a rotation). The *Procrustes form average* minimises

$$\sum_{i=1}^{n} \| \mathbf{U}_i \mathbf{R}_i - \boldsymbol{\mu} \|^2 \tag{14.10.4}$$

over $\mathbf{R}_i$ in $SO(m)$ and $\boldsymbol{\mu}$ in $S^{m(k-1)-1}$. In practice, the solution is obtained iteratively.

Let the $m \times (k-1)$ matrix γ be a preshape. Then a convenient projection (Dryden & Mardia, 1993) of the shape space Σ_m^k onto the tangent space to Σ_m^k at $[\gamma]$ can be described as follows. A preshape $\mathbf{Z}$ is mapped to

$$\mathbf{v} = \{\mathbf{I}_{km-m} - \text{vec}(\gamma)\text{vec}(\gamma)^T\}\text{vec}(\mathbf{Z}\boldsymbol{\Gamma}),$$

where the rotation matrix $\boldsymbol{\Gamma}$ minimizes

$$\text{tr}(\gamma - \boldsymbol{\Gamma}\mathbf{Z})^T(\gamma - \boldsymbol{\Gamma}\mathbf{Z}), \tag{14.10.5}$$

i.e. $\boldsymbol{\Gamma} = \mathbf{U}\mathbf{V}^T$, where $\gamma^T\mathbf{Z} = \mathbf{V}\boldsymbol{\Lambda}\mathbf{U}^T$ is the singular value decomposition. Note that $\mathbf{v}^T\{\text{vec}(\gamma)\} = 0$. Here, we have given an exposition of the field of shape analysis to indicate how directional statistics and shape analysis are related but there are various other developments in shape analysis which do not fit into this framework. In particular, tangent projections to the shape space are found to be adequate for practical purposes. For various other developments, we refer to Dryden & Mardia (1998).

Appendix 1

SPECIAL FUNCTIONS

We list below some important formulae involving Bessel functions and Kummer functions.

1. The *modified Bessel function* I_p of the *first kind* and *order* p can be defined by

$$I_p(\kappa) = \frac{1}{2\pi}\int_0^{2\pi} \cos p\theta \, e^{\kappa\cos\theta}\, d\theta \tag{A.1}$$

(Abramowitz & Stegun, 1965, p. 376, 9.6.19);

2.

$$I_p(\kappa) = \sum_{r=0}^{\infty} \frac{1}{\Gamma(p+r+1)\Gamma(r+1)}\left(\frac{\kappa}{2}\right)^{2r+p} \tag{A.2}$$

(Abramowitz & Stegun, 1965, p. 375, 9.6.10);

3.

$$I_p(\kappa) = \frac{(\kappa/2)^p}{\Gamma(p+\frac{1}{2})\Gamma(\frac{1}{2})}\int_{-1}^{1} e^{\kappa t}(1-t^2)^{p-\frac{1}{2}}\, dt \tag{A.3}$$

(Abramowitz & Stegun, 1965, p. 376, 9.6.18).

4. For large κ,

$$\begin{aligned} I_p(\kappa) &= (2\pi\kappa)^{-\frac{1}{2}} e^{\kappa}\left\{1 - \frac{4p^2-1}{8\kappa} + \frac{(4p^2-1)(4p^2-9)}{2(8\kappa)^2}\right. \\ &\quad \left. - \frac{(4p^2-1)(4p^2-9)(4p^2-25)}{2(8\kappa)^3}\right\} + O\left(\frac{1}{\kappa^4}\right) \end{aligned} \tag{A.4}$$

(Abramowitz & Stegun, 1965, p. 377, 9.7.1).

5. The *Neumann addition formula* is

$$I_0(\{\kappa_1^2 + \kappa_2^2 + 2\kappa_1\kappa_2\cos\theta\}^{1/2}) = I_0(\kappa_1)I_0(\kappa_2) + 2\sum_{p=1}^{\infty} I_p(\kappa_1)I_p(\kappa_2)\cos p\theta \tag{A.5}$$

(G. N. Watson, 1948, Section 11.2).

6.

$$I'_p(\kappa) = \tfrac{1}{2}\{I_{p-1}(\kappa) + I_{p+1}(\kappa)\} \quad (A.6)$$
$$I'_0(\kappa) = I_1(\kappa) \quad (A.7)$$

(Abramowitz & Stegun, 1965, p. 376, 9.6.26–27).

7.

$$\kappa I'_p(\kappa) = pI_p(\kappa) + \kappa I_{p+1}(\kappa) \quad (A.8)$$
$$= \kappa I_{p-1}(\kappa) - pI_p(\kappa) \quad (A.9)$$

(G. N. Watson, 1948, p. 79; Abramowitz & Stegun, 1965, p. 376, 9.6.26).

8. In (9.3.8) we defined A_p by

$$A_p(\kappa) = \frac{\int_{-1}^{1} t^2 e^{\kappa t}(1-t^2)^{(p-3)/2} dt}{\int_{-1}^{1} e^{\kappa t}(1-t^2)^{(p-3)/2} dt}. \quad (A.10)$$

9.

$$A_p(\kappa) = \frac{I_{p/2}(\kappa)}{I_{p/2-1}(\kappa)}. \quad (A.11)$$

10. For small κ, (A.2) gives

$$A_p(\kappa) = \frac{1}{p}\kappa - \frac{1}{p^2(p+2)}\kappa^3 + O\left(\kappa^5\right) \quad (A.12)$$

(Schou, 1978, (5)).

11. For large κ, (A.4) gives

$$A_p(\kappa) = 1 - \frac{p-1}{2}\frac{1}{\kappa} + \frac{(p-1)(p-3)}{8}\frac{1}{\kappa^2} + O\left(\frac{1}{\kappa^3}\right) \quad (A.13)$$

(Schou, 1978, (6)).

12. From (A.8) and (A.9),

$$A'_p(\kappa) = 1 - A_p(\kappa)^2 - \frac{p-1}{\kappa}A_p(\kappa) \quad (A.14)$$

(Schou, 1978, (3)).

13. The *Bessel function* J_p of the first kind and *order* p can be defined by

$$J_p(\kappa) = \sum_{r=0}^{\infty}(-1)^r \frac{1}{\Gamma(p+r+1)\Gamma(r+1)}\left(\frac{\kappa}{2}\right)^{2r+p} \quad (A.15)$$

(Abramowitz & Stegun, 1965, p. 360, 9.1.10).

14.
$$J_p(\kappa) = e^{\frac{1}{2}p\pi i} I_p(e^{-\frac{1}{2}p\pi i}\kappa) \tag{A.16}$$

(Abramowitz & Stegun, 1965, p. 375, 9.6.3).

15. The *Kummer function* $M(a, b, \cdot)$ can be defined by

$$M(a, b, \kappa) = B\,(a, b-a)^{-1} \int_{-1}^{1} e^{\kappa t^2} t^{2a-1} (1-t^2)^{b-a-1} dt \tag{A.17}$$

(Abramowitz & Stegun, 1965, p. 505, 13.2.1). The Kummer function $M(a, b, \cdot)$ is also known as the *confluent hypergeometric function* ${}_1F_1(a, b, \cdot)$.

16.
$$M(a, b, \kappa) = \sum_{n=0}^{\infty} \frac{\Gamma(a+n)\Gamma(b)}{\Gamma(a)\Gamma(b+n)} \frac{\kappa^n}{n!} \tag{A.18}$$

(Abramowitz & Stegun, 1965, p. 504, 13.1.2).

17. For κ real with $|\kappa|$ large,

$$M(a, b, \kappa) = \frac{\Gamma(b)}{\Gamma(b-a)}(-\kappa)^{-a} M_- + \frac{\Gamma(b)}{\Gamma(a)} e^{\kappa} \kappa^{a-b} M_+, \tag{A.19}$$

where

$$M_- = 1 - \frac{a(a-b+1)}{\kappa} + \frac{a(a+1)(a-b+1)(a-b+2)}{2\kappa^2} + O\left(\frac{1}{|\kappa|^3}\right)$$
$$M_+ = 1 + \frac{(b-a)(1-a)}{\kappa} + \frac{(b-a)(b-a+1)(1-a)(2-a)}{2\kappa^2} + O\left(\frac{1}{|\kappa|^3}\right)$$

(Abramowitz & Stegun, 1965, p. 508, 13.5.1).

18.
$$M'(a, b, \kappa) = \frac{a}{b} M(a+1, b+1, \kappa) \tag{A.20}$$

(Abramowitz & Stegun, 1965, p. 507, 13.4.8).

19. In (10.3.32) we defined D_p by

$$D_p(\kappa) = \frac{\int_0^1 t^2 e^{\kappa t^2} (1-t^2)^{(p-3)/2} dt}{\int_0^1 e^{\kappa t^2} (1-t^2)^{(p-3)/2} dt}. \tag{A.21}$$

20.
$$D_p(\kappa) = \frac{M(3/2, p/2+1, \kappa)}{pM(1/2, p/2, \kappa)}. \tag{A.22}$$

21. For small κ, (A.18) gives

$$D_p(\kappa) = \frac{1}{p}\left\{1 + \frac{2(p-1)}{p(p+2)}\kappa\right\} + O\left(\kappa^2\right). \tag{A.23}$$

22. For κ real with κ large, (A.19) gives

$$D_p(\kappa) = 1 - \frac{p-1}{2\kappa} + O\left(\frac{1}{\kappa^2}\right), \quad \kappa \to \infty \tag{A.24}$$

$$D_p(\kappa) = \frac{1}{2|\kappa|} + O\left(\frac{1}{|\kappa|^2}\right), \quad \kappa \to -\infty. \tag{A.25}$$

23. The *hypergeometric function* ${}_1F_1\,(r/2, p/2; \cdot)$ of matrix argument can be defined by

$${}_1F_1\left(\frac{r}{2}, \frac{p}{2}; \mathbf{A}\right) = \int_{V_r(\mathbb{R}^p)} e^{\mathbf{X}^T\mathbf{A}\mathbf{X}} d\mathbf{X}, \tag{A.26}$$

where $\mathbf{A}$ is a symmetric $p \times p$ matrix and the integration is with respect to the uniform distribution on the Stiefel manifold $V_r(\mathbb{R}^p)$ (Muirhead, 1982, p. 288, Ex. 7.8).

In particular,

$${}_1F_1\left(\frac{1}{2}, \frac{p}{2}; \mathbf{A}\right) = \int_{S^{p-1}} e^{\mathbf{x}^T\mathbf{A}\mathbf{x}} d\mathbf{x}, \tag{A.27}$$

where the integration is with respect to the uniform distribution on S^{p-1}.

Definitions of the general hypergeometric functions ${}_pF_q$ of matrix argument are given in Muirhead (1982, Section 7.3).

24. The *hypergeometric function* ${}_0F_1\,(p/2; \cdot)$ of matrix argument can be defined by

$${}_0F_1\left(\frac{p}{2}; \frac{1}{4}\mathbf{F}^T\mathbf{F}\right) = \int_{V_r(\mathbb{R}^p)} e^{\mathrm{tr}(\mathbf{F}^T\mathbf{X})} d\mathbf{X}, \tag{A.28}$$

where $\mathbf{F}$ is a $p \times r$ matrix and the integration is with respect to the uniform distribution on the Stiefel manifold $V_r(\mathbb{R}^p)$ (Muirhead, 1982, p. 262, Theorem 7.4.1).

Appendix 2

TABLES AND CHARTS FOR THE CIRCULAR CASE

The tables and charts in this appendix are presented in the same order in which they were first cited in the text.

Appendix 2.1. The cumulative distribution function $F(\alpha) = F(\alpha; \pi, \kappa)$ of the von Mises distribution $M(\pi, \kappa)$ transferred to the linear interval $[0, 2\pi]$ by cutting the unit circle at 0.

α	$F(\alpha)$									
(degrees)	$\kappa = 0$	$\kappa = 0.2$	$\kappa = 0.4$	$\kappa = 0.6$	$\kappa = 0.8$	$\kappa = 1.0$	$\kappa = 1.2$	$\kappa = 1.4$	$\kappa = 1.6$	$\kappa = 1.8$
0	0.00000	0.00000	0.00000	0.00000	0.00000	0.00000	0.00000	0.00000	0.00000	0.00000
5	0.01389	0.01126	0.00895	0.00699	0.00536	0.00404	0.00301	0.00221	0.00161	0.00116
10	0.02778	0.02254	0.01793	0.01400	0.01074	0.00811	0.00604	0.00444	0.00323	0.00233
15	0.04167	0.03385	0.02697	0.02108	0.01620	0.01225	0.00913	0.00672	0.00490	0.00353
20	0.05556	0.04522	0.03608	0.02826	0.02175	0.01647	0.01230	0.00907	0.00662	0.00479
25	0.06944	0.05665	0.04531	0.03557	0.02744	0.02083	0.01559	0.01153	0.00843	0.00611
30	0.08333	0.06816	0.05467	0.04304	0.03329	0.02535	0.01903	0.01411	0.01035	0.00753
35	0.09722	0.07978	0.06420	0.05071	0.03936	0.03007	0.02266	0.01686	0.01241	0.00906
40	0.11111	0.09152	0.07392	0.05861	0.04567	0.03504	0.02650	0.01981	0.01465	0.01073
45	0.12500	0.10338	0.08386	0.06679	0.05228	0.04029	0.03062	0.02299	0.01709	0.01259
50	0.13889	0.11540	0.09405	0.07527	0.05921	0.04587	0.03505	0.02647	0.01978	0.01466
55	0.15278	0.12757	0.10452	0.08409	0.06653	0.05184	0.03985	0.03028	0.02278	0.01699
60	0.16667	0.13992	0.11529	0.09331	0.07428	0.05825	0.04059	0.03450	0.02614	0.01965
65	0.18056	0.15246	0.12639	0.10295	0.08251	0.06517	0.05082	0.03919	0.02994	0.02270
70	0.19444	0.16520	0.13784	0.11306	0.09128	0.07265	0.05711	0.04442	0.03425	0.02620
75	0.20833	0.17815	0.14968	0.12368	0.10064	0.08078	0.06407	0.05030	0.03915	0.03027
80	0.22222	0.19132	0.16192	0.13485	0.11066	0.08962	0.07176	0.05690	0.04477	0.03501
85	0.23611	0.20471	0.17460	0.14662	0.12139	0.09925	0.08028	0.06436	0.05122	0.04053
90	0.25000	0.21834	0.18772	0.15901	0.13289	0.10975	0.08974	0.07277	0.05863	0.04699
95	0.26389	0.23222	0.20130	0.17206	0.14522	0.12122	0.10025	0.08228	0.06714	0.05455
100	0.27778	0.24633	0.21537	0.18582	0.15844	0.13372	0.11191	0.09302	0.07693	0.06339
105	0.29167	0.26069	0.22992	0.20030	0.17260	0.14734	0.12483	0.10514	0.08816	0.07370
110	0.30556	0.27529	0.24498	0.21554	0.18774	0.16217	0.13913	0.11876	0.10101	0.08571
115	0.31944	0.29014	0.26054	0.23154	0.20392	0.17825	0.15491	0.13405	0.11566	0.09962

120	0.33333	0.30522	0.27659	0.24832	0.22114	0.19566	0.17226	0.15112	0.13228	0.11565
125	0.34722	0.32053	0.29314	0.26587	0.23944	0.21444	0.19125	0.17009	0.15103	0.13402
130	0.36111	0.33606	0.31017	0.28420	0.25882	0.23460	0.21194	0.19106	0.17206	0.15491
135	0.37500	0.35180	0.32766	0.30327	0.27926	0.25616	0.23435	0.21408	0.19545	0.17847
140	0.38889	0.36774	0.34559	0.32306	0.30073	0.27909	0.25849	0.23918	0.22127	0.20479
145	0.40278	0.38385	0.36392	0.34353	0.32319	0.30334	0.28431	0.26633	0.24951	0.23390
150	0.41667	0.40013	0.38263	0.36463	0.34656	0.32883	0.31172	0.29544	0.28010	0.26575
155	0.43056	0.41655	0.40166	0.38628	0.37077	0.35546	0.34060	0.32638	0.31290	0.30020
160	0.44444	0.43309	0.42098	0.40841	0.39570	0.38309	0.37079	0.35897	0.34769	0.33701
165	0.45833	0.44973	0.44053	0.43095	0.42122	0.41155	0.40208	0.39294	0.38418	0.37585
170	0.47222	0.46644	0.46025	0.45379	0.44722	0.44066	0.43423	0.42800	0.42201	0.41630
175	0.48611	0.48321	0.48009	0.47684	0.47353	0.47022	0.46696	0.46381	0.46077	0.45786
180	0.50000	0.50000	0.50000	0.50000	0.50000	0.50000	0.50000	0.50000	0.50000	0.50000

	$\kappa = 2.0$	$\kappa = 2.2$	$\kappa = 2.4$	$\kappa = 2.6$	$\kappa = 2.8$	$\kappa = 3.0$	$\kappa = 3.2$	$\kappa = 3.4$	$\kappa = 3.6$	$\kappa = 3.8$
0	0.00000	0.00000	0.00000	0.00000	0.00000	0.00000	0.00000	0.00000	0.00000	0.00000
5	0.00083	0.00059	0.00041	0.00029	0.00020	0.00014	0.00010	0.00007	0.00005	0.00003
10	0.00167	0.00118	0.00084	0.00059	0.00041	0.00029	0.00020	0.00014	0.00010	0.00007
15	0.00253	0.00180	0.00127	0.00090	0.00063	0.00044	0.00031	0.00021	0.00015	0.00010
20	0.00344	0.00245	0.00174	0.00123	0.00086	0.00060	0.00042	0.00029	0.00020	0.00014
25	0.00440	0.00314	0.00223	0.00158	0.00111	0.00078	0.00055	0.00038	0.00027	0.00019
30	0.00543	0.00389	0.00278	0.00197	0.00139	0.00098	0.00069	0.00048	0.00034	0.00024
35	0.00656	0.00472	0.00338	0.00241	0.00171	0.00121	0.00085	0.00060	0.00042	0.00029
40	0.00781	0.00564	0.00406	0.00290	0.00207	0.00147	0.00104	0.00074	0.00052	0.00037
45	0.00920	0.00669	0.00483	0.00348	0.00249	0.00178	0.00127	0.00090	0.00064	0.00046
50	0.01078	0.00788	0.00574	0.00416	0.00300	0.00216	0.00155	0.00111	0.00079	0.00057
55	0.01259	0.00927	0.00679	0.00496	0.00361	0.00262	0.00190	0.00137	0.00099	0.00071

Reproduced from Batschelet (1965) by permission of the publisher, Amer. Inst. Biol. Sci.

Appendix 2.1. *(continued)*

α	$F(\alpha)$									
(degrees)	$\kappa = 2.0$	$\kappa = 2.2$	$\kappa = 2.4$	$\kappa = 2.6$	$\kappa = 2.8$	$\kappa = 3.0$	$\kappa = 3.2$	$\kappa = 3.4$	$\kappa = 3.6$	$\kappa = 3.8$
60	0.01467	0.01089	0.00805	0.00593	0.00435	0.00319	0.00233	0.00170	0.00124	0.00090
65	0.01709	0.01281	0.00956	0.00711	0.00527	0.00390	0.00288	0.00213	0.00157	0.00116
70	0.01993	0.01509	0.01138	0.00856	0.00642	0.00481	0.00360	0.00269	0.00201	0.00150
75	0.02328	0.01782	0.01360	0.01035	0.00786	0.00596	0.00452	0.00343	0.00260	0.00197
80	0.02723	0.02111	0.01631	0.01258	0.00969	0.00746	0.00573	0.00441	0.00339	0.00261
85	0.03193	0.02508	0.01965	0.01537	0.01201	0.00938	0.00733	0.00573	0.00448	0.00350
90	0.03752	0.02988	0.02376	0.01887	0.01498	0.01189	0.00944	0.00750	0.00596	0.00475
95	0.04418	0.03571	0.02882	0.02325	0.01876	0.01514	0.01223	0.00988	0.00800	0.00648
100	0.05210	0.04276	0.03506	0.02875	0.02359	0.01936	0.01591	0.01308	0.01077	0.00888
105	0.06150	0.05127	0.04273	0.03563	0.02972	0.02482	0.02075	0.01737	0.01456	0.01222
110	0.07263	0.06152	0.05212	0.04419	0.03749	0.03185	0.02710	0.02308	0.01968	0.01681
115	0.08574	0.07379	0.06355	0.05477	0.04727	0.04084	0.03534	0.03062	0.02657	0.02308
120	0.10109	0.08840	0.07736	0.06777	0.05946	0.05223	0.04595	0.04048	0.03571	0.03154
125	0.11895	0.10564	0.09391	0.08360	0.07452	0.06652	0.05946	0.05323	0.04771	0.04282
130	0.13954	0.12581	0.11356	0.10264	0.09290	0.08421	0.07644	0.06947	0.06322	0.05760
135	0.16308	0.14917	0.13662	0.12530	0.11507	0.10583	0.09746	0.08986	0.08294	0.07665
140	0.18970	0.17592	0.16335	0.15188	0.14141	0.13183	0.12305	0.11499	0.10756	0.10071
145	0.21948	0.20618	0.19393	0.18264	0.17222	0.16259	0.15367	0.14538	0.13767	0.13048
150	0.25239	0.23996	0.22841	0.21766	0.20766	0.19832	0.18959	0.18140	0.17371	0.16647
155	0.28829	0.27713	0.26668	0.25688	0.24769	0.23903	0.23088	0.22317	0.21586	0.20893
160	0.32694	0.31744	0.30848	0.30004	0.29206	0.28450	0.27732	0.27050	0.26399	0.25776
165	0.36795	0.36047	0.35339	0.34667	0.34029	0.33421	0.32841	0.32287	0.31755	0.31244
170	0.41087	0.40570	0.40079	0.39612	0.39166	0.38740	0.38332	0.37941	0.37563	0.37200
175	0.45509	0.45246	0.44994	0.44755	0.44526	0.44306	0.44095	0.43893	0.43697	0.43508
180	0.50000	0.50000	0.50000	0.50000	0.50000	0.50000	0.50000	0.50000	0.50000	0.50000

	$\kappa = 4.0$	$\kappa = 4.2$	$\kappa = 4.4$	$\kappa = 4.6$	$\kappa = 4.8$	$\kappa = 5.0$	$\kappa = 5.2$	$\kappa = 5.4$	$\kappa = 5.6$	$\kappa = 5.8$
0	0.00000	0.00000	0.00000	0.00000	0.00000	0.00000	0.00000	0.00000	0.00000	0.00000
5	0.00002	0.00002	0.00001	0.00001	0.00001	0.00000	0.00000	0.00000	0.00000	0.00000
10	0.00005	0.00003	0.00002	0.00001	0.00001	0.00001	0.00000	0.00000	0.00000	0.00000
15	0.00007	0.00005	0.00003	0.00002	0.00002	0.00001	0.00001	0.00001	0.00000	0.00000
20	0.00010	0.00007	0.00005	0.00003	0.00002	0.00002	0.00001	0.00001	0.00000	0.00000
25	0.00013	0.00009	0.00006	0.00004	0.00003	0.00002	0.00001	0.00001	0.00001	0.00000
30	0.00016	0.00011	0.00008	0.00005	0.00004	0.00003	0.00002	0.00001	0.00001	0.00001
35	0.00021	0.00014	0.00010	0.00007	0.00005	0.00003	0.00002	0.00002	0.00001	0.00001
40	0.00026	0.00018	0.00013	0.00009	0.00006	0.00004	0.00003	0.00002	0.00001	0.00001
45	0.00032	0.00023	0.00016	0.00011	0.00008	0.00006	0.00004	0.00003	0.00002	0.00001
50	0.00040	0.00029	0.00021	0.00015	0.00010	0.00007	0.00005	0.00004	0.00003	0.00002
55	0.00051	0.00037	0.00027	0.00019	0.00014	0.00010	0.00007	0.00005	0.00004	0.00003
60	0.00066	0.00048	0.00035	0.00025	0.00018	0.00013	0.00010	0.00007	0.00005	0.00004
65	0.00085	0.00063	0.00046	0.00034	0.00025	0.00018	0.00014	0.00010	0.00007	0.00005
70	0.00112	0.00084	0.00062	0.00047	0.00035	0.00026	0.00019	0.00015	0.00011	0.00008
75	0.00149	0.00113	0.00086	0.00065	0.00049	0.00037	0.00028	0.00022	0.00016	0.00013
80	0.00201	0.00155	0.00119	0.00092	0.00071	0.00055	0.00042	0.00033	0.00025	0.00020
85	0.00274	0.00215	0.00168	0.00132	0.00104	0.00082	0.00064	0.00051	0.00040	0.00031
90	0.00378	0.00301	0.00240	0.00192	0.00154	0.00123	0.00098	0.00079	0.00063	0.00051
95	0.00525	0.00426	0.00346	0.00282	0.00229	0.00187	0.00152	0.00124	0.00102	0.00083
100	0.00733	0.00606	0.00501	0.00415	0.00344	0.00286	0.00237	0.00197	0.00164	0.00137
105	0.01027	0.00864	0.00727	0.00613	0.00517	0.00437	0.00370	0.00313	0.00265	0.00224
110	0.01437	0.01231	0.01055	0.00905	0.00777	0.00668	0.00575	0.00495	0.00426	0.00367
115	0.02008	0.01748	0.01524	0.01330	0.01162	0.01016	0.00889	0.00778	0.00682	0.00597
120	0.02789	0.02470	0.02189	0.01941	0.01724	0.01531	0.01362	0.01212	0.01079	0.00961
125	0.03847	0.03460	0.03114	0.02806	0.02530	0.02284	0.02062	0.01863	0.01685	0.01524
130	0.05253	0.04795	0.04382	0.04007	0.03667	0.03358	0.03077	0.02821	0.02588	0.02375
135	0.07089	0.06563	0.06081	0.05638	0.05232	0.04857	0.04512	0.04194	0.03900	0.03628
140	0.09438	0.08852	0.08309	0.07804	0.07334	0.06897	0.06488	0.06107	0.05751	0.05418
145	0.12377	0.11748	0.11159	0.10605	0.10084	0.09594	0.09132	0.08695	0.08283	0.07893

Appendix 2.1. *(continued)*

α (degrees)	$F(\alpha)$ $\kappa = 4.0$	$\kappa = 4.2$	$\kappa = 4.4$	$\kappa = 4.6$	$\kappa = 4.8$	$\kappa = 5.0$	$\kappa = 5.2$	$\kappa = 5.4$	$\kappa = 5.6$	$\kappa = 5.8$
150	0.15964	0.15319	0.14708	0.14129	0.13578	0.13055	0.12558	0.12083	0.11631	0.11199
155	0.20234	0.19606	0.19006	0.18433	0.17884	0.17358	0.16853	0.16368	0.15902	0.15453
160	0.25180	0.24608	0.24059	0.23530	0.23020	0.22528	0.22053	0.21593	0.21148	0.20716
165	0.30752	0.30277	0.29818	0.29374	0.28944	0.28527	0.28122	0.27728	0.27344	0.26971
170	0.36848	0.36508	0.36177	0.35857	0.35545	0.35241	0.34944	0.34655	0.34372	0.34096
175	0.43324	0.43146	0.42973	0.42805	0.42641	0.42481	0.42324	0.42171	0.42021	0.41873
180	0.50000	0.50000	0.50000	0.50000	0.50000	0.50000	0.50000	0.50000	0.50000	0.50000

	$\kappa = 6.0$	$\kappa = 6.2$	$\kappa = 6.4$	$\kappa = 6.6$	$\kappa = 6.8$	$\kappa = 7.0$	$\kappa = 7.2$	$\kappa = 7.4$	$\kappa = 7.6$	$\kappa = 7.8$
30	0.00000	0.00000	0.00000	0.00000	0.00000	0.00000	0.00000	0.00000	0.00000	0.00000
35	0.00001	0.00000	0.00000	0.00000	0.00000	0.00000	0.00000	0.00000	0.00000	0.00000
40	0.00001	0.00001	0.00000	0.00000	0.00000	0.00000	0.00000	0.00000	0.00000	0.00000
45	0.00001	0.00001	0.00000	0.00000	0.00000	0.00000	0.00000	0.00000	0.00000	0.00000
50	0.00001	0.00001	0.00001	0.00000	0.00000	0.00000	0.00000	0.00000	0.00000	0.00000
55	0.00002	0.00001	0.00001	0.00001	0.00000	0.00000	0.00000	0.00000	0.00000	0.00000
60	0.00003	0.00002	0.00001	0.00001	0.00001	0.00001	0.00000	0.00000	0.00000	0.00000
65	0.00004	0.00003	0.00002	0.00002	0.00001	0.00001	0.00001	0.00000	0.00000	0.00000
70	0.00006	0.00005	0.00003	0.00003	0.00002	0.00001	0.00001	0.00001	0.00001	0.00000
75	0.00010	0.00007	0.00006	0.00004	0.00003	0.00002	0.00002	0.00001	0.00001	0.00001
80	0.00015	0.00012	0.00009	0.00007	0.00006	0.00004	0.00003	0.00003	0.00002	0.00002
85	0.00025	0.00020	0.00015	0.00012	0.00010	0.00008	0.00006	0.00005	0.00004	0.00003

90	0.00041	0.00033	0.00026	0.00021	0.00017	0.00014	0.00011	0.00009	0.00007	0.00006
95	0.00068	0.00056	0.00046	0.00037	0.00031	0.00025	0.00021	0.00017	0.00014	0.00011
100	0.00114	0.00095	0.00079	0.00066	0.00055	0.00046	0.00038	0.00032	0.00027	0.00023
105	0.00190	0.00161	0.00137	0.00116	0.00099	0.00084	0.00071	0.00061	0.00052	0.00044
110	0.00317	0.00274	0.00236	0.00204	0.00176	0.00153	0.00132	0.00114	0.00099	0.00086
115	0.00524	0.00460	0.00404	0.00355	0.00312	0.00274	0.00241	0.00212	0.00187	0.00164
120	0.00857	0.00764	0.00681	0.00608	0.00543	0.00485	0.00434	0.00388	0.00346	0.00310
125	0.01379	0.01249	0.01131	0.01025	0.00929	0.00843	0.00765	0.00694	0.00630	0.00572
130	0.02181	0.02003	0.01841	0.01693	0.01556	0.01432	0.01318	0.01213	0.01117	0.01029
135	0.03377	0.03144	0.02928	0.02728	0.02543	0.02371	0.02211	0.02063	0.01925	0.01796
140	0.05106	0.04814	0.04540	0.04283	0.04041	0.03815	0.03602	0.03402	0.03213	0.03036
145	0.07525	0.07175	0.06844	0.06530	0.06233	0.05950	0.05681	0.05426	0.05184	0.04953
150	0.10787	0.10392	0.10015	0.09654	0.09308	0.08976	0.08658	0.08353	0.08061	0.07780
155	0.15020	0.14604	0.14202	0.13814	0.13440	0.13078	0.12728	0.12390	0.12063	0.11746
160	0.20298	0.19893	0.19499	0.19116	0.18744	0.18382	0.18029	0.17686	0.17352	0.17027
165	0.26606	0.26251	0.25904	0.25566	0.25235	0.24911	0.24594	0.24285	0.23981	0.23684
170	0.33826	0.33561	0.33302	0.33047	0.32798	0.32553	0.32312	0.32076	0.31844	0.31616
175	0.41729	0.41588	0.41449	0.41312	0.41178	0.41045	0.40915	0.40787	0.40661	0.40537
180	0.50000	0.50000	0.50000	0.50000	0.50000	0.50000	0.50000	0.50000	0.50000	0.50000

Appendix 2.1. *(continued)*

α	$F(\alpha)$										
(degrees)	$\kappa = 8.0$	$\kappa = 8.2$	$\kappa = 8.4$	$\kappa = 8.6$	$\kappa = 8.8$	$\kappa = 9.0$	$\kappa = 9.2$	$\kappa = 9.4$	$\kappa = 9.6$	$\kappa = 9.8$	$\kappa = 10.0$
70	0.00000	0.00000	0.00000	0.00000	0.00000	0.00000	0.00000	0.00000	0.00000	0.00000	0.00000
75	0.00001	0.00000	0.00000	0.00000	0.00000	0.00000	0.00000	0.00000	0.00000	0.00000	0.00000
80	0.00001	0.00001	0.00001	0.00001	0.00000	0.00000	0.00000	0.00000	0.00000	0.00000	0.00000
85	0.00002	0.00002	0.00002	0.00001	0.00001	0.00001	0.00001	0.00000	0.00000	0.00000	0.00000
90	0.00005	0.00004	0.00003	0.00003	0.00002	0.00002	0.00001	0.00001	0.00001	0.00001	0.00001
95	0.00009	0.00008	0.00006	0.00005	0.00004	0.00004	0.00003	0.00002	0.00002	0.00002	0.00001
100	0.00019	0.00016	0.00013	0.00011	0.00009	0.00008	0.00007	0.00005	0.00005	0.00004	0.00003
105	0.00038	0.00032	0.00027	0.00023	0.00020	0.00017	0.00014	0.00012	0.00011	0.00009	0.00008
110	0.00074	0.00064	0.00056	0.00048	0.00042	0.00036	0.00032	0.00027	0.00024	0.00021	0.00018
115	0.00145	0.00127	0.00112	0.00099	0.00087	0.00077	0.00068	0.00060	0.00053	0.00047	0.00041
120	0.00277	0.00248	0.00222	0.00199	0.00178	0.00160	0.00143	0.00128	0.00115	0.00103	0.00092
125	0.00519	0.00472	0.00429	0.00390	0.00354	0.00322	0.00293	0.00266	0.00242	0.00221	0.00201
130	0.00948	0.00873	0.00805	0.00742	0.00684	0.00631	0.00582	0.00537	0.00495	0.00457	0.00422
135	0.01677	0.01566	0.01462	0.01366	0.01276	0.01193	0.01115	0.01042	0.00974	0.00911	0.00852
140	0.02869	0.02712	0.02564	0.02425	0.02293	0.02169	0.02052	0.01942	0.01838	0.01740	0.01647
145	0.04733	0.04524	0.04325	0.04136	0.03955	0.03783	0.03619	0.03463	0.03314	0.03171	0.03036
150	0.07510	0.07251	0.07001	0.06762	0.06531	0.06310	0.06096	0.05891	0.05693	0.05503	0.05319
155	0.11440	0.11143	0.10855	0.10576	0.10305	0.10043	0.09789	0.09542	0.09302	0.09069	0.08843
160	0.16710	0.16400	0.16099	0.15804	0.15517	0.15236	0.14962	0.14694	0.14433	0.14177	0.13927
165	0.23393	0.23107	0.22827	0.22552	0.22283	0.22019	0.21759	0.21504	0.21253	0.21007	0.20765
170	0.31391	0.31170	0.30952	0.30738	0.30527	0.30319	0.30115	0.29913	0.29714	0.29517	0.29323
175	0.40414	0.40293	0.40174	0.40056	0.39940	0.39825	0.39712	0.39600	0.39489	0.39380	0.39271
180	0.50000	0.50000	0.50000	0.50000	0.50000	0.50000	0.50000	0.50000	0.50000	0.50000	0.50000

Reproduced from Batschelet (1965) by permission of the publisher, Amer. Inst. Biol. Sci.

Appendix 2.2. Quantiles (in degrees) of the von Mises distribution $M(0, \kappa)$ transferred to the linear interval $[-180°, 180°]$ by cutting the unit circle at 180°. The lower tail area $\Pr(-180° < \theta < -180° + \delta)$ is $\alpha/2$. The upper tail area $\Pr(180° - \delta < \theta < 180°)$ is $\alpha/2$. For values of κ greater than those shown in the table, approximate quantiles δ are given by $\delta = 180° - (180/\pi)° z_\alpha \kappa^{-\frac{1}{2}}$.

κ	$\alpha = 0.001$	0.01	0.05	0.1	κ	$\alpha = 0.001$	0.01	0.05	0.1
0.0	0.1	0.6	2.9	5.7	4.6	71.1	102.4	123.4	133.2
0.1	0.2	2.0	10.0	19.9	4.7	73.2	103.5	124.1	133.8
0.2	0.2	2.2	11.1	22.1	4.8	75.2	104.6	124.8	134.3
0.3	0.2	2.5	12.4	24.6	4.9	77.1	105.6	125.5	134.9
0.4	0.3	2.8	13.9	27.5	5.0	78.8	106.6	126.2	135.4
0.5	0.3	3.2	15.7	30.8	5.2	82.0	108.4	127.4	136.4
0.6	0.4	3.6	17.7	34.5	5.4	84.9	110.1	128.5	137.3
0.7	0.4	4.1	20.1	38.7	5.6	87.5	111.7	129.6	138.2
0.8	0.5	4.7	22.9	43.3	5.8	89.8	113.2	130.6	139.0
0.9	0.5	5.4	26.0	48.3	6.0	92.0	114.5	131.5	139.7
1.0	0.6	6.2	29.6	53.5	6.2	94.0	115.8	132.4	140.5
1.1	0.7	7.2	33.6	58.9	6.4	95.8	117.0	133.3	141.1
1.2	0.8	8.3	38.1	64.3	6.6	97.6	118.2	134.1	141.8
1.3	1.0	9.6	42.9	69.6	6.8	99.2	119.2	134.8	142.4
1.4	1.1	11.2	47.9	74.8	7.0	100.7	120.3	135.5	143.0
1.5	1.3	13.1	53.1	79.6	7.2	102.1	121.2	136.2	143.6
1.6	1.6	15.3	58.4	84.1	7.4	103.5	122.2	136.9	144.1
1.7	1.8	17.9	63.5	88.3	7.6	104.7	123.0	137.5	144.6
1.8	2.2	20.8	68.4	92.1	7.8	105.9	123.9	138.1	145.1
1.9	2.6	24.2	73.0	95.6	8.0	107.1	124.7	138.7	145.6
2.0	3.0	28.0	77.3	98.8	8.2	108.2	125.5	139.2	146.0
2.1	3.6	32.1	81.3	101.7	8.4	109.2	126.2	139.8	146.5
2.2	4.3	36.6	84.9	104.3	8.6	110.2	126.9	140.3	146.9
2.3	5.1	41.2	88.3	106.7	8.8	111.2	127.6	140.8	147.3
2.4	6.0	46.0	91.3	109.0	9.0	112.1	128.2	141.2	147.7
2.5	7.2	50.7	94.1	111.0	9.2	113.0	128.9	141.7	148.0
2.6	8.5	55.2	96.7	112.9	9.4	113.8	129.0	142.1	148.4
2.7	10.1	59.6	99.1	114.6	9.6	114.7	130.1	142.6	148.8
2.8	12.1	63.6	101.3	116.2	9.8	115.4	130.6	143.0	149.1
2.9	14.3	67.4	103.3	117.7	10.0	116.2	131.2	143.4	149.4
3.0	16.9	70.9	105.1	119.1	10.5	118.0	132.5	144.3	150.2
3.1	19.9	74.2	106.9	120.4	11.0	119.6	133.7	145.2	150.9
3.2	23.2	77.1	108.5	121.6	11.5	121.1	134.8	146.0	151.6
3.3	26.9	79.9	110.0	122.8	12.0	122.5	135.8	146.7	152.2
3.4	30.8	82.4	111.4	123.8	12.5	123.8	136.8	147.5	152.8
3.5	34.9	84.8	112.7	124.9	13.0	125.0	137.7	148.1	153.4
3.6	39.1	86.9	114.0	125.8	14	127.2	139.3	149.3	154.4
3.7	43.2	89.0	115.2	126.7	15	129.2	140.8	150.4	155.3
3.8	47.1	90.8	116.3	127.6	20	136.5	146.3	154.5	158.7
3.9	50.9	92.6	117.3	128.4	30	144.9	152.7	159.3	162.7
4.0	54.5	94.3	118.3	129.2	40	149.7	156.4	162.1	165.0
4.1	57.8	95.8	119.3	129.9	50	153.0	159.0	164.0	166.6
4.2	60.9	97.3	120.2	130.7	100	161.1	165.2	168.7	170.6
4.3	63.7	98.7	121.0	131.3		$z_\alpha = 3.291$	2.576	1.960	1.645
4.4	66.3	100.0	121.9	132.0					
4.5	68.8	101.2	122.7	132.6					

Appendix 2.3. The function A, defined by $A(\kappa) = I_1(\kappa)/I_0(\kappa)$. The mean resultant length ρ of the von Mises distribution $M(\mu, \kappa)$ is $\rho = A(\kappa)$.

κ	$A(\kappa)$	κ	$A(\kappa)$	κ	$A(\kappa)$
0.0	0.000	3.5	0.841	7.0	0.926
0.1	0.050	3.6	0.846	7.1	0.927
0.2	0.100	3.7	0.851	7.2	0.928
0.3	0.148	3.8	0.855	7.3	0.929
0.4	0.196	3.9	0.860	7.4	0.930
0.5	0.242	4.0	0.864	7.5	0.931
0.6	0.287	4.1	0.867	7.6	0.932
0.7	0.330	4.2	0.871	7.7	0.933
0.8	0.371	4.3	0.874	7.8	0.934
0.9	0.410	4.4	0.877	7.9	0.934
1.0	0.446	4.5	0.880	8.0	0.935
1.1	0.481	4.6	0.883	8.1	0.936
1.2	0.513	4.7	0.886	8.2	0.937
1.3	0.543	4.8	0.889	8.3	0.938
1.4	0.570	4.9	0.891	8.4	0.938
1.5	0.596	5.0	0.893	8.5	0.939
1.6	0.620	5.1	0.896	8.6	0.940
1.7	0.642	5.2	0.898	8.7	0.941
1.8	0.662	5.3	0.900	8.8	0.941
1.9	0.681	5.4	0.902	8.9	0.942
2.0	0.698	5.5	0.904	9.0	0.943
2.1	0.714	5.6	0.906	9.2	0.944
2.2	0.728	5.7	0.907	9.4	0.945
2.3	0.741	5.8	0.909	9.6	0.946
2.4	0.754	5.9	0.911	9.8	0.948
2.5	0.765	6.0	0.912	10	0.949
2.6	0.775	6.1	0.914	12	0.957
2.7	0.785	6.2	0.915	15	0.966
2.8	0.794	6.3	0.917	20	0.975
2.9	0.802	6.4	0.918	24	0.979
3.0	0.810	6.5	0.920	30	0.983
3.1	0.817	6.6	0.921	40	0.987
3.2	0.824	6.7	0.922	60	0.992
3.3	0.830	6.8	0.923	120	0.996
3.4	0.836	6.9	0.924	∞	1.000

Based on Table C of Batschelet (1965),
with permission of the Amer. Inst. Biol. Sci.

Appendix 2.4. The function A^{-1}. The maximum likelihood estimate $\hat{\kappa}$ of the concentration parameter κ of the von Mises distribution $M(\mu, \kappa)$ is $\hat{\kappa} = A^{-1}(\bar{R})$.

x	$A^{-1}(x)$	x	$A^{-1}(x)$	x	$A^{-1}(x)$
0.00	0.000	0.35	0.748	0.70	2.014
0.01	0.020	0.36	0.772	0.71	2.077
0.02	0.040	0.37	0.797	0.72	2.144
0.03	0.060	0.38	0.823	0.73	2.214
0.04	0.080	0.39	0.848	0.74	2.289
0.05	0.100	0.40	0.874	0.75	2.369
0.06	0.120	0.41	0.900	0.76	2.455
0.07	0.140	0.42	0.927	0.77	2.547
0.08	0.161	0.43	0.954	0.78	2.646
0.09	0.181	0.44	0.982	0.79	2.754
0.10	0.201	0.45	1.010	0.80	2.871
0.11	0.221	0.46	1.039	0.81	3.000
0.12	0.242	0.47	1.068	0.82	3.143
0.13	0.262	0.48	1.098	0.83	3.301
0.14	0.283	0.49	1.128	0.84	3.479
0.15	0.303	0.50	1.159	0.85	3.680
0.16	0.324	0.51	1.191	0.86	3.911
0.17	0.345	0.52	1.223	0.87	4.177
0.18	0.366	0.53	1.257	0.88	4.489
0.19	0.387	0.54	1.291	0.89	4.859
0.20	0.408	0.55	1.326	0.90	5.305
0.21	0.430	0.56	1.362	0.91	5.852
0.22	0.451	0.57	1.398	0.92	6.540
0.23	0.473	0.58	1.436	0.93	7.426
0.24	0.495	0.59	1.475	0.94	8.610
0.25	0.516	0.60	1.516	0.95	10.272
0.26	0.539	0.61	1.557	0.96	12.766
0.27	0.561	0.62	1.600	0.97	16.927
0.28	0.584	0.62	1.645	0.98	25.252
0.29	0.606	0.64	1.691	0.99	50.242
0.30	0.629	0.65	1.739		
0.31	0.652	0.66	1.790		
0.32	0.676	0.67	1.842		
0.33	0.700	0.68	1.896		
0.34	0.724	0.69	1.954		

Based on Table B of Batschelet (1965) by permission of the Amer. Inst. Biol. Sci., and on Table 2 of Gumbel, Greenwood & Durand (1953) by permission of the authors and publisher of *J. Amer. Statist. Assoc.*

Appendix 2.5. The marginal maximum likelihood estimate $\hat{\kappa}$ of the concentration parameter κ of the von Mises distribution $M(\mu, \kappa)$. $\bar{R}$ denotes the sample mean resultant length.

n	$\bar{R} = 0.10$	0.15	0.20	0.25	0.30	0.35	0.40	0.45	0.50	0.55	0.60	0.65	0.70	0.75	0.80	0.85	0.90	0.95
5	0	0	0	0	0	0	0	0.15	0.67	0.94	1.18	1.41	1.68	2.00	2.44	3.10	4.39	8.33
6	0	0	0	0	0	0	0	0.56	0.83	1.04	1.25	1.48	1.74	2.07	2.51	3.20	4.54	8.66
7	0	0	0	0	0	0	0.38	0.69	0.90	1.10	1.30	1.52	1.78	2.11	2.56	3.27	4.65	8.89
8	0	0	0	0	0	0	0.53	0.76	0.95	1.13	1.33	1.55	1.81	2.15	2.60	3.32	4.73	9.06
9	0	0	0	0	0	0.31	0.61	0.80	0.98	1.16	1.35	1.57	1.84	2.17	2.63	3.36	4.79	9.19
10	0	0	0	0	0	0.42	0.65	0.83	1.00	1.18	1.37	1.59	1.86	2.19	2.66	3.39	4.84	9.30
11	0	0	0	0	0	0.48	0.69	0.85	1.02	1.19	1.38	1.61	1.87	2.21	2.68	3.42	4.89	9.39
12	0	0	0	0	0.23	0.53	0.71	0.87	1.03	1.20	1.40	1.62	1.88	2.22	2.69	3.44	4.92	9.46
13	0	0	0	0	0.32	0.56	0.73	0.88	1.04	1.21	1.41	1.63	1.89	2.23	2.71	3.46	4.95	9.53
14	0	0	0	0	0.37	0.58	0.74	0.89	1.05	1.22	1.41	1.63	1.90	2.24	2.72	3.47	4.98	9.58
15	0	0	0	0	0.41	0.60	0.75	0.90	1.06	1.23	1.42	1.64	1.91	2.25	2.73	3.49	5.00	9.63
20	0	0	0	0.30	0.50	0.65	0.79	0.93	1.09	1.26	1.45	1.67	1.94	2.28	2.76	3.53	5.07	9.79
25	0	0	0	0.38	0.54	0.67	0.81	0.95	1.10	1.27	1.46	1.68	1.95	2.30	2.79	3.56	5.12	9.88
30	0	0	0.22	0.42	0.56	0.69	0.82	0.96	1.11	1.28	1.47	1.69	1.96	2.31	2.80	3.58	5.15	9.95
35	0	0	0.27	0.44	0.57	0.70	0.83	0.97	1.12	1.29	1.48	1.70	1.97	2.32	2.81	3.60	5.17	9.99
40	0	0	0.31	0.45	0.58	0.70	0.83	0.97	1.12	1.29	1.48	1.70	1.98	2.33	2.82	3.61	5.19	10.03
45	0	0.04	0.33	0.46	0.58	0.71	0.84	0.98	1.13	1.30	1.49	1.71	1.98	2.33	2.82	3.62	5.20	10.06
50	0	0.14	0.34	0.47	0.59	0.71	0.84	0.98	1.13	1.30	1.49	1.71	1.98	2.34	2.83	3.62	5.21	10.08
100	0	0.26	0.38	0.49	0.61	0.73	0.86	1.00	1.15	1.31	1.50	1.73	2.00	2.35	2.85	3.65	5.26	10.18
150	0.18	0.28	0.39	0.50	0.62	0.74	0.86	1.00	1.15	1.32	1.51	1.73	2.00	2.36	2.86	3.66	5.27	10.21
200	0.19	0.29	0.40	0.51	0.62	0.74	0.87	1.00	1.15	1.32	1.51	1.73	2.01	2.36	2.86	3.67	5.28	10.22
∞	0.20	0.30	0.41	0.52	0.63	0.75	0.87	1.01	1.16	1.33	1.52	1.74	2.01	2.37	2.87	3.68	5.31	10.27

Reproduced from Table D1 of Batschelet (1981) by permission of the publisher.

Appendix 2.6. Upper quantiles of $\bar{C}$ from the uniform distribution.

n	$\alpha =$ 0.10	0.05	0.025	0.01
5	0.413	0.522	0.611	0.709
6	0.376	0.476	0.560	0.652
7	0.347	0.441	0.519	0.607
8	0.324	0.412	0.486	0.569
9	0.305	0.388	0.459	0.538
10	0.289	0.368	0.436	0.512
11	0.275	0.351	0.416	0.489
12	0.264	0.336	0.398	0.468
13	0.253	0.323	0.383	0.451
14	0.244	0.311	0.369	0.435
15	0.235	0.301	0.357	0.420
16	0.228	0.291	0.345	0.407
17	0.221	0.282	0.335	0.395
18	0.215	0.274	0.326	0.384
19	0.209	0.267	0.317	0.374
20	0.204	0.260	0.309	0.365
21	0.199	0.254	0.302	0.356
22	0.194	0.248	0.295	0.348
23	0.190	0.243	0.288	0.341
24	0.186	0.238	0.282	0.334
25	0.182	0.233	0.277	0.327
30	0.17	0.21	0.25	0.30
35	0.15	0.20	0.23	0.28
40	0.14	0.18	0.22	0.26
45	0.14	0.17	0.21	0.25
50	0.13	0.16	0.20	0.23
$(2n)^{1/2}\bar{C} \dot{\sim} N(0,1)$	1.282	1.645	1.960	2.326

Based on Table 3 of Stephens (1969d) by permission of the *J. Amer. Statist. Assoc*

Appendix 2.7. Quantiles of the Hodges–Ajne statistic m under uniformity. The quantiles m_0 have been chosen to make the tail probabilities $\Pr(m \leq m_0)$ closest to the nominal level α.

n	$\alpha = 0.10$	0.05	0.025	0.01
9	0	0	0	0
10	1	0	0	0
11	1	0	0	0
12	1	1	0	0
13	1	1	1	0
14	2	1	1	0
15	2	2	1	1
16	2	2	1	1
17	3	2	2	1
18	3	3	2	2
19	3	3	2	2
20	4	3	3	2
21	4	4	3	2
22	5	4	3	3
23	5	4	4	3
24	5	5	4	3
25	6	5	4	4
30	7	7	6	5
35	9	9	8	7
40	11	10	10	9
50	15	14	13	12

Compiled from Hodges (1955) with permission of the publisher

Appendix 2.8. Quantiles (in degrees) of the circular range w under uniformity.

n	$\alpha = 0.005$	0.01	0.05	0.10
4	38.8	48.9	83.6	105.3
5	64.0	76.1	113.8	135.4
6	87.2	100.2	138.2	158.7
7	107.6	120.8	158.0	177.3
8	125.5	138.5	174.4	192.5
9	141.1	153.8	188.1	205.1
10	154.7	167.1	199.8	215.8
11	166.8	178.7	209.9	225.0
12	177.4	189.0	218.7	233.0
13	187.0	198.1	226.5	240.0
14	195.5	206.2	233.4	246.2
15	203.2	213.5	239.5	251.7
16	210.2	220.1	245.1	256.7
17	216.6	226.2	250.1	261.2
18	222.4	231.6	254.7	265.3
19	227.7	236.7	258.8	269.1
20	232.7	241.3	262.7	272.5
21	237.2	245.6	266.2	275.6
22	241.4	249.5	269.4	278.6
23	245.4	253.2	272.5	281.3
24	249.0	256.7	275.3	283.8
25	252.5	259.9	277.9	286.1
26	255.7	262.9	280.4	288.3
27	258.7	265.7	282.7	290.4
28	261.5	268.3	284.8	292.3
29	264.2	270.8	286.9	294.1
30	266.7	273.2	288.8	295.9

Reproduced from Laubscher & Rudolph (1968) by permission of the authors.

Appendix 2.9. Quantiles (in degrees) of the equal-spacings statistic L under uniformity.

n	$\alpha = 0.01$	0.05	0.10
4	221.0	186.5	171.7
5	212.0	183.6	168.8
6	206.0	180.7	166.3
7	202.7	177.8	164.9
8	198.4	175.7	163.4
9	195.1	173.5	162.4
10	192.2	172.1	161.3
11	189.7	170.3	160.2
12	187.6	169.2	159.2
13	185.8	167.8	158.4
14	184.0	166.7	157.7
15	182.2	165.6	157.0
16	180.7	164.9	156.6
17	179.6	164.2	155.9
18	178.2	163.1	155.2
19	177.1	162.4	154.8
20	176.0	161.6	154.4
25	171.9	158.9	152.7
30	168.8	156.7	151.4
35	166.4	155.0	150.3
40	164.4	153.6	149.5
45	162.7	152.4	148.7
50	161.2	151.4	148.1
100	152.8	146.8	143.7
200	146.8	142.6	140.4

Reproduced from Rao (1969) by permission of the Indian Statistical Institute.

Appendix 2.10. Upper quantiles of the statistic U^2 for testing goodness of fit of a von Mises distribution.

κ	Significance level α							
	0.500	0.25	0.15	0.10	0.05	0.025	0.01	0.005
(a) Case 0: both parameters known								
All κ	0.069	0.105	0.131	0.152	0.187	0.222	0.268	0.304
(b) Case 1: shape parameter κ known								
0.0	0.047	0.071	0.089	0.105	0.133	0.163	0.204	0.235
0.50	0.048	0.072	0.091	0.107	0.135	0.165	0.205	0.237
1.00	0.051	0.076	0.095	0.111	0.139	0.169	0.209	0.241
1.50	0.053	0.080	0.099	0.115	0.144	0.173	0.214	0.245
2.00	0.055	0.082	0.102	0.119	0.147	0.177	0.217	0.248
4.00	0.058	0.086	0.107	0.124	0.153	0.183	0.224	0.255
∞	0.059	0.089	0.110	0.127	0.157	0.187	0.228	0.259
(c) Case 2: μ known								
0.0	0.047	0.071	0.089	0.105	0.133	0.163	0.204	0.235
0.50	0.048	0.072	0.091	0.107	0.135	0.165	0.205	0.237
1.00	0.051	0.076	0.095	0.111	0.139	0.169	0.209	0.241
1.50	0.053	0.080	0.100	0.116	0.144	0.174	0.214	0.245
2.00	0.055	0.082	0.103	0.119	0.148	0.177	0.218	0.249
4.00	0.057	0.085	0.106	0.122	0.151	0.181	0.221	0.253
∞	0.057	0.085	0.105	0.122	0.151	0.180	0.221	0.252
(d) Case 3: neither parameter known								
0.0	0.030	0.040	0.046	0.052	0.061	0.069	0.081	0.090
0.50	0.031	0.042	0.050	0.056	0.066	0.077	0.090	0.100
1.00	0.035	0.049	0.059	0.066	0.079	0.092	0.110	0.122
1.50	0.039	0.056	0.067	0.077	0.092	0.108	0.128	0.144
2.00	0.043	0.061	0.074	0.084	0.101	0.119	0.142	0.159
4.00	0.047	0.067	0.082	0.093	0.113	0.132	0.158	0.178
∞	0.048	0.069	0.084	0.096	0.117	0.137	0.164	0.184

For $\kappa > 4$ use linear interpolation in $1/\kappa$. For cases 2 and 3 enter the table at the estimate of κ.

Reproduced from Table 1 of Lockhart & Stephens (1985) by permission of the Biometrika Trust.

Appendix 2.11. Confidence intervals for κ.

Appendix 2.11a. Chart for obtaining a 90% confidence interval for the concentration parameter κ of the von Mises distribution $M(\mu, \kappa)$. n denotes the sample size. The curves in the upper part give the upper confidence limit κ_u; those in the lower part give the lower confidence limit κ_l.

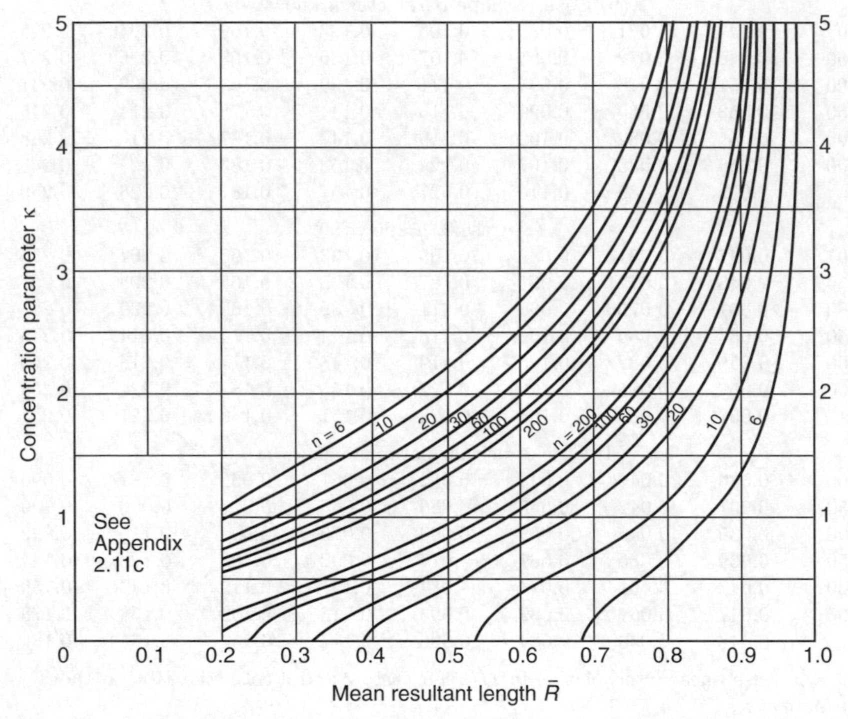

Reproduced from Figure 5.3.1 of Batschelet (1981) by permission of the publisher.

Appendix 2.11b. Chart for obtaining a 98% confidence interval for the concentration parameter κ of the von Mises distribution $M(\mu, \kappa)$. n denotes the sample size. The curves in the upper part give the upper confidence limit κ_u; those in the lower part give the lower confidence limit κ_l.

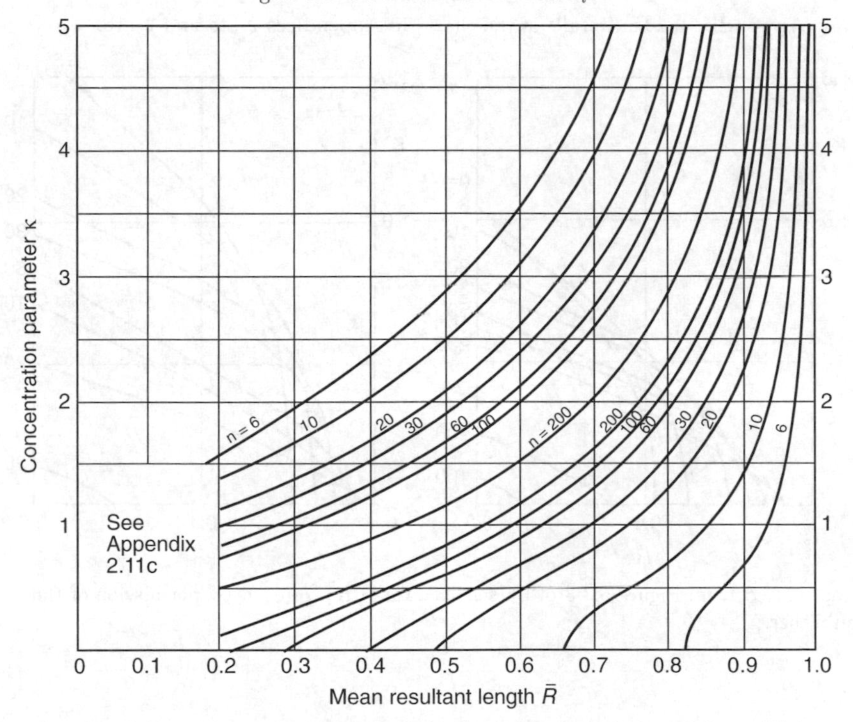

Reproduced from Figure 5.3.2 of Batschelet (1981) by permission of the publisher.

Appendix 2.11c. Details of the charts in Appendices 2.11a and 2.11b.

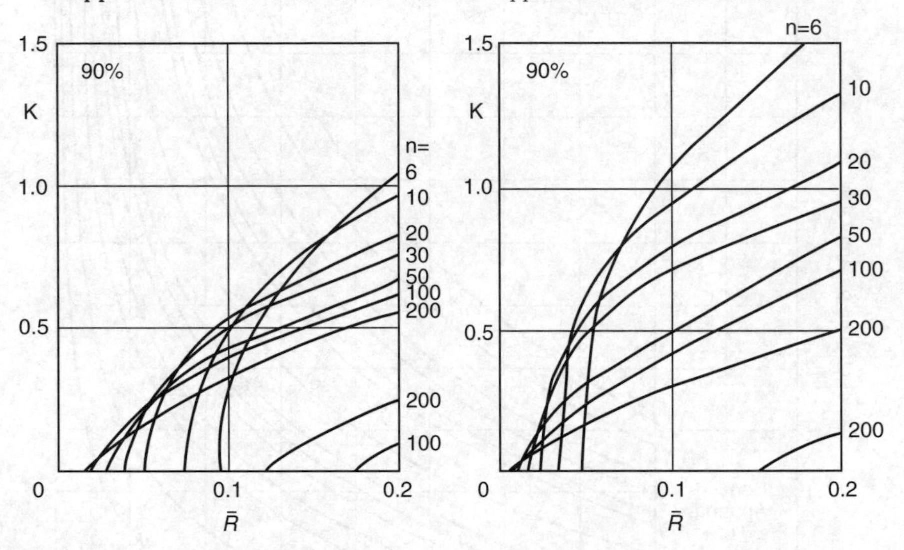

Reproduced from Figure 5.3.3 of Batschelet (1981) reproduced by permission of the publisher.

Appendix 2.12. Upper quantiles of the two-sample Watson–Williams test.

Appendix 2.12a. Chart for obtaining the upper 5% quantile of $\bar{R}' = (R_1 + R_2)/n$ conditional on the mean resultant length $\bar{R}$ of the combined sample (Equal sample sizes: $n_1 = n_2$).

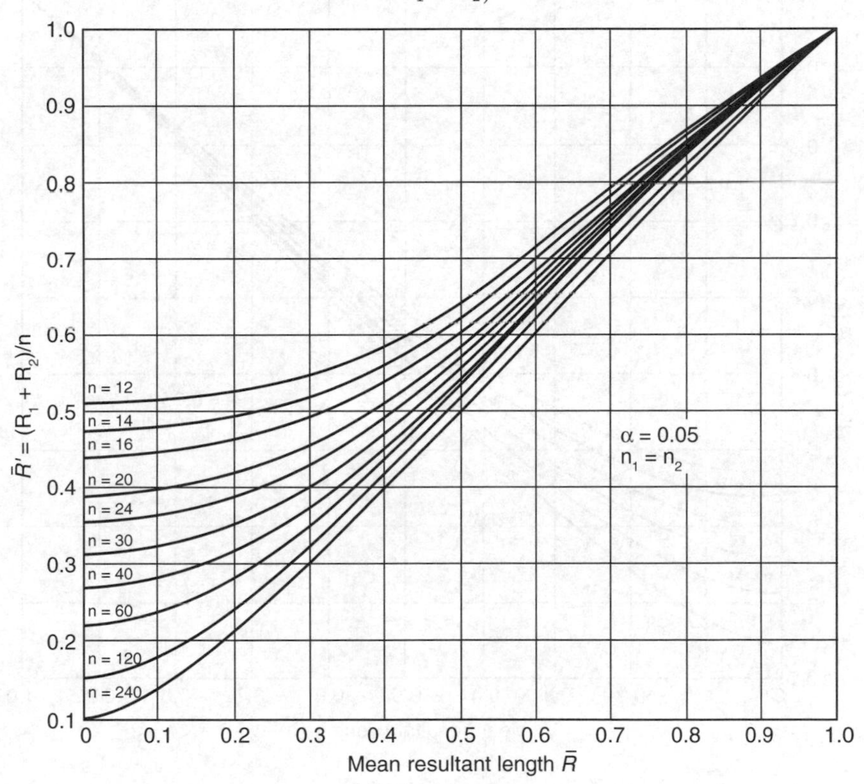

Appendix 2.12b. Chart for obtaining the upper 5% quantile of $\bar{R}' = (R_1 + R_2)/n$ conditional on the mean resultant length $\bar{R}$ of the combined sample ($n_2 = 2n_1$).

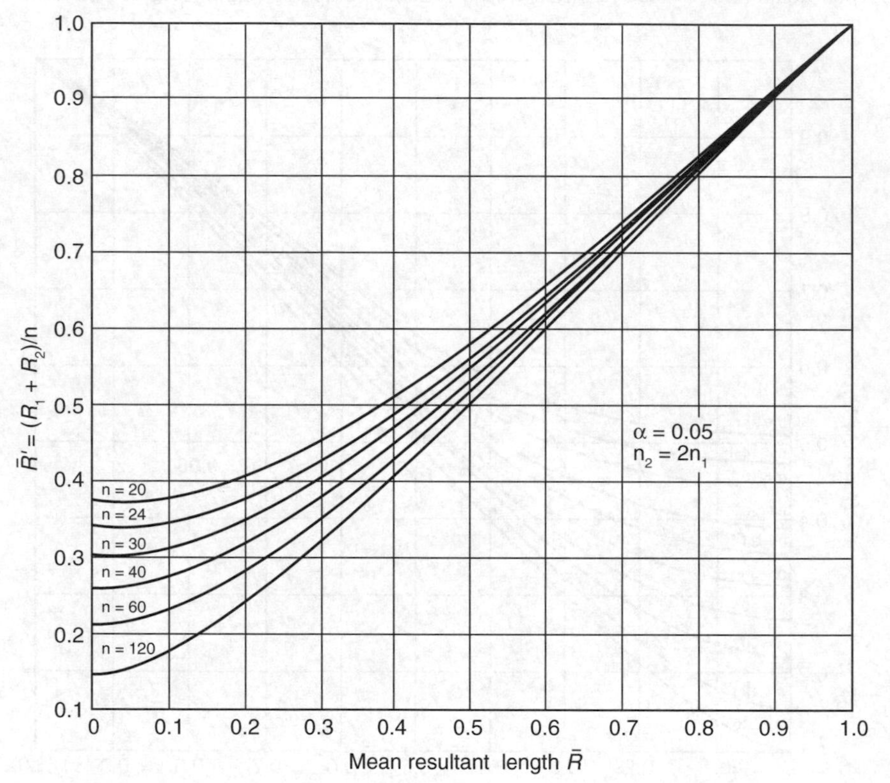

Appendix 2.13. Upper quantiles of R_1^2 for the two-sample uniform scores test. The quantiles $R_{1,0}^2$ have been chosen to make the tail probabilities $\Pr(R_1^2 \geq R_{1,0}^2)$ closest to the nominal level α.

n	n_1	$\alpha = 0.001$	0.01	0.05	0.10
8	4				6.83
9	3				6.41
	4			8.29	4.88
10	3				6.85
	4			9.47	6.24
	5			10.47	6.85
11	3			7.20	5.23
	4			10.42	7.43
	5		12.34	8.74	6.60
12	3			7.46	5.73
	4		11.20	8.46	7.46
	5		13.93	10.46	7.46
	6		14.93	11.20	7.46
13	3			7.68	6.15
	4		11.83	9.35	7.03
	5		15.26	10.15	7.39
	6		17.31	10.42	8.04
14	3			7.85	6.49
	4		12.34	9.30	7.60
	5		16.39	10.30	7.85
	6	19.20	15.59	12.21	7.94
	7	20.20	16.39	11.65	8.85
15	3			7.99	6.78
	4		12.78	8.74	7.91
	5	17.35	14.52	10.36	7.91
	6	20.92	17.48	11.61	9.12
	7	22.88	16.14	11.57	9.06
16	3			8.11	5.83
	4		13.14	9.44	7.38
	5	18.16	15.55	10.44	9.03
	6	22.43	16.98	11.54	9.11
	7	25.27	18.16	12.66	9.78

Appendix 2.13. (*continued*)

n	n_1	$\alpha = 0.001$	0.01	0.05	0.10
17	3		8.21	7.23	6.14
	4	13.44	11.76	9.74	7.64
	5	18.86	16.44	11.03	8.76
	6	23.73	17.76	12.21	9.41
	7	27.40	17.98	12.63	10.11
	8	29.37	19.11	13.36	10.15
18	2				3.88
	3		8.29	7.41	6.41
	4	13.70	12.17	9.94	8.06
	5	19.46	16.05	11.45	8.76
	6	24.87	17.40	12.25	9.94
	7	29.28	19.46	13.41	10.29
	8	28.40	20.11	13.82	10.60
	9	29.28	20.23	13.99	11.04
19	2				3.89
	3		8.36	7.56	6.48
	4	13.93	12.52	9.69	7.54
	5	19.98	15.88	11.29	8.96
	6	25.87	18.19	12.57	9.87
	7	27.71	19.34	13.54	10.55
	8	31.04	21.12	14.29	11.12
	9	29.46	21.07	14.58	11.37
20	2				3.90
	3		8.42	7.70	6.70
	4	14.12	12.83	9.87	7.80
	5	20.43	16.29	11.49	9.08
	6	26.75	18.64	12.93	9.98
	7	29.36	20.43	14.05	11.03
	8	30.08	21.77	14.77	11.47
	9	32.44	22.99	15.45	11.97
	10	33.26	22.67	15.39	12.19
	$R^* = \chi^2_{2;\alpha}$	13.816	9.210	5.991	4.605

Appendix 2.14. Upper quantiles of Watson's two-sample statistic $U^2_{n_1,n_2}$. The quantiles $U^2_{n_1,n_2,0}$ have been chosen to make the tail probabilities $\Pr(U^2_{n_1,n_2} \geq U^2_{n_1,n_2,0})$ closest to the nominal level α.

n	n_1	$\alpha = 0.001$	0.01	0.05	0.10
9	4			0.204	
10	4			0.217	
	5			0.225	
11	4		0.169	0.227	
	5		0.242	0.182	
12	4		0.236	0.181	0.163
	5		0.257	0.200	0.171
	6		0.264	0.206	0.171
13	4		0.244	0.192	0.175
	5		0.269	0.189	0.165
	6		0.282	0.190	0.154
14	4		0.250	0.186	0.157
	5	0.280	0.229	0.191	0.159
	6	0.298	0.246	0.186	0.161
	7	0.304	0.251	0.199	0.158
15	4		0.256	0.195	0.156
	5	0.289	0.241	0.177	0.161
	6	0.311	0.262	0.188	0.156
	7	0.322	0.239	0.182	0.156
16	4	0.260	0.217	0.182	0.156
	5	0.297	0.251	0.188	0.156
	6	0.323	0.248	0.190	0.156
	7	0.339	0.245	0.182	0.156
	8	0.344	0.250	0.184	0.156
17	4	0.265	0.224	0.185	0.154
	5	0.304	0.239	0.186	0.155
	6	0.333	0.253	0.185	0.157
	7	0.353	0.247	0.185	0.155
	8	0.363	0.248	0.186	0.155
$U^2_{\infty,\infty}$		0.385	0.268	0.187	0.152

Compiled from Burr (1964) with permission of the publisher.

Appendix 2.15. Upper quantiles of the total number r of runs. The quantiles r_0 have been chosen to make the tail probabilities $\Pr(r \geq r_0)$ closest to the nominal level α.

n_2 \ $n_1 =$	3	4	5	6	7	8	9	10	11	12	13	14	15	16	17	18	19	20
										$\alpha = 0.05$								
4		2																
5		2	2															
6	2	2	2	2														
7	2	2	2	4	4													
8	2	2	2	4	4	4												
9	2	2	2	4	4	4	4											
10	2	2	4	4	4	4	6	6										
11	2	2	4	4	4	6	6	6	6									
12	2	2	4	4	4	6	6	6	6	8								
13	2	2	4	4	4	6	6	6	8	8	8							
14	2	2	4	4	6	6	6	8	8	8	8	8						
15	2	2	4	4	6	6	6	8	8	8	8	10	10					
16	2	2	4	4	6	6	6	8	8	8	10	10	10	10				
17	2	4	4	4	6	6	8	8	8	10	10	10	10	12	12			
18	2	4	4	6	6	6	8	8	8	10	10	10	12	12	12	12		
19	2	4	4	6	6	6	8	8	8	10	10	10	12	12	12	12	14	
20	2	4	4	6	6	8	8	8	10	10	10	12	12	12	12	14	14	14
n_2										$\alpha = 0.01$								
5			2															
6		2	2	2														
7		2	2	2	2													
8		2	2	2	2	2												
9		2	2	2	2	2	4											
10	2	2	2	2	2	4	4	4										
11	2	2	2	2	4	4	4	4	4									
12	2	2	2	2	4	4	4	4	6	6								
13	2	2	2	2	4	4	4	6	6	6	6							
14	2	2	2	4	4	4	4	6	6	6	6	8						
15	2	2	2	4	4	4	6	6	6	6	8	8	8					
16	2	2	2	4	4	4	6	6	6	6	8	8	8	8				
17	2	2	2	4	4	4	6	6	6	8	8	8	8	10	10			
18	2	2	2	4	4	6	6	6	6	8	8	8	10	10	10	10		
19	2	2	2	4	4	6	6	6	8	8	8	8	10	10	10	10	12	
20	2	2	4	4	4	6	6	6	8	8	10	10	10	10	10	12	12	12

Compiled from Asano (1965) with permission of Kluwer Academic Publishers, Tokyo.

Appendix 2.16. Upper quantiles of W for the three-sample uniform scores test. The quantiles W_0 have been chosen to make the tail probabilities $\Pr(W \geq W_0)$ closest to the nominal level α.

n_1	n_2	n_3	$\alpha = 0.01$	0.05	0.10
3	3	3	12.82	9.45	9.06
4	3	2	11.95	9.06	8.02
4	3	3	10.89	9.40	7.97
4	4	1		10.29	8.59
4	4	2	10.47	9.05	8.05
4	4	3	11.53	9.36	8.21
4	4	4	12.20	9.60	8.23
5	2	2	10.38	8.48	7.85
5	3	1		9.59	7.90
5	3	2	12.38	9.14	7.65
5	3	3	11.78	9.22	7.85
5	4	1	10.92	9.31	6.97
5	4	2	11.48	9.00	7.72
5	4	3	11.50	9.30	8.05
5	4	4	11.82	9.52	8.19
5	5	1	11.87	8.99	7.14
5	5	2	11.14	8.79	7.68
5	5	3	11.87	9.25	7.99
5	5	4	12.00	9.46	8.20
	$\chi^2_{4;\alpha}$		13.277	9.488	7.779

Appendix 2.17. Upper quantiles (upper entries) of the linear–circular rank correlation coefficient U_n. The lower entries give the exact levels.

	α		
n	0.10	0.05	0.01
5	3.97	–	–
	0.083	–	–
6	4.57	4.67	4.95
	0.083	0.067	0.017
7	4.30	4.90	5.75
	0.100	0.047	0.008
8	4.488	5.17	6.15
	0.010	0.050	0.010
9	4.50	5.34	6.68
	0.010	0.050	0.010
10	4.52	5.48	6.68
11	4.55	5.5	7.2
12	4.57	5.6	7.5
15	4.59	5.7	7.9
20	4.60	5.8	8.3
30	4.60	5.9	8.7
40	4.60	5.9	8.8
50	4.61	6.0	8.9
100	4.61	6.0	9.1

Adapted from Mardia (1976).

Appendix 2.18. Upper quantiles (upper entries) of the circular–circular rank correlation coefficient r_0. The lower entries give the exact levels.

	α		
n	0.10	0.05	0.01
5	1.00	–	–
	0.083	–	–
6	0.694	–	1.00
	0.117	–	0.017
7	0.522	0.616	0.796
	0.100	0.061	0.022
8	0.432	0.534	0.729
	0.098	0.053	0.012
9	0.37†	0.627	6.68
	–	0.052	0.010
10	0.33†	0.41†	0.59†

Adapted from Mardia (1975a).

† Values calculated from approximation (11.2.17).

Appendix 3

TABLES FOR THE SPHERICAL CASE

The tables in this appendix are presented in the same order in which they were first cited in the text.

Appendix 3.1. Quantiles (in degrees) of colatitude θ in the Fisher distribution $F(\boldsymbol{\mu}, \kappa)$. For values of κ greater than those shown in the table, approximate quantiles δ are given by $\delta = a\kappa^{-\frac{1}{2}}$.

κ	$\alpha = 0.001$	0.01	0.05	0.10
0.0	176.4	168.5	154.2	143.1
0.1	176.2	167.9	152.9	141.4
0.2	176.0	167.3	151.5	139.5
0.3	175.8	166.6	150.0	137.4
0.4	175.5	165.8	148.4	135.3
0.5	175.3	165.0	146.6	133.1
0.6	175.0	164.1	144.8	130.7
0.7	174.6	163.1	142.8	128.3
0.8	174.3	162.1	140.8	125.7
0.9	173.9	161.0	138.6	123.1
1.0	173.5	159.7	136.3	120.4
1.1	173.1	158.4	133.9	117.7
1.2	172.6	157.0	131.4	114.9
1.3	172.1	155.5	128.9	112.2
1.4	171.5	153.8	126.2	109.4
1.5	170.9	152.1	123.6	106.8
1.6	170.2	150.2	120.9	104.1
1.7	169.5	148.3	118.3	101.5
1.8	168.7	146.2	115.6	99.0
1.9	167.8	144.0	113.0	96.6
2.0	166.9	141.8	110.4	94.3
2.1	165.9	139.4	107.9	92.1
2.2	164.8	137.0	105.4	89.9
2.3	163.6	134.6	103.1	87.9
2.4	162.3	132.1	100.8	86.0
2.5	160.9	129.6	98.6	84.1
2.6	159.4	127.1	96.5	82.4
2.7	157.9	124.7	94.5	80.7
2.8	156.2	122.2	92.6	79.1
2.9	154.4	119.8	90.8	77.6
3.0	152.5	117.5	89.0	76.1
3.1	150.6	115.2	87.4	74.7
3.2	148.5	113.0	85.8	73.4
3.3	146.4	110.9	84.3	72.2
3.4	144.3	108.9	82.8	71.0
3.5	142.1	106.9	81.4	69.9
3.6	139.8	105.0	80.1	68.8
3.7	137.6	103.2	78.8	67.7
3.8	135.3	101.5	77.6	66.7
3.9	133.1	99.8	76.5	65.8
4.0	130.9	98.2	75.4	64.8
4.1	128.7	96.7	74.3	64.0
4.2	126.6	95.2	73.3	63.1
4.3	124.6	93.8	72.3	62.3

Appendix 3.1. (*continued*)

κ	$\alpha = 0.001$	0.01	0.05	0.10
4.4	122.6	92.5	71.3	61.5
4.5	120.6	91.2	70.4	60.8
4.6	118.7	89.9	69.6	60.0
4.7	116.9	88.7	68.7	59.3
4.8	115.2	87.6	67.9	58.6
4.9	113.5	86.5	67.1	58.0
5.0	111.9	85.4	66.4	57.3
5.2	108.8	83.4	64.9	56.1
5.4	106.0	81.5	63.6	55.0
5.6	103.4	79.8	62.3	53.9
5.8	100.9	78.1	61.1	52.9
6.0	98.6	76.6	60.0	52.0
6.2	96.5	75.1	58.9	51.1
6.4	94.5	73.7	57.9	50.2
6.6	92.7	72.4	56.9	49.4
6.8	90.9	71.2	56.0	48.6
7.0	89.2	70.0	55.1	47.9
7.2	87.7	68.9	54.3	47.1
7.4	86.2	67.8	53.5	46.5
7.6	84.8	66.8	52.7	45.8
7.8	83.4	65.8	52.0	45.2
8.0	82.2	64.9	51.3	44.6
8.2	80.9	64.0	50.6	44.0
8.4	79.8	63.1	50.0	43.5
8.6	78.7	62.3	49.3	42.9
8.8	77.6	61.5	48.7	42.4
9.0	76.6	60.8	48.2	41.9
9.2	75.6	60.0	47.6	41.4
9.4	74.6	59.3	47.1	41.0
9.6	73.7	58.6	46.5	40.5
9.8	72.8	58.0	46.0	40.1
10.0	72.0	57.4	45.5	39.7
10.5	70.0	55.8	44.4	38.7
11.0	68.2	54.5	43.3	37.8
11.5	66.5	53.2	42.3	36.9
12.0	64.9	52.0	41.4	36.1
12.5	63.4	50.8	40.5	35.3
13.0	62.1	49.8	39.7	34.6
14.0	59.6	47.9	38.2	33.3
15.0	57.4	46.1	36.8	32.2
20.0	49.1	39.7	31.8	27.8
30.0	39.7	32.2	25.8	22.6
40.0	34.2	27.8	22.3	19.5
50.0	30.5	24.8	19.9	17.5
100	21.3	17.4	14.0	12.3
$\delta = a\kappa^{-1/2}$	$a = 212.9$	173.9	140.2	123.0

Appendix 3.2. The function A_3^{-1}. The maximum likelihood estimate $\hat{\kappa}$ of the concentration parameter κ of the Fisher distribution $F(\boldsymbol{\mu}, \kappa)$ is $\hat{\kappa} = A_3^{-1}(\bar{R})$.

x	$A_3^{-1}(x)$	x	$A_3^{-1}(x)$	x	$A_3^{-1}(x)$
0.00	0.000	0.35	1.137	0.70	3.304
0.01	0.030	0.36	1.176	0.71	3.423
0.02	0.060	0.37	1.215	0.72	3.551
0.03	0.090	0.38	1.255	0.73	3.687
0.04	0.120	0.39	1.295	0.74	3.832
0.05	0.150	0.40	1.336	0.75	3.989
0.06	0.180	0.41	1.378	0.76	4.158
0.07	0.211	0.42	1.421	0.77	4.341
0.08	0.241	0.43	1.464	0.78	4.541
0.09	0.271	0.44	1.508	0.79	4.759
0.10	0.302	0.45	1.554	0.80	4.998
0.11	0.332	0.46	1.600	0.81	5.262
0.12	0.363	0.47	1.647	0.82	5.555
0.13	0.394	0.48	1.696	0.83	5.882
0.14	0.425	0.49	1.746	0.84	6.250
0.15	0.456	0.50	1.797	0.85	6.667
0.16	0.488	0.51	1.849	0.86	7.143
0.17	0.519	0.52	1.903	0.87	7.692
0.18	0.551	0.53	1.958	0.88	8.333
0.19	0.583	0.54	2.015	0.89	9.091
0.20	0.615	0.55	2.074	0.90	10.000
0.21	0.647	0.56	2.135	0.91	11.111
0.22	0.680	0.57	2.198	0.92	12.500
0.23	0.713	0.58	2.263	0.93	14.286
0.24	0.746	0.59	2.330	0.94	16.667
0.25	0.780	0.60	2.401	0.95	20.000
0.26	0.814	0.61	2.473	0.96	25.000
0.27	0.848	0.62	2.549	0.97	33.333
0.28	0.883	0.63	2.628	0.98	50.000
0.29	0.918	0.64	2.711	0.99	100.000
0.30	0.953	0.65	2.798		
0.31	0.989	0.66	2.888		
0.32	1.025	0.67	2.984		
0.33	1.062	0.68	3.085		
0.34	1.100	0.69	3.191		

Appendix 3.3. The marginal maximum likelihood estimate $\breve{\kappa}$ of the concentration parameter κ of the Fisher distribution $F(\boldsymbol{\mu}, \kappa)$. $\bar{R}$ denotes the sample mean resultant length.

n	$\bar{R} = 0.10$	0.15	0.20	0.25	0.30	0.35	0.40	0.45	0.50	0.55	0.60	0.65	0.70	0.75	0.80	0.85	0.90	0.95
5	0	0	0	0	0	0	0	0.216	0.941	1.344	1.712	2.104	2.565	3.154	3.986	5.332	8.000	16.000
6	0	0	0	0	0	0	0	0.778	1.170	1.505	1.848	2.232	2.696	3.297	4.156	5.554	8.333	16.667
7	0	0	0	0	0	0	0.536	0.970	1.294	1.605	1.938	2.320	2.786	3.398	4.277	5.713	8.571	17.143
8	0	0	0	0	0	0	0.740	1.079	1.374	1.673	2.001	2.383	2.853	3.473	4.368	5.833	8.750	17.500
9	0	0	0	0	0	0.429	0.853	1.151	1.431	1.723	2.049	2.431	2.905	3.531	4.438	5.925	8.889	17.778
10	0	0	0	0	0	0.587	0.927	1.203	1.473	1.762	2.087	2.470	2.946	3.578	4.494	6.000	9.000	18.000
11	0	0	0	0	0	0.681	0.979	1.242	1.507	1.793	2.117	2.501	2.979	3.615	4.540	6.060	9.091	18.182
12	0	0	0	0	0.324	0.744	1.019	1.273	1.534	1.818	2.142	2.526	3.006	3.647	4.579	6.111	9.167	18.333
13	0	0	0	0	0.445	0.791	1.050	1.298	1.556	1.840	2.163	2.548	3.030	3.673	4.611	6.153	9.231	18.462
14	0	0	0	0	0.520	0.827	1.076	1.319	1.575	1.857	2.180	2.566	3.049	3.696	4.639	6.190	9.286	18.571
15	0	0	0	0	0.574	0.856	1.097	1.337	1.591	1.873	2.196	2.582	3.067	3.716	4.663	6.222	9.333	18.667
20	0	0	0	0.422	0.713	0.942	1.164	1.396	1.646	1.925	2.248	2.637	3.126	3.784	4.746	6.333	9.500	19.000
25	0	0	0	0.540	0.774	0.987	1.202	1.429	1.677	1.956	2.279	2.669	3.162	3.825	4.797	6.400	9.600	19.200
30	0	0	0.308	0.596	0.810	1.015	1.226	1.451	1.698	1.976	2.300	2.691	3.186	3.853	4.830	6.444	9.667	19.333
35	0	0	0.388	0.630	0.833	1.034	1.243	1.466	1.712	1.990	2.314	2.706	3.203	3.872	4.854	6.476	9.714	19.429
40	0	0	0.434	0.653	0.850	1.048	1.255	1.478	1.723	2.001	2.325	2.718	3.215	3.887	4.872	6.500	9.750	19.550
45	0	0.065	0.463	0.670	0.863	1.058	1.264	1.486	1.732	2.009	2.334	2.727	3.225	3.898	4.886	6.518	9.778	19.556
50	0	0.190	0.483	0.682	0.873	1.067	1.272	1.493	1.738	2.016	2.341	2.734	3.233	3.907	4.897	6.533	9.800	19.600
100	0	0.373	0.557	0.734	0.914	1.103	1.304	1.524	1.768	2.045	2.371	2.766	3.268	3.948	4.948	6.600	9.900	19.800
150	0.204	0.405	0.578	0.750	0.928	1.115	1.315	1.534	1.777	2.055	2.381	2.776	3.280	3.962	4.964	6.622	9.933	19.867
200	0.238	0.419	0.588	0.758	0.934	1.120	1.320	1.539	1.782	2.060	2.386	2.782	3.286	3.969	4.973	6.633	9.950	19.900
∞	0.302	0.456	0.615	0.780	0.953	1.137	1.336	1.554	1.797	2.074	2.401	2.798	3.304	3.989	4.998	6.667	10.000	20.000

Appendix 3.4. The function D_3^{-1} on $(0, \frac{1}{3})$. The maximum likelihood estimate $\hat{\kappa}$ of the concentration parameter κ of the girdle Watson distribution $W(\boldsymbol{\mu}, \kappa)$ is $\hat{\kappa} = D_3^{-1}(\bar{t}_3)$.

t	$D_3^{-1}(t)$	t	$D_3^{-1}(t)$	t	$D_3^{-1}(t)$
0.001	−500.0	0.115	−4.196	0.230	−1.357
0.005	−100.0	0.120	−3.993	0.235	−1.279
0.010	−50.00	0.125	−3.802	0.240	−1.202
0.015	−33.33	0.130	−3.624	0.245	−1.127
0.020	−25.00	0.135	−3.457	0.250	−1.053
0.025	−20.00	0.140	−3.298	0.255	−0.982
0.030	−16.67	0.145	−3.148	0.260	−0.911
0.035	−14.29	0.150	−3.006	0.265	−0.842
0.040	−12.25	0.155	−2.870	0.270	−0.774
0.045	−11.11	0.160	−2.741	0.275	−0.708
0.050	−9.992	0.165	−2.617	0.280	−0.642
0.055	−9.087	0.170	−2.499	0.285	−0.578
0.060	−8.327	0.175	−2.385	0.290	−0.514
0.065	−7.681	0.180	−2.275	0.295	−0.452
0.070	−7.126	0.185	−2.170	0.300	−0.390
0.075	−6.641	0.190	−2.068	0.305	−0.330
0.080	−6.215	0.195	−1.970	0.310	−0.270
0.085	−5.836	0.200	−1.874	0.315	−0.211
0.090	−5.495	0.205	−1.782	0.320	−0.152
0.095	−5.188	0.210	−1.692	0.325	−0.095
0.100	−4.908	0.215	−1.605	0.330	−0.038
0.105	−4.651	0.220	−1.520	0.333	−0.004
0.110	−4.415	0.225	−1.438		

Appendix 3.5. The function D_3^{-1} on $(\frac{1}{3}, 1)$. The maximum likelihood estimate $\hat{\kappa}$ of the concentration parameter κ of the bipolar Watson distribution $W(\boldsymbol{\mu}, \kappa)$ is $\hat{\kappa} = D_3^{-1}(\bar{t}_1)$.

t	$D_3^{-1}(t)$	t	$D_3^{-1}(t)$	t	$D_3^{-1}(t)$
0.34	0.075	0.57	2.392	0.80	5.797
0.35	0.184	0.58	2.496	0.81	6.063
0.36	0.292	0.59	2.602	0.82	6.354
0.37	0.398	0.60	2.709	0.83	6.676
0.38	0.503	0.61	2.819	0.84	7.035
0.39	0.606	0.62	2.930	0.85	7.438
0.40	0.708	0.63	3.044	0.86	7.897
0.41	0.809	0.64	3.160	0.87	8.426
0.42	0.909	0.65	3.280	0.88	9.043
0.43	1.008	0.66	3.402	0.89	9.776
0.44	1.106	0.67	3.529	0.90	10.654
0.45	1.204	0.68	3.659	0.91	11.746
0.46	1.302	0.69	3.764	0.92	13.112
0.47	1.399	0.70	3.934	0.93	14.878
0.48	1.497	0.71	4.079	0.94	17.242
0.49	1.594	0.72	4.231	0.95	20.560
0.50	1.692	0.73	4.389	0.96	25.546
0.51	1.790	0.74	4.556	0.97	33.866
0.52	1.888	0.75	4.731	0.98	50.521
0.53	1.987	0.76	4.917	0.99	100.510
0.54	2.087	0.77	5.115	0.99	100.510
0.55	2.188	0.78	5.326	0.995	200.5
0.56	2.289	0.79	5.552	0.999	1000.5

Appendix 3.6. Upper quantiles $\bar{R}_0$ of $\bar{R}$ for testing that the concentration parameter of a Fisher distribution has a prescribed value κ. $\Pr(\bar{R} > \bar{R}_0) = \alpha$. $\alpha = 0.01$ (upper entry) and $\alpha = 0.05$ (lower entry).

n	$\kappa = 0.0$	0.5	1.0	1.5	2.0	2.5	3.0	3.5	4.0	4.5	5.0
5	0.805	0.828	0.867	0.898	0.920	0.935	0.945	0.953	0.959	0.963	0.967
	0.700	0.729	0.784	0.833	0.868	0.892	0.909	0.922	0.932	0.939	0.945
6	0.747	0.777	0.828	0.868	0.896	0.916	0.929	0.939	0.947	0.953	0.957
	0.642	0.676	0.724	0.799	0.841	0.870	0.891	0.906	0.918	0.927	0.934
7	0.699	0.735	0.795	0.843	0.876	0.899	0.915	0.927	0.936	0.943	0.949
	0.597	0.635	0.708	0.773	0.819	0.852	0.876	0.894	0.907	0.917	0.925
8	0.658	0.699	0.767	0.821	0.859	0.885	0.903	0.917	0.927	0.935	0.942
	0.560	0.602	0.681	0.750	0.802	0.838	0.864	0.883	0.897	0.909	0.918
9	0.624	0.669	0.743	0.802	0.843	0.872	0.893	0.908	0.919	0.928	0.935
	0.529	0.575	0.659	0.732	0.786	0.825	0.853	0.874	0.889	0.902	0.912
10	0.594	0.643	0.722	0.785	0.830	0.861	0.884	0.900	0.912	0.922	0.930
	0.503	0.552	0.640	0.716	0.774	0.815	0.844	0.866	0.883	0.896	0.906
12	0.546	0.601	0.687	0.757	0.808	0.843	0.868	0.887	0.901	0.912	0.920
	0.460	0.515	0.609	0.691	0.753	0.797	0.830	0.853	0.872	0.886	0.897
16	0.476	0.541	0.638	0.717	0.774	0.815	0.845	0.867	0.883	0.896	0.906
	0.400	0.463	0.567	0.655	0.723	0.773	0.809	0.835	0.856	0.872	0.884
20	0.428	0.500	0.604	0.688	0.751	0.796	0.828	0.852	0.871	0.885	0.897
	0.358	0.428	0.538	0.631	0.703	0.755	0.794	0.823	0.845	0.862	0.876
30	0.355	0.436	0.550	0.643	0.713	0.764	0.802	0.829	0.850	0.867	0.880
	0.295	0.374	0.493	0.594	0.671	0.729	0.771	0.803	0.827	0.846	0.862
40	0.307	0.398	0.517	0.616	0.690	0.745	0.785	0.815	0.838	0.856	0.870
	0.255	0.343	0.467	0.572	0.653	0.713	0.758	0.791	0.817	0.837	0.853
60	0.251	0.353	0.479	0.583	0.663	0.722	0.765	0.798	0.823	0.842	0.858
	0.208	0.307	0.437	0.546	0.631	0.694	0.742	0.777	0.805	0.826	0.844
100	0.194	0.308	0.441	0.550	0.635	0.698	0.745	0.780	0.807	0.828	0.845
	0.161	0.272	0.407	0.521	0.609	0.676	0.725	0.763	0.792	0.815	0.834
∞	0.000	0.164	0.313	0.438	0.537	0.613	0.672	0.716	0.751	0.778	0.800

From M. A. Stephens in *Biometrika*, Vol. 54, pp. 211–223, 1967. Reproduced by permission of the publisher.

Appendix 3.7. Upper quantiles of the two-sample Watson–Williams test.

Appendix 3.7a. Upper α quantiles of $\bar{R}' = (R_1 + R_2)/n$ conditional on the mean resultant length $\bar{R}$ of the combined sample. $\alpha = 0.01$ (upper entry) and $\alpha = 0.05$ (lower entry): equal sample sizes: $n_1 = n_2$.

n	$\bar{R} = 0.05$	0.10	0.15	0.20	0.25	0.30	0.35	0.40	0.45	0.50	0.55	0.60	0.65	0.70
12	0.571	0.573	0.557	0.583	0.592	0.605	0.622	0.642	0.666	0.692	0.720	0.749	0.780	0.811
	0.482	0.486	0.492	0.501	0.514	0.531	0.552	0.577	0.606	0.637	0.670	0.704	0.741	0.778
16	0.493	0.496	0.502	0.511	0.524	0.542	0.563	0.589	0.617	0.648	0.680	0.714	0.749	0.784
	0.415	0.419	0.427	0.440	0.457	0.478	0.505	0.535	0.568	0.603	0.640	0.679	0.718	0.758
20	0.440	0.444	0.452	0.464	0.480	0.500	0.525	0.555	0.586	0.620	0.655	0.692	0.730	0.768
	0.369	0.375	0.385	0.400	0.420	0.466	0.475	0.509	0.545	0.583	0.622	0.663	0.704	0.746
24	0.401	0.407	0.416	0.429	0.448	0.471	0.499	0.531	0.565	0.610	0.639	0.677	0.716	0.757
	0.337	0.343	0.355	0.372	0.395	0.423	0.455	0.491	0.529	0.569	0.601	0.652	0.695	0.738
30	0.359	0.365	0.376	0.392	0.414	0.441	0.474	0.507	0.544	0.581	0.621	0.662	0.703	0.746
	0.300	0.308	0.322	0.341	0.368	0.400	0.435	0.473	0.514	0.555	0.598	0.642	0.686	0.731
40	0.311	0.319	0.332	0.352	0.378	0.409	0.444	0.481	0.521	0.562	0.604	0.647	0.691	0.734
	0.260	0.270	0.286	0.310	0.340	0.376	0.414	0.455	0.492	0.542	0.586	0.631	0.677	0.723
60	0.255	0.265	0.282	0.307	0.339	0.375	0.414	0.455	0.498	0.542	0.586	0.632	0.677	0.723
	0.213	0.225	0.247	0.276	0.312	0.351	0.393	0.437	0.482	0.528	0.574	0.621	0.668	0.715
120	0.182	0.197	0.223	0.258	0.297	0.339	0.383	0.428	0.474	0.521	0.568	0.616	0.664	0.711
	0.153	0.171	0.201	0.239	0.281	0.326	0.372	0.419	0.466	0.514	0.562	0.610	0.659	0.708
240	0.133	0.155	0.189	0.230	0.274	0.320	0.367	0.415	0.463	0.511	0.559	0.608	0.657	0.706
	0.113	0.139	0.177	0.220	0.267	0.314	0.363	0.410	0.458	0.507	0.556	0.605	0.654	0.704
∞	0.05	0.10	0.15	0.20	0.25	0.30	0.35	0.40	0.45	0.50	0.55	0.60	0.65	0.70

From M. A. Stephens in *Biometrika*, Vol. 56, pp. 169–181, 1969. Reproduced by permission of the publisher.

Appendix 3.7b. Upper α quantiles of $\bar{R}' = (R_1 + R_2)/n$ conditional on the mean resultant length $\bar{R}$ of the combined sample. $\alpha = 0.01$ (upper entry) and $\alpha = 0.05$ (lower entry): $n_2 = 2n_1$.

n	$\bar{R} = 0.10$	0.15	0.20	0.25	0.30	0.35
20	0.420	0.436	0.454	0.474	0.497	0.525
	0.355	0.371	0.392	0.414	0.444	0.475
24	0.385	0.403	0.422	0.445	0.469	0.499
	0.325	0.345	0.366	0.392	0.422	0.455
30	0.350	0.366	0.387	0.412	0.441	0.472
	0.296	0.314	0.338	0.367	0.399	0.435
40	0.307	0.325	0.350	0.378	0.409	0.444
	0.261	0.281	0.309	0.340	0.376	0.414
60	0.257	0.278	0.307	0.339	0.375	0.414
	0.220	0.244	0.275	0.312	0.351	0.393
120	0.195	0.223	0.258	0.297	0.339	0.383
	0.169	0.201	0.239	0.281	0.326	0.372
∞	0.10	0.15	0.20	0.25	0.30	0.35

From M. A. Stephens in *Biometrika*, Vol. 56, pp. 169–181, 1969. Reproduced by permission of the publisher.

Appendix 4

NOTATION

Relations

$\simeq$	is approximately equal to
$\sim$	is distributed as
$\dot{\sim}$	is approximately distributed as

Sample spaces

S^1	unit circle
S^{p-1}	unit sphere in $\mathbb{R}^p$
$\mathbb{R}P^{p-1}$	real $(p-1)$-dimensional projective space
$O(p)$	group of orthogonal $p \times p$ matrices
$SO(p)$	group of $p \times p$ rotation matrices
$U(p)$	group of unitary $p \times p$ matrices
$V_r(\mathbb{R}^p)$	Stiefel manifold of orthonormal r-frames in $\mathbb{R}^p$
$G_r(\mathbb{R}^p)$	Grassmann manifold of r-dimensional subspaces of $\mathbb{R}^p$
H^{p-1}	unit hyperboloid in $\mathbb{R}^p$
$\mathbb{C}P^{k-1}$	complex $(k-1)$-dimensional projective space
Σ_m^k	space of shapes of k labelled points in $\mathbb{R}^m$

Population quantities

μ	mean direction on S^1
$\boldsymbol{\mu}$	mean direction on S^{p-1}
κ	concentration parameter
ρ	mean resultant length
$\phi_p = \alpha_p + \beta_p$	pth component of characteristic function of distribution on S^1

Sample quantities

$\bar{x}$	mean direction on S^1
$\bar{\mathbf{x}}_0$	mean direction on S^{p-1}
R	resultant length
$\bar{R}$	mean resultant length
$\bar{C}$	component along $\boldsymbol{\mu}$ (or along x-axis if $p=2$) of vector mean
$\bar{S}$	component normal to $\boldsymbol{\mu}$ (or along y-axis if $p=2$) of vector mean
$\bar{\mathbf{T}}$	scatter matrix of observations on S^{p-1}
a_p, b_p	pth trigonometric moments of observations on S^1
$\hat{\mu}$	maximum likelihood estimate of μ
$\hat{\boldsymbol{\mu}}$	maximum likelihood estimate of $\boldsymbol{\mu}$
$\hat{\kappa}$	maximum likelihood estimate of κ
$\breve{\kappa}$	marginal maximum likelihood estimate of κ

Distributions

$M(\mu, \kappa)$	von Mises
$C(\mu, \rho)$	cardioid
$WN(\mu, \rho)$	wrapped normal
$WC(\mu, \rho)$	wrapped Cauchy
$M_p(\boldsymbol{\mu}, \kappa)$	von Mises–Fisher
$F(\boldsymbol{\mu}, \kappa)$	$M_3(\boldsymbol{\mu}, \kappa)$, Fisher
$BM_p(\boldsymbol{\mu}, \kappa)$	Brownian motion
$PN_p(\boldsymbol{\mu}, \boldsymbol{\Sigma})$	projected normal
$W_p(\boldsymbol{\mu}, \kappa)$	Watson
$B(\mathbf{A})$	Bingham
$ACG(\mathbf{A})$	angular central Gaussian
$\mathbb{C}W_{k-2}(\boldsymbol{\mu}, \kappa)$	complex Watson
$\mathbb{C}B_{k-2}(\mathbf{A})$	complex Bingham

Special functions

I_p	modified Bessel, defined in (A.1)
J_p	Bessel, defined in (A.15)
A_p	defined in (A.10)
$A = A_2$	
D_p	defined in (A.21)

Shape

$d_F(\mathbf{z}_1, \mathbf{z}_2)$	full Procrustes distance between $\mathbf{z}_1$ and $\mathbf{z}_2$ $(0 \leq d_F \leq 1)$
$d_\rho(\mathbf{z}_1, \mathbf{z}_2)$	Procrustes distance between $\mathbf{z}_1$ and $\mathbf{z}_2$ $(0 \leq d_\rho \leq 1)$
$\mathbf{H}$	Helmertised sub-matrix $((k-1) \times k)$
$\mathbf{z}^0$	vector of raw landmarks ($k \times 1$ vector)
$\mathbf{z}_H$	vector of Helmertised landmarks ($(k-1) \times 1$ vector)
$\mathbf{z}$	Helmertised preshape
$[\mathbf{z}]$	shape of $\mathbf{z}$
$\mathbf{z}_C$	centred preshape
$\mathbf{z}^*$	transpose of complex conjugate of $\mathbf{z}$
$\mathbf{u}$	centred landmarks
$\mathbf{X}_H = \mathbf{HX}$	Helmertised landmark coordinates ($(k-1) \times m$ matrix)
$\mathbf{Z} = \mathbf{X}_H / \lVert\mathbf{X}_H\rVert$	preshape ($(k-1) \times m$ matrix)

References and Author Index

Abrahamson, I. G. (1967). Exact Bahadur efficiencies for the Kolmogorov–Smirnov and Kuiper one- and two-sample statistics. *Ann. Math. Statist.*, **38**, 1475–1490. (150)

Abramowitz, M. & Stegun, I. A. (1965). *Handbook of Mathematical Functions.* Dover, New York. (50, 349, 350, 351)

Accardi, L., Cabrera, J. & Watson, G. S. (1987). Some stationary Markov processes in discrete time for unit vectors. *Metron*, **45**, 115–133. (265)

Aitchison, J. (1986). *The Statistical Analysis of Compositional Data*, Chapman & Hall, London. (307)

Ajne, B. (1968). A simple test for uniformity of a circular distribution. *Biometrika*, **55**, 343–354. (106, 107, 109)

Anderson, C. M. & Wu, C. F. J. (1995). Measuring location effects from factorial experiments with a directional response. *Internat. Statist. Rev.*, **63**, 345–363. (136, 141, 227)

Anderson, T. W. & Stephens, M. A. (1972). Tests for randomness of directions against equatorial and bimodal alternatives. *Biometrika*, **43**, 613–621. (233)

Andrews, D. F. (1974). A robust method for multiple linear regression. *Technometrics*, **16**, 523–531. (276)

Arnold, K. J. (1941). *On Spherical Probability Distributions.* Ph.D. thesis, Massachusetts Institute of Technology. (170, 171, 183)

Arsham, H. (1988). Kuiper's *P*-value as a measuring tool and decision procedure for the goodness-of-fit test. *J. Appl. Statist.*, **15**, 131–135. (102)

Asano, C. (1965). Runs test for a circular distribution and a table of probabilities. *Ann. Inst. Statist. Math.*, **17**, 331–346. (153, 378)

Bagchi, P. (1994). Empirical Bayes estimation in directional data. *J. Appl. Statist.*, **21**, 317–326. (279)

Bagchi, P. & Guttman, I. (1988). Theoretical considerations of the multivariate von Mises-Fisher distribution. *J. Appl. Statist.*, **15**, 149–169. (279)

Bagchi, P. & Guttman, I. (1990). Spuriosity and outliers in directional data. *J. Appl. Statist.*, **17**, 341–350. (269, 279)

Bagchi, P. & Kadane, J. B. (1991). Laplace approximations to posterior moments and marginal distributions on circles, spheres, and cylinders. *Canad. J. Statist.*, **19**, 67–77. (279)

Bai, Z., Rao, C. & Zhao, L. (1988). Kernel estimators of density functions of directional data. *J. Multivariate Anal.*, **27**, 24–39. (278)

Ball, F. & Blackwell, P. (1992). A finite form for the wrapped Poisson distribution. *Adv. Appl. Probab.*, **24**, 221–222. (49)

Barndorff-Nielsen, O. E. (1978a). *Information and Exponential Families.* Wiley, Chichester. (33, 88, 120, 211)

Barndorff-Nielsen, O. E. (1978b). Hyperbolic distributions and distributions on hyperbolae. *Scand. J. Statist.*, **5**, 151–157. (298)

Barndorff-Nielsen, O. E. (1988). *Parametric Statistical Models and Likelihood,* Lecture Notes in Statistics **50**. Springer-Verlag, Heidelberg. (33)

Barndorff-Nielsen, O. E., Blæsild, P. & Eriksen, P. S. (1989). *Decomposition and Invariance of Measures, and Statistical Transformation Models,* Lecture Notes in Statistics **58**. Springer-Verlag, New York. (33, 300)

Barndorff-Nielsen, O. E. & Cox, D. R. (1989). *Asymptotic Techniques for Use in Statistics.* Chapman and Hall, London. (143)

Barndorff-Nielsen, O. E., Blæsild, P., Jensen, J. L., & Jørgensen, B. (1982). Exponential transformation models. *Proc. Roy. Soc. London Ser. A*, **379**, 41–65. (33)

Barnett, V. & Lewis, T. (1994). *Outliers in Statistical Data, 3rd ed.* Wiley, Chichester. (269, 274)

Bartels, R. (1984). Estimation in a bidirectional mixture of von Mises distributions. *Biometrics*, **40**, 777–784. (91)

Barton, D. E. & David, F. N. (1958). Runs in a ring. *Biometrika*, **45**, 572–578. (153, 158)

Batschelet, E. (1965). *Statistical Methods for the Analysis of Problems in Animal Orientation and Certain Biological Rhythms.* Amer. Inst. Biol. Sci., Washington. (38, 156, 355, 360, 362, 363)

Batschelet, E. (1971). Recent statistical methods for orientation data. In *Animal Orientation, Symposium 1970 on Wallops Island*, Amer. Inst. Biol. Sci., Washington. (106)

Batschelet, E. (1981). *Circular Statistics in Biology.* Academic Press, London. (xx, 18, 55, 150, 364, 370, 371, 372)

Batschelet, E., Hillman, D., Smolensky, M. & Halberg, F. (1973). Angular-linear correlation coefficient for rhythmometry and circannually changing human birth rates at different geographic latitudes. *Int. J. Chronobiol.*, **1**, 183–202. (55)

Beckmann, P. (1959). The probability distribution of the vector sum of n unit vectors with arbitrary phase distributions. *Acta Technica*, **4**, 323–335. (9)

Benford, F. (1938). The law of anomalous numbers. *Proc. Amer. Phil. Soc.*, **78**, 551–572. (64)

Beran, R. J. (1968). Testing for uniformity on a compact homogeneous space. *J. Appl. Probab.*, **5**, 177–195. (110, 112, 209, 243)

Beran, R. J. (1969a). Asymptotic theory of a class of tests for uniformity of a circular distribution. *Ann. Math. Statist.*, **40**, 1196–1206. (110, 114, 115)

Beran, R. J. (1969b). The derivation of nonparametric two-sample tests for uniformity of a circular distribution. *Biometrika*, **56**, 561–570. (152, 154, 155)

Beran, R. J. (1979). Exponential models for directional data. *Ann. Statist.*, **7**, 1162–1178. (174, 175)

Beran, R. J. & Fisher, N. I. (1998). Nonparametric comparison of mean directions or mean axes. *Ann. Statist*, **26**, 472–493. (220, 228, 240)

Bernoulli, D. (1735). Quelle est la cause physique de l'inclination des plans des orbites des planètes?. In *Recueil des pièces qui ont remporté le prix de l'Académie Royale des Sciences de Paris 1734, 93–122.* Académie Royale des Sciences de Paris, Paris. Reprinted in *Daniel Bernoulli, Werke, Vol. 3*, 226–303. Birkhäuser, Basel (1982). (11, 208)

Best, D. J. & Fisher, N. I. (1979). Efficient simulation of the von Mises distribution. *Appl. Statist.*, **24**, 152–157. (43)

Best, D. J. & Fisher, N. I. (1981). The bias of the maximum likelihood estimators of the von Mises-Fisher concentration parameters. *Comm. Statist. Simulation Comput.*, **10**, 493–502. (87, 200)

Best, D. J. & Fisher, N. I. (1986). Goodness of fit and discordancy tests for samples from the Watson distribution on the sphere. *Austral. J. Statist.*, **28**, 13–31. (197, 268, 272)

Bhattacharyya, G. K. & Johnson, R. A. (1969). On Hodges's bivariate sign test and a test for uniformity of a circular distribution. *Biometrika*, **56**, 446–449. (98, 106)

Bingham, C. (1964). *Distributions on the Sphere and on the Projective Plane.* Ph.D. thesis, Yale University. (182, 203)

Bingham, C. (1974). An antipodally symmetric distribution on the sphere. *Ann. Statist.*, **2**, 1201–1225. (182, 232, 234)

Bingham, C., Chang, T. & Richards, D. (1992). Approximating the matrix Fisher and Bingham distributions: applications to spherical regression and Procrustes analysis. *J. Multivariate Anal.*, **41**, 314–337. (192, 260, 284, 292, 293, 330)

Bingham, C. & Mardia, K. V. (1978). A small circle distribution on the sphere. *Biometrika*, **65**, 379–389. (177)

Bingham, M. S. (1971). Stochastic processes with independent increments taking values in an Abelian group. *Proc. London Math. Soc.*, **22**, 507–530. (51)

Bingham, M. S. & Mardia, K. V. (1975). Maximum likelihood characterization of the von Mises distribution. In G. P. Patil, S. Kotz & J. K. Ord (eds), *Statistical Distributions in Scientific Work, 3*, pp. 387–398, Reidel, Dordrecht. (171, 179)

Blake, A. & Marinos, C. (1990). Shape from texture – estimation, isotropy and moments. *Artificial Intelligence*, **45**, 323–380. (10, 47, 90)

Bookstein, F. L. (1986). Size and shape spaces for landmark data in two dimensions (with discussion). *Statist. Sci.*, **1**, 181–242. (306, 307)

Bookstein, F. L. (1991). *Morphometric Tools for Landmark Data: Geometry and Biology.* Cambridge Univ. Press, Cambridge. (306, 313, 326)

Bookstein, F. L. (1996). Biometrics, biomathematics and the morphometric synthesis. *Bull. Math. Biol.*, **58**, 313–365. (340, 341)

Boulerice, B. & Ducharme, G. R. (1994). Decentered directional data. *Ann. Inst. Statist. Math.*, **46**, 573–586. (179)

Boulerice, B. & Ducharme, G. R. (1997). Smooth tests of goodness-of-fit for directional and axial data. *J. Multivariate Anal.*, **60**, 154–175. (274)

Bowman, A. W. (1992). Density based tests for goodness of fit. *J. Statist. Comput. Simulation*, **40**, 1–13. (274)

Breckling, J. (1989). *The Analysis of Directional Time Series: Applications to Wind Speed and Direction*, Lecture Notes in Statistics **61**. Springer-Verlag, Berlin. (263, 264, 265)

Breitenberger, E. (1963). Analogues of the normal distribution on the circle and the sphere. *Biometrika*, **50**, 81–88. (171)

Brown, B. M. (1994). Grouping corrections for circular goodness-of-fit tests. *J. Roy. Statist. Soc. Ser. B*, **56**, 275–283. (117, 158)

Brunner, L. & Lo, A. Y. (1994). Nonparametric Bayes methods for directional data. *Canad. J. Statist.*, **22**, 401–412. (279)

Burr, E. J. (1964). Small-sample distributions of the two-sample Cramér–von Mises W^2 and Watson's U^2. *Ann. Math. Statist.*, **35**, 1091–1098. (151, 152, 377)

Cabrera, J. & Watson, G. S. (1990). On a spherical median related distribution. *Comm. Statist. Theory Methods*, **19**, 1973–1986. (180)

Cairns, M. B. (1975). *A Structural Model for the Analysis of Directional Data.* Ph.D. thesis, University of Toronto. (47, 92)

Chan, Y. M. & He, X. (1993). On median-type estimators of direction for the von Mises–Fisher distribution. *Biometrika*, **80**, 869–875. (275)

Chang, T. (1986). Spherical regression. *Ann. Statist.*, **14**, 907–924. (259, 260, 284)

Chang, T. (1987). On the statistical properties of estimated rotations. *J. Geophys. Res.*, **92** (B7), 6319–6329. (261)

Chang, T. (1988). Estimating the relative rotation of two tectonic plates from boundary crossings. *J. Amer. Statist. Assoc.*, **83**, 1178–1183. (261)

Chang, T. (1989). Spherical regression with errors in variables. *Ann. Statist.*, **17**, 293–306. (260)

Chang, T. (1993). Spherical regression and the statistics of tectonic plate reconstruction. *Internat. Statist. Rev.*, **61**, 299–316. (6, 259)

Chang, T. & Ko, D. (1995). *M*-estimates of rigid body motion on the sphere and in Euclidean space. *Ann. Statist.*, **23**, 1823–1847. (276)

Chapman, G. R., Chen, G. & Kim, P. T. (1995). Assessing geometric integrity through spherical regression techniques. *Statist. Sinica*, **5**, 173–220. (259)

Chikuse, Y. (1990a). Distributions of orientations on Stiefel manifolds. *J. Multivariate Anal.*, **33**, 247–264. (286)

Chikuse, Y. (1990b). The matrix angular central Gaussian distribution. *J. Multivariate Anal.*, **33**, 265–274. (296)

Chikuse, Y. (1991a). High dimensional limit theorems and matrix decompositions on the Stiefel manifold. *J. Multivariate Anal.*, **36**, 145–162. (292)

Chikuse, Y. (1991b). Asymptotic expansions for distributions of the large-sample matrix resultant and related statistics on the Stiefel manifold. *J. Multivariate Anal.*, **39**, 270–283. (292)

Chikuse, Y. (1993a). High dimensional asymptotic expansions for the matrix Langevin distributions on the Stiefel manifold. *J. Multivariate Anal.*, **44**, 82–101. (286, 292)

Chikuse, Y. (1993b). Asymptotic theory for the concentrated Langevin distributions on the Grassmann manifold. In K. Matsuita *et al.* (eds), *Statistical Science and Data Analysis: Proceedings of the 3rd Pacific Area Statistical Conference*, pp. 237–245, VSP, Utrecht. (295)

Chikuse, Y. (1993c). Invariant measures on Stiefel manifolds with applications to multivariate analysis. In T. W. Anderson, K. T. Fang & I. W. Olkin (eds), *Multivariate Analysis and its Applications: Proceedings of the International Symposium of Multivariate Analysis and its Applications*, pp. 177–193. IMS Lecture Notes Monograph Series 24, Inst. Math. Statist., Hayward, CA. (292)

Chikuse, Y. (1998). Density estimation on the Stiefel manifold. *J. Multivariate Anal.*, **66**, 188–206. (278)

Chikuse, Y. & Watson, G. S. (1995). Large sample asymptotic theory of tests for uniformity on the Grassmann manifold. *J. Multivariate Anal.*, **53**, 18–31. (296, 297)

Chou, R. J. (1986). Small sample theory of the Langevin distribution. *Austral. J. Statist.*, **28**, 335–344. (212)

Choulakian, V., Lockhart, R. A. & Stephens, M. A. (1994). Cramér–von Mises statistics for discrete distributions. *Canad. J. Statist.*, **22**, 125–137. (117)

Clark, R. M. (1983). Estimation of parameters in the marginal Fisher distribution. *Austral. J. Statist.*, **25**, 227–237. (180)

Coles, S. G. (1998). Inference for circular distributions and processes. *Statist. Comput.*, **8**, 105–113. (264)

Collett, D. (1980). Outliers in circular data. *Appl. Statist.*, **29**, 50–57. (268)

Cordeiro, G. M. & Ferrari, S. (1991). A modified score test statistic having chi-squared distribution to order n^{-1}. *Biometrika*, **78**, 573–582. (95)

Cordeiro, G. M., Paula, G. A. & Botter, D. A. (1994). Improved likelihood ratio tests for dispersion models. *Internat. Statist. Rev.*, **62**, 257–274. (136)

Cox, D. R. (1975). Contribution to discussion of Mardia (1975a). *J. Roy. Statist. Soc. Ser. B*, **37**, 380–381. (45, 143, 273)

Cox, D. R. & Hinkley, D. V. (1974). *Theoretical Statistics.* Chapman & Hall, London. (34, 86, 95)

Creer, K. M., Irving, E. & Nairn, A. E. M. (1959). Palaeomagnetism of the Great Whin Sill. *Geophys. J. Roy. Astron. Soc.*, **2**, 306–323. (177)

Curray, J. R. (1956). The analysis of two-dimensional orientation data. *J. Geol.*, **64**, 117–131. (7)

Dagpunar, J. (1990). Sampling from the von Mises distribution via a comparison of random numbers. *J. Appl. Statist.*, **17**, 165–168. (44)

Daniels, H. E. (1954). A distribution-free test for regression parameters. *Ann. Math. Statist.*, **25**, 499–513. (106)

Darling, D. A. (1953). On a class of problems related to the random division of an interval. *Ann. Math. Statist.*, **24**, 239–253. (108)

Darling, D. A. (1983). On the asymptotic distribution of Watson's statistic. *Ann. Statist.*, **11**, 1263–1266. (105)

David, F. N. & Barton, D. E. (1962). *Combinatorial Chance.* Griffin, London. (153)

David, H. A. (1970). *Order Statistics.* Wiley, New York. (107)

David, H. A. & Newell, D. J. (1965). The identification of annual peak periods for a disease. *Biometrics*, **21**, 645–650. (96)

de Hass-Lorentz, G. L. (1913). *Die Brownsche Bewegung und einige verwandte Erscheinungen.* Frieder, Vieweg und Sohn, Brunswick. (51)

de Waal, D. J. (1979). On the normalizing constant for the Bingham–von Mises–Fisher matrix distribution. *South African Statist. J.*, **13**, 103–112. (175, 292–293)

Diggle, P. J. & Fisher, N. I. (1985). SPHERE – a contouring program for spherical data. *Comput. Geosci.*, **11**, 725–766. (194, 278)

Diggle, P. J., Fisher, N. I. & Lee, A. J. (1985). A comparison of tests of uniformity for spherical data. *Austral. J. Statist.*, **27**, 53–69. (210, 234)

Dimroth, E. (1962). Untersuchungen zum Mechanismus von Blastesis und syntexis in Phylliten and Hornfelsen des südwestlichen Fichtelgebirges I. Die statistische Auswertung einfacher Gürteldiagramme. *Tscherm. Min. Petr. Mitt.*, **8**, 248–274. (181)

Dimroth, E. (1963). Fortschritte der Gefügestatistik. *N. Jb. Min. Mh.*, **13**, 186–192. (181)

Dobson, A. J. (1978). Simple approximations to the von Mises concentration statistic. *Appl. Statist.*, **27**, 345–346. (86)

Downs, T. D. (1966). Some relationships among the von Mises distributions of different dimensions. *Biometrika*, **53**, 269–272. (173)

Downs, T. D. (1972). Orientation statistics. *Biometrika*, **59**, 665–676. (284, 289, 291, 292)

Downs, T. D. (1974). Rotational angular correlation. In M. Ferin, F. Halberg & L. van der Wiele (eds), *Biorhythms and Human Reproduction*, pp. 97–104, Wiley, New York. (249, 256)

Downs, T. D. & Gould, A. L. (1967). Some relationships between the normal and von Mises distributions. *Biometrika*, **54**, 684–687. (173)

Downs, T. D. & Liebman, J. (1969). Statistical methods for vectorcardiographic directions. *IEEE Trans. Bio-med. Engr.*, **16**, 87–94. (10)

Dryden, I. L. (1991). Contribution to discussion of Goodall (1991). *J. Roy. Statist. Soc. Ser. B*, **53**, 327–328. (345)

Dryden, I. L., Faghihi, M. & Taylor, C. C. (1995). Investigating regularity in spatial point patterns using shape analysis. In K. V. Mardia & C. A. Gill (eds), *Current Issues in Statistical Shape Analysis*, pp. 40–48, University of Leeds Press, Leeds. (321)

Dryden, I. L. & Mardia, K. V. (1991). General shape distributions in a plane. *Adv. Appl. Prob.*, **23**, 259–276. (325, 343, 345)

Dryden, I. L. & Mardia, K. V. (1993). Multivariate shape analysis. *Sanhkyā Ser. A*, **55**, 460–480. (347)

Dryden, I. L. & Mardia, K. V. (1998). *Statistical Shape Analysis.* Wiley, Chichester. (303, 304, 308, 309, 311, 319, 325, 333, 341, 345, 346, 347)

Ducharme, G. R. & Milasevic, P. (1987a). Spatial median and directional data. *Biometrika*, **74**, 212–215. (275)

Ducharme, G. R. & Milasevic, P. (1987b). Some asymptotic properties of the circular median. *Comm. Statist. Theory Methods*, **16**, 659–664. (275)

Ducharme, G. R. & Milasevic, P. (1990). Estimating the concentration of the Langevin distribution. *Canad. J. Statist.*, **18**, 163–169. (276)

Ducharme, G. R., Jhun, M., Romano, J. & Truong, K. N. (1985). Bootstrap confidence cones for directional data. *Biometrika*, **72**, 637–645. (277)

Durand, D. & Greenwood, J. A. (1957). Random unit vectors II: Usefulness of Gram–Charlier and related series in approximating distributions. *Ann. Math. Statist.*, **28**, 978–985. (67, 99)

Efron, B. & Tibshirani, R. (1993). *An Introduction to the Bootstrap.* Chapman and Hall, London. (277)

Eplett, W. J. R. (1979). The small sample distribution of a Mann–Whitney type statistic for circular data. *Ann. Statist.*, **7**, 446–453. (156)

Eplett, W. J. R. (1982). Two Mann-Whitney type rank tests. *J. Roy. Statist. Soc. Ser. B*, **44**, 270–286. (156)

Epp, R. J., Tukey, J. W. & Watson, G. S. (1971). Testing unit vectors for correlation. *J. Geophys. Res.*, **76**, 8480–8483. (263)

Feller, W. (1966). *An Introduction to Probability Theory and its Applications, Vol. 2.* Wiley, New York. (27, 54, 57, 64)

Fisher, N. I. (1982). Robust estimation of the concentration parameter of Fisher's distribution on the sphere. *Appl. Statist.*, **31**, 152–154. (276)

Fisher, N. I. (1985). Spherical medians. *J. Roy. Statist. Soc. Ser. B*, **47**, 342–348. (275)

Fisher, N. I. (1986a). Robust comparison of dispersions for samples of directional data. *Austral. J. Statist.*, **28**, 213–219. (276)

Fisher, N. I. (1986b). Robust tests for comparing the dispersions of several Fisher or Watson distributions on the sphere. *Geophys. J. Roy. Astr. Soc.*, **85**, 563–572. (276)

Fisher, N. I. (1993). *Statistical Analysis of Circular Data.* Cambridge University Press, Cambridge. (xx, 86, 139, 248, 250, 254, 258, 264, 277)

Fisher, N. I. & Best, D. J. (1984). Goodness-of-fit tests for Fisher's distribution on the sphere. *Austral. J. Statist.*, **26**, 142–150. (270)

Fisher, N. I. & Hall, P. (1989). Bootstrap confidence regions for directional data. *J. Amer. Statist. Assoc.*, **84**, 996–1002; correction (1990), **85**, 608. (277)

Fisher, N. I. & Hall, P. (1992). Bootstrap methods for directional data. In K. V. Mardia (ed.), *The Art of Statistical Science: A Tribute to G.S. Watson*, pp. 47–63. Wiley, Chichester. (277)

Fisher, N. I. & Lee, A. J. (1981). Non-parametric measures of angular-linear association. *Biometrika*, **68**, 629–636. (248)

Fisher, N. I. & Lee, A. J. (1982). Non-parametric measures of angular-angular association. *Biometrika*, **69**, 315–321. (250, 252, 254)

Fisher, N. I. & Lee, A. J. (1983). A correlation coefficient for circular data. *Biometrika*, **70**, 327–332. (250, 252, 256)

Fisher, N. I. & Lee, A. J. (1986). Correlation coefficients for random variables on a sphere or hypersphere. *Biometrika*, **73**, 159–164. (255, 256, 257)

Fisher, N. I. & Lee, A. J. (1992). Regression models for an angular response. *Biometrics*, **48**, 665–677. (258)

Fisher, N. I. & Lee, A. J. (1994). Time series analysis of circular data. *J. Roy. Statist. Soc. Ser. B*, **56**, 327–339. (264)

Fisher, N. I. & Lewis, T. (1983). Estimating the common mean direction of several circular or spherical distributions with differing dispersions. *Biometrika*, **70**, 333–341. (215)

Fisher, N. I. & Lewis, T. (1985). A note on spherical splines. *J. Roy. Statist. Soc. Ser. B*, **47**, 482–488. (281)

Fisher, N. I., Lewis, T. & Embleton, B. J. J. (1987). *Statistical Analysis of Spherical Data.* Cambridge Univ. Press, Cambridge. 1st paperback edition (with corrections) (1993). (xx, 194, 215, 270, 272, 277)

Fisher, N. I., Lewis, T. & Willcox, M. E. (1981). Tests of discordancy for samples from Fisher's distribution on the sphere. *Appl. Statist.*, **30**, 230–237. (268)

Fisher, N. I., Lunn, A. D. & Davies, S. J. (1993). Spherical median axes. *J. Roy. Statist. Soc. Ser. B*, **55**, 117–124. (276)

Fisher, N. I. & Willcox, M. E. (1978). A useful decomposition of the resultant length for samples from the von Mises–Fisher distributions. *Comm. Statist. Simulation Comput.*, **7**, 257–267. (190)

Fisher, N. I., Hall, P., Jing, B.-Y., & Wood, A. T. A. (1996). Improved pivotal methods for constructing confidence regions with directional data. *J. Amer. Statist. Assoc.*, **91**, 1062–1070. (277)

Fisher, R. A. (1929). Tests of significance in harmonic analysis. *Proc. Roy. Soc. London Ser. A*, **125**, 54–59. (107)

Fisher, R. A. (1953). Dispersion on a sphere. *Proc. Roy. Soc. London Ser. A*, **217**, 295–305. (166, 168, 170, 186, 216)

Fisher, R. A. (1959). *Statistical Methods and Scientific Inference, 2nd ed.* Oliver and Boyd, Edinburgh. (120)

Fraser, D. A. S. (1979). *Inference and Linear Models.* McGraw-Hill, New York. (47, 92)

Fraser, M. D., Hsu, Y.-S. & Walker, J. J. (1981). Identifiability of finite mixtures of von Mises distributions. *Ann. Statist.*, **9**, 1130–1131. (90)

Freedman, L. S. (1979). The use of a Kolmogorov–Smirnov type statistic in testing hypotheses about seasonal variation. *J. Epidem. Comm. Health*, **33**, 223–228. (103)

Freedman, L. S. (1981). Watson's U_N^2 statistic for a discrete distribution. *Biometrika*, **68**, 708–711. (117)

Fujikoshi, Y. & Watamori, Y. (1992). Tests for the mean direction of the Langevin distribution with large concentration parameter. *J. Multivariate Anal.*, **42**, 210–225. (217)

Gadsden, R. J. & Kanji, G. K. (1980). Sequential analysis for angular data. *Statistician*, **30**, 119–129. (282)

Gadsden, R. J. & Kanji, G. K. (1982). Sequential analysis applied to circular data. *Sequential Anal.*, **1**, 305–314. (282)

Geary, R. C. (1936). The distribution of "Student's" ratio for non-normal samples. *J. Roy. Statist. Soc. Ser. B*, **3**, 178–184. (66)

Giné, M. E. (1975). Invariant tests for uniformity on compact Riemannian manifolds based on Sobolev norms. *Ann. Statist.*, **3**, 1243–1266. (209, 210, 233, 234, 240, 241, 243, 300)

Goodall, C. R. (1991). Procrustes methods in the statistical analysis of shape (with discussion). *J. Roy. Statist. Soc. Ser. B*, **53**, 285–339. (310, 339)

Goodall, C. R. & Mardia, K. V. (1993). Multivariate aspects of shape theory. *Ann. Statist*, **21**, 848–866. (325, 345)

Gordon, A. D., Jupp, P. E. & Byrne, R. W. (1989). Construction and assessment of mental maps. *British J. Math. Statist. Psych.*, **42**, 169–182. (9)

Gordon, L. & Hudson, M. (1977). A characterization of the von Mises distribution. *Ann. Statist.*, **5**, 813–814. (43)

Gould, A. L. (1969). A regression technique for angular variates. *Biometrics*, **25**, 683–700. (10, 257)

Greenwood, J. A. & Durand, D. (1955). The distribution of length and components of the sum of n random unit vectors. *Ann. Math. Statist.*, **26**, 233–246. (67, 69, 71, 95)

Gumbel, E. J., Greenwood, J.A. and Durand, D. (1953). The circular normal distribution: theory and tables. *J. Amer. Statist. Assoc.*, **48**, 131–152. (363)

Guttorp, P. & Lockhart, R. A. (1988). Finding the location of a signal – a Bayesian analysis. *J. Amer. Statist. Assoc.*, **83**, 322–330. (279)

Hájek, J. & Šidák, Z. (1967). *Theory of Rank Tests.* Academia, Prague. (247, 251)

Hall, P., Watson, G. S. & Cabrera, J. (1987). Kernel density estimation with spherical data. *Biometrika*, **74**, 751–762. (278)

Hansen, K. M. & Mount, V. S. (1990). Smoothing and extrapolation of crustal stress orientation measurements. *J. Geophys. Res.*, **95** (B2), 1155–1165. (282)

Hanson, B., Klink, K., Matsuura, K., Robeson, S. M. & Willmott, C. J. (1992). Vector correlation: Review, exposition and geographic application. *Ann. Assoc. Amer. Geographers*, **82**, 103–116. (256)

Harrison, D. & Kanji, G. K. (1988). The development of analysis of variance for circular data. *J. Appl. Statist.*, **15**, 197–224. (138, 141, 227)

Harrison, D., Kanji, G. K., & Gadsden, R. J. (1986). Analysis of variance for circular data. *J. Appl. Statist.*, **13**, 123–138. (138, 139, 141, 227)

Hartman, P. & Watson, G. S. (1974). "Normal" distribution functions on spheres and the modified Bessel functions. *Ann. Probab.*, **2**, 593–607. (173)

Hayakawa, T. (1990). On tests for the mean direction of the Langevin distribution. *Ann. Inst. Statist. Math.*, **42**, 359–376. (210, 212, 213)

Hayakawa, T. & Puri, M. L. (1985). Asymptotic expansions of the distributions of some test statistics. *Ann. Inst. Statist. Math.*, **37**, 95–108. (213)

He, X. (1992). Robust statistics of directional data: a survey. In A. K. Md. E. Saleh (ed.), *Proceedings of the International Symposium of Nonparametric Statistics and Related Topics*, pp. 87–96, North Holland, New York. (274)

He, X. & Simpson, D. G. (1992). Robust direction estimation. *Ann. Statist.*, **20**, 351–369. (274)

Healy, D. M., Hendriks, H., & Kim, P. T. (1998). Spherical deconvolution. *J. Multivariate Anal.*, **67**, 1–22. (279)

Healy, D. M. & Kim, P. T. (1996). An empirical Bayes approach to directional data and efficient computation on the sphere. *Ann. Statist.*, **24**, 232–254. (279)

Hendriks, H. (1990). Non-parametric estimation of a probability density on a Riemannian manifold using Fourier expansions. *Ann. Statist.*, **18**, 832–849. (278)

Hendriks, H. (1991). A Cramér–Rao type lower bound for estimators with values in a manifold. *J. Multivariate Anal.*, **38**, 245–261. (84, 300)

Hendriks, H., Janssen, J. H. M. & Ruymgaart, F. H. (1993). Strong uniform convergence of density estimators on compact Euclidean manifolds. *Statist. Probab. Lett.*, **16**, 301–305. (278)

Hendriks, H. & Landsman, Z. (1996a). Asymptotic behavior of sample mean location for manifolds. *Statist. Probab. Lett.*, **26**, 169–178. (301)

Hendriks, H. & Landsman, Z. (1996b). Asymptotic tests for mean location on manifolds. *C. R. Acad. Sci. Paris Sér. I Math.*, **322**, 773–778. (301)

Hendriks, H. & Landsman, Z. (1998). Mean location and sample mean location on manifolds: asymptotics, tests, confidence regions. *J. Multivariate Anal.*, **67**, 227–243. (301)

Hendriks, H., Landsman, Z. & Ruymgaart, F. (1996). Asymptotic behavior of sample mean direction for spheres. *J. Multivariate Anal.*, **59**, 141–152. (77, 188)

Hermans, M. & Rasson, J. P. (1985). A new Sobolev test for uniformity on the circle. *Biometrika*, **72**, 698–702. (109)

Hill, G. W. (1976). New approximations to the von Mises distribution. *Biometrika*, **63**, 673–676. (41)

Hill, G. W. (1977). Algorithm 518: Incomplete Bessel function $I_0(x)$; the von Mises distribution. *ACM Trans. Math. Software*, **3**, 279–284. (41)

Hill, T. P. (1995). A statistical derivation of the significant-digit law. *Statist. Sci.*, **10**, 354–363. (65)

Hodges, J. L. (1955). A bivariate sign test. *Ann. Math. Statist.*, **26**, 523–527. (106, 366)

Hospers, J. (1955). Rock magnetism and polar wandering. *J. Geol.*, **63**, 59–74. (216)

Huber, P. (1964). Robust estimation of a location parameter. *Ann. Math. Statist.*, **35**, 73–101. (275)

IMSL (1991). *The IMSL Libraries. Edition 2.0.* IMSL Inc., Sugar Land, TX. (43)

Jammalamadaka, S. R. (1984). Nonparametric methods in directional data analysis. In P. R. Krishnaiah & P. K. Sen (eds), *Handbook of Statistics, Vol 4: Nonparametric Methods*, pp. 755–770. Elsevier, Amsterdam. (145)

Jaynes, E. T. (1957). Information theory and statistical mechanics. *Phys. Rev. Ser. A*, **106**, 620–630. (43)

Jaynes, E. T. (1963). Information theory and statistical mechanics. In *Statistical*

Physics: Brandeis University Summer Institute Lectures in Physics, 1963, Vol. 3, pp. 81–218, Benjamin, New York. (43)

Jeffreys, H. (1948). *Theory of Probability, 2nd ed*, Oxford Univ. Press, Oxford. (48)

Jensen, J. L. (1981). On the hyperboloid distribution. *Scand. J. Statist.*, **8**, 193–206. (297, 298, 299)

Jensen, J. L. (1995). *Saddlepoint Approximations.* Oxford University Press, Oxford. (68, 207)

Johnson, R. A. & Wehrly, T. (1977). Measures and models for angular correlation and angular-linear correlation. *J. Roy. Statist. Soc. Ser. B*, **39**, 222–229. (246, 256)

Johnson, R. A. & Wehrly, T. (1978). Some angular-linear distributions and related regression models. *J. Amer. Statist. Assoc.*, **73**, 602–606. (257, 258, 261, 262, 263)

Jones, T. A. & James, W. R. (1969). Analysis of bimodal orientation data. *Math. Geol.*, **1**, 129–135. (90)

Jørgensen, B. (1987). Small dispersion asymptotics. *Rev. Bras. Probab. Estatist. (Brazilian J. Probab. Statist.)*, **1**, 59–90. (189, 192, 237, 238, 292)

Jupp, P. E. (1987). A non-parametric correlation coefficient and a two-sample test for random vectors or directions. *Biometrika*, **74**, 887–890. (255)

Jupp, P. E. (1988). Residuals for directional data. *J. Appl. Statist.*, **15**, 137–147. (261)

Jupp, P. E. (1995). Some applications of directional statistics to astronomy. In E.-M. Tiit, T. Kollo & H. Niemi (eds), *New Trends in Probability and Statistics. Vol. 3. Multivariate Statistics and Matrices in Statistics*, pp. 123–133, VSP, Utrecht. (11)

Jupp, P. E. (1999). Modifications of the Rayleigh and Bingham tests for uniformity of directions. *Submitted for publication.* (207, 233, 287, 295, 323)

Jupp, P. E. & Kent, J. T. (1987). Fitting smooth paths to spherical data. *Appl. Statist.*, **36**, 34–46. (280)

Jupp, P. E. & Mardia, K. V. (1979). Maximum likelihood estimation for the matrix von Mises–Fisher and Bingham distributions. *Ann. Statist.*, **7**, 599–606. (283, 288)

Jupp, P. E. & Mardia, K. V. (1980). A general correlation coefficient for directional data and related regression problems. *Biometrika*, **67**, 163–173; correction (1981), **68**, 738. (55, 249, 256, 262)

Jupp, P. E. & Mardia, K. V. (1989). A unified view of the theory of directional statistics, 1975-1988. *Internat. Statist. Rev.*, **57**, 261–294. (xviii)

Jupp, P. E. & Spurr, B. D. (1983). Sobolev tests for symmetry of directional data. *Ann. Statist.*, **11**, 1225–1231. (146, 236, 243, 288, 300)

Jupp, P. E. & Spurr, B. D. (1985). Sobolev tests for independence of directions. *Ann. Statist.*, **13**, 1140–1155. (241, 243, 253, 257, 300)

Jupp, P. E. & Spurr, B. D. (1989). Statistical estimation of a shock centre: Slate islands astrobleme. *J. Math. Geol.*, **21**, 191–198. (261)

Jupp, P. E., Spurr, B. D., Nichols, G. J. & Hirst, J. P. P. (1987). Statistical estimation of the apex of a sedimentary distributary system. *J. Math. Geol.*, **19**, 319–333. (261)

Kagan, A. M., Linnik, Y. V., & Rao, C. R. (1973). *Characterization Problems in Mathematical Statistics.* Wiley-Interscience, New York. (43)

Kanatani, K. (1993). *Geometric Computation for Machine Vision.* Clarendon Press, Oxford. (259)

Keilson, J., Petrondas, D., Sumita, U. & Wellner, J. (1983). Significance points for some tests of uniformity on the sphere. *J. Statist. Comput. Simulation*, **17**, 195–218. (209, 233)

Kelker, D. G. & Langenberg, C. W. (1982). A mathematical model for orientation data from macroscopic conical folds. *J. Math. Geol.*, **14**, 289–307. (183)

Kendall, D. G. (1977). The diffusion of shape. *Adv. Appl. Probab.*, **9**, 428–430. (303)

Kendall, D. G. (1981), in V. Barnett (ed.), *Interpreting Multivariate Data*, pp. 75–80, Wiley, Chichester.

Kendall, D. G. (1983). The shape of Poisson-Delaunay triangles. In H. C. Demetrescu & M. Iosifescu (eds), *Studies in Probability and Related Topics in Honour of Octav Onicescu*, pp. 321–330, Nagard, Montreal. (318)

Kendall, D. G. (1984). Shape manifolds, Procrustean metrics and complex projective spaces. *Bull. London Math. Soc.*, **16**, 81–121. (309, 317, 325, 345)

Kendall, D. G. (1989). A survey of the statistical theory of shape (with discussion). *Statist. Sci.*, **4**, 87–120. (321)

Kendall, M. G. (1946). Discussion on 'The statistical study of infectious diseases' by M. J. Greenwood. *J. Roy. Statist. Soc.*, **109**, 103–105. (108)

Kent, J. T. (1975). Contribution to discussion of Mardia (1975a). *J. Roy. Statist. Soc. Ser. B*, **37**, 377–378. (43)

Kent, J. T. (1977). The infinite divisibility of the von Mises–Fisher distribution for all values of the parameter in all dimensions. *Proc. London Math. Soc.*, **35**, 359–384. (173)

Kent, J. T. (1978). Limiting behaviour of the von Mises–Fisher distribution. *Math. Proc. Cambridge Phil. Soc.*, **84**, 531–536. (38, 174)

Kent, J. T. (1982). The Fisher–Bingham distribution on the sphere. *J. Roy. Statist. Soc. Ser. B*, **44**, 71–80. (175, 177, 205, 206, 273, 274)

Kent, J. T. (1983a). Identifiability of finite mixtures for directional data. *Ann. Statist.*, **11**, 954–988. (90)

Kent, J. T. (1983b). Information gain and a general measure of correlation. *Biometrika*, **70**, 163–173. (255)

Kent, J. T. (1987). Asymptotic expansions for the Bingham distribution. *Appl. Statist.*, **36**, 139–144. (182, 204)

Kent, J. T. (1992). New directions in shape analysis. In K. V. Mardia (ed.), *The Art of Statistical Science: A Tribute to G. S. Watson*, pp. 115–127. Wiley, Chichester. (309, 310, 313, 314, 344)

Kent, J. T. (1994). The complex Bingham distribution and shape analysis. *J. Roy. Statist. Soc. Ser. B*, **56**, 285–299. (182, 310, 312, 315, 317, 324, 327, 332, 333)

Kent, J. T. (1997). Data analysis for shapes and images. *J. Statist. Plan. Inference*, **57**, 181–193. (343)

Kent, J. T. & Mardia, K. V. (1994). Statistical shape methodology. In O. Ying-Lie, A. Toet, D. Foster & P. Meer (eds), *Shape in Picture*, pp. 443–452. Springer-Verlag, Berlin. (313)

Kent, J. T. & Mardia, K. V. (1997). Consistency of Procrustes estimators. *J. Roy. Statist. Soc. Ser. B*, **59**, 281–290. (179)

Kent, J. T., Briden, J. C. & Mardia, K. V. (1983). Linear and planar structure in ordered multivariate data as applied to progressive demagnetization of palaeomagnetic resonance. *Geophys. J. Roy. Astron. Soc.*, **75**, 593–621. (281)

Kent, J. T., Mardia, K. V., & Rao, J. S. (1979). A characterization of the uniform distribution on the circle. *Ann. Statist.*, **7**, 882–889. (66, 184)

Kent, J. T. & Tyler, D. E. (1988). Maximum likelihood estimation for the wrapped Cauchy distribution. *J. Appl. Statist.*, **15**, 247–254. (52, 90)

Khatri, C. G. & Mardia, K. V. (1977). The von Mises–Fisher distribution in orientation statistics. *J. Roy. Statist. Soc. Ser. B*, **39**, 95–106. (290, 291, 292)

Kim, P. T. (1991). Decision theoretic analysis of spherical regression. *J. Multivariate Anal.*, **38**, 233–240. (279)

Kim, P. T. (1998). Deconvolution density estimation on $SO(N)$. *Ann. Statist.*, **26**, 1083–1102. (279)

Kimber, A. C. (1985). A note on the detection and accommodation of outliers relative to Fisher's distribution on the sphere. *Appl. Statist.*, **34**, 169–172. (268, 276)

Klotz, J. (1959). Null distribution of the Hodges bivariate sign test. *Ann. Math. Statist.*, **30**, 1029–1033. (106)

Kluyver, J. C. (1906). A local probability theorem. *Ned. Akad. Wet. Proc. Ser. A*, **8**, 341–350. (68)

Ko, D. (1992). Robust estimation of the concentration parameter of the von Mises–Fisher distribution. *Ann. Statist.*, **20**, 917–928. (276)

Ko, D. & Chang, T. (1993). Robust *M*-estimators on spheres. *J. Multivariate Anal.*, **45**, 104–136. (274, 276)

Ko, D. & Guttorp, P. (1988). Robustness of estimators for directional data. *Ann. Statist.*, **16**, 609–618. (274, 275)

Kuhn, W. & Grün, F. (1942). Beziehungen zwischen elastischen Konstanten and Dehnungs doppelbrechung hochelastischer Stoffe. *Kolloid. Z.*, **101**, 248–271. (9, 170)

Kuiper, N. H. (1960). Tests concerning random points on a circle. *Ned. Akad. Wet. Proc. Ser. A*, **63**, 38–47. (100, 102, 150)

Langevin, P. (1905). Magnétisme et théorie des électrons. *Ann. Chim. Phys.*, **5**, 71–127. (170)

Laubscher, N. F. & Rudolph, G. J. (1968). *A distribution arising from random points on the circumference of a circle.* Technical Report 268, CSIR, Pretoria. (107, 367)

Laycock, P. J. (1975). Optimal design: regression models for directions. *Biometrika*, **62**, 305–311. (257, 282)

Lee, J. H. A. (1963). (Correspondence). *British Med. J.*, **10**, 623. (9, 97)

Lee, J. M. & Ruymgaart, F. H. (1996). Nonparametric curve estimation on Stiefel manifolds. *J. Nonparametr. Statist.*, **6**, 57–68. (278)

Lehmann, E. L. (1959). *Testing Statistical Hypotheses.* Wiley, New York. (96)

Lenth, R. V. (1981a). Robust measures of location for directional data. *Technometrics*, **23**, 77–81. (275)

Lenth, R. V. (1981b). On finding the source of a signal. *Technometrics*, **23**, 149–154. (261)

Lévy, P. (1939). L'addition des variables aléatoires définies sur une circonférence. *Bull. Soc. Math. France*, **67**, 1–41. (49, 52)

Lewis, T. & Fisher, N. I. (1982). Graphical methods for investigating the fit of a Fisher distribution to spherical data. *Geophys. J. R. Astr. Soc.*, **69**, 1–13. (195)

Lewis, T. & Fisher, N. I. (1995). Estimating the angle between the mean directions of two spherical distributions. *Austral. J. Statist.*, **37**, 179–191. (220)

Liddell, I. G. & Ord, J. K. (1978). Linear-circular correlation coefficients: some further results. *Biometrika*, **65**, 448–450. (246)

Lo, A. & Cabrera, J. (1987). Bayes procedures for rotationally symmetric models on the sphere. *Ann. Statist.*, **15**, 1257–1268. (279)

Lockhart, R. A. & Stephens, M. A. (1985). Tests of fit for the von Mises distribution. *Biometrika*, **72**, 647–652. (116, 369)

Lord, R. D. (1948). A problem on random vectors. *Phil. Mag. (7)*, **39**, 66–71. (71)

Lord, R. D. (1954). The use of the Hankel transform in statistics. I, General theory and examples. *Biometrika*, **41**, 44–55. (60)

Lukacs, E. (1970). *Characteristic Functions, 2nd ed.* Griffin, London. (52, 57, 61)

Maag, U. R. (1966). A k-sample analogue of Watson's U^2 statistic. *Biometrika*, **53**, 579–583. (158)

Maag, U. R. & Dicaire, G. (1971). On Kolmogorov–Smirnov type one-sample statistics. *Biometrika*, **58**, 653–656. (102)

Mackenzie, J. K. (1957). The estimation of an orientation relationship. *Acta. Crystallog.*, **10**, 61–62. (255, 256, 259)

Maling, D. H. (1992). *Coordinate Systems and Map Projections, 2nd ed.* Pergamon Press, Oxford. (161)

Mardia, K. V. (1967). A non-parametric test for the bivariate two-sample location problem. *J. Roy. Statist. Soc. Ser. B*, **29**, 320–342. (148, 376)

Mardia, K. V. (1968). Small sample power of a non-parametric test for the bivariate two-sample location problem in the normal case. *J. Roy. Statist. Soc. Ser. B*, **30**, 83–92. (148)

Mardia, K. V. (1969a). On the null distribution of a non-parametric test for the bivariate two-sample problem. *J. Roy. Statist. Soc. Ser. B*, **31**, 98–102. (148, 376)

Mardia, K. V. (1969b). On Wheeler and Watson's two-sample test on a circle. *Sanhkyā*, *Ser. A* **31**, 177–190. (148, 149, 155)

Mardia, K. V. (1970). A bivariate non-parametric c-sample test. *J. Roy. Statist. Soc. Ser. B*, **32**, 74–87. (157, 379)

Mardia, K. V. (1972a). *Statistics of Directional Data.* Academic Press, London. (xviii, 43, 44, 68, 70, 71, 81, 82, 84, 96, 121, 131, 137, 153, 211)

Mardia, K. V. (1972b). A multisample uniform scores test on a circle and its parametric competitor. *J. Roy. Statist. Soc. Ser. B*, **34**, 102–113. (157)

Mardia, K. V. (1975a). Statistics of directional data (with discussion). *J. Roy. Statist. Soc. Ser. B*, **37**, 349–393. (55, 174, 251, 256, 262, 268, 283, 380)

Mardia, K. V. (1975b). Distribution theory for the von Mises–Fisher distribution and its application. In G. P. Patil, S. Kotz & J. K. Ord (eds), *Statistical Distributions in Scientific Work, 1*, pp. 113–130. Reidel, Dordrecht. (185)

Mardia, K. V. (1975c). Characterizations of directional distributions. In G. P. Patil, S. Kotz & J. K. Ord (eds), *Statistical Distributions in Scientific Work, 3*, pp. 365–385. Reidel, Dordrecht. (55, 171, 262)

Mardia, K. V. (1976). Linear-circular correlation coefficients and rhythmometry. *Biometrika*, **63**, 403–405. (246, 247, 256, 257, 380)

Mardia, K. V. (1989). Shape analysis of triangles through directional techniques. *J. Roy. Statist. Soc. Ser. B*, **51**, 449–458. (321, 322, 325)

Mardia, K. V. (1995). Shape advances and future perspectives. In K. V. Mardia & C. A. Gill (eds), *Current Issues in Statistical Shape Analysis*, pp. 57–75, University of Leeds Press, Leeds. (313)

Mardia, K. V. (1999). Directional statistics and shape analysis. *J. Appl. Statist.*, **26**, 949–957. (317)

Mardia, K. V. & Dryden, I. L. (1989a). Statistical analysis of shape data. *Biometrika*, **76**, 271–281. (324, 325)

Mardia, K. V. & Dryden, I. L. (1989b). Shape distributions for landmark data. *Adv. Appl. Probab.*, **21**, 742–755. (317, 324, 325)

Mardia, K. V. & Dryden, I. L. (1999). The complex Watson distribution and shape analysis. *J. Roy. Statist. Soc. Ser. B*, **61**, 913–926. (334, 341, 342)

Mardia, K. V. & Edwards, R. (1982). Weighted distributions and rotating caps. *Biometrika*, **69**, 323–330. (180)

Mardia, K. V., Edwards, R. & Puri, M. L. (1977). Analysis of central place theory (with discussion). *Bull. Internat. Statist. Inst.*, **47**, 93–110, 139–146. (307, 320, 321)

Mardia, K. V. & El-Atoum, S. A. M. (1976). Bayesian inference for the von Mises–Fisher distribution. *Biometrika*, **63**, 203–205. (279)

Mardia, K. V. & Gadsden, R. J. (1977). A circle of best fit for spherical data and areas of vulcanism. *Appl. Statist.*, **26**, 238–245. (177)

Mardia, K. V., Goodall, C. R. & Walder, A. (1996). Distributions of projective invariants and model-based machine vision. *Adv. Appl. Probab.*, **28**, 641–661. (10)

Mardia, K. V., Holmes, D. & Kent, J. T. (1984). A goodness-of-fit test for the von Mises–Fisher distribution. *J. Roy. Statist. Soc. Ser. B*, **46**, 72–78. (272)

Mardia, K. V., Kent, J. T. & Bibby, J. M. (1979). *Multivariate Analysis.* Academic Press, London. (184, 245, 249, 310, 344)

Mardia, K. V. & Khatri, C. G. (1977). Uniform distribution on a Stiefel manifold. *J. Multivariate Anal.*, **7**, 468–473. (294)

Mardia, K. V. & Puri, M. L. (1978). A robust spherical correlation coefficient against scale. *Biometrika*, **65**, 391–396. (256)

Mardia, K. V., Southworth, H. R. & Taylor, C. C. (1999). On bias in maximum likelihood estimators. *J. Statist. Plann. Inference*, **76**, 31–39. (88, 200)

Mardia, K. V. & Sutton, T. W. (1975). On the modes of a mixture of two von Mises distributions. *Biometrika*, **62**, 699–701. (91)

Mardia, K. V. & Sutton, T. W. (1978). A model for cylindrical variables with applications. *J. Roy. Statist. Soc. Ser. B*, **40**, 229–233. (55, 257, 262)

Mardia, K. V. & Walder, A. N. (1988). *On Kendall's Spherical Blackboard.* Technical report, Dept. of Statist., University of Leeds. (319)

Mardia, K. V. & Zemroch, P. J. (1975a). Algorithm AS 81: Circular statisics. *Appl. Statist.*, **24**, 147–150. (86)

Mardia, K. V. & Zemroch, P. J. (1975b). Algorithm A-s 86: The von Mises distribution function. *Appl. Statist.*, **24**, 268–272. (41)

Mardia, K. V. & Zemroch, P. J. (1977). Table of maximum likelihood estimates for the Bingham distribution. *J. Statist. Comput. Simulation*, **6**, 29–34. (204)

Mardia, K. V., Kent, J. T., Goodall, C. R. & Little, J. A. (1996). Kriging and splines with derivative information. *Biometrika*, **83**, 207–221. (281)

Mardia, K. V., Baczkowski, A., Feng, X. & Hainsworth, T. J. (1997). Statistical methods for automatic interpretation of digitally scanned finger prints. *Pattern Recognition Lett.*, **18**, 1197–1203. (10)

Markov, A. A. (1912). *Wahrscheinlichkeitsrechnung.* Taubner, Leipzig. (68)

Marshall, A. W. & Olkin, I. (1961). Game theoretic proof that Chebyshev inequalities are sharp. *Pacific J. Math.*, **11**, 1421–1429. (31)

Matthews, G. V. T. (1961). "Nonsense" orientation in mallard *anas platyrhynchos* and its relation to experiments on bird navigation. *Ibis*, **103a**, 211–230. (3)

McCullagh, P. (1996). Möbius transformation and Cauchy parameter estimation. *Ann. Statist.*, **24**, 787–808. (52)

Mendoza, C. E. (1986). Smoothing unit vector fields. *J. Math. Geol.*, **18**, 307–322. (281)

Moran, P. A. P. (1975). Quaternions, Haar measure and the estimation of a palaeomagnetic rotation. In J. Gani (ed.), *Perspectives in Probability and Statistics*, pp. 295–301. Applied Probability Trust, Sheffield. (286)

Morris, J. E. & Laycock, P. J. (1974). Discriminant analysis of directional data. *Biometrika*, **61**, 335–341. (282)

Muirhead, R. J. (1982). *Aspects of Multivariate Statistical Theory.* Wiley, New York. (352)

Newcomb, S. (1881). Note on the frequency of use of the different digits in natural numbers. *Amer. J. Math.*, **4**, 39–40. (64)

Okabe, A., Boots, B. & Sugihara, K. (1992). *Spatial Tessellations: Concepts and Applications of Voronoi Diagrams.* Wiley, Chichester. (320)

Oller, J. (1993). On an intrinsic analysis of statistical estimation. In C. M. Cuadras & C. R. Rao (eds), *Multivariate Analysis: Future Directions 2*, pp. 421–437. North-Holland, Amsterdam. (300)

Oller, J. M. & Corcuera, J. M. (1995). Intrinsic analysis of statistical estimation. *Ann. Statist.*, **23**, 1562–1581. (300)

Pearson, E. S. & Stephens, M. A. (1962). The goodness-of-fit tests based on W_N^2 and U_N^2. *Biometrika*, **49**, 397–402. (104)

Pearson, K. (1905). The problem of the random walk. *Nature*, **72**, 294–342. (65, 68)

Pearson, K. (1906). A mathematical theory of random migration. *Drapers' Company Research Memoirs, Biometric Series, III*, **15**. (67, 68, 95)

Perrin, F. (1928). Etude mathématique du mouvement brownien de rotation. *Ann. Sci. Ec. Norm. Sup. Paris*, **45**, 1–51. (174)

Persson, T. (1979). A new way to obtain Watson's U^2. *Scand. J. Statist.*, **6**, 119–122. (152)

Pincus, H. J. (1953). The analysis of aggregates of orientation data in the earth sciences. *J. Geol.*, **61**, 482–509. (7)

Poincaré, H. (1912). Chance. *Monist*, **22**, 31–52. (63)

Pólya, G. (1919). Zur Statistik der sphärischen Verteilung der Fixsterne. *Ast. Nachr.*, **208**, 175–180. (11)

Prentice, M. J. (1978). On invariant tests of uniformity for directions and orientations. *Ann. Statist.*, **6**, 169–176; correction (1979), **7**, 926. (209, 233, 234)

Prentice, M. J. (1982). Antipodally symmetric distributions for orientation statistics. *J. Statist. Plann. Inference*, **6**, 205–214. (292)

Prentice, M. J. (1984). A distribution-free method of interval estimation for unsigned directional data. *Biometrika*, **71**, 147–154. (235)

Prentice, M. J. (1986). Orientation statistics without parametric assumptions. *J. Roy. Statist. Soc. Ser. B*, **48**, 214–222. (286, 289)

Prentice, M. J. (1987). Fitting smooth paths to rotation data. *Appl. Statist.*, **36**, 325–331. (281, 286)

Prentice, M. J. (1989). Spherical regression on matched pairs of orientation statistics. *J. Roy. Statist. Soc. Ser. B*, **51**, 241–248. (260)

Pukkila, T. M. & Rao, C. R. (1988). Pattern recognition based on scale invariant discriminant functions. *Inform. Sci.*, **45**, 379–389. (178, 307)

Purkayastha, S. (1991). A rotationally symmetric directional distribution obtained through maximum likelihood characterization. *Sankhyā Ser. A*, **53**, 70–83. (180)

Purkayastha, S. & Mukerjee, R. (1992). Maximum likelihood characterization of the von Mises–Fisher matrix distribution. *Sankhyā Ser. A*, **54**, 123–127. (291)

Rao, C. R. (1973). *Linear Statistical Inference and its Applications, 2nd. ed.* Wiley, New York. (76, 171)

Rao, J. S. (1969). *Some Contributions to the Analysis of Circular Data.* Ph.D. thesis, Indian Statist. Inst., Calcutta. (75, 107, 108, 115, 153, 368)

Rao, J. S. (1972a). Bahadur efficiencies of some tests for uniformity on the circle. *Ann. Math. Statist.*, **43**, 468–479. (108)

Rao, J. S. (1972b). Some variants of chi-square for testing uniformity on the circle. *Z. Wahrsch. verw. Geb.*, **22**, 33–44. (109, 112)

Rao, J. S. (1976). Some tests based on arc length for the circle. *Sankhyā Ser. B*, **38**, 329–338. (108, 153)

Rao, J. S. & Sengupta, S. (1970). An optimum hierarchical sampling procedure for cross-bedding data. *J. Geol.*, **78**, 533–534. (136)

Rayleigh, Lord (1880). On the resultant of a large number of vibrations of the same pitch and of arbitrary phase. *Phil. Mag.*, **10**, 73–78. (65)

Rayleigh, Lord (1905). The problem of the random walk. *Nature*, **72**, 318. (65)

Rayleigh, Lord (1919). On the problem of random vibrations, and of random flights in one, two, or three dimensions. *Phil. Mag., (6)*, **37**, 321–347. (9, 68, 189, 206)

Richardus, P. & Adler, R. K. (1972). *Map Projections for Geodesists, Cartographers & Geographers.* North-Holland, London. (161)

Rivest, L.-P. (1982). Some statistical methods for bivariate circular data. *J. Roy. Statist. Soc. Ser. B*, **44**, 81–90. (256, 257)

Rivest, L.-P. (1984). On the information matrix for symmetric distributions on the hypersphere. *Ann. Statist.*, **12**, 1085–1089. (176)

Rivest, L.-P. (1986). Modified Kent's statistics for testing the goodness-of-fit of the Fisher distribution in small concentrated samples. *Statist. Probab. Lett.*, **4**, 1–4. (273)

Rivest, L.-P. (1988). A distribution for dependent unit vectors. *Comm. Statist. Theory Methods*, **17**, 461–483. (262)

Rivest, L.-P. (1989). Spherical regression for concentrated Fisher–von Mises distributions. *Ann. Statist.*, **17**, 307–317. (260, 261, 269)

Rivest, L.-P. (1997). A decentred predictor for circular-circular regression. *Biometrika*, **84**, 717–726. (261)

Roberts, P. H. & Ursell, H. D. (1960). Random walk on a sphere and on a Riemannian manifold. *Phil. Trans. Roy. Soc. London Ser. A*, **252**, 317–356. (174)

Roberts, P. H. & Winch, D. E. (1984). On random rotations. *Adv. Appl. Probab.*, **16**, 638–655. (293)

Ronchetti, E. (1992). Optimal robust estimators for the concentration parameter of a Fisher–von Mises distribution. In K. V. Mardia (ed.), *The Art of Statistical Science: A Tribute to G. S. Watson*, pp. 65–74. Wiley, Chichester. (276)

Ross, H. E., Crickmar, S. D., Sills, N. V. & Owen, E. P. (1969). Orientation to the vertical in free divers. *Aerospace Med.*, **40**, 728–732. (9)

Rothman, E. D. (1971). Tests of coordinate independence for a bivariate sample on a torus. *Ann. Math. Statist.*, **42**, 1962–1969. (253)

Rothman, E. D. (1972). Tests for uniformity of a circular distribution. *Sankhyā Ser. A*, **34**, 23–32. (109, 112)

Russell, G. S. & Levitin, D. J. (1996). An expanded table of probability values for Rao's spacing test. *Comm. Statist. Simulation Comput.*, **24**, 879–888. (108)

Ruymgaart, F. H. (1989). Strong uniform convergence of density estimators on spheres. *J. Statist. Plann. Inference*, **23**, 45–52. (278)

Saw, J. G. (1978). A family of distributions on the m-sphere and some hypothesis tests. *Biometrika*, **65**, 69–73. (179)

Saw, J. G. (1983). Dependent unit vectors. *Biometrika*, **70**, 665–671. (179)

Saw, J. G. (1984). Ultraspherical polynomials and statistics on the m-sphere. *J. Multivariate Anal.*, **14**, 105–113. (179)

Schach, S. (1967). *Nonparametric Tests of Location for Circular Distributions.* Technical Report 95, Dept. of Statist., Minnesota University. (154, 155)

Schach, S. (1969a). Nonparametric symmetry tests for circular distributions. *Biometrika*, **56**, 571–577. (146)

Schach, S. (1969b). On a class of nonparametric two-sample tests for circular distributions. *Ann. Math. Statist.*, **40**, 1791–1800. (149, 155)

Schaeben, H. (1993). Towards statistics of crystal orientations in quantitative texture analysis. *J. Appl. Cryst.*, **26**, 112–121. (289)

Schmidt-Koenig, K. (1958). Experimentelle Einflussnahme auf die 24-Stunden-Periodik bei Brieftauben und deren Auswirkungen unter besonderer Berücksichtigung des Heimfindevermögens.. *Z. Tierpsychol.*, **15**, 301–331. (131)

Schmidt-Koenig, K. (1963). On the role of the loft, the distance and site of release in pigeon homing (the "cross-loft experiment"). *Biol. Bull.*, **125**, 154–164. (93, 125)

Schmidt-Koenig, K. (1965). Current problems in bird orientation. In D. Lehrman*et al.* (eds), *Advances in the Study of Behaviour*, pp. 217–278, Academic Press, New York. (8)

Schou, G. (1978). Estimation of the concentration parameter in von Mises–Fisher distributions. *Biometrika*, **65**, 369–377. (87, 88, 89, 200, 201, 350)

Selby, B. (1964). Girdle distributions on a sphere. *Biometrika*, **51**, 381–392. (183)

Sengupta, S. & Rao, J. S. (1966). Statistical analysis of crossbedding azimuths from the Kamthi formation around Bheemaram, Pranhita–Godavari valley. *Sankhyā Ser. B*, **28**, 165–174. (8)

Sherman, B. (1950). A random variable related to the spacing of sample value. *Ann. Math. Statist.*, **21**, 339–351. (108)

Siegel, S. (1956). *Nonparametric Statistics for the Behavioural Sciences.* McGraw–Hill, New York. (146)

Spurr, B. D. (1981). On estimating the parameters in mixtures of circular normal distributions. *J. Math. Geol.*, **13**, 163–173. (91)

Spurr, B. D. & Koutbeiy, M. A. (1991). A comparison of various methods for estimating the parameters in mixtures of von Mises distributions. *Comm. Statist. Simulation Comput.*, **20**, 725–741. (90)

Stam, A. J. (1982). Limit theorems for uniform distributions on spheres in high dimensional Euclidean spaces. *J. Appl. Probab.*, **19**, 221–229. (192)

Steck, G. P. (1969). The Smirnov two-sample tests as rank tests. *Ann. Math. Statist.*, **40**, 1449–1466. (150)

Stephens, M. A. (1962a). Exact and approximate tests for directions I. *Biometrika*, **49**, 463–477. (67, 124)

Stephens, M. A. (1962b). Exact and approximate tests for directions II. *Biometrika*, **49**, 547–552. (185, 213)

Stephens, M. A. (1963). Random walk on a circle. *Biometrika*, **50**, 385–390. (38, 44, 51)

Stephens, M. A. (1965). Significance points for the two-sample statistic $U^2_{M,N}$. *Biometrika*, **52**, 661–663. (152)

Stephens, M. A. (1967). Tests for the dispersion and the modal vector of a distribution on a sphere. *Biometrika*, **54**, 211–223. (190, 388)

Stephens, M. A. (1969a). Tests for the von Mises distribution. *Biometrika*, **56**, 149–160. (79, 127)

Stephens, M. A. (1969b). A goodness-of-fit statistic for the circle, with some comparisons. *Biometrika*, **56**, 161–168. (109, 115)

Stephens, M. A. (1969c). Multi-sample tests for the Fisher distribution for directions. *Biometrika*, **56**, 169–181. (389, 390)

Stephens, M. A. (1969d). Tests for randomness of directions against two circular alternatives. *J. Amer. Statist. Assoc.*, **64**, 280–289. (365)

Stephens, M. A. (1969e). *Techniques for Directional Data.* Technical Report 150, Dept. of Statist., Stanford Univ. (10)

Stephens, M. A. (1970). Use of the Kolmogorov–Smirnov, Cramér–von Mises and related statistics without extensive tables. *J. Roy. Statist. Soc. Ser. B*, **32**, 115–122. (102, 104)

Stephens, M. A. (1972). *Multi-sample Tests for the von Mises Distribution*, Technical Report 190, Dept. of Statist., Stanford Univ. (130, 135)

Stephens, M. A. (1974). EDF statistics for goodness of fit and some comparisons. *J. Amer. Statist. Assoc.*, **69**, 730–737. (270)

Stephens, M. A. (1979). Vector correlation. *Biometrika*, **66**, 41–48. (255, 256)

Stephens, M. A. (1982). Use of the von Mises distribution to analyse continuous proportions. *Biometrika*, **69**, 197–203. (141, 227)

Stephens, M. A. (1992). On Watson's ANOVA for directions. In K. V. Mardia (ed.), *The Art of Statistical Science: A Tribute to G. S. Watson*, pp. 75–85, Wiley, Chichester. (190)

Stuart, A. & Ord, J. K. (1987). *Kendall's Advanced Theory of Statistics, Vol. 1* (5th edition), Griffin, London. (23)

Stuart, A. & Ord, J. K. (1991). *Kendall's Advanced Theory of Statistics, Vol. 2* (5th edition), Edward Arnold, London. (79, 120, 121, 140, 226)

Theobald, C. M. (1975). An inequality for the trace of the product of two symmetric matrices. *Proc. Cambridge Phil. Soc.*, **77**, 265–267. (203)

Tiku, M. L. (1965). Chi-square approximations for the distributions of goodness-of-fit statistics U^2_N and W^2_N. *Biometrika*, **52**, 630–633. (104)

Titchmarsh, E. C. (1958). *The Theory of Functions, 2nd ed.* Oxford University Press, Oxford. (58, 61, 63)

Titterington, D. M., Smith, A. F. M. & Makov, U. E. (1985). *Statistical Analysis of Finite Mixture Distributions.* Wiley, Chichester. (90)

Tyler, D. (1987). Statistical analysis for the angular central Gaussian distribution on the sphere. *Biometrika*, **74**, 579–589. (182, 206, 343)

Upton, G. J. G. (1973). Single-sample tests for the von Mises distribution. *Biometrika*, **60**, 87–99. (122)

Upton, G. J. G. (1974). New approximations to the distribution of certain angular statistics. *Biometrika*, **61**, 369–373. (41)

Upton, G. J. G. & Fingleton, B. (1989). *Spatial Data Analysis by Example, Vol. 2.* Wiley, Chichester. (xx)

Vistelius, A. B. (1966). *Structural Diagrams.* Pergamon, London. (219, 220)

von Mises, R. (1918). Über die "Ganzzahligkeit" der Atomgewichte und verwandte Fragen. *Phys. Z.*, **19**, 490–500. (9, 36, 42, 99)

Wahba, G. (1966). Section on problems and solutions: A least squares estimate of satellite attitude. *SIAM Rev.*, **8**, 384–385. (259)

Wahba, G. (1990). *Spline Models for Observational Data.* Society for Industrial and Applied Mathematics, Philadelphia. (281)

Watamori, Y. (1992). Tests for a given linear structure of the mean direction of the Langevin distribution. *Ann. Inst. Statist. Math.*, **44**, 147–156. (217, 218)

Waterman, T. H. (1963). The analysis of spatial orientation. *Ergeb. Biol.*, **26**, 98–117. (12)

Watson, G. N. (1948). *A Treatise on the Theory of Bessel Functions, 2nd ed.* Cambridge University Press, Cambridge. (349, 350)

Watson, G. S. (1956a). Analysis of dispersion on a sphere. *Monthly Notices Roy. Astr. Soc., Geophys. Suppl.*, **7**, 153–159. (190, 191, 223)

Watson, G. S. (1956b). A test for randomness of directions. *Monthly Notices Roy. Astr. Soc., Geophys. Suppl.*, **7**, 160–161. (206)

Watson, G. S. (1960). More significance tests on the sphere. *Biometrika*, **47**, 87–91. (231)

Watson, G. S. (1961). Goodness-of-fit tests on a circle. *Biometrika*, **48**, 109–114. (103, 105)

Watson, G. S. (1962). Goodness-of-fit tests on a circle, II. *Biometrika*, **49**, 57–63. (149, 151, 152)

Watson, G. S. (1965). Equatorial distributions on a sphere. *Biometrika*, **52**, 193–201. (180)

Watson, G. S. (1967). Another test for the uniformity of a circular distribution. *Biometrika*, **54**, 675–677. (109)

Watson, G. S. (1970). Orientation statistics in the earth sciences. *Bull. Geol. Inst. Univ. Uppsala*, **2**, 73–89. (7, 208, 209)

Watson, G. S. (1976). Optimal invariant tests for uniformity. In E. J. Williams (ed.), *Studies in Probability and Statistics*, pp. 121–127, North-Holland, Amsterdam. (105)

Watson, G. S. (1983a). Statistics on Spheres. *University of Arkansas Lecture Notes in the Mathematical Sciences*, **6**. (xx, 178, 179, 180, 187, 188, 190, 212, 215, 218, 228, 229, 239, 277)

Watson, G. S. (1983b). Large sample theory of the Langevin distribution. *J. Stat. Plann. Inference*, **8**, 245–256. (142, 187, 188, 215, 228, 229, 231)

Watson, G. S. (1983c). Large sample theory for distributions on the hypersphere with rotational symmetries. *Ann. Inst. Statist. Math.*, **35**, 303–319. (180, 187, 188, 215)

Watson, G. S. (1983d). Limit theorems on high dimensional spheres and Stiefel manifolds. In S. Karlin, T. Amemiya & L. A. Goodman (eds), *Studies in Econometrics, Time Series and Multivariate Statistics*, pp. 559–570, Academic Press, New York. (192, 292)

Watson, G. S. (1983e). The computer simulation treatment of directional data. In *Proceedings of the Geological Conference, Kharagpur, India, December 1983.* Reprinted in *Indian J. Earth Sci.*, **12**, 19–23. (277)

Watson, G. S. (1984). The theory of concentrated Langevin distributions. *J. Multivariate Anal.*, **14**, 74–82. (229)

Watson, G. S. (1985). Interpolation and smoothing of directed and undirected line data. In P. R. Krishnaiah (ed.), *Multivariate Analysis VI*, pp. 613–625, Academic Press, New York. (280, 281)

Watson, G. S. (1986). Some estimation theory on the sphere. *Ann. Inst. Statist. Math.*, **38**, 263–275. (274)

Watson, G. S. (1988). The Langevin distribution on high dimensional spheres. *J. Appl. Statist.*, **15**, 123–130. (192)

Watson, G. S. (1994). Distribution of angles between unit vectors and the multiple comparison problem for unit vectors. *J. Appl. Statist.*, **21**, 327–333. (226)

Watson, G. S. (1995). U_n^2 test for uniformity of discrete distributions. *J. Appl. Statist.*, **22**, 273–276. (105)

Watson, G. S. & Beran, R. J. (1967). Testing a sequence of unit vectors for serial correlation. *J. Geophys. Res.*, **72**, 5655–5659. (256, 263)

Watson, G. S. & Debiche, M. G. (1992). Testing whether two Fisher distributions have the same center. *Geophys. J.*, **109**, 225–232. (228)

Watson, G. S. & Williams, E. (1956). On the construction of significance tests on the circle and the sphere. *Biometrika*, **43**, 344–352. (75, 80, 123, 129, 130, 170, 185, 214, 219)

Wehrly, T. & Johnson, R. A. (1980). Bivariate models for dependence of angular observations and a related Markov process. *Biometrika*, **67**, 255–256. (265)

Wehrly, T. & Shine, E. P. (1981). Influence curves for directional data. *Biometrika*, **68**, 634–635. (274)

Wellner, J. A. (1979). Permutation tests for directional data. *Ann. Statist.*, **7**, 929–943. (243, 300)

Wheeler, S. & Watson, G. S. (1964). A distribution-free two-sample test on a circle. *Biometrika*, **51**, 256–257. (148)

Wilkie, D. (1983). Rayleigh test for randomness of circular data. *Appl. Statist.*, **32**, 311–312. (95)

Wintner, A. (1947). On the shape of the angular case of Cauchy's distribution curves. *Ann. Math. Statist.*, **18**, 589–593. (52)

Wood, A. T. A. (1982). A bimodal distribution for the sphere. *Appl. Statist.*, **31**, 52–58. (178)

Wood, A. T. A. (1988). Some notes on the Fisher–Bingham family on the sphere. *Comm. Statist. Theory Methods*, **17**, 3881–3897. (175, 176)

Wood, A. T. A. (1994). Simulation of the von Mises–Fisher distribution. *Comm. Statist. Simulation Comput.*, **23**, 157–164. (172)

Yamamoto, E. & Yanagimoto, T. (1995). A modified likelihood ratio test for the mean direction in the von Mises distribution. *Comm. Statist. Theory Methods*, **24**, 2659–2678. (122, 211)

Yfantis, E. A. & Borgman, L. E. (1982). An extension of the von Mises distribution. *Comm. Statist. Theory Methods*, **11**, 1695–1706. (45)

Zygmund, A. (1959). *Trigonometric Series.* Cambridge University Press, *Vol. 1*, Cambridge. (27)

Index

WILEY SERIES IN PROBABILITY AND STATISTICS

ESTABLISHED BY WALTER A. SHEWHART AND SAMUEL S. WILKS

Probability and Statistics Section

*ANDERSON · The Statistical Analysis of Time Series
ARNOLD, BALAKRISHNAN, and NAGARAJA · A First Course in Order Statistics
BACCELLI, COHEN, OLSDER, and QUADRAT · Synchronization and Linearity: An Algebra for Discrete Event Systems
BARNETT · Comparative Statistical Inference, *Third Edition*
BASILEVSKY · Statistical Factor Analysis and Related Methods: Theory and Applications
BERNARDO and SMITH · Bayesian Statistical Concepts and Theory
BILLINGSLEY · Convergence of Probability Measures
BOROVKOV · Asymptotic Methods in Queuing Theory
BOROVKOV · Ergodicity and Stability of Stochastic Processes
BRANDT, FRANKEN, and LISEK · Stationary Stochastic Models
CAINES · Linear Stochastic Systems
CAIROLI and DALANG · Sequential Stochastic Optimization
CONSTANTINE · Combinatorial Theory and Statistical Design
COOK · Regression Graphics
COVER and THOMAS · Elements of Information Theory
CSÖRGŐ and HORVÁTH · Weighted Approximation in Probability Statistics
CSÖRGŐ and HORVÁTH · Limit Theorems in Change Point Analysis
DETTE and STUDDEN · The Theory of Canonical Moments with Applications in Statistics, Probability, and Analysis
*DOOB · Stochastic Processes
DRYDEN and MARDIA · Statistical Analysis of Shape
DUPUIS and ELLIS · A Weak Convergence Approach to the Theory of Large Deviations
ETHIER and KURTZ · Markov Processes: Characterization and Convergence
FELLER · An Introduction to Probability Theory and Its Applications, Volume 1, *Third Edition*, Revised; Volume II, *Second Edition*
FULLER · Introduction to Statistical Time Series, *Second Edition*
FULLER · Measurement Error Models
GHOSH, MUKHOPADHYAY, and SEN · Sequential Estimation
GIFI · Nonlinear Multivariate Analysis
GUTTORP · Statistical Inference for Branching Processes
HALL · Introduction to the Theory of Coverage Processes

*Now available in a lower priced paperback edition in the Wiley Classics Library.

Probability and Statistics Section (Continued)

HAMPEL · Robust Statistics: The Approach Based on Influence Functions
HANNAN and DEISTLER · The Statistical Theory of Linear Systems
HUBER · Robust Statistics
HUŠKOVÁ, BERAN, and DUPAČ · Collected Works of Jaroslav Hájek—With Commentary
IMAN and CONOVER · A Modern Approach to Statistics
JUREK and MASON · Operator-Limit Distributions in Probability Theory
KASS and VOS · The Geometrical Foundations of Asymptotic Inference
KAUFMAN and ROUSSEEUW · Finding Groups in Data: An Introduction to Cluster Analysis
KELLY · Probability, Statistics, and Optimization
LINDVALL · Lectures on the Coupling Method
McFADDEN · Management of Data in Clinical Trials
MANTON, WOODBURY, and TOLLEY · Statistical Applications Using Fuzzy Sets
MORGENTHALER and TUKEY · Configural Polysampling: A Route to Practical Robustness
MUIRHEAD · Aspects of Multivariate Statistical Theory
OLIVER and SMITH · Influence Diagrams, Belief Nets, and Decision Analysis
*PARZEN · Modern Probability Theory and Its Applications
PRESS · Bayesian Statistics: Principles, Models, and Applications
PUKELSHEIM · Optimal Experimental Design
RAO · Asymptotic Theory of Statistical Inference
RAO · Linear Statistical Inference and Its Applications, *Second Edition*
RAO and SHANBHAG · Choquet-Deny Type Functional Equations with Applications to Stochastic Models
ROBERTSON, WRIGHT, and DYKSTRA · Order Restricted Statistical Inference
ROGERS and WILLIAMS · Diffusions, Markov Processes, and Martingales, Volume I: Foundations, *Second Edition*, Volume II: Itô Calculus
RUBINSTEIN and SHAPIRO · Discrete Event Systems: Sensitivity Analysis and Stochastic Optimization by the Score Function Method
RUZSA and SZEKELEY · Algebraic Probability Theory
SCHEFFÉ · The Analysis of Variance
SEBER · Linear Regression Analysis
SEBER · Multivariate Observations
SEBER and WILD · Nonlinear Regression
SERFLING · Approximation Theorems of Mathematical Statistics
SHORACK and WELLNER · Empirical Processes with Applications to Statistics
SMALL and McLEISH · Hilbert Space Methods in Probability and Statistical Inference
STAPLETON · Linear Statistical Models
STAUDTE and SHEATHER · Robust Estimation and Testing
STOYANOV · Counterexamples in Probability
TANAKA · Time Series Analysis: Nonstationary and Noninvertible Distribution Theory
THOMPSON and SEBER · Adaptive Sampling
WELSH · Aspects of Statistical Inference
WHITTAKER · Graphical Models in Applied Multivariate Statistics
YANG · The Construction Theory of Denumerable Markov Processes

---

*Now available in a lower priced paperback edition in the Wiley Classics Library.

Applied Probability and Statistics Section

ABRAHAM and LEDOLTER · Statistical Methods for Forecasting
AGRESTI · Analysis of Ordinal Categorical Data
AGRESTI · Categorical Data Analysis
ANDERSON, AUQUIER, HAUCH, OAKES, VANDAELE, and WEISBERG · Statistical Methods for Comparative Studies
ARMITAGE and DAVID (editors) · Advances in Biometry
*ARTHANARI and DODGE · Mathematical Programming in Statistics
ASMUSSEN · Applied Probability and Queues
*BAILEY · The Elements of Stochastic Processes with Applications to the Natural Sciences
BARNET and LEWIS · Outliers in Statistical Data, *Third Edition*
BARTHOLOMEW, FORBES, and McLEAN · Statistical Techniques for Manpower Planning, *Second Edition*
BATES and WATTS · Nonlinear Regression Analysis and Its Applications
BECHHOFER, SANTNER, and GOLDSMAN · Design and Analysis of Experiments for Statistical Selection, Screening, and Multiple Comparisons
BELSLEY · Conditioning Diagnostics: Collinearity and Weak Data in Regression
BELSLEY, KUH, and WELSCH · Regression Diagnostics: Identifying Influential Data and Sources of Collinearity
BHAT · Elements of Applied Stochastic Processes, *Second Edition*
BHATTACHARYA and WAYMIRE · Stochastic Processes with Applications
BIRKES and DODGE · Alternative Methods of Regression
BLOOMFIELD · Fourier Analysis of Time Series: An Introduction
BOLLEN · Structural Equations with Latent Variables
BOULEAU · Numerical Methods for Stochastic Processes
BOX · Bayesian Inference in Statistical Analysis
BOX and DRAPER · Empirical Model-Building and Response Surfaces
BOX and DRAPER · Evolutionary Operation: A Statistical Method for Process Improvement
BUCKLEW · Large Deviation Techniques in Decision, Simulation, and Estimation
BUNKE and BUNKE · Nonlinear Regression, Functional Relations, and Robust Methods: Statistical Methods of Model Building
CHATTERJEE and HADI · Sensitivity Analysis in Linear Regression
CHOW and LIU · Design and Analysis of Clinical Trials
CLARKE and DISNEY · Probability and Random Processes: A First Course with Applications, *Second Edition*
*COCHRAN and COX · Experimental Designs, *Second Edition*
CONOVER · Practical Nonparametric Statistics, *Second Edition*
CORNELL · Experiments with Mixtures, Designs, Models, and the Analysis of Mixture Data, *Second Edition*
*COX · Planning of Experiments
CRESSIE · Statistics for Spatial Data, *Revised Edition*
DANIEL · Applications of Statistics to Industrial Experimentation
DANIEL · Biostatistics: A Foundation for Analysis in the Health Sciences, *Sixth Edition*
DAVID · Order Statistics, *Second Edition*
*DEGROOT, FIENBERG, and KADANE · Statistics and the Law
*DODGE · Alternative Methods of Regression
DOWDY and WEARDEN · Statistics for Research, *Second Edition*

---

*Now available in a lower priced paperback edition in the Wiley Classics Library.

Applied Probability and Statistics (Continued)

DUNN and CLARK · Applied Statistics: Analysis of Variance and Regression, *Second Edition*
ELANDT-JOHNSON and JOHNSON · Survival Models and Data Analysis
EVANS, PEACOCK, and HASTINGS · Statistical Distributions, *Second Edition*
FLEISS · The Design and Analysis of Clinical Experiments
FLEISS · Statistical Methods for Rates and Proportions, *Second Edition*
FLEMING and HARRINGTON · Counting Processes and Survival Analysis
GALLANT · Nonlinear Statistical Models
GLASSERMAN and YAO · Monotone Structure in Discrete-Event Systems
GNANADESIKAN · Methods for Statistical Data Analysis of Multivariate Observations. *Second Edition*
GOLDSTEIN and LEWIS · Assessment: Problems, Development, and Statistical Issues
GREENWOOD and NIKULIN · A Guide to Chi-squared Testing
*HAHN · Statistical Models in Engineering
HAHN and MEEKER · Statistical Intervals: A Guide for Practitioners
HAND · Construction and Assessment of Classification Rules
HAND · Discrimination and Classification
HEIBERGER · Computation for the Analysis of Designed Experiments
HINKELMAN and KEMPTHORNE · Design and Analysis of Experiments, Volume 1: Introduction to Experimental Design
HOAGLIN, MOSTELLER, and TUKEY · Exploratory Approach to Analysis of Variance
HOAGLIN, MOSTELLER, and TUKEY · Exploring Data Tables, Trends and Shapes
HOAGLIN, MOSTELLER, and TUKEY · Understanding Robust and Exploratory Data Analysis
HOCHBERG and TAMHANE · Multiple Comparison Procedures
HOCKING · Methods and Applications of Linear Models: Regression and the Analysis of Variables
HOGG and KLUGMAN · Loss Distributions
HOLLANDER and WOLFE · Nonparametric Statistical Methods
HOSMER and LEMESHOW · Applied Logistic Regression
HØYLAND and RAUSAND · System Reliability Theory: Models and Statistical Methods
HUBERTY · Applied Discriminant Analysis
JACKSON · A User's Guide to Principal Components
JOHN · Statistical Methods in Engineering and Quality Assurance
JOHNSON · Multivariate Statistical Simulation
JOHNSON & KOTZ · Distributions in Statistics

Continuous Multivarious Distributions

JOHNSON, KOTZ, and BALAKRISHNAN · Continuous Univariate Distributions, Volume 1, *Second Edition*
JOHNSON, KOTZ, and BALAKRISHNAN · Continuous Univariate Distributions, Volume 2, *Second Edition*
JOHNSON, KOTZ, and BALAKRISHNAN · Discrete Multivariate Distributions
JOHNSON, KOTZ, and KEMP · Univariate Discrete Distributions, *Second Edition*
JUREČKOVÁ and SEN · Robust Statistical Procedures: Asymptotics and Interrelations
KADANE · Bayesian Methods and Ethics in a Clinical Trial Design
KADANE and SCHUM · A Probabilistic Analysis of the Sacco and Vanzetti Evidence
KALBFLEISCH and PRENTICE · The Statistical Analysis of Failure Time Data

---

*Now available in a lower priced paperback edition in the Wiley Classics Library.

Applied Probability and Statistics (Continued)

KELLY · Reversibility and Stochastic Networks
KHURI, MATHEW, and SINHA · Statistical Tests for Mixed Linear Models
KLUGMAN, PANJER, and WILLMOT · Loss Models: From Data to Decisions
KLUGMAN, PANJER, and WILLMOT · Solutions Manual to Accompany Loss Models: From Data to Decisions
KOVALENKO, KUZNETZOV, and PEGG · Mathematical Theory of Reliability of Time-Dependent Systems with Practical Applications
LAD · Operational Subjective Statistical Methods: A Mathematical, Philosophical, and Historical Introduction
LANGE, RYAN, BILLARD, BRILLINGER, CONQUEST, and GREENHOUSE · Case Studies in Biometry
LAWLESS · Statistical Models and Methods for Lifetime Data
LEE · Statistical Methods for Survival Data Analysis, *Second Edition*
LEPAGE and BILLARD · Exploring the Limits of Bootstrap
LINHART and ZUCCHINI · Model Selection
LITTLE and RUBIN · Statistical Analysis with Missing Data
MAGNUS and NEUDECKER · Matrix Differential Calculus with Applications in Statistics and Econometrics, *Revised Edition*
MALLER and ZHOU · Survival Analysis with Long Term Survivors
MANN, SCHAFER, and SINGPURWALLA · Methods for Statistical Analysis of Reliability and Life Data
McLACHLAN and KRISHNAN · The EM Algorithm and Extensions
McLACHLAN · Discriminant Analysis and Statistical Pattern Recognition
McNEIL · Epidemiological Research Methods
MILLER · Survival Analysis
MONTGOMERY and PECK · Introduction to Linear Regression Analysis, *Second Edition*
MYERS and MONTGOMERY · Response Surface Methodology: Process and Product in Optimization Using Designed Experiments
NELSON · Accelerated Testing, Statistical Models, Test Plans, and Data Analyses
NELSON · Applied Life Data Analysis
OCHI · Applied Probability and Stochastic Processes in Engineering and Physical Sciences
OKABE, BOOTS, and SUGIHARA · Spatial Tesselations: Concepts and Applications of Voronoi Diagrams
PANKRATZ · Forecasting with Dynamic Regression Models
PANKRATZ · Forecasting with Univariate Box–Jenkins Models: Concepts and Cases
PIANTADOSI · Clinical Trials: A Methodologic Perspective
PORT · Theoretical Probability for Applications
PUTERMAN · Markov Decision Processes: Discrete Stochastic Dynamic Programming
RACHEV · Probability Metrics and the Stability of Stochastic Models
RÉNYI · A Diary on Information Theory
RIPLEY · Spatial Statistics
RIPLEY · Stochastic Simulation
ROUSSEEUW and LEROY · Robust Regression and Outlier Detection
RUBIN · Multiple Imputation for Nonresponse in Surveys
RUBINSTEIN · Simulation and the Monte Carlo Method
RUBINSTEIN and MELAMED · Modern Simulation and Modeling
RYAN · Statistical Methods for Quality Improvement
SCHUSS · Theory and Applications of Stochastic Differential Equations

---

*Now available in a lower priced paperback edition in the Wiley Classics Library.

Applied Probability and Statistics (Continued)

SCOTT · Multivariate Density Estimation: Theory, Practice, and Visualization
*SEARLE · Linear Models
SEARLE · Linear Models for Unbalanced Data
SEARLE, CASELLA, and McCULLOCH · Variance Components
STOYAN, KENDALL, and MECKE · Stochastic Geometry and Its Applications, *Second Edition*
STOYAN and STOYAN · Fractals, Random Shapes, and Point Fields: Methods of Geometrical Statistics
THOMPSON · Empirical Model Building
THOMPSON · Sampling
TIJMS · Stochastic Modeling and Analysis: A Computational Approach
TIJMS · Stochastic Models: An Algorithmic Approach
TITTERINGTON, SMITH, and MARKOV · Statistical Analysis of Finite Mixture Distributions
UPTON and FINGLETON · Spatial Data Analysis by Example, Volume 1: Point Pattern and Quantitative Data
UPTON and FINGLETON · Spatial Data Analysis by Example, Volume II: Categorical and Directional Data
VAN RIJKEVORSEL and DE LEEUW · Component and Correspondence Analysis
WEISBERG · Applied Linear Regression, *Second Edition*
WESTFALL and YOUNG · Resampling-Based Multiple Testing: Examples and Methods for *p*-Value Adjustment
WHITTLE · Systems in Stochastic Equilibrium
WOODING · Planning Pharmaceutical Clinical Trials: Basic Statistical Principles
WOOLSON · Statistical Methods for the Analysis of Biomedical Data
*ZEELNER · An Introduction to Bayesian Inference in Econometrics

Texts and References Section

AGRESTI · An Introduction to Categorical Data Analysis
ANDERSON · An Introduction to Multivariate Statistical Analysis, *Second Edition*
ANDERSON and LOYNES · The Teaching of Practical Statistics
ARMITAGE and COLTON · Encyclopedia of Biostatistics. 6 Volume set
BARTOSZYNSKI and NIEWIADOMSKA-BUGAJ · Probability and Statistical Inference
BERRY, CHALONER, and GEWEKE · Bayesian Analysis in Statistics and Econometrics: Essays in Honor of Arnold Zellner
BHATTACHARYA and JOHNSON · Statistical Concepts and Methods
BILLINGSLEY · Probability and Measure, *Second Edition*
BOX · R. A. Fisher, the Life of a Scientist
BOX, HUNTER, and HUNTER · Statistics for Experimenters: An Introduction to Design, Data Analysis, and Model Building
BOX and LUCEÑO · Statistical Control by Monitoring and Feedback Adjustment
BROWN and HOLLANDER · Statistics: A Biomedical Introduction
CHATTERJEE and PRICE · Regression Analysis by Example, *Second Edition*
COOK and WEISBERG · An Introduction to Regression Graphics

*Now available in a lower priced paperback edition in the Wiley Classics Library.

Texts and References Section (Continued)

COX · A Handbook of Introductory Statistical Methods
DILLON and GOLDSTEIN · Multivariate Analysis: Methods and Applications
DODGE and ROMIG · Sampling Inspection Tables, *Second Edition*
DRAPER and SMITH · Applied Regression Analysis, *Third Edition*
DUDEWICZ and MISHRA · Modern Mathematical Statistics
DUNN · Basic Statistics: A Primer for the Biomedical Sciences, *Second Edition*
FISHER and VAN BELLE · Biostatistics: A Methodology for the Health Sciences
FREEMAN and SMITH · Aspects of Uncertainty: A Tribute to D. V. Lindley
GROSS and HARRIS · Fundamentals of Queueing Theory, *Third Edition*
HALD · A History of Probability and Statistics and their Applications Before 1750
HALD · A History of Mathematical Statistics from 1750 to 1930
HELLER · MACSYMA for Statisticians
HOEL · Introduction to Mathematical Statistics, *Fifth Edition*
JOHNSON and BALAKRISHNAN · Advances in the Theory and Practice of Statistics: A Volume in Honor of Samuel Kotz
JOHNSON and KOTZ (editors) · Leading Personalities in Statistical Sciences: From the Seventeenth Century to the Present
JUDGE, GRIFFITHS, HILL, LÜTKEPOHL, and LEE · The Theory and Practice of Econometrics, *Second Edition*
KHURI · Advanced Calculus with Applications in Statistics
KOTZ and JOHNSON (editors) · Encyclopedia of Statistical Sciences. Volumes 1 to 9 with Index
KOTZ and JOHNSON (editors) · Encyclopedia of Statistical Sciences: Supplement Volume
KOTZ, REED, and BANKS (editors) · Encyclopedia of Statistical Sciences: Update Volume 1
KOTZ, REED, and BANKS (editors) · Encyclopedia of Statistical Sciences: Update Volume 2
LAMPERTI · Probability: A Survey of the Mathematical Theory, *Second Edition*
LARSON · Introduction to Probability Theory and Statistical Inference, *Third Edition*
LE · Applied Survival Analysis
MALLOWS · Design, Data, and Analysis by Some Friends of Cuthbert Daniel
MARDIA · The Art of Statistical Science: A Tribute to G. S. Watson
MASON, GUNST, and HESS · Statistical Design and Analysis of Experiments with Applications to Engineering and Science
MURRAY · X-STAT 2.0 Statistical Experimentation, Design Data Analysis, and Nonlinear Optimization
PURI, VILAPLANA, and WERTZ · New Perspectives in Theoretical and Applied Statistics
RENCHER · Methods of Multivariate Analysis
RENCHER · Multivariate Statistical Inference with Applications
ROSS · Introduction to Probability and Statistics for Engineers and Scientists
ROHATGI · An Introduction to Probability Theory and Mathematical Statistics
RYAN · Modern Regression Methods
SCHOTT · Matrix Analysis for Statistics
SEARLE · Matrix Algebra Useful for Statistics
STYAN · The Collected Papers of T. W. Anderson: 1943–1985
TIERNEY · LISP-STAT: An Object-Oriented Environment for Statistical Computing and Dynamic Graphics
WONNACOTT and WONNACOTT · Econometrics, *Second Edition*

---

*Now available in a lower priced paperback edition in the Wiley Classics Library.

WILEY SERIES IN PROBABILITY AND STATISTICS

ESTABLISHED BY WALTER A. SHEWHART AND SAMUEL S. WILKS

Editors
Robert M. Groves, Graham Kalton, J. N. K. Rao, Norbert Schwarz, Christopher Skinner

Survey Methodology Section

BIEMER, GROVES, LYBERG, MATHIOWETZ, and SUDMAN · Measurement Errors in Surveys
COCHRAN · Sampling Techniques, *Third Edition*
COX, BINDER, CHINNAPPA, CHRISTIANSON, COLLEDGE, and KOTT (editors) · Business Survey Methods
*DEMING · Sample Design in Business Research
DILLMAN · Mail and Telephone Surveys: The Total Design Method
GROVES · Survey Errors and Survey Costs
GROVES and COUPER · Nonresponse in Household Interview Surveys
GROVES, BIEMER, LYBERG, MASSEY, NICHOLLS, and WAKSBERG · Telephone Survey Methodology
*HANSEN, HURWITZ, and MADOW · Sample Survey Methods and Theory, Volume I: Methods and Applications
*HANSEN, HURWITZ, and MADOW · Sample Survey Methods and Theory, Volume II: Theory
KASPRZYK, DUNCAN, KALTON, and SINGH · Panel Surveys
KISH · Statistical Design for Research
*KISH · Survey Sampling
LESSLER and KALSBEEK · Nonsampling Error in Surveys
LEVY and LEMESHOW · Sampling of Populations: Methods and Applications
LYBERG, BIEMER, COLLINS, de LEEUW, DIPPO, SCHWARZ, TREWIN (editors) · Survey Measurement and Process Quality
SKINNER, HOLT and SMITH · Analysis of Complex Surveys

---

*Now available in a lower priced paperback edition in the Wiley Classics Library.